Computational Electrodynamics
A Gauge Approach with Applications in Microelectronics

RIVER PUBLISHERS SERIES IN ELECTRONIC MATERIALS AND DEVICES

Indexing: All books published in this series are submitted to Thomson Reuters Book Citation Index (BkCI), CrossRef and to Google Scholar.

The "River Publishers Series in Electronic Materials and Devices" is a series of comprehensive academic and professional books which focus on the theory and applications of advanced electronic materials and devices. The series focuses on topics ranging from the theory, modeling, devices, performance and reliability of electron and ion integrated circuit devices and interconnects, insulators, metals, organic materials, micro-plasmas, semiconductors, quantum-effect structures, vacuum devices, and emerging materials. Applications of devices in biomedical electronics, computation, communications, displays, MEMS, imaging, micro-actuators, nanoelectronics, optoelectronics, photovoltaics, power ICs and micro-sensors are also covered.

Books published in the series include research monographs, edited volumes, handbooks and textbooks. The books provide professionals, researchers, educators, and advanced students in the field with an invaluable insight into the latest research and developments.

Topics covered in the series include, but are by no means restricted to the following:

- Integrated circuit devices
- Interconnects
- Insulators
- Organic materials
- Semiconductors
- Quantum-effect structures
- Vacuum devices
- Biomedical electronics
- Displays and imaging
- MEMS
- Sensors and actuators
- Nanoelectronics
- Optoelectronics
- Photovoltaics
- Power ICs

For a list of other books in this series, visit www.riverpublishers.com

Computational Electrodynamics
A Gauge Approach with Applications in Microelectronics

Wim Schoenmaker
MAGWEL, Belgium

Published, sold and distributed by:
River Publishers
Alsbjergvej 10
9260 Gistrup
Denmark

River Publishers
Lange Geer 44
2611 PW Delft
The Netherlands

Tel.: +45369953197
www.riverpublishers.com

ISBN: 978-87-93519-84-8 (Hardback)
978-87-93519-83-1 (Ebook)

Contents

Preface xv

Acknowledgments xix

List of Figures xxi

List of Tables xxxix

List of Symbols xli

List of Abbreviations xlv

PART I: Introduction to Electromagnetism

1 Introduction 3

2 The Microscopic Maxwell Equations 7

2.1 Definition of the Electric Field 7

2.2 Definition of the Magnetic Field 8

2.3 The Microscopic Maxwell Equations in Integral and Differential Form 9

2.4 Conservation Laws 12

2.4.1 Conservation of Charge – The Continuity Equation 12

2.4.2 Conservation of Energy – Poynting's Theorem . . 13

2.4.3 Conservation of Linear Momentum – The Electromagnetic Field Tensor 14

2.4.4 Angular Momentum Conservation 15

3 Potentials and Fields and the Lagrangian 17
3.1 The Scalar and Vector Potential 17
3.2 Gauge Invariance 19
3.3 Lagrangian for an Electromagnetic Field Interacting with Charges and Currents 19

4 The Macroscopic Maxwell Equations 23
4.1 Constitutive Equations 23
4.2 Boltzmann Transport Equation 24
4.3 Currents in Metals 26
4.4 Charges in Metals 29
4.5 Semiconductors 30
4.6 Currents in Semiconductors 31
4.7 Insulators 36
4.8 Dielectric Media 37
4.9 Magnetic Media 41

5 Wave Guides and Transmission Lines 45
5.1 TEM Modes 47
5.2 TM Modes 49
5.3 TE Modes 49
5.4 Transmission Line Theory – S Parameters 50
5.5 Classical Ghosts Fields 54
5.6 The Static Approach and Dynamic Parts 56
5.7 Interface Conditions 58
5.8 Boundary Conditions 59

6 Energy Calculations and the Poynting Vector 69
6.1 Static Case 69
6.2 High-Frequency Case 70

7 From Macroscopic Field Theory to Electric Circuits 73
7.1 Kirchhoff's Laws 73
7.2 Circuit Rules 78
7.3 Inclusion of Time Dependence 80

8 Gauge Conditions 87
8.1 The Coulomb Gauge 89
8.2 The Lorenz Gauge 90
8.3 The Landau Gauge 91

8.4 The Temporal Gauge 94
8.5 The Axial Gauge 95
8.6 The 't Hooft Gauge 95

9 The Geometry of Electrodynamics **97**
9.1 Gravity as a Gauge Theory 98
9.2 The Geometrical Interpretation of Electrodynamics 104

10 Integral Theorems **107**
10.1 Vector Identities 113

PART II: Discretization Methods for Sources and Fields

11 The Finite Difference Method **117**

12 The Finite Element Method **121**
12.1 Trial Solutions 121
12.2 The Element Concept 122

13 The Finite Volume Method and Finite Surface Method **129**
13.1 Differential Operators in Cartesian Grids 132
13.2 Discretized Equations 134
13.3 The No-Ghost Approach 134
13.4 Current Continuity Equation 139
13.5 Computational Details of the Hole Transport Equation . . . 141
13.5.1 Scaling 144
13.6 Computational Details of the Electron Transport Equation 151
13.6.1 Couplings 152
13.7 The Poisson Equation 156
13.8 Maxwell-Ampere Equation 162
13.9 Using Gauge Conditions to Decrease Matrix Fill-In 164
13.9.1 Poisson System 165
13.9.2 Metals 166
13.9.3 Dielectrics 168
13.9.4 Maxwell-Ampere System 170
13.9.5 "Standard" Implementation 171
13.9.6 Decoupling Implementation 171

13.10 The Generalized Coulomb Gauge 172
13.10.1 Implementation Details of the Ampere-Maxwell System 173
13.11 The *EV* Solver 174
13.11.1 Boundary Conditions for the *EV* System 176
13.11.2 Implementation Details of the *EV* System 177
13.11.3 Solution Strategy of the *EV* System 179
13.12 The Scharfetter-Gummel Discretization 179
13.12.1 The Static and Dynamic Parts 181
13.13 Using Unstructured Grids 183

14 Finite Volume Method and the Transient Regime 187
14.1 The Electromagnetic Drift-Diffusion Solver in the Time Domain 188
14.2 Gauge Conditions 191
14.3 Semiconductor Treatment 194
14.4 Implementation of Numerical Methods for Solving the Equations 197
14.5 Spatial Discretization 197
14.6 Discretization of Gauss' Law 198
14.7 Boundary Conditions for Gauss' Discretized Law 199
14.8 Discretization of the Maxwell-Ampere System 202
14.9 Boundary Conditions for the Maxwell-Ampere Equation 207
14.10 Generalized Boundary Conditions for the Maxwell-Ampere Equation 211
14.11 Discretization of the Gauge Condition 213
14.12 Temporal Discretization 214
14.13 BDF for DAEs 215
14.14 State-Space Matrices and Linking Harmonic to Transient Analysis 216
14.15 A Technical Detail: Link Orientations 221
14.16 Scaling 222
14.16.1 Scaling the Poisson Equation 222
14.16.2 Scaling the Current-Continuity Equations 223
14.16.3 Scaling the Maxwell-Ampere Equation 224
Summary 226

PART III: Applications

15 Simple Test Cases 229
15.1 Examples 229
15.1.1 Crossing Wires 229
15.1.2 Square Coaxial Cable 229
15.1.3 Spiral Inductor 231
15.2 S-Parameters, Y-Parameters, Z-Parameters 233
15.3 A Simple Conductive Rod 235
15.4 Strip Line above a Conductive Plate 239
15.4.1 Finite t_M Results 246
15.5 Running the Adapter 247
15.6 Simulations with Opera – VectorFields 247
15.7 Coax Configuration 256
15.8 Inductor with Grounded Guard Ring 258
15.9 Inductor with Narrow Winding above a Patterned Semiconductor Layer 265
Summary 280

16 Evaluation of Coupled Inductors 281
16.1 Scaling Rules for the Maxwell Equations 282
16.2 Discretization 283
16.3 The EV Solver 285
16.3.1 Boundary Conditions 287
16.4 Scattering Parameters 288
16.5 Application to Compute the Coupling of Inductors 290

17 Coupled Electromagnetic-TCAD Simulation for High Frequencies 295
17.1 Review of A-V Formulation 298
17.1.1 A-V Formulation of the Coupled System 298
17.2 Origin of the High-Frequency Breakdown of the A-V Solver 300
17.3 E-V Formulation 301
17.3.1 Redundancy in Coupled System 303
17.3.2 Issues of Material Properties 304
17.3.3 Boundary Conditions 305
17.3.4 Implementation Details 306
17.3.5 Matrix Permutation 307

17.4 Numerical Results 308
17.4.1 Accuracy of E-V Solver 308
17.4.2 Spectral Analyses 311
17.4.3 Performance Comparisons 314
Summary 316

18 EM-TCAD Solving from 0–100 THz 317
18.1 From AV to EV 317
18.2 Discretization 319
18.3 Simplified EV Schemes 320
18.4 Combination of AV and EV Solvers 321
18.5 Numerical Experiments 321
18.6 Best Practices for Iterative Solving 325

19 Large Signal Simulation of Integrated Inductors on Semi-Conducting Substrates 327
19.1 Need for Mimetic Formulation 328
19.2 Field Equations 329
19.3 Application to An Octa-Shaped Inductor 332
Summary 339

20 Inclusion of Lorentz Force Effects in TCAD Simulations 341
20.1 Steady-State Equations 342
20.2 Discretization of the Lorentz Current Densities 344
20.3 Static Skin Effects in Conducting Wires 347
20.4 Self-Induced Lorentz Force Effects in Metallic Wires 348
20.5 Self-Induced Lorentz Force Effects in Silicon Wires 349
20.6 External Fields 349
Summary 351

21 Self-Induced Magnetic Field Effects, the Lorentz Force and Fast-Transient Phenomena 353
21.1 Time-Domain Formulation of EM-TCAD Problem 356
21.2 Inclusion of the Lorentz Force 358
21.3 Discretization of the Lorentz Current Densities 360
21.4 Applications 366
Summary 377

22 EM Analysis of ESD Protection for Advanced CMOS Technology **379**
22.1 Simulation of a Metallic Wire 380
22.2 In-depth Simulation of the Full ESD Structure 383
22.3 Negative Stress with Active Diode 387
22.4 Diode SCR 389
22.5 Comparison with TLP Measurements 391
Summary 392

23 Coupled Electromagnetic-TCAD Simulation for Fast-Transient Systems **395**
23.1 Time-Domain A-V formulation 397
23.2 Analysis of Fast-Transient Breakdown 400
23.3 Time-Domain E-V Formulation 402
23.4 Numerical Results 404
Summary 407

24 A Fast Time-Domain EM-TCAD Coupled Simulation Framework via Matrix Exponential with Stiffness Reduction **409**
24.1 Time-Domain Formulation of EM-TCAD Problem 411
24.2 Time-Domain Simulation with Matrix Exponential Method 415
24.3 Error Control and Adaptivity 420
24.4 E-V Formulation of EM-TCAD for MEXP Method 421
24.5 Numerical Results 424
24.6 Validity Proof of Regularization with Differentiated Gauss' Law 431
24.7 Fast Computation of Mx in E-V Formulation 432
Summary 433

PART IV: Advanced Topics

25 Surface-Impedance Approximation to Solve RF Design Problems **437**
25.1 Surface Impedance Approximation 437
25.2 Formulation of the BISC in Potentials 440
25.3 Scaling Considerations 442
25.4 One-Dimensional Test Example 444

26 Using the Ghost Method for Floating Domains in Electromagnetic Field Solvers **455**
26.1 Problem Description 456
26.2 Proposed Solution 458
26.3 Example 1: Metal Blocks Embedded in Insulator 459
26.4 Example 2: A Transformer System 460
26.5 Initial Guess 462
26.6 High-Frequency Problems 462
26.7 Floating Semiconductor Regions 468

27 Integrating Factors for Discretizing the Maxwell-Ampere Equation **477**
27.1 Review of the Scharfetter-Gummel Discretization 478
27.2 Observations 479
27.3 Maxwell Equations 481
27.4 Discretization of the Curl-Curl Operator 482
27.5 Discretization of the Divergence Operator 484
27.6 Discretization of Poisson-Type Operators 489
27.7 Equivalence 490
27.8 High-Frequency Maxwell Equations 491
27.9 Integrating Factors for Unstructured Grids 493
27.10 Implementation Details 494
27.11 Effect of the Inclusion of the Integrating Factor 495
27.12 Simulation Set Up and Results 495
Summary 501

28 Stability Analysis of the Transient Field Solver **503**
28.1 Impact of the Gauge Condition 509
28.2 Magnetic Neumann Boundary Conditions 513
28.3 Results for Larger Values of the Conductance 513
28.4 Yet Another Experiment 518
28.5 Inductor Experiments 518
28.6 Results for a Metal Loop 524
28.7 Results for a Twisted Bar 526
28.8 Corner Example 530
28.9 Returning to the Original Problem 534
28.10 Revisiting the Equations 538
28.11 Redoing the Corner Structure 543
28.12 Simple Test Structure for the Stability Problem 549

28.13 Results for a Single Line 555
28.14 Some Theoretical Considerations 558
28.15 The Impact of the Meshing 559
28.16 Final Summary of Stability Study 561

29 Summary of the Numerical Techniques 563
29.1 Equations 563
29.2 Boundary conditions 566
29.3 Spatial Discretization 567

References 575

Index 585

About the Author 595

Preface

Why another book on computing and electrodymanics? There are many good books around that certainly cover important aspects of the computation of electromagnetic phenomena. In order to justify the effort of writing or reading another book on the subject it should contain material that is not easily obtained in other domains of the literature. This is indeed the case for the present work. The main motivation for writing this book is that it provides space to approach the subject in a more tutorial manner such that subtle issues can be explained in sufficient depth. In peer-reviewed journals contributions such material is usually lacking since – once it is explained – it is considered as evident and a waste of paper. The motivation to allocate room for a more detailed approach is the view point that deviates from the mainstream set up of the topic. In electronic engineering, the starting point for describing electro-magnetic effects are the Maxwell equations which give direct relations between the electric and magnetic fields and their sources. The engineer's intuition is fully tuned to understand these equations as describing the mutual forces between charges. The concept of force goes back to Newton and therefore, there is a direct connection between the entities that are contained in the Maxwell equations and our intuitive idea corresponding to a 'force'. On the other hand, in the physics community the force concept, although still being used in elementary computations, the concept of energy is considered the more fundamental entity and the force is a derived concept. An intuitive notion of energy is also less evident. One could say that energy is, loosely speaking, the ability to impose change. The Maxwell equations can be reformulated such that their variables represent energy. The electric voltage is one of them. So far, that is not a big deal and there is no dispute about what the 'voltage' represents. However, if the forces, voltages or other electromagnetic variables change rapidly in time, one can arrive at contradictory results if one does not define precisely the meaning of variables. In this book special attention will be paid to define the concepts of electromagnetism carefully. In particular the connection between the concepts and the transition to quantities that can be submitted to a computation algorithm will be addressed. The notion of energy as a starting point for formulating a computational scheme implies a deeper level of abstraction. As is known from mechanical engineering identical situations can be described by different setting of the zero-point energy, in other words, energy is not specified by a uniquely defined number. In electrodynamics this translates to having potentials that are neither quantified by a unique numerical value: one may redefine the potential without changing the physical content. In the last century this observation has resulted into the standard model based on gauge theories that is capable of describing nature. Gauge theories are the mathematical frames to describe physical observations without reference to uniquely defined entities. The deeper origin of non-uniqueness is rooted in symmetry and it is actually the presence of symmetry that is responsible for numerical non-uniqueness. Now one may adapt the view point in computational electrodynamics to get away from the non-uniqueness at the earliest stage and focus on the variables that do have a unique numerical values (electric field, magnetic field, charge) but in this way insight is lost in the mathematical structure of the gauge formulation and as a consequence shuts off exploiting this structure in a computational approach. The purpose of this book is to present an approach that does exploit the gauge formulation of electrodynamics in a computational setting. The terminology of the gauge principle will be explained in depth such that the words are not just a matter of tautology. It is the hope of the author that the reader will develop a valuable intuition of the concepts. The tools to realize this outcome is to present the concepts using pictorial illustrations. Advanced mathematical rigor is replaced by using more elementary notations for the purpose of obtaining an operational understanding.

The applications that are described in this work are all rooted in micro-electronic design. The reason for this is quite simple: this is the author's field of expertise. However, I believe that the computational methods that are presented here may be used in very diverse areas. After all, electromagnetism plays an important role in the description of nature ranging from astrophysics, chemistry and (sub-) nuclear physics. Whereas the connection with sub-nuclear physics is rather direct, e.g. here the gauge principle is used as a starting point rather than an interesting observation down the road, I encourage to apply these techniques elsewhere.

This book consists of four parts. In Part I, we set up the scene of electromagnetism with some focus on the application in microelectrnic device engineering. Semiconductors are of pivotal importance in microelectronic engineering and the transport of holes and electrons in semiconductors is discussed. Part II deals with the construction of computational algorithms that come with the scene setting of Part I. In Part II, the novelties of the gauge theoretical aspects will be presented. Next in Part III we discuss numerous applications of the computational schemes that are developed in Part II. Every chapter in Part III is self-contained in the sense that the extensive discussions of Part II are briefly summarized. Chapters 16–24 in Part III are reprints from published work. Finally Part IV deals with more advanced topics. Here the word advanced can be read as unfinished, disputable or challenging. The idea is that the chapters in Part IV trigger the reader to form his/her own opinion what will be the next steps in computational electrodynamics.

Acknowledgments

The work described in this book is the result of a few decades of analyzing electromagnetic field problems, coding and comparing the computed results with measurements. It is needless to say that such an effort can not be achieved by one person. I would like to acknowledge here all the fruitful collaborations that were established in EU funded projects (Codestar, CHAMELEON-RF, pullNano, SQWIRE, ICESTARS, nanoCOPS) as well as national projects supported by the Flanders Agency VLAIO. Some special thanks I owe to some people for sharpening and challenging my thoughts on the subject of computational electrodynamics. In particular, I would like to thank Wim Magnus and Peter Meuris as co-workers realizing the geometrical picture of the magnetic vector potential including its numerical implementation. I also owe a special thanks to Philippe Galy (ST microelectronics) and Quan (Alex) Chen (Hong-Kong University). Philippe is gratefully acknowledge in supporting the development of the chapters dealing with the Lorentz force, whereas Alex' help was indispensable in clarifying the correct formulation of the EV solver. For the chapter dealing with the stability analysis, I owe Hans-Georg Brachtendorf and Kai Bittner (University of Applied Science Upper Austrian). In the projects there were produced many measurement results that turned out to be very important in the code development. I am very grateful for getting the data from Walter Pflanz, Alexander Steinmayer and Ehrenfried Seebacher (AMS), Rick Janssen (NXP), Pascal Reynier (ACCO) and many more. Furthermore, I enjoyed long-term relations with the mathematics community. I like to mention explicitly Henk Van der Vorst (Utrecht Univerisity), Wil Schilders (TU/e), Caren Tischendorf (Humbolt University Berlin), Karl Meerbergen (KU Leuven). Last but not least, I express my gratitude to Vee, whose continuous support was critical in many stages of the activities that finally led to the content of this work.

List of Figures

Figure 4.1 Band edge modulation by doping. 30
Figure 5.1 Contours for evaluating voltage drops and currents of a two-conductor system in a TEM mode. 48
Figure 5.2 Ground-Signal-Ground (GSG) probe. The tip and the connector to the coax cable are shown. Photo of Infinity Probe ® courtesy of Cascade Microtech, a FormFactor company. 51
Figure 5.3 Set up of the s-matrix simulation. 52
Figure 5.4 Two port s-parameter set up. 53
Figure 7.1 Closed electric circuit containing a resistor connected to a DC power supply through two resistive leads. 75
Figure 7.2 The electric circuit of Figure 7.1, taking into account the spatial extension of the leads. Γ is a circuit loop, i.e. an internal, closed loop encircling the "hole" of the circuit. 75
Figure 7.3 The electric circuit of Figure 7.2 with a helix-shaped "resistor". 81
Figure 8.1 Effective confinement potential in a Hall bar (shaded area). The bare confinement is invoked by a "hard wall" restricting the lateral motion to the interval $|y| < W/2$. 94
Figure 9.1 Parallel displacement in the locally Euclidean coordinate system. 101
Figure 9.2 The covariant derivative of a vector field. 102
Figure 9.3 Determination of the curvature from a round trip along a closed loop. 103
Figure 9.4 Local frames for the phase of a wave function. . . . 104
Figure 12.1 Function with domain split in elements. 123
Figure 12.2 Base functions for one element of the domain. . . . 124
Figure 12.3 Weighted sum of base functions for one element of the domain. 124

Figure 13.1 The fundamental variables on the Cartesian grid. . 130
Figure 13.2 The assembly of the $\nabla \times \nabla \times$-operator using 12 contributions of neighboring links. 133
Figure 13.3 The assembly of the $\nabla \cdot \nabla$-operator using 6 contributions of neighboring nodes. 133
Figure 13.4 Identification of link direction. 139
Figure 13.5 Labeling of the front (tail) and back (head) nodes of a link. 168
Figure 13.6 Labeling of the front (tail) and back (head) nodes of a link. 177
Figure 13.7 A collection of oriented links. 178
Figure 13.8 Visualization of the solution procedure. 179
Figure 13.9 Layout of a spiral inductor. 183
Figure 13.10 Zoom-in of an unstructured grid. 184
Figure 13.11 Inductances obtained on a structured and unstructured grid. 185
Figure 14.1 Illustration of the use of Gauss' law. 194
Figure 14.2 Correction factor as a function of the acuteness of the discretization cells. 204
Figure 14.3 A collection of oriented links. 205
Figure 14.4 Sketch of the boundary nodes and links. 210
Figure 14.5 Set up for direct simulation of S-parameters. 212
Figure 14.6 Illustration of the justification for using internal contacts. (VNA – vector network analyser, e.g. frequency domain measurement equipment.). . . . 213
Figure 14.7 Illustration of the currents at a contact node, i. The nodes j are field nodes. 219
Figure 14.8 A collection of oriented links. 221
Figure 15.1 Layout of two crossing wires in insulating environment. 230
Figure 15.2 Layout of the square coax structure. 230
Figure 15.3 Layout of the spiral inductor structure. 231
Figure 15.4 Magnetic field strength in the plane of the spiral inductor. 232
Figure 15.5 Set up of the S-parameter evaluation: 1 port is excited and all others are floating. 235
Figure 15.6 Illustration of the wire in the MAGWEL editor; the red square is a contact area. 236
Figure 15.7 Illustration of the mesh in the MAGWEL editor. . . 237
Figure 15.8 Zoom-in of the mesh in the MAGWEL editor. . . . 238

Figure 15.9 View of the magnetic field strength. 238
Figure 15.10 View of the magnetic field strength. 239
Figure 15.11 Layout of the parallel strip above a conductive plate. 240
Figure 15.12 Illustration of the contacts. 241
Figure 15.13 Layout of the parallel strip above a conductive plate. 242
Figure 15.14 Layout of the parallel strip above a conductive plate. 243
Figure 15.15 Separation in two domains of the electric field. red: $E > 0.01E_{max}$, blue: $E < 0.01E_{max}$. 244
Figure 15.16 Illustration of the mesh. 244
Figure 15.17 Illustration of the mesh. 245
Figure 15.18 Illustration of the model with X and Y at different scales. 248
Figure 15.19 Illustration of the potential with X and Y at the same scale. 248
Figure 15.20 Illustration of the electric field with X and Y at the same scale. 248
Figure 15.21 Mesh1: 436 elements, evaluated error 18.21371%. 250
Figure 15.22 Mesh2: 532 elements, evaluated error 15.04066%. 250
Figure 15.23 Mesh3: 671 elements, evaluated error 17.95522%. 250
Figure 15.24 Mesh4: 881 elements, evaluated error 10.40742%. 251
Figure 15.25 Mesh5: 1079 elements, evaluated error 8.484277%. 251
Figure 15.26 Mesh6: 1411 elements, evaluated error 7.563303% and with zoom-in. 251
Figure 15.27 Mesh7: 1955 elements, evaluated error 6.604616% and with zoom-in. 252
Figure 15.28 Mesh8: 2706 elements, evaluated error 4.726004% and with zoom-in. 252
Figure 15.29 Mesh9: 3647 elements, evaluated error 3.773939%. 253
Figure 15.30 Mesh10: 4967 elements, evaluated error 3.268908%. 253

Figure 15.31 Mesh11: 7100 elements, evaluated error 2.078323% and with zoom-in. 253
Figure 15.32 Mesh12: 9894 elements, evaluated error 2.127898%. 254
Figure 15.33 Mesh13: 13752 elements, evaluated error 1.517278%. 254
Figure 15.34 Mesh14: 18945 elements, evaluated error 1.209618%. 255
Figure 15.35 Mesh15: 24736 elements, evaluated error 0.908549%. 255
Figure 15.36 Illustration of the coax wire in the MAGWEL editor. 256
Figure 15.37 Simulation results and analytical MAGWEL results for a Manhattan mesh. 257
Figure 15.38 View of the magnetic field strength. 258
Figure 15.39 Inductor layout of an inductor with a closed guard ring. 259
Figure 15.40 log(abs(Re(Y11)) + abs(Im((Y11))) and log(abs(Re(Y12)) + abs(Im(Y12))). 260
Figure 15.41 Re (S11) and Im (S11). 261
Figure 15.42 Re (S12) and Im (S12). 262
Figure 15.43 Re (22) and Im (S22). 263
Figure 15.44 The inductance, resistance and Q factor of the inductor with grounded closed-guard ring. . 264
Figure 15.45 View of the inductor (no details). 265
Figure 15.46 2D View of the Nwell pattern. 266
Figure 15.47 View of the design using vertical contacts. 267
Figure 15.48 View of the small contacts. 267
Figure 15.49 Re(S11). 268
Figure 15.50 Im(S11). 268
Figure 15.51 Re(S12). 269
Figure 15.52 Im(S12). 269
Figure 15.53 Re(S22). 270
Figure 15.54 Im(S22). 270
Figure 15.55 L using Y12. 272
Figure 15.56 L using Y11. 273
Figure 15.57 R using Y12. 273
Figure 15.58 R using Y11. 274
Figure 15.59 Q factor. 275
Figure 15.60 Re(Y11). 276

Figure 15.61 Im(Y11). 276
Figure 15.62 Re(Y12). 277
Figure 15.63 Im(Y12). 277
Figure 15.64 L using Y12 and L using Y11. 278
Figure 15.65 R using Y12 and R using Y11. 279
Figure 15.66 Q-factor. 280
Figure 16.1 Discretized version of the regularized curl-curl operator acting on a vector field. 285
Figure 16.2 Set up of the s-parameter evaluation: 1 port is excited and all others are floating. 290
Figure 16.3 View on the coupled spiral inductor using the Virtuosa design environment. 291
Figure 16.4 View on the coupled spiral inductor using the MAGWEL editor. 291
Figure 16.5 Comparison of the experiment and simulation results for s_{11}. 292
Figure 16.6 Comparison of the experiment and simulation results for s_{12}. 292
Figure 16.7 Comparison of the experimental and simulation results for the gain. 293
Figure 17.1 A typical on-chip structure consisting of metallic interconnects, semiconductor devices and substrate. 296
Figure 17.2 Cross wire structure. Simulation domain is $10 \times 10 \times 10 \mu m^3$ and the cross sections of metal wires are $2 \times 2 \mu m^2$. $\sigma = 5.96 \times 10^7 S/m$. FVM discretization generates 1400 nodes and 3820 links. 308
Figure 17.3 Metal plug structure. Simulation domain is $10 \times 10 \times 10 \mu m^3$ the cross section of metal plug is $4 \times 4 \mu m^2$. $\sigma = 3.37 \times 10^7 S/m$. A uniform doping of $N_D = 1 \times 10^{24}$ is used. FVM discretization generates 1300 nodes and 3540 links. 309
Figure 17.4 Substrate noise isolation structure. A deep n-well (DNW) (pink region) is implanted in the p-type substrate to isolate analog circuits from digital noise sources. Simulation domain is $100 \times 50 \times 11 \mu m^3$. $\sigma = 3.37 \times 10^7$ S/m. A user-defined doping profile is adopted. FVM discretization generates 6300 nodes and 13540 links. 309

Figure 17.5 Differences between the A-V and E-V solvers for the testing structures (with direct solver). 310
Figure 17.6 Current density at the middle layer of the substrate of SNI structure (shown in log10 scale). 310
Figure 17.7 Eigenvalue distribution of the preconditioned Jacobian matrices of A-V solver at different frequencies. 311
Figure 17.8 Eigenvalue distribution of the preconditioned Jacobian matrices of E-V solver at different frequencies. 311
Figure 17.9 Eigenvalue distribution of the preconditioned Jacobian matrices of A-V solver at different frequencies (no metal). 313
Figure 17.10 Eigenvalue distribution of the preconditioned Jacobian matrices of E-V solver at different frequencies (no metal). 313
Figure 18.1 Lay out of an integrated inductor problem. 322
Figure 18.2 Voltage in the inductor plane at 10 GHz using the AV solver. 322
Figure 18.3 Voltage in the inductor plane at 10 GHz using the EV Gauss solver. 323
Figure 18.4 Magnitude of the electric field in the inductor plane at 10 GHz using the AV solver. 323
Figure 18.5 Magnitude of the electric field in the inductor plane at 10 GHz using the EV Gauss solver. 324
Figure 18.6 Condition number Vs frequency for the integrated inductor problem. 325
Figure 19.1 'Artist impression' of the Gauss' law-induced constraint for the time evolution of the full wave variables. 328
Figure 19.2 Illustration of the discrete variables in one mesh cell. 331
Figure 19.3 Octagonal (left) and 8-shaped (right) VCO coil. . . 333
Figure 19.4 Measurement of spur level of octagonal and octa-shaped coil. 333
Figure 19.5 Inductance and quality factor of octagonal (red) and 8-shaped (green) coil. 333
Figure 19.6 View of the integrated 8-shaped inductor from above. The vertical direction is stretched. . . . 334

Figure 19.7 View of the full simulation domain. 334
Figure 19.8 Value of currents at the left and right contact of the inductor. 335
Figure 19.9 Value of the current in the ground plane contact. A transient overshoot is observed. 335
Figure 19.10 Logarithm of the absolute values of the current in the ground-plane contact. Two time scales are observed. 336
Figure 19.11 Set up of a compact model for the transient results. 337
Figure 19.12 Results of a compact model for the transient simulation using a step magnitude of 1Volt. 337
Figure 19.13 Current through the first contact of the 8-shaped inductor after transient simulation with variable order and variable time step size. 338
Figure 19.14 Current through the second contact of the 8-shaped inductor after transient simulation with variable order and variable time step size. 339
Figure 19.15 Sum of all simulated currents (including the substrate current) through the 8-shaped inductor after transient simulation with variable order and variable time step size. 339
Figure 20.1 ESD protection structure whose position requires an accurate knowledge of substrate current flow. . . 342
Figure 20.2 Mesh element ingredients that illustrate the decomposition of the Lorentz force vector product. d_{ij} is the (partial) dual area which is enclosed by the vectors $\mathbf{s}$, $\mathbf{t}$, $\mathbf{u}$ and $\mathbf{v}$. The distance between node i and j is $\langle h_{ij} \rangle$. 345
Figure 20.3 Locations of the angles α, that participate in the computation of the components K_u and K_v. 346
Figure 20.4 Illustration of the Lorentz force in a conducting wire. 347
Figure 20.5 Radial voltage at mid cut of the conducting wire. 348
Figure 20.6 Radial voltage at mid cut of the silicon wire. 349
Figure 20.7 Radial voltage at mid cut of the silicon wire with an external field in the y-direction (arbitrary units). 350

Figure 20.8 Radial voltage at mid cut of the silicon wire with an external field in the y-direction. The bias voltage is 0.5 V and the external field is taken 1 T. 350
Figure 21.1 Mesh element illustrating the ingredients of the decomposition of the Lorentz force vector product. Part of the dual area, d_{ij} (see text) of the link $\langle ij \rangle$ is enclosed by the vectors $\mathbf{u}$, $\mathbf{v}$, $\mathbf{s}$, $\mathbf{t}$. The distance between node i and j is h_{ij}. 362
Figure 21.2 Illustration of the decomposition of the Lorentz force vector product in the local coordinate bases. 363
Figure 21.3 Typical ESD current pulse used in the simulation of the fast-transient signals. 368
Figure 21.4 2D cross section of the Silicon wire. 368
Figure 21.5 ESD voltage pulse computed in the simulation of the fast-transient signals at the injection contact. 369
Figure 21.6 Current density at 0.30×10^{-9} sec. Some reduction is observed in the center due to the skin effect. . . . 369
Figure 21.7 Electric field intensity at 0.30×10^{-9} sec. Some reduction in the center is observed due to the skin effect and the value is $\sim 10^3$ V/m. 370
Figure 21.8 Magnitude of the ∇V at 0.30×10^{-9} sec. The maximum value is $\sim 10^3$ V/m. 370
Figure 21.9 Magnitude of the pseudo-canonical momentum $\mathbf{\Pi} = \partial_t \mathbf{A}$ at 0.30×10^{-9}. The maximum value is $\sim 10^3$ V/m. 371
Figure 21.10 Magnitude of the vector potential $\mathbf{A}$ at 0.30×10^{-9} sec. The maximum value is $3 \sim \times 10^{-7}$ Vsec/m. 371
Figure 21.11 Magnitude of the magnetic induction, $\mathbf{B}$ at 0.30×10^{-9} sec. The maximum value is $\sim 10^{-2}$ T. 372
Figure 21.12 Relative change in the voltage of the current injection contact due to the Lorentz force. 372
Figure 21.13 Circuit layout for use of a Silicon-Controlled Rectifier (SCR). The location of the SCR is encircled and is presented in more detail in Figure 21.14. . . 373

Figure 21.14 Device implementation the SCR. The left picture is a stretched view of the actual implementation that is presented in Figure 21.15. 373
Figure 21.15 Actual implementation of the SCR structure. . . . 374
Figure 21.16 Change in the voltage of the current injection contact due to magnetic effects. 375
Figure 21.17 Change in the gate current due to magnetic effects. 375
Figure 21.18 Change in the gate current and change in the voltage of the current injection contact (LVDS) due to switch-on of the Lorentz force for the majority carriers. 376
Figure 21.19 Change in the voltage of the current injection contact (LVDS, dashed line) and change in the gate current (continuous line) due to an additional switch-on of the Lorentz force for the minority carriers. . . . 376
Figure 21.20 Relative change in the voltage of the current injection contact (LVDS) due to an additional switch-on of the Lorentz force for the minority carriers. 377
Figure 22.1 Basic electrical schematic of the local protection with SCR and diode. 380
Figure 22.2 SCR schematic with its top view layout and 3D TCAD structure. 380
Figure 22.3 IV curves for (E, EM, EMLF) simulations and calculation of $r(t)$ for E, EM and EMLF. . . . 382
Figure 22.4 IV curves of SCR during ESD stress and typical point identification. 384
Figure 22.5 IGN(t) curves of N gate current during surge and current density extraction at A. 387
Figure 22.6 Typical IV diode curve. 388
Figure 22.7 Front/Back side distribution magnetic field for the diode and SCR for 2.5A. 390
Figure 22.8 IGN(t) curves as collector current during surge and current density extraction at α & γ. 391
Figure 23.1 Conductor current in the XWR structure obtained by A-V and E-V solvers. 405

Figure 23.2 Eigenvalue distribution of A-V solver with different step sizes (preconditioned Jacobian matrix from the XWR case). 406
Figure 23.3 Eigenvalue distribution of E-V solver with different step sizes (preconditioned Jacobian matrix from the XWR case). 406
Figure 24.1 Flows of traditional methods and MEXP. 420
Figure 24.2 Generalized eigenvalues of the systems using the original and the differentiated Gauss law. Only the part near the origin is plotted. 425
Figure 24.3 3D view of the SWR case with one silicon wire in the middle and two copper leads having a cross section of $4 \times 4\mu m^2$, surrounded by an $8\mu m$ thick oxide layer (not shown). The lengths of the three parts are respectively $8\mu m$, $16\mu m$ and $8\mu m$. The silicon part has an n-type doping of $10^{21}m^{-3}$. 425
Figure 24.4 Square wave input. 425
Figure 24.5 Current through the left contact obtained by GR and MEXP-AV and MEXP-EV for the SWR example. 426
Figure 24.6 Sparsity pattern of the Jacobian matrices in GR2 and MEXP-AV solvers (VCO example). 427
Figure 24.7 Error vs. step size for MEXP-AV, MEXP-EV and MEXP-EVBG (SWR case with fixed $m = 80, h = 0.25ps$. 428
Figure 24.8 Left-contact current computed by GR2, MEXP-AV and MEXP-EVBG with adaptive time-stepping. 429
Figure 25.1 Various components of the electric field and magnetic induction. 439
Figure 25.2 Various components of the electric field and magnetic induction at large but finite metallic conducutance. 441
Figure 25.3 Variables in the assembling of the SIBC. 442
Figure 25.4 Two-dimensional cross section of the test structure with mesh. 444
Figure 25.5 Three-dimensional test structure for the surface impedance approximation. 444

Figure 25.6 The Poisson potential in metal for the standard calculation (left panel) and SIBC (right panel). . . . 445
Figure 25.7 Component of the B-field that is reported in Figures 25.8–25.10. 446
Figure 25.8 Real part of the magnetic induction tangential to the metal surface. 446
Figure 25.9 Imaginary part of the magnetic induction tangential to the metal surface. 447
Figure 25.10 Ratio of the real and imaginary parts in the range 1–10 GHz. 447
Figure 25.11 Real part of the magnetic induction tangential to the metal surface. 448
Figure 25.12 Imaginary part of the magnetic induction tangential to the metal surface. 448
Figure 25.13 Ratio of the real and imaginary parts in the range 1GHz. 449
Figure 25.14 Illustration of contact current calculation. 450
Figure 25.15 Comparison of the real parts of the contact currents from 1–10 GHz. 450
Figure 25.16 Comparison of the imaginary parts of the contact currents from 1–10 GHz. 451
Figure 25.17 Voltage vs. time plot of the transient test set up. . . 452
Figure 25.18 Current vs. time plot of the transient test set up. . . 453
Figure 26.1 Grid of a one-dimensional structure of a metallic region squeezed between two insulating regions. . . 457
Figure 26.2 Two floating metallic regions embedded into an insulating volume and two contacts. 459
Figure 26.3 The Poisson potential along a line from the bottom to the top contact. 460
Figure 26.4 A 3D view of the geometry of a ring transformer used to show the validity of the method. 460
Figure 26.5 The electric potential along the line AA' in the cross-section. The applied potentials on the ports are also shown. 461
Figure 26.6 Design of a resonator structure. The floating ring (light-gray colored) is resonating with a narrow stripe transmission line (blue colored). The right view is a stretched version of the left unstretched view. 463

Figure 26.7 S11 parameter. The resonance is observed in the measurement and in the simulation set that does not put $E_C = 0$. 464
Figure 26.8 S21 parameter. The resonance is observed in the measurement and in the simulation set that does not put $E_C = 0$. 464
Figure 26.9 Voltage over the resonating ring at 6.85 GHz. . . . 465
Figure 26.10 Voltage over the resonating ring at 7.75 GHz. . . . 465
Figure 26.11 Vector potential over the resonating ring at 6.85 GHz. 466
Figure 26.12 Vector potential over the resonating ring at 7.75 GHz. 466
Figure 26.13 Structure with a stack of two floating regions and the used mesh. 467
Figure 26.14 The resulting voltage along a vertical cut through the stack of Figure 26.13. 468
Figure 26.15 Structure with a semiconductor floating region between two metallic blocks. 469
Figure 26.16 Hole density profile in the floating floating semiconductor regions. The applied bias is 0.5 volt. 470
Figure 26.17 Electron density profile in the floating floating semiconductor regions. The applied bias is 0.5 volt. 470
Figure 26.18 Poisson potential profile in the floating floating semiconductor regions. The applied bias is 0.5 volt. 471
Figure 26.19 Hole Fermi level in the floating semiconductor regions. The applied bias is 0.5 volt. 471
Figure 26.20 Electron Fermi level in the floating semiconductor regions. The applied bias is 0.5 volt. 472
Figure 26.21 Structure of a floating metal/semiconductor stack. The lower part is semiconductor (pink). 472
Figure 26.22 Poisson potential profile in the vertical direction through Semi/Metal floating stack. Applied bias at top is 0.5 volt. 473
Figure 26.23 Hole density profile in the vertical direction through Semi/Metal floating stack. Applied bias at top is 0.5 volt. 473

Figure 26.24 Electron density profile in the vertical direction through Semi/Metal floating stack. Applied bias at top is 0.5 volt. 474
Figure 26.25 Hole Fermi level profile in the vertical direction through Semi/Metal floating stack. Applied bias at top is 0.5 volt. 474
Figure 26.26 Electron Fermi level profile in the vertical direction through Semi/Metal floating stack. Applied bias at top is 0.5 volt. 475
Figure 27.1 Links involved in the discretization of the curl-curl operator for the central link 1. 483
Figure 27.2 Links involved in the discretization of the divergence operator. 485
Figure 27.3 Symbolic representation of the equation $\nabla \cdot \mathbf{A} = 0$. 486
Figure 27.4 Symbolic equation extracted from the equation $\nabla \cdot \mathbf{A} = 0$. 486
Figure 27.5 Links involved in the discretization of the curl-curl operator for the central link 1 and the transition to the Laplace operator for this link. 487
Figure 27.6 Links contributing to the Laplacian representation of the discretized curl-curl operator for the central link 1. 488
Figure 27.7 Laplace nodes and corresponding links central link and node 1. 489
Figure 27.8 Inductor layout of an inductor with a closed guard ring. 495
Figure 27.9 log(abs(Re(Y11))+abs(Im(Y11))) and log(abs(Re(Y12))+abs(Im(Y12))). 496
Figure 27.10 Re (S11) and Im (S11). 497
Figure 27.11 Re (S12) and Im (S12). 498
Figure 27.12 Re (22) and Im (S22). 499
Figure 27.13 The inductance, resistance and Q factor of the inductor with grounded closed-guard ring. . . 500
Figure 28.1 Test structure to study the stability problem. 504
Figure 28.2 Eigenvalue plot of $A^{-1}J$. Both permittivities are equal one. The Lorenz gauge is used. 504

Figure 28.3 Eigenvalue plot of $A^{-1}J$. We have different permittivities and a moderately low conductance. (Lorenz gauge). 505
Figure 28.4 Eigenvalue plot of $A^{-1}J$. We have equal permittivities with value equal 4. (Lorenz gauge). . 506
Figure 28.5 Structure with different metals. 506
Figure 28.6 Eigenvalue plot of $A^{-1}J$. Using different conductances but same permittivities (Lorenz gauge). 507
Figure 28.7 Zoom in to the eigenvalue spectrum for different conductances but equal permittivities. (Lorenz gauge). 507
Figure 28.8 Zoom in to the eigenvalue spectrum for different conductances but equal permittivities. (Lorenz gauge). 508
Figure 28.9 The eigenvalue spectrum for metal permittivity taken equal to the oxide permittivity. (Lorenz gauge). 509
Figure 28.10 Eigenvalues for the set up belonging to Figure 28.3 but computed in the Coulomb gauge. 510
Figure 28.11 Lorenz gauge eigenvalues. 510
Figure 28.12 Coulomb gauge eigenvalues. 511
Figure 28.13 Eigenvalue distribution in the Lorenz gauge after different implementation of Equation (28.4). 512
Figure 28.14 Eigenvalue when applying magnetic boundary conditions and $\varepsilon_{\text{oxide}} \neq \varepsilon_{\text{metal}}$. 513
Figure 28.15 $\varepsilon_{\text{oxide}} = \varepsilon_{\text{metal}}$ and Neumann boundary conditions and Lorenz gauge. 514
Figure 28.16 $\varepsilon_{\text{oxide}} = \varepsilon_{\text{metal}}$ and Neumann boundary conditions and Coulomb gauge. 514
Figure 28.17 A straight line structure and Neumann boundary conditions and Coulomb gauge. 515
Figure 28.18 Global view at the eigenvalue spectrum $\varepsilon_{\text{oxide}} = 10,\ \varepsilon_{\text{metal}} = 0$ and Lorenz gauge. 515
Figure 28.19 Zoomed view at the eigenvalue spectrum $\varepsilon_{\text{oxide}} = 10,\ \varepsilon_{\text{metal}} = 0$ and Lorenz gauge. 516
Figure 28.20 Zoomed view at the eigenvalue spectrum $\varepsilon_{\text{oxide}} = 10,\ \varepsilon_{\text{metal}} = 5$ and Lorenz gauge. 516
Figure 28.21 Zoomed view at the eigenvalue spectrum $\varepsilon_{\text{oxide}} = 10,\ \varepsilon_{\text{metal}} = 8$ and Lorenz gauge. 517

Figure 28.22 Zoomed view at the eigenvalue spectrum $\varepsilon_{\text{oxide}} = 10,\ \varepsilon_{\text{metal}} = 10$ and Lorenz gauge. 517
Figure 28.23 Global view at the eigenvalue spectrum $\varepsilon_{\text{oxide}} = 5,\ \varepsilon_{\text{metal}} = 10$ and Lorenz gauge. 518
Figure 28.24 Zoomed view at the eigenvalue spectrum $\varepsilon_{\text{oxide}} = 5,\ \varepsilon_{\text{metal}} = 10$ and Lorenz gauge. 519
Figure 28.25 Further zoomed view at the eigenvalue spectrum $\varepsilon_{\text{oxide}} = 5,\ \varepsilon_{\text{metal}} = 10$ and Lorenz gauge. 519
Figure 28.26 Inductor design. 520
Figure 28.27 Global view at the eigenvalue spectrum $\varepsilon_{\text{oxide}} = 3.9,\ \varepsilon_{\text{metal}} = 10$ and Lorenz gauge. . . . 520
Figure 28.28 Global view at the eigenvalue spectrum $\varepsilon_{\text{oxide}} = 3.9,\ \varepsilon_{\text{metal}} = 10$ and Lorenz gauge. . . . 521
Figure 28.29 Global view at the eigenvalue spectrum $\varepsilon_{\text{oxide}} = 3.9,\ \varepsilon_{\text{metal}} = 10$ and Lorenz gauge. . . . 521
Figure 28.30 Mesh and structure of the 3D folded conductor. . . 522
Figure 28.31 Same structure and mesh under a different viewing angle. 522
Figure 28.32 Eigenvalue spectrum (full range). 523
Figure 28.33 Eigenvalue spectrum (zoom around zero). 523
Figure 28.34 Eigenvalue spectrum (zoom around zero). 524
Figure 28.35 Metal loop layout and mesh. 524
Figure 28.36 Eigenvalue spectrum results for the inductor case. . 525
Figure 28.37 Zoom of the results for the inductor case. 525
Figure 28.38 Impulse response of the inductor showing the instability. 526
Figure 28.39 Spectrum results for a finer mesh. 526
Figure 28.40 Domain size: 20x20x12. 527
Figure 28.41 Domain size: 20x20x20. 527
Figure 28.42 Domain size: 20x20x36. 528
Figure 28.43 Spectrum for the domain size: 20x20x12. 528
Figure 28.44 Spectrum for the domain size: 20x20x20. 529
Figure 28.45 Spectrum for the domain size: 20x20x36. 529
Figure 28.46 2D view of the metal corner embedding in oxide. The green block is oxide with $\varepsilon_{\text{oxide}} = 5$. The yellow part is oxide with $\varepsilon_{\text{oxide}} = 1$. 530
Figure 28.47 3D view of the metal corner embedding in Oxide. The background oxide ($\varepsilon_{\text{oxide}} = 1$) is not shown but present in all layers. 531

Figure 28.48 Zoomed view of the spectrum with Ax, Ay and Az included in the calculation. 531
Figure 28.49 Global view of the spectrum with only A_x and A_y included in the calculation. 532
Figure 28.50 Zoomed view at the spectrum with only A_x and A_y included in the calculation. 532
Figure 28.51 Spectrum for $\tau = 10^{-15}$. With an adapted time scaling we obtain a spectrum that corresponds to a much more stable equation. 533
Figure 28.52 Zoom in to the spectrum around zero. 533
Figure 28.53 Original structure used in this appendix. $\varepsilon_{\text{metal}} = 1$ and $\varepsilon_{\text{oxide}} = 5$. 534
Figure 28.54 Full-range spectrum using the new time scaling factor. 534
Figure 28.55 Spectrum zoom in around Re(λ) is 0. The spread is limited to (–10, 10) using time-scaling 10^{-12} seconds. 535
Figure 28.56 Spectrum zoom in around Re(λ) is 0. The time-scaling 10^{-14} seconds. 535
Figure 28.57 Spectrum zoom in around Re(λ) is 0. The time-scaling 10^{-13} seconds. 536
Figure 28.58 Result (no zoom) in the Coulomb gauge for the time-scaling $\tau = 10^{-14}$ sec. $\varepsilon_{\text{oxide}} = 5$, $\varepsilon_{\text{metal}} = 1$ and mesh as shown in Figure 28.53. 537
Figure 28.59 Result (zoom) in the Coulomb gauge for the time-scaling $\tau = 10^{-14}$ sec. $\varepsilon_{\text{oxide}} = 5$, $\varepsilon_{\text{metal}} = 1$ and mesh as shown in Figure 28.53. 537
Figure 28.60 Result (extra zoom) in the Coulomb gauge for the time-scaling $\tau = 10^{-14}$ sec. $\varepsilon_{\text{oxide}} = 5$, $\varepsilon_{\text{metal}} = 1$ and mesh as shown in Figure 28.53. . . 538
Figure 28.61 Zoom-in to the eigenvalue spectrum around the real axis. 539
Figure 28.62 Spectrum of all eigenvalues for the structure in Figure 28.53. The Lorenz gauge is used and the new implementation of mixed space-time terms is applied. 540
Figure 28.63 Eigenvalue spectrum with new implementation of the transient voltage terms in the Maxwell-Ampere equation. (Zoom in around zero). 541

Figure 28.64 Electric field plot. 541
Figure 28.65 A-field in the middle plane of the structure. 542
Figure 28.66 Π-field in the middle plane of the structure. 542
Figure 28.67 Π-field in the middle plane of the structure. 543
Figure 28.68 Results for the corner using the old implementation. 543
Figure 28.69 Spectrum zoom-in for the calculation including A_x, A_y (no A_z) containing the wrong-sign real part eigenvalues. 544
Figure 28.70 Eigenvalue spectrum for the calculation including A_x, A_y and A_z containing the wrong-sign real part eigenvalues. 544
Figure 28.71 Zoom-in to the spectrum for the bar structure using a course mesh. 545
Figure 28.72 Further zoom-in to the spectrum for the bar structure using a course mesh. 546
Figure 28.73 Spectrum for a calculation with the E-field switched on but the vector potential is not activated. 546
Figure 28.74 Eigenvalue spectrum zoom-in with lowly conducting oxide: $\sigma_{\mathrm{Ox}} = 1$ S/m. 547
Figure 28.75 Eigenvalue spectrum for a system without conducting material. 548
Figure 28.76 2D view in the mid plane of the structure including the mesh. 549
Figure 28.77 3D view of the structure. The mesh is also shown. . 549
Figure 28.78 Eigenvalue plot (no zoom). 550
Figure 28.79 Zoomed view of the eigenvalue spectrum of the U-shape. 551
Figure 28.80 Spectrum corresponding the $A_z = 0$. 553
Figure 28.81 Damping Pi using delta= 0.01. 554
Figure 28.82 Using Pi damp with loss = 2000. 555
Figure 28.83 Using Pi damp with loss = 1000. 556
Figure 28.84 The bar of metal. 556
Figure 28.85 3D view + mesh of the bar of metal. 557
Figure 28.86 The eigenvalue spectrum around zero. 558
Figure 28.87 Test structure: 2D view. 560
Figure 28.88 Test structure: 3D view. 560

List of Tables

Table 15.1 Some characteristic results for two crossing wires . . 230
Table 15.2 Some characteristic results for a square coaxial cable 231
Table 15.3 Some characteristic results for the spiral inductor . . 232
Table 17.1 Norms of L^{-1} and U^{-1} computed for the A-V and E-V solvers (ILUT(10^{-6})) 312
Table 17.2 Iterative performance of A-V and E-V solvers for the cross wire structure. (ILUT(10^{-6}), time unit: second) 314
Table 17.3 Iterative performance of A-V and E-V solvers for the metal plug structure. (ILUT(10^{-6}), time unit: second) 314
Table 17.4 Iterative performance of A-V and E-V solvers for the SNI structure. (ILUT(10^{-4}), time unit: second) 315
Table 18.1 Condition numbers of the **A**-V and **E**-V- solvers . . 324
Table 22.1 3D TCAD extraction of E, B and J in E, EM and EMLF simulation cases 381
Table 22.2 Resistance values 383
Table 22.3 Extraction of main significant point on IV curve response 384
Table 22.4 Extraction of magnetic field B in the full structure and in silicon region for A, B, C, D points 385
Table 22.5 Extraction of electric field E in the full structure and in silicon region for A, B, C, D points 385
Table 22.6 Extraction of current density $\mathbf{J} \sim; 8 \times 10^{10} A/m^2$, $\mathbf{J}_\mathrm{n} \sim 1 \times 10^{11} A/m^2$, $\mathbf{J}_\mathrm{p} \sim 2 \times 10^{10} A/m^2$ in the full structure and in the silicon for A, B, C, D points . . . 386
Table 22.7 Main values for physical parameters extracted on points A, B, C, D 386

Table 22.8 Extraction of Magnetic field B in the full structure and in silicon region for the α, β, γ points 389
Table 22.9 Extraction of current density $\mathbf{J}$, $\mathbf{J}_\mathrm{n}$ and $\mathbf{J}_\mathrm{p}$ in the structure and in silicon for α, β, γ points 389
Table 22.10 Main values for extracted variables at the points α, β and γ 391
Table 22.11 Comparison between measurements and simulation for Vt1 and Vhold on full ESD structure 392
Table 23.1 Specifications of test structures 404
Table 23.2 Performance of iterative solution in A-V and E-V solvers 405
Table 24.1 Specifications of test structures 424
Table 24.2 Runtime breakdown of GR2, MEXP-AV and MEXP-EV for the SWR case using $h = 0.25ps$ (time unit: sec) 426
Table 24.3 Performance using adaptive step size (time unit: sec) 428
Table 24.4 Effect of breakdown of linear/nonlinear nodes (unit: sec) 430
Table 28.1 Comparison of Lorenz and Coulomb gauge results 512

List of Symbols

$\mathbf{A}$	vector potential
$\mathbf{A}_{\mathrm{EX}}$	external vector potential
$\mathbf{A}_{\mathrm{IN}}$	induced vector potential
$\mathbf{B}, \mathbf{B}_{\mathrm{IN}}$	magnetic induction
C	capacitance
c	speed of light, concentration
$\mathbf{dr}$	line element
$\mathbf{dS}$	surface element
$\mathrm{d}s$	elementary distance in Riemannian geometry
$\mathrm{d}\tau$	volume element
D_{n}	electron diffusion coefficient
D_{p}	hole diffusion coefficient
$\mathbf{D}$	electric displacement vector
E	energy
E_{F}	Fermi energy
$E_{\alpha\mathbf{k}}(W)$	electron energy
e	elementary charge
$\mathbf{E}$	electric field
$\mathbf{E}_{\mathrm{C}}$	conservative electric field
$\mathbf{E}_{\mathrm{EX}}$	external electric field
$\mathbf{E}_{\mathrm{IN}}$	induced electric field
$\mathbf{E}_{\mathrm{NC}}$	non-conservative electric field
$\mathbf{e}_{\mathrm{z}}$	unit vector along z-axis
$\mathbf{e}_{\phi}$	azimuthal unit vector
$F_{\mu\nu}$	electromagnetic field tensor
$f, f_{\mathrm{n}}, f_{\mathrm{p}}$	(Boltzmann) distribution function
G	conductance, generation rate
G_{Q}	quantized conductance
$g_{\mu\nu}$	metric tensor
H	Hamiltonian
$H_{\mathbf{p}'\mathbf{p}}$	Hamiltonian scatttering matrix element

h	Planck's constant
$\hbar$	reduced Planck constant ($h/2\pi$)
I	electric current
i	imaginary unit
J_G	gate leakage current
$\mathbf{J}, \mathbf{J}_\mathrm{n}, \mathbf{J}_\mathrm{p}$	electric current density
$\mathbf{H}$	magnetic field intensity
k	wavenumber
k_B	Boltzmann's constant
$\mathbf{k}$	electron wave vector
L	inductance, Lagrangian, length
L_x, L_y	length
$\mathbf{L}$	total angular momentum
l	subband index, angular momentum quantum number, length
m, m_n	charge carrier effective mass
m_0	free electron mass
$m_\mathrm{n}, m_{\alpha x}, m_{\alpha y}, m_{\alpha z}$	electron effective masse
m_p	hole effective masse
$\mathbf{M}$	magnetization vector
$\mathbf{m}$	magnetic moment
N	number of particles, coordinates or modes
N_A	acceptor doping density
n	electron concentration
$\mathbf{n}$	unit vector
$p, p_i, \mathbf{p}, \mathbf{p}_i, P, P_i, \mathbf{P}, \mathbf{P}_i, \ldots$	generalized momenta
p	hole concentration
$\mathbf{P}$	total momentum, electric polarization vector
$\mathbf{p}$	momentum, electric dipole moment
$q, q_i, \mathbf{q}, \mathbf{q}_i, Q, Q_i, \mathbf{Q}, \mathbf{Q}_i, \ldots$	generalized coordinates
Q	electric charge
q_n	carrier charge
Q_A	electric charge residing in active area
R	resistance, recombination rate
R_H	Hall resistance
R_K	von Klitzing resistance
R_L	lead resistance

R_{Q}	quantized resistance
$R^{\mu}_{\rho\lambda\sigma}$	Riemann tensor
$\mathbb{R}$	set of real numbers
(r, θ, ϕ)	spherical coordinates
$\mathbf{r}, \mathbf{r}_n$	position vector
S	action, entropy
$S(\mathbf{p}, \mathbf{p}')$	transition rate
$\mathbf{S}$	Poynting vector
$\mathbf{S}_{\mathrm{n}}, \mathbf{S}_{\mathrm{p}}$	energy flux vector
t	time
T	lattice temperature
T_{n}	electron temperature
T_{p}	hole temperature
$\mathbf{T}, \mathbf{T}_{\alpha\beta}$	EM energy momentum tensor
U_{E}	electric energy
U_{M}	magnetic energy
U_{EM}	EM energy
$U(y), U(z)$	potential energy
u_{EM}	EM energy density
V	scalar electric potential
V_{H}	Hall voltage
V_{G}	gate voltage
$\mathbf{v}_n$	carrier velocity
$\mathbf{v}_{\mathrm{n}}, \mathbf{v}_{\mathrm{p}}$	drift velocity
$\mathbf{v}$	drift velocity, velocity field
$W, W_{\alpha l}(W)$	subband energy
$w, w_{\mathrm{n}}, w_{\mathrm{p}}$	carrier energy density
(x, y, z)	Cartesian coordinates
Y	admittance
Z	impedance
α	summation index, valley index, variational parameters
β	summation index, $1/k_{\mathrm{B}}T$
$\partial\Omega$	boundary surface of Ω
$\partial\Omega_{\infty}$	boundary surface of Ω_{∞}
ϵ_0	electric permittivity of vacuum
ϵ	electric permittivity
$\epsilon_{\mathrm{r}}, \epsilon_{\mathrm{S}}$	relative electric permittivity

Γ	closed curve inside a circuit
$\Gamma_{\alpha l}$	resonance width
$\Gamma^{\alpha}_{\mu\nu}$	affine connection
κ	wavenumber
$\kappa_{\mathrm{n}}, \kappa_{\mathrm{p}}$	thermal conductivity
Λ_{EM}	EM angular momentum density
μ	magnetic permeability, chemical potential
μ_0	magnetic permeability of vacuum
μ, μ_n, μ_p	carrier mobility
μ_{r}	relative magnetic permeability
Ω	connected subset of $\mathbb{R}^3$, volume, circuit region, Ohm
Ω_{∞}	all space
ω	angular frequency
$\boldsymbol{\pi}_{\mathrm{EM}}$	EM momentum density
$\mathbf{\Pi}$	pseudo-canonical momentum
ρ	electric charge density
(ρ, ϕ, z)	cylindrical coordinates
σ	electrical conductivity, spin index
$\tau, \tau_0, \tau_{\mathrm{e}}, \tau_{\mathbf{p}}, \tau_{\mathrm{en}}, \tau_{\mathrm{ep}}$	relaxation time
$\tau_{\alpha l}$	resonance lifetime
χ	gauge function
χ_{e}	electric susceptibility
$\chi_k(y)$	wave function
χ_{m}	magnetic susceptibility
Φ_{D}	electric flux (displacement)
Φ_{E}	electric flux (electric field)
Φ_{ex}	external magnetic flux
Φ, Φ_{M}	magnetic flux
$\psi(\mathbf{r})$	wave function
$\boldsymbol{\nabla}$	gradient
$\boldsymbol{\nabla}\cdot$	divergence
$\boldsymbol{\nabla}\times$	curl
$\boldsymbol{\nabla}^{\mathbf{2}}$	vectorial Laplace operator
∇^2	Laplace operator
$\mathcal{L}$	Lagrange density, inductance per unit length
V_{ϵ}	electromotive force

List of Abbreviations

AK	anode/cathode (K from German: katode)
AMD	approximate minimum degree
AV	vector potential A, voltage potential V (solver)
B	back (link orientation, also head)
BDF	backward-differentiation formula
BE	backward Euler
BEM	boundary element method
BEOL	back-end of line (device architecture of interconnect)
BiCGStab	biconjugate gradient stabilized (method)
CAD	computer-aided design
CGS	conjugate gradient squared
CMOS	complementary metal-oxide-semiconductor
COLAMD	column approximate minimum degree (permutation)
CPU	central processing unit
DAE	differential-algebraic equation
DC	direct current
DD	drift-diffusion
devEM	device electromagnetic (solver)
DOF	degree of freedom
DS	direct solver
DUT	device under test
E	electric
EHF	extreme high frequency
EM	electromagnetic
EMC	electromagnetic compatibility
EMF	electromotive force
EMLF	electromagnetic Lorentz force
EQS	electro-quasi static
ESD	electrostatic discharge
EV	electric field E, voltage potential V (solver)
EVBG	electric-field voltage-field balanced gauge

EXP	experimental (data)
F	front (link orientation, also tail)
FD	finite difference scheme
FDTD	finite difference time domain
FEM	finite element method
FEOL	front-end of line (device architectures in silicon)
FIT	finite integration technique
FMM	fast multi-pole method
FSM	finite surface method
FVM	finite volume method
GMRES	generalized minimal residue (method)
GR2	Gear's second order (method)
HFFS	High-frequency structural simulator (Ansys)
HJ	Hammerstad and Jensen (authors)
IGN	ground current
ILU	incomplete lower upper
ILUT	incomplete lower-upper with threshold
IO	input/output
IQHE	integer quantum Hall effect
IR	infrared
L	left (link orientation)
LF	Lorentz force
LHS (lhs)	left-hand side
LMS	linear multi-step
LNA	low-noise amplifier
LTE	local truncation error
MAE	Maxwell-Ampere equation
MAGWEL	**M**axwell-**G**host-solver-for **W**ell-designed **El**ectronics
MEXP	matrix exponential
MGW	MAGWEL
MMIC	monolithic microwave integrated circuit
MOM (MoM)	method of moments
MOSFET	metal-oxide semiconductor field effect transistor
MQS	magneto-quasi static
NBC	Neumann boundary conditions
NC	non-convergence
ODE	ordinary differential equation
OPERA	electromagnetic finite-element simulation software (Cobham)

PDE	partial-differential equation
PEEC	partial element electrical circuit
Q	Quality (factor)
QM	quantum-mechanical
R	right (link orientation)
RAM	random access memory
RF	radio frequency
RHS (rhs)	right-hand side
S (parameter)	scattering (parameter)
SAMG	algebraic multi-grid methods for Systems
SCR	silicon-controlled rectifier
SG	Scharfetter-Gummel
SGS	signal-ground-signal
SG2	second order Gear's (method)
SIBC	surface impedance boundary condition
SNI	substrate noise isolation
SOI	silicon on insulator
solvEM	batch-mode version of devEM simulation software
SRH	Shockley-Read-Hall (recombination)
SWR	silicon wire
TCAD	technology computer aided design
TEM	tranversal electromagnetic (wave)
TLP	transmission line pulse
TSV	through-silicon via
UV	ultraviolet
VCO	voltage controlled oscillator
VLAIO	Vlaanderen Agentschap innoveren & ondernemen
VNA	vector network analyzer
XWR	cross wire

PART I

Introduction to Electromagnetism

In this part the derivation and illustrations of the Maxwell equations are given with specific attention to the goal of transfering these equations to computational algorithms that will be discussed in detail in Part II.

1

Introduction

Electromagnetism, formulated in terms of the Maxwell equations, and quantum mechanics, formulated in terms of the Schrödinger equation, constitute the physical laws by which the bulk of natural experiences are described. Apart from the gravitational forces, nuclear forces and weak decay processes, the description of the physical facts starts with these underlying microscopic theories. However, knowledge of these basic laws is only the beginning of the process to apply these laws in realistic circumstances and to determine their quantitative consequences. With the advent of powerful computer resources, it has become feasible to extract information from these basic laws with unprecedented accuracy. In particular, the complexity of realistic systems manifests itself in the non-trivial boundary conditions, such that without computers, reliable calculation are beyond reach.

The ambition of physicists, chemists and engineers, to provide tools for performing calculations, does not only boost progress in technology but also has a strong impact on the formulation of the equations that represent the physics knowledge and hence provides a deeper understanding of the underlying physics laws. As such, computational physics has become a cornerstone of theoretical physics and we may say that without a computational recipe, a physics law is void or at least incomplete. Contrary to what is sometimes claimed, that after having found the unifying theory for gravitation and quantum theory, there is nothing left to investigate, we believe that physics has just started to flourish and there are wide fields of research waiting for exploration.

This volume is dedicated to the study of electrodynamic problems. The Maxwell equations appear in the form

$$\Delta(\textit{field}) = \textit{source} \tag{1.1}$$

where Δ describes the near-by field variable correlation of the field that is induced by a source or field disturbance. Near-by correlations can be mathematically expressed by differential operators that probe changes going from one location to a neighboring one. It should be emphasized that "near-by" refers to space and time.

One could "easily" solve these equations by construction the inverse of the differential operator. Such an inverse is usually known as a Green function.

There are two main reasons that prevent a straightforward solution of the Maxwell equations.

First of all, realistic structure boundaries may be very irregular, and therefore the corresponding boundary conditions cannot be implemented analytically.

Secondly, the sources themselves may depend on the values of the fields and will turn the problem in a highly non-linear one, as may be seen from Equation (1.1) that should be read as

$$\Delta(\mathit{field}) = \mathit{source}(\mathit{field}) \tag{1.2}$$

The bulk of this volume is dedicated to find solutions to equations of this kind. In particular, the Parts II, III and IV are dealing with above type of equations. A considerable amount of work deals with obtaining the details of the right-hand side of Equation (1.2), namely how the source terms, being charges and currents depend in detail on the values of the field variables.

Whereas, the microscopic equation describe the physical processes in great detail, i.e. at every space-time point field and source variables are declared, it may be profitable to collect a whole bunch of these variables into a single basket and to declare for each basket a few representative variables as the appropriate values for the fields and the sources. This kind of reduction of parameters is the underlying strategy of circuit modeling. Here, the Maxwell equations are replaced by Kirchhoff's network equations.

The "basket" containing a large collection of fundamental degrees of freedom of field and source variables, should not be filled at random. Physical intuition suggests that we put together in one basket degrees of freedom that are "alike". Field and source variables at near-by points are candidates for being grabbed together, since physical continuity implies that a all elements in the basket will have similar values[1].

The baskets are not only useful for simplifying the continuous equations. They are vital to the discretization schemes. Since any computer has only

---

[1]It should be emphasized that such a picture works at the classical level. Quantum physics implies that near-by field point may take any value and the continuity of fields is not required.

a finite memory storage, the continuous or infinite collection of degrees of freedom must be mapped onto a finite subset. This may be accomplished by appropriately positioning and sizing of all the baskets. This procedure is named "grid generation" and the construction of a good grid is often of great importance to obtain accurate solutions.

After having mapped the continuous problem onto a finite grid one may establish a set of algebraic equations connecting the grid variables (basket representatives) and explicitly reflecting the non-linearity of the original differential equations. The solution of large systems of non-linear algebraic equations is based on Newton's iterative method. To find the solution of the set of non-linear equations $\mathbf{F}(\mathbf{x}) = \mathbf{0}$, an initial guess is made: $\mathbf{x} = \mathbf{x}_{\text{init}} = \mathbf{x}_0$. Next the guess is (hopefully) improved by looking at the equation:

$$\mathbf{F}(\mathbf{x}_0 + \Delta\mathbf{x}) \simeq \mathbf{F}(\mathbf{x}_0) + \mathbf{A} \cdot \Delta\mathbf{x}, \tag{1.3}$$

where the matrix $\mathbf{A}$ is

$$\mathbf{A}_{ij} = \left(\frac{\partial F_i(\mathbf{x})}{\partial x_j}\right)_{\mathbf{x}_0}. \tag{1.4}$$

In particular, by assuming that the correction brings us close to the solution, i.e. $\mathbf{x}_1 = \mathbf{x}_0 + \Delta\mathbf{x} \simeq \mathbf{x}^*$, where $\mathbf{F}(\mathbf{x}^*) = 0$, we obtain that

$$\begin{aligned} 0 &= \mathbf{F}(\mathbf{x}_0) + \mathbf{A} \cdot \Delta\mathbf{x} \qquad \text{or} \\ \Delta\mathbf{x} &= -\mathbf{A}^{-1} \cdot \mathbf{F}(\mathbf{x}_0) \end{aligned} \tag{1.5}$$

Next we repeat this procedure, untill convergence is reached. A series of vectors, $\mathbf{x}_{\text{init}} = \mathbf{x}_0, \mathbf{x}_1, \mathbf{x}_2, \ldots \mathbf{x}_{n-1}, \mathbf{x}_n = \mathbf{x}_{\text{final}}$, is generated, such that $|\mathbf{F}(\mathbf{x}_{\text{final}})| < \epsilon$, where ϵ is some prescribed error criterion. In each iteration a large linear matrix problem of the type $\mathbf{A}|\mathbf{x}\rangle = |\mathbf{b}\rangle$ needs to be solved. 'Solving' means here: find the vector $|\mathbf{x}\rangle$ that satisfies this equation. It does not inply: find the inverse of $\mathbf{A}$ and perform the multiplicaton $\mathbf{A}^{-1}|\mathbf{b}\rangle = |\mathbf{x}\rangle$. In fact, there exist numerous techniques to find $|\mathbf{x}\rangle$ avoiding the computation of $\mathbf{A}^{-1}$. These methods are exploited in the third part of this book.

2

The Microscopic Maxwell Equations

2.1 Definition of the Electric Field

Static electric phenomena are time-independent manifestations of a repulsive or attractive forces that are attributed to the presence of net charges located at material bodies. Electric forces were originally observed by rubbing amber with a piece of tissues made of silk. (The Greek word elektron - $\epsilon\lambda\epsilon\kappa\tau\rho\omega\nu$ - is amber). More experiments led to the conclusion that objects can be electrified or being in an electric state which lead to mutually attractive or repulsive forces. The states are next marked as being positively charged or negatively charged, where equal charge sign corresponds to repulsion and opposite charge sign corresponds to attraction. Further experiments led to the conclusion that electric states correspond to a *quantitative* amount of electricity or charge being present on/in the material bodies. The amount of charge in the above experiments is nowadays understood as an excess or lack of electrons and each electron has a fixed unit of charge attached to it. Another crucial piece of information was gathered by experiments showing that the total amount of charge is conserved: by putting two equal-sized metal spheres in contact with each other it was shown that if one sphere had no electric effect before the contacting, after the touch the electric effect was equally distributed and with half of the size compared to the effect corresponding to the charged sphere before putting the spheres in contact. The strength of the electric force was made explicit by Coulomb and is summarized in Coulomb's law: The force between two charges Q_1 and Q_2 varies proportional with the magnitude of each charge, varies inversely with square of the distance between the two charges, is oriented along the line joining the two charges and is attractive for opposite charges and repulsive for equal type of charges. When more than two bodies with charges are involved the resulting force is composed as a vector sum of the participating two-body forces.

$$\mathbf{F} = k\frac{Q_1 Q_2}{r^2} \tag{2.1}$$

The constant k is a proportionality constant that is also written as

$$k = \frac{1}{4\pi\epsilon_0} \tag{2.2}$$

The unit of charge is "Coulomb" and one electron has a charge of 1.6210^{-19} Coulomb. The permittivity ϵ_0 corresponds to the situation that the force law is applied in vacuum and has a numerical value equal to $8.854187817\ 10^{-12}$ Newton $\text{meter}^2/\text{Coulomb}^2 = \text{Farad}/\text{meter}$.

A major step forward was made by imaging that a charge changes its environment by creating a surrounding *field*. Thus, given a charge Q, another charge q will sense the presence of Q indirectly by a field that is due to the presence of a change in the properties of the surrounding volume around Q. This quality of the surrounding space around Q is referred to as the *electric field* and does not depend on the probing charge q that is used to detect the presence of the electric field. Of course, the probing charge brings with it its own electric field such the total electric field is a combination of the two fields corresponding to the two fields if the charges are separated by a large distance.
From Coulomb's law, the electric field is given by

$$\mathbf{E} = \frac{1}{4\pi\epsilon_0}\frac{Q}{r^2}\hat{\mathbf{r}} \tag{2.3}$$

where $\hat{\mathbf{r}}$ is a unit vector pointing away from the the charge Q being located in the origin.

2.2 Definition of the Magnetic Field

Magnetic effects have been known at least as long as electric effects, but their understanding turned out to be more difficult than the electric ones. The reason being that magnetic materials have intricate alignment behavior. It was only after this discovery that currents of electric charge have a magnetic effect. Just as with the electric fields, the field concept was also exploited to get a handle on magnetic effects. Thus, we say that a current creates a magnetic field in its surrounding and the strength of the magnetic field can be found by measurement of the torque force on a magnetic test particle. In fact, the magnetic particle responds similar to a pair of opposite charges. There alignment to the magnetic fields suggests that magnetic particles are *dipoles*. Contrary to electric charge, magnetic monopoles have never been observed.

The magnetic force law was formulated by Biot and Savard. For a current I in a wire element of length Δl located at the origin, the contribution to the magnetic-flux density $\Delta \mathbf{B}$ is

$$\Delta \mathbf{B} = kI\frac{\Delta \mathbf{l} \times \mathbf{x}}{r^3} \tag{2.4}$$

where k is some constant to be determined soon. Since currents of charges have a magnetic effect, one may consider two currents instead of one current and one magnetic dipole. Indeed it is found that two parallel currents also exert a mutual force, proportional to the currents in the wires and inversely proportional to the distance at least for two long parallel wires. In order to use the magnetic field concept one can deduce this force as the result of a cumulative force due to all current elements that constitute the wires. The force element ΔF acting on a current element Δl due to the presence of a magnetic intensity flux is

$$\Delta \mathbf{F} = k' I \Delta \mathbf{l} \times B \tag{2.5}$$

where k' is a constant. We are now in the position to determine k and k', Combining both equations, the mutual force acting on two current loops is

$$\mathbf{F}_{12} = kk' I_1 I_2 \oint \oint \frac{d\mathbf{l_1} \times d\mathbf{l_1} \times \mathbf{x}_{12}}{|x_{12}|^3} \tag{2.6}$$

The constant kk' has dimension $[\mathrm{N/A^2}]$ and is written as $\mu_0/2\pi$. The value of $\mu_0 = 4\pi \times 10^{-7}$ and this value of course depends on the unit charge current which is the Ampere [A]. The ampere is the current corresponding to a throughput of one Coulomb charge per second. Experimentally it turned out in the past that it is more convenient to select the Ampere as a starting unit such that the Coulomb becomes a derived unit being 1 A.sec.

We will close our introductory section here concerning the origin of the static electric and magnetic fields and present in the next chapter a review of the electrodynamics of space and time dependent electromagnetic fields.

2.3 The Microscopic Maxwell Equations in Integral and Differential Form

In general, any electromagnetic field can be described and characterized on a microscopic scale by two vector fields $\mathbf{E}(\mathbf{r}, t)$ and $\mathbf{B}(\mathbf{r}, t)$ specifying respectively the electric field and the magnetic induction in an arbitrary space point $\mathbf{r}$ at an arbitrary time t. All dynamical features of these vector fields are contained in the well-known Maxwell equations (Maxwell [1954] Jackson [1975] Feynman et al. [1964])

$$\nabla \cdot \mathbf{E} = \frac{\rho}{\epsilon_0} \tag{2.7}$$

$$\nabla \cdot \mathbf{B} = 0 \tag{2.8}$$

$$\nabla \times \mathbf{E} = -\frac{\partial \mathbf{B}}{\partial t} \tag{2.9}$$

$$\nabla \times \mathbf{B} = \mu_0 \mathbf{J} + \epsilon_0 \mu_0 \frac{\partial \mathbf{E}}{\partial t}. \tag{2.10}$$

They describe the spatial and temporal behavior of the electromagnetic field vectors and relate them to the sources of electric charge and current that may be present in the region of interest. Within the framework of a microscopic description, the electric charge density ρ and the electric current density $\mathbf{J}$ are considered spatially localized distributions residing in vacuum. As such they represent both mobile charges giving rise to macroscopic currents in solid-state devices, chemical solutions, plasmas etc., and bound charges that are confined to the region of an atomic nucleus. In turn, the Maxwell equations in the above presented form explicitly refer to the values taken by $\mathbf{E}$ and $\mathbf{B}$ in vacuum and, accordingly, the electric permittivity ϵ_0 and the magnetic permeability μ_0 appearing in Equations (2.7) and (2.10) correspond to vacuum.

From the mathematical point of view, the solution of the differential Equations (2.7), (2.8), (2.9) and (2.10) together with appropriate boundary conditions in space and time, should in principle unequivocally determine the fields $\mathbf{E}(\mathbf{r}, t)$ and $\mathbf{B}(\mathbf{r}, t)$. In practice however, analytic solutions may be achieved only in a limited number of cases and, due to the structural and geometrical complexity of modern electronic devices, one has to adopt advanced numerical simulation techniques to obtain reliable predictions of electromagnetic field profiles. In this light, the aim is to solve Maxwell's equations on a discrete set of mesh points using suitable discretization techniques which are often taking advantage of integral form of Maxwell's equations. The latter may be derived by a straightforward application of Gauss' and Stokes' theorems. In particular, one may integrate Equations (2.7) and (2.7) over a simply connected region $\Omega \in \mathbb{R}^3$ bounded by a closed surface $\partial\Omega$ to obtain

$$\int_{\partial\Omega} \mathbf{E}(\mathbf{r}, t) \cdot \mathbf{dS} = \frac{1}{\epsilon_0} Q(t) \tag{2.11}$$

$$\int_{\partial\Omega} \mathbf{B}(\mathbf{r}, t) \cdot \mathbf{dS} = 0, \tag{2.12}$$

where $Q(t)$ denotes the instantaneous charge residing in the volume Ω, i.e.

$$Q(t) = \int_{\Omega} \rho(\mathbf{r}, t) \, \mathrm{d}\tau. \tag{2.13}$$

Equation (2.11) is nothing but Gauss' law stating that the total outward flux of the electric field threading the surface $\partial\Omega$ equals the total charge contained in the volume Ω up to a factor ϵ_0 whereas Equation (2.12) reflects the absence of magnetic monopoles.

Similarly, introducing an arbitrary, open and simply connected surface Σ bounded by a simple, closed curve Γ, one may extract the induction law of Faraday and Ampère's law by integrating respectively Equations (2.9) and (2.10) over Σ:

$$\oint_{\Gamma} \mathbf{E}(\mathbf{r}, t) \cdot \mathbf{dr} = -\frac{\mathrm{d}\Phi_{\mathrm{M}}(t)}{\mathrm{d}t} \tag{2.14}$$

$$\oint_{\Gamma} \mathbf{B}(\mathbf{r}, t) \cdot \mathbf{dr} = \mu_0 \left(I(t) + \epsilon_0 \frac{\mathrm{d}\Phi_{\mathrm{E}}(t)}{\mathrm{d}t} \right). \tag{2.15}$$

The variables $\Phi_{\mathrm{E}}(t)$ and $\Phi_{\mathrm{M}}(t)$ are representing the time-dependent electric and magnetic fluxes piercing the surface Σ and are defined as:

$$\Phi_{\mathrm{E}}(t) = \int_{\Sigma} \mathbf{E}(\mathbf{r}, t) \cdot \mathbf{dS} \tag{2.16}$$

$$\Phi_{\mathrm{M}}(t) = \int_{\Sigma} \mathbf{B}(\mathbf{r}, t) \cdot \mathbf{dS}, \tag{2.17}$$

while the circulation of the electric field around Γ is the instantaneous electromotive force $V_\epsilon(t)$ along Γ is:

$$V_\epsilon(t) = \oint_{\Gamma} \mathbf{E}(\mathbf{r}, t) \cdot \mathbf{dr}. \tag{2.18}$$

The right-hand side of Equation (2.15) consists of the total current flowing through the surface Σ

$$I(t) = \int_{\Sigma} \mathbf{J}(\mathbf{r}, t) \cdot \mathbf{dS} \tag{2.19}$$

and the so-called displacement current which is proportional to the time derivative of the electric flux.

The sign of the above line integrals depends on the orientation of the closed loop Γ, the positive traversal sense of which is uniquely defined by the orientation of the surface Σ imposed by the vectorial surface element $\mathbf{dS}$. Apart from this restriction it should be noted that the surface Σ can be chosen freely so as to extract meaningful physical information from the corresponding Maxwell equation. In particular, though being commonly labeled by the symbol Σ, the surfaces appearing in Faraday's and Ampère's laws (Equations (2.14)–(2.15) will generally be chosen in a different way as can be illustrated by the example of a simple electric circuit. In the case of Faraday's law, one usually wants $\Phi_{\mathrm{M}}(t)$ to be the magnetic flux threading the circuit and therefore Σ would be chosen to "span" the circuit while Γ would be located in the interior of the circuit area. On the other hand, in order to exploit Ampère's law, the surface Σ should be pierced by the current density in the circuit in order to make $I(t)$ the current flowing through the circuit.

2.4 Conservation Laws

Although a complete description of the electromagnetic field requires the full solution of the Maxwell equations in their differential form, one may extract a number of conservation laws may by simple algebraic manipulation. The differential form of the conservation laws takes the generic form

$$\boldsymbol{\nabla} \cdot \mathbf{F} + \frac{\partial \mathbf{G}}{\partial t} = \mathbf{K}, \tag{2.20}$$

where $\mathbf{F}$ is the generalized flow tensor associated with the field $\mathbf{G}$ and $\mathbf{K}$ is related to any possible external sources or sinks.

2.4.1 Conservation of Charge – The Continuity Equation

Taking the divergence of Equation (2.10) and the time derivative of Equation (2.7) and combining the resulting equations, one easily obtains the charge-current continuity equation expressing the conservation of electric charge:

$$\boldsymbol{\nabla} \cdot \mathbf{J} + \frac{\partial \rho}{\partial t} = 0. \tag{2.21}$$

Integration over a closed volume Ω yields

$$\int_{\partial\Omega} \mathbf{J} \cdot \mathrm{d}\mathbf{S} = -\frac{\partial}{\partial t}\int_{\Omega} \rho \, \mathrm{d}\tau, \tag{2.22}$$

which states that the total current flowing through the bounding surface $\partial\Omega$ equals the time rate of change of all electric charge residing within Ω.

2.4.2 Conservation of Energy – Poynting's Theorem

The electromagnetic energy flow generated by a time dependent electromagnetic field is most adequately represented by the well-known Poynting vector given by

$$\mathbf{S} = \frac{1}{\mu_0}\mathbf{E} \times \mathbf{B}. \tag{2.23}$$

Calculating the divergence of $\mathbf{S}$ and using the Maxwell equations, one may relate the Poynting vector to the electromagnetic energy density u_{EM} through the energy conservation law

$$\boldsymbol{\nabla} \cdot \mathbf{S} + \frac{\partial\, u_{\mathrm{EM}}}{\partial t} = -\mathbf{J} \cdot \mathbf{E}, \tag{2.24}$$

which is also known as the Poynting theorem. The energy density u_{EM} is given by

$$u_{\mathrm{EM}} = \frac{1}{2}\left(\epsilon_0 E^2 + \frac{B^2}{\mu_0}\right). \tag{2.25}$$

The energy conservation expressed in Equation (2.24) refers to the total energy of the electromagnetic field and all charged particles contributing to the charge and current distributions. In particular, denoting the mechanical energy of the charged particles residing in the volume Ω by E_{MECH} one may derive for both classical and quantum mechanical systems that the work done per unit time by the electromagnetic field on the charged volume is given by

$$\frac{\mathrm{d}E_{\mathrm{MECH}}}{\mathrm{d}t} = \int_{\Omega} \mathbf{J} \cdot \mathbf{E} \, \mathrm{d}\tau. \tag{2.26}$$

Introducing the total electromagnetic energy associated with the volume Ω as $E_{\mathrm{EM}} = \int_{\Omega} u_{\mathrm{EM}} \, \mathrm{d}\tau$ one may integrate Poynting's theorem to arrive at

$$\frac{\mathrm{d}}{\mathrm{d}t}(E_{\mathrm{MECH}} + E_{\mathrm{EM}}) = -\int_{\partial\Omega} \mathbf{S} \cdot \mathrm{d}\mathbf{S}. \tag{2.27}$$

It should be emphasized that the above result also covers most of the common situations where the energy of the charged particles is relaxed to the environment through dissipative processes. The latter may be accounted for by invoking appropriate constitutive equations expressing the charge and current densities as linear or non-linear responses to the externally applied electromagnetic fields and other driving force fields. As an example, we mention Ohm's law, proposing a linear relation between the macroscopic electric current density and the externally applied electric field in a non-ideal conductor:

$$\mathbf{J}_{\mathrm{M}} = \sigma \mathbf{E}_{\mathrm{EXT}}. \tag{2.28}$$

Here, the conductivity σ is assumed to give an adequate characterization of all microscopic elastic and inelastic scattering processes that are responsible for the macroscopically observable electric resistance.

The derivation of constitutive equations will be discussed in greater detail in Section 2.4.

2.4.3 Conservation of Linear Momentum – The Electromagnetic Field Tensor

In an analogous way, an appropriate linear momentum density $\boldsymbol{\pi}_{\mathrm{EM}}$ may be assigned to the electromagnetic field, which differs from the Poynting vector merely by a factor $\epsilon_0\mu_0 = 1/c^2$:

$$\begin{aligned} \boldsymbol{\pi}_{\mathrm{EM}} &= \epsilon_0 \mathbf{E} \times \mathbf{B} \\ &= \frac{1}{c^2}\mathbf{S}. \end{aligned} \tag{2.29}$$

The time evolution of $\boldsymbol{\pi}_{\mathrm{EM}}$ is not only connected to the rate of change of the mechanical momentum density giving rise to the familiar Lorentz force term, but also involves the divergence of a second rank tensor $\mathbf{T}$ which is usually called the Maxwell stress tensor (Jackson [1975] Landau and Lifshitz [1962]). The latter is defined most easily by its Cartesian components

$$\mathbf{T}_{\alpha\beta} = \epsilon_0 \left(\frac{1}{2}|\mathbf{E}|^2 - E_\alpha E_\beta \right) + \frac{1}{\mu_0}\left(\frac{1}{2}|\mathbf{B}|^2 - B_\alpha B_\beta \right) \tag{2.30}$$

with $\alpha, \beta = x, y, z$.

A straightforward calculation yields:

$$\frac{\partial \boldsymbol{\pi}_{\mathrm{EM}}}{\partial t} = -\rho\, \mathbf{E} - \mathbf{J} \times \mathbf{B} - \boldsymbol{\nabla} \cdot \mathbf{T}. \tag{2.31}$$

2.4.4 Angular Momentum Conservation

The angular momentum density of the electromagnetic field and its corresponding flux may be defined respectively by the relations

$$\begin{aligned} \boldsymbol{\Lambda}_{\mathrm{EM}} &= \mathbf{r} \times \boldsymbol{\pi}_{\mathrm{EM}} \\ \boldsymbol{\Gamma} &= \mathbf{r} \times \mathbf{T}. \end{aligned} \tag{2.32}$$

The conservation law that governs the angular momentum, reads

$$\frac{\partial \boldsymbol{\Lambda}_{\mathrm{EM}}}{\partial t} = -\mathbf{r} \times (\rho\, \mathbf{E} + \mathbf{J} \times \mathbf{B}) - \boldsymbol{\nabla} \cdot \boldsymbol{\Gamma}. \tag{2.33}$$

3

Potentials and Fields and the Lagrangian

Not only the Maxwell equations themselves but also all related conservation laws have been expressed with the help of two key observables describing the microscopic electromagnetic field, namely $\mathbf{E}$ and $\mathbf{B}$. Strictly speaking, all relevant physics involving electromagnetic phenomena can be described correctly and completely in terms of the variables $\mathbf{E}$ and $\mathbf{B}$ solely, and from this point of view there is absolutely no need of defining auxiliary potentials akin to $\mathbf{E}$ and $\mathbf{B}$. Nevertheless, it proves quite beneficial to introduce the scalar potential $V(\mathbf{r}, t)$ and the vector potential $\mathbf{A}(\mathbf{r}, t)$ as alternative electrodynamical degrees of freedom.

3.1 The Scalar and Vector Potential

From the Maxwell equation $\boldsymbol{\nabla} \cdot \mathbf{B} = 0$ and Helmholtz' theorem it follows that, within a simply connected region Ω, there exists a regular vector field $\mathbf{A}$ – called vector potential – such that

$$\mathbf{B} = \boldsymbol{\nabla} \times \mathbf{A}, \tag{3.1}$$

which allows us to rewrite Faraday's law (2.14) as

$$\boldsymbol{\nabla} \times \left(\mathbf{E} + \frac{\partial \mathbf{A}}{\partial t} \right) = 0. \tag{3.2}$$

The scalar potential V emerges from the latter equation and Helmholtz' theorem stating that, in a simply connected region Ω there must exist a regular scalar function V such that

$$\mathbf{E} = -\boldsymbol{\nabla} V - \frac{\partial \mathbf{A}}{\partial t}. \tag{3.3}$$

Although V and $\mathbf{A}$ do not add new physics, there are at least three good reasons to introduce them anyway. First, it turns out that (Jackson [1975]

Feynman et al. [1964]) the two potentials greatly facilitate the mathematical treatment of classical electrodynamics in many respects. For instance, the choice of an appropriate gauge[2] allows one to convert the Maxwell equations into convenient wave equations for V and $\mathbf{A}$ for which analytic ssolutions can be derived occasionally. Moreover, the scalar potential V provides an natural link to the concept of macroscopic potential differences that are playing a crucial role in conventional simulations of electric circuits.

Next, most quantum mechanical treatments directly invoke the "potential" picture to deal with the interaction between a charged particle and an electromagnetic field. In particular, adopting the path integral approach, one accounts for the presence of electric and magnetic fields by correcting the action functional S related to the propagation from $(\mathbf{r}_0, t_0)$ to $(\mathbf{r}_1, t_1)$ along a world line, according to

$$S[V, \mathbf{A}] = S[0, \mathbf{0}] + q\left(\int_{\mathbf{r}_1}^{\mathbf{r}_2} \mathbf{A} \cdot \mathbf{dr} - \int_{t_0}^{t_1} \mathrm{d}t\, V(\mathbf{r}, t)\right), \tag{3.4}$$

while the field-dependent Hamiltonian term appearing in the non-relativistic, one-particle Schrödinger equation $\mathrm{i}\hbar\,(\partial\psi/\partial t) = H\,\psi$, takes the form

$$H = \frac{1}{2m}\,(\mathbf{p} - q\mathbf{A})^2 + qV \tag{3.5}$$

with $\mathbf{p} = -\mathrm{i}\hbar\boldsymbol{\nabla}$. Furthermore, the canonical quantization of the electromagnetic radiation field leads to photon modes corresponding to the quantized transverse modes of the vector potential.

Finally, the third motivation for adopting scalar and vector potentials lies in the perspective of developing new numerical simulation techniques. For example, it was observed (Schoenmaker et al. [2002a]) that the magnetic field generated by a steady current distribution may alternatively be extracted from the fourth Maxwell equation (Ampère's law),

$$\boldsymbol{\nabla}\times\boldsymbol{\nabla}\times\mathbf{A} = \mu_0\mathbf{J} \tag{3.6}$$

by assigning discretized vector potential variables to the *links* connecting adjacent nodes. This will be discussed in Section 9.

[2]Gauge transformations will extensively be treated in Chapter 8.

3.2 Gauge Invariance

In contrast to the electric field and the magnetic induction, neither the scalar nor the vector potential are uniquely defined. Indeed, performing a so-called gauge transformation

$$\begin{aligned} \mathbf{A}'(\mathbf{r},t) &= \mathbf{A}(\mathbf{r},t) + \boldsymbol{\nabla}\chi(\mathbf{r},t) \\ V'(\mathbf{r},t) &= V(\mathbf{r},t) - \frac{\partial\chi(\mathbf{r},t)}{\partial t}, \end{aligned} \tag{3.7}$$

where the gauge field $\chi(\mathbf{r},t)$ is an arbitrary regular, real scalar field, one clearly observes that the potentials are modified while the electromagnetic fields $\mathbf{E}(\mathbf{r},t)$ and $\mathbf{B}(\mathbf{r},t)$ remain unchanged. Similarly, any quantum mechanical wave function $\psi(\mathbf{r},t)$ transforms according to

$$\begin{aligned} \psi'(\mathbf{r},t) &= \psi(\mathbf{r},t)\exp\left(\mathrm{i}q\chi(\mathbf{r},t)\right) \\ \psi'^{*}(\mathbf{r},t) &= \psi^{*}(\mathbf{r},t)\exp\left(-\mathrm{i}q\chi(\mathbf{r},t)\right), \end{aligned}$$

whereas the quantum mechanical probability density $|\psi(\mathbf{r},t)|^2$ and other observable quantities are invariant under a gauge transformation, as required.

3.3 Lagrangian for an Electromagnetic Field Interacting with Charges and Currents

While the Maxwell equations are the starting point in the so-called *inductive approach*, one may alternatively adopt the *deductive approach* and try to "derive" the Maxwell equations from a proper variational principle. As a matter of fact it is possible indeed to postulate a Lagrangian density $\mathcal{L}(\mathbf{r},t)$ and an action functional $S[\mathcal{L},t_0,t_1] = \int_{t_0}^{t_1}\mathcal{L}(\mathbf{r},t)\mathrm{d}\tau$ such that the Maxwell equations emerge as the Euler-Lagrange equations that make the action

$$\delta S = 0. \tag{3.8}$$

stationary. While such a "derivation" is of utmost importance for the purpose of basic understanding from the theoretical point of view, the Lagrangian and Hamiltonian formulation of electromagnetism may look redundant when it comes to numerical computations. However, we have quoted the Lagrangian density of the electromagnetic field not only for the sake of completeness but also to illustrate the numerical potential of the underlying variational principle.

The Lagrangian density for the interacting electromagnetic field is conventionally postulated as a quadratic functional of the scalar and vector potential and their derivatives:

$$\mathcal{L} = \frac{1}{2}\epsilon_0 \left| \nabla V + \frac{\partial \mathbf{A}}{\partial t} \right|^2 - \frac{1}{2\mu_0} |\nabla \times \mathbf{A}|^2 + \mathbf{J} \cdot \mathbf{A} - \rho V, \qquad (3.9)$$

where the field variables V and $\mathbf{A}$ are linearly coupled to the charge and current distribution ρ and $\mathbf{J}$.

It is now straightforward to obtain the Maxwell equations as the Euler-Lagrange equations corresponding to Equations (3.9) provided that the set of field variables is chosen to be either V or A_α. The first possibility gives rise to

$$\sum_{\beta=x,y,z} \frac{\partial}{\partial x_\beta} \left[\frac{\partial \mathcal{L}}{\partial \left(\frac{\partial V}{\partial x_\beta} \right)} i \right] + \frac{\partial}{\partial t} \left[\frac{\partial \mathcal{L}}{\partial \left(\frac{\partial V}{\partial t} \right)} \right] = \frac{\partial \mathcal{L}}{\partial V}. \qquad (3.10)$$

Inserting all non-zero derivatives, we arrive at

$$\epsilon_0 \sum_\beta \left(\frac{\partial V}{\partial x_\beta} + \frac{\partial A_\beta}{\partial t} \right) = -\rho, \qquad (3.11)$$

which clearly reduces to the first Maxwell equation

$$\epsilon_0 \nabla \cdot \mathbf{E} = \rho \qquad \text{(Gauss' law)}. \quad (3.12)$$

Similarly, the three Euler-Lagrange equations

$$\sum_{\beta=x,y,z} \frac{\partial}{\partial x_\beta} \left[\frac{\partial \mathcal{L}}{\partial \left(\frac{\partial A_\alpha}{\partial x_\beta} \right)} \right] + \frac{\partial}{\partial t} \left[\frac{\partial \mathcal{L}}{\partial \left(\frac{\partial A_\alpha}{\partial t} \right)} \right] = \frac{\partial \mathcal{L}}{\partial A_\alpha}, \qquad \alpha = x, y, z \qquad (3.13)$$

lead to the fourth Maxwell equation

$$\frac{1}{\mu_0} \left(\nabla \times \mathbf{B} - \epsilon_0 \frac{\partial \mathbf{E}}{\partial t} \right) = \mathbf{J} \qquad \text{(Ampère-Faraday's law)} \qquad (3.14)$$

It should be noted that, within the deductive approach, the electric and magnetic field vectors are *defined* by the equations

$$\mathbf{E} = -\nabla V - \frac{\partial \mathbf{A}}{\partial t}, \quad \mathbf{B} = \nabla \times \mathbf{A}, \qquad (3.15)$$

whereas the latter are directly resulting from the Maxwell equations in the inductive approach. Mutates mutandis, the two remaining Maxwell equations $\boldsymbol{\nabla}\cdot\mathbf{B} = 0$ and $\boldsymbol{\nabla}\times\mathbf{E} = -\partial\mathbf{B}/\partial t$ are a direct consequence of the operation of the vector identities (10.33) and (10.34) on Equations (3.15). It should also be noted that the Lagrangian density may be written as

$$\mathcal{L} = \frac{1}{2}\epsilon_0\mathbf{E}^2 - \frac{1}{2\mu_0}\mathbf{B}^2 \tag{3.16}$$

So far, we have considered the Maxwell equations from the perspective that the charge and the current densities are given and the fields should be determined. However, as was already mentioned in the introduction, the charge and current densities may also be influenced by the fields. In order to illustrate the opposite cause-effect relation, we consider the Lagrangian of N charged particles moving in an electromagnetic field. The Lagragian is

$$L = \sum_{n=1}^{N}\frac{1}{2}m_n v_n^2 + \frac{1}{2}\int \mathrm{d}\tau\left(\epsilon_0\mathbf{E}^2 - \frac{1}{\mu_0}\mathbf{B}^2\right) - \int \mathrm{d}\tau\rho V + \int \mathrm{d}\tau\mathbf{J}\cdot\mathbf{A}, \tag{3.17}$$

where we defined the charge and current densities as

$$\begin{aligned}\rho(\mathbf{r},t) &= \sum_{n=1}^{N} q_n\delta(\mathbf{r}-\mathbf{r}_n),\\ \mathbf{J}(\mathbf{r},t) &= \sum_{n=1}^{N} q_n\mathbf{v}_n\delta(\mathbf{r}-\mathbf{r}_n)\end{aligned} \tag{3.18}$$

and the particles' velocities as $\mathbf{v}_n = \mathrm{d}\mathbf{r}_n/\mathrm{d}t$. Applying the Euler-Lagrange equations:

$$\frac{\mathrm{d}}{\mathrm{d}t}\left(\frac{\partial L}{\partial\mathbf{v}_n}\right) - \frac{\partial L}{\partial\mathbf{r}_n} = 0, \tag{3.19}$$

gives

$$m_n\frac{\mathrm{d}^2\mathbf{r}_n}{\mathrm{d}t^2} = q_n\mathbf{E}(\mathbf{r}_n,t) + q_n\mathbf{v}_n\times\mathbf{B}(\mathbf{r}_n,t). \tag{3.20}$$

The last term is recognized as the Lorentz force.

4

The Macroscopic Maxwell Equations

4.1 Constitutive Equations

The Maxwell equations contain source terms being the charge densities and the currents. In this chapter we will present the physics behind these terms and derive their precise form. We will see that the charge and current formulas depend very much on the medium in which these charges and currents are present. For solid media we can distinguish between insulators, semiconductors and conductors. The corresponding expressions differ considerably for the different materials. Furthermore, in the gas phase or the liquid phase again other expressions will be found. In the latter case, we enter the realm of plasma physics and magnetohydrodynamics. These topics are beyond the present scope.

Before starting to derive the constitutive equations we need to address another machinery, namely statistical physics. From a philosophical point of view, statistical physics is a remarkable part of natural science. It does not contribute to a deeper understanding of the fundamental forces of nature, yet it introduces a fundamental constant of nature, the Boltzmann constant $k_{\rm B} = 1.3805 \times 10^{23} J/K$. Furthermore, there has been a discussion over several generations of physicists, debating the reality of irreversibility. The dispute in a nutshell is whether the idea of entropy increase is a sensible one, considering the fact that the microscopic dynamics is time-reversal invariant. As has been demonstrated in (Magnus and Schoenmaker [1993]) the time reversal invariance is broken in the limit of infinitely many degrees of freedom. In practice, 'infinity' is already reached for 30 degrees of freedom in the study of (Magnus and Schoenmaker [1993]). Therefore, we believe that the dispute is settled and statistical physics is 'solid as a rock'.

4.2 Boltzmann Transport Equation

In this section, we will consider the assumptions that lead to the Boltzmann transport equation. This equation serves as the starting point for deriving the formules for the constitutive equation for the currents in metals, semiconductors and insulators.

When describing the temporal evolution of many particles, one is not interested in the detailed trajectory of each individual particle in space and time. First of all, the particles are identical and therefore their trajectories are interchangeable. Secondly, the individual trajectories exhibit stochastic motion on a short time scale that is irrelevant on a larger time scale. In a similar way, the detailed knowledge at a short length scale is also not of interest for understanding the behavior at larger length scales. Thus, we must obtain a procedure for eliminating the short-distance fluctuations from the description of the many particle systems. In fact, to arrive at a manageable set of equations such a procedure should also reduce the number of variables for which the evolution equations need to be formulated.

There are a number of schemes that allow for such a reduction. All methods apply some kind of coarse graining, i.e. a number of microscopic variables are bundled and are represented by a single effective variable. In this section, we discuss the method that is due to Boltzmann and that leads to the Boltzmann transport equation.

Consider N particles with generalized coordinates $\mathbf{q}_i, i = 1, \ldots, N$ and generalized momenta $\mathbf{p}_i, i = 1, \ldots, N$. Each particle can be viewed as a point of the so-called μ-space, a six-dimensional space, spanned by the coordinates $\mathbf{q}, \mathbf{p}$. In this light, the N particles will trace out N curves in phase space as time evolves. Let us now subdivide the phase space into cells of size $\Delta\Omega = \Delta q^3\ \Delta p^3$. Each cell can be labeled by a pair of coordinates $\mathbf{Q}_i$ and momenta $\mathbf{P}_i$. The number of particles that is found in the cell Ω_i is given by $f(\mathbf{P}_i, \mathbf{Q}_i, t)$. We can illustrate the role of the cell size setting $\Delta\Omega$. The function $f(\mathbf{P}_i, \mathbf{Q}_i, t)$ is given by

$$f(\mathbf{P}_i, \mathbf{Q}_i, t) = \sum_{i=1}^{N} \int_{\Delta\Omega} \mathrm{d}^3p\, \mathrm{d}^3q\, \boldsymbol{\delta}(\mathbf{p} - \mathbf{p}_i(t))\, \boldsymbol{\delta}(\mathbf{q} - \mathbf{q}_i(t)). \tag{4.1}$$

We can illustrate the role of the coarse-graining scaling parameter $\Delta\Omega$. If we take the size of the cell arbitrary small then we will occasionally find a particle in the cell. Such a choice of $\Delta\Omega$ corresponds to a fully microscopic description of the mechanical system and we will not achieve a reduction in degrees of freedom.

On the other hand, if we choose $\Delta\Omega$ arbitrary large, then all degrees of freedom are represented by one (static) point f, and we have lost all knowledge of the system. Therefore, $\Delta\Omega$ must be chosen such that it acts as the "communicator" between the microscopic and macroscopic worlds. This connection can be obtained by setting the size of the cell large enough such that each cell contains a number of particles. Within each cell the particles are considered to be in a state of thermal equilibrium. Thus for each cell a temperature T_i and a chemical potential μ_i can be given. The (local) thermal equilibrium is realized if there occurs a thermalization, i.e. within the cell collisions should occur within a time interval Δt. Therefore, the cell should be chosen such that its size exceeds at least a few mean-free path lengths.

On the macroscopic scale, the cell labels $\mathbf{P}_i$ and $\mathbf{Q}_i$ are smooth variables. The cell size is the denoted by the differential $d\Omega = \mathrm{d}^3p\ \mathrm{d}^3q$. Then we may denote the distribution functions as $f(\mathbf{P}, \mathbf{Q}, t) \equiv f(\mathbf{p}, \mathbf{q}, t)$. From the distribution function $f(\mathbf{p}, \mathbf{q}, t)$, the particle density function can be obtained from

$$\int \mathrm{d}^3p\ f(\mathbf{p}, \mathbf{q}, t) = \rho(\mathbf{q}, t). \tag{4.2}$$

As time progresses from t to $t + \delta t$, all particles in a cell at $\mathbf{p}, \mathbf{q}$ will be found in a cell at $\mathbf{p}', \mathbf{q}'$, provided that no collisions occurred. Hence

$$f(\mathbf{p}, \mathbf{q}, t)\ \mathrm{d}^3p\ \mathrm{d}^3q = f(\mathbf{p} + \mathbf{F}\delta t,\ \mathbf{q} + \mathbf{v}\delta t,\ t + \delta t)\ \mathrm{d}^3p\prime\ \mathrm{d}^3q\prime. \tag{4.3}$$

According to Liouville's theorem (Fowler [1936] Huang [1963]), the two volume elements $\mathrm{d}^3p\ \mathrm{d}^3q$ and $\mathrm{d}^3p\prime\ \mathrm{d}^3q\prime$ are equal, which may appear evident if there are no external forces. If there are forces that do not explicitly depend on time, any cubic element deforms into a parallelepiped but with the same volume as the original cube. Taking also into account the effect of collisions that may kick particles in or out of the cube in the time interval δt, we arrive at the following equation for the distribution function

$$\left(\frac{\partial}{\partial t} + \frac{\mathbf{p}}{m} \cdot \boldsymbol{\nabla}_{\mathrm{q}} + \mathbf{F} \cdot \boldsymbol{\nabla}_{\mathrm{p}}\right) f(\mathbf{p}, \mathbf{q}, t) = \left(\frac{\partial f}{\partial t}\right)_{\mathrm{c}}, \tag{4.4}$$

where the "collision term" $(\partial f/\partial t)_{\mathrm{c}}$ *defines* the effects of scattering. A quantitative estimate of this term is provided by studying the physical mechanisms that contribute to this term. As carriers traverse, their motion is frequently disturbed by scattering due to collisions with impurity atoms, phonons, crystal defects, other carriers or even with foreign particles (cosmic rays). The frequency at which such events occur can be estimated by assuming that these events take place in an uncorrelated way; in other words two such events are statistically independent. Each physical mechanism is described by an interaction Hamiltonian or potential function, $U_S(\mathbf{r})$ that describes the details of the scattering process. The matrix element that describes the transition from a carrier in a state with momentum $|\mathbf{p}\rangle$ to a state with momentum $|\mathbf{p}'\rangle$ is

$$H_{\mathbf{p}'\mathbf{p}} = \frac{1}{\Omega}\int \mathrm{d}\tau \; \mathrm{e}^{-\frac{\mathrm{i}}{\hbar}\mathbf{p}'\cdot\mathbf{r}} \, U_S(\mathbf{r}) \; \mathrm{e}^{\frac{\mathrm{i}}{\hbar}\mathbf{p}\cdot\mathbf{r}}, \tag{4.5}$$

where Ω is a box that is used to count the number of momentum states. This box is of the size Δq^3 as defined above.

The evaluation of the transition amplitude relies on Fermi's Golden Rule. The transition rate then becomes

$$S(\mathbf{p}', \mathbf{p}) = \frac{2\pi}{\hbar}|H_{\mathbf{p}'\mathbf{p}}|^2 \delta(E(\mathbf{p}') - E(\mathbf{p}) - \Delta E), \tag{4.6}$$

where ΔE is the change in energy related to the transition. If $\Delta E = 0$, the collision is *elastic* The collision term is the result of the balance between kick-in and kick-out of the transitions that take place per unit time:

$$\left(\frac{\partial f}{\partial t}\right)_{\mathrm{c}} = \sum_{\mathbf{p}'} \left(S(\mathbf{p}', \mathbf{p}) \; f(\mathbf{q}, \mathbf{p}', t) - S(\mathbf{p}, \mathbf{p}') \; f(\mathbf{q}, \mathbf{p}, t) \right). \tag{4.7}$$

Once more it should be emphasized that although this balance picture is heuristic, looks reasonable and leads to a description of irreversibility it does not explain the latter. The collision term can be further fine-tuned to mimic the consequences of Pauli's exclusion principle by suppression of multiple occupation of states:

$$\begin{aligned}\left(\frac{\partial f}{\partial t}\right)_{\mathrm{c}} = \sum_{\mathbf{p}'} \big[& S(\mathbf{p}', \mathbf{p}) \; f(\mathbf{q}, \mathbf{p}', t)(1 - f(\mathbf{q}, \mathbf{p}, t)) \\ & - S(\mathbf{p}', \mathbf{p}) \; f(\mathbf{q}, \mathbf{p}, t)(1 - f(\mathbf{q}, \mathbf{p}', t)) \big].\end{aligned} \tag{4.8}$$

4.3 Currents in Metals

In many materials, the conduction current that flows due to the presence of an electric field, $\mathbf{E}$, is proportional to $\mathbf{E}$, so that

$$\mathbf{J} = \sigma \mathbf{E}, \tag{4.9}$$

where the electrical conductivity σ is a material parameter. In metallic materials, Ohm's law, Equation (4.9) is accurate. However, a fast generalization should be allowed for anisotropic conducting media. Moreover, the conductivity may depend on the frequency mode such that we arrive at

$$\mathbf{J}_i(\omega) = \boldsymbol{\sigma}_{ij}(\omega)\mathbf{E}_j(\omega) \tag{4.10}$$

and $\boldsymbol{\sigma}$ is a second-rank tensor. The derivation of Ohm's law from the Boltzmann transport equation was initiated by Drude. In Drude's model (Drude [1900a,b]), the electrons move as independent particles in the metallic region suffering from scattering during their travel from the cathode to the anode. The distribution function is assumed to be of the following form:

$$f(\mathbf{q}, \mathbf{p}, t) = f_{\mathrm{O}}(\mathbf{q}, \mathbf{p}, t) + f_{\mathrm{A}}(\mathbf{q}, \mathbf{p}, t), \tag{4.11}$$

where f_{O} is the equilibrium distribution function, being symmetric in the momentum variable $\mathbf{p}$, and f_{A} is a perturbation due to an external field that is anti-symmetric in the momentum variable. The collision term in Drude's model is crudely approximated by the following assumptions:

- only kick-out,
- all $S(\mathbf{p}, \mathbf{p}')$ are equal,
- no Pauli exclusion principle,
- no carrier heating, i.e. low-field transitions.

The last assumption implies that only the anti-symmetric part participates in the collision term (Lundstrom [1999]). Defining a characteristic time $\tau_{\mathbf{p}}$, the momentum-relaxation time, we find that

$$\left(\frac{\partial f}{\partial t}\right)_{\mathrm{c}} = -\frac{f_{\mathrm{A}}}{\tau_{\mathbf{p}}} \quad \text{and} \quad 1/\tau_{\mathbf{p}} = \sum_{\mathbf{p}'} S(\mathbf{p}, \mathbf{p}'). \tag{4.12}$$

Furthermore, assuming a constant electric field $\mathbf{E}$ and a spatially uniform charge electron distribution, the Boltzmann transport equation becomes

$$-q\,\mathbf{E} \cdot \boldsymbol{\nabla}(f_{\mathrm{O}} + f_{\mathrm{A}}) = -\frac{f_{\mathrm{A}}}{\tau_{\mathbf{p}}}. \tag{4.13}$$

Finally, if we assume that $f \simeq f_{\mathrm{O}} \propto \exp\left(-p^2/2mk_{\mathrm{B}}T\right)$ then

$$f_{\mathrm{A}} = q\tau_{\mathbf{p}}\mathbf{E} \cdot \boldsymbol{\nabla}_{\mathbf{p}} f_{\mathrm{O}} = \frac{q\tau_{\mathbf{p}}}{k_{\mathrm{B}}T}\,\mathbf{E} \cdot \mathbf{v} f_{\mathrm{O}}. \tag{4.14}$$

Another way of looking at this result is to consider $f = f_0 + f_A$ as a Taylor series for f_0:

$$f(\mathbf{p}) = f_0(\mathbf{p}) + (q\tau_{\mathbf{p}}\mathbf{E}) \cdot \nabla_{\mathbf{p}} f_0(\mathbf{p}) + \cdots = f_0(\mathbf{p} + q\tau_{\mathbf{p}}\mathbf{E}). \tag{4.15}$$

This is a *displaced* Maxwellian distribution function in the direction opposite to the applied field $\mathbf{E}$. The current density is $\mathbf{J} = qn\mathbf{v}$ follows from the averaged velocity

$$\mathbf{J} = qn \frac{\int \mathrm{d}^3 p \, (\mathbf{p}/m) \, f(\mathbf{p})}{\int \mathrm{d}^3 p \;\; f(\mathbf{p})} = \frac{q^2 \tau_{\mathbf{p}}}{m} n\mathbf{E}. \tag{4.16}$$

The electron mobility, μ_n, is defined as the proportionality constant in the constitutive relation $\mathbf{J} = q\mu_n n\mathbf{E}$, such that

$$\mu_n = \frac{q\tau_{\mathbf{p}}}{m}. \tag{4.17}$$

So we have been able to "deduce" Ohm's law from the Boltzmann transport equation.

It is a remarkable fact that Drude's model is quite accurate, given the fact that no reference was made to Pauli's exclusion principle and the electron waves do not scatter while traveling in a perfect crystal lattice. Indeed, it was recognized by Sommerfeld that ignoring these effects will give rise to errors in the calculation of the order of 10^2, but both these errors cancel. Whereas Drude's model explains the existence of resistance, more advanced models are needed to accommodate for the non-linear current-voltage characteristics, the frequency dependence and the anisotropy of the conductance for some materials. A "modern" approach to derive conductance properties was initiated by (Kubo [1957]). His theory naturally leads to the inclusion of anisotropy, non-linearity and frequency dependence. Kubo's approach also serves as the starting point to calculate transport properties in the quantum theory of many particles at finite temperature (Mahan [1981]). These approaches start from the quantum-Liouville equation and the Gibb's theory of assembles on phase space. The latter has a more transparent generalization to the many-particle Hilbert space of quantum states.

Instead of reproducing here text book presentations of these various domains of physics, we intend to give the reader some sense of alertness, that the validity of some relations is limited. In order to push back the restrictions, one needs to re-examine the causes of the limitations. Improved models can be *guessed* by widening the defining expression as in the foregoing case where the scalar σ was substituted by the conductivity tensor $\boldsymbol{\sigma}$. The consequences of these guesses can be tested in simulation experiments. Therefore, simulation plays an important role to obtain improved models.

In the process of purchasing model improvements a few guidelines will be of help. First of all, the resulting theory should respect some fundamental physical principles. The *causality* principle is an important example. It states that there is a retarded temporal relation between cause and effect. The causality principle is a key ingredient to derive the Kramers-Kronig relations, that put severe limitations on the real and imaginary parts of the material parameters. Yet these relationships are not sufficient to determine the models completely, but one needs to include additional physical models.

4.4 Charges in Metals

Metallic materials are characterized as having an appreciable conductivity. Any excess free charge distribution in the metal will decay exponentially to zero in a small time. Combining Gauss' law with the current continuity equation

$$\begin{aligned} \nabla \cdot (\epsilon \mathbf{E}) &= \rho \\ \nabla \cdot (\sigma \mathbf{E}) &= \frac{\partial \rho}{\partial t} \end{aligned} \tag{4.18}$$

and considering ϵ and σ constant, we find

$$\begin{aligned} \frac{\partial \rho}{\partial t} &= -\frac{\sigma}{\epsilon}\rho \\ \rho &= \rho_0 \exp\left(-\frac{\sigma}{\epsilon}t\right). \end{aligned} \tag{4.19}$$

In metallic materials, the decay time $\tau = \epsilon/\sigma$ is of the order of 10^{-18} seconds, such that $\rho = 0$ at any instant.

For conducting materials one usually assumes $\nabla \cdot \mathbf{D} = 0$ and for constant ϵ and ρ, the electric field $\mathbf{E}$ and current density $\mathbf{J}$ are constant (Collin [1960]). A subtlety arises when ϵ and ρ are varying in space. Considering the steady-state version of above set of equations, we obtain

$$\begin{aligned} \nabla \cdot (\epsilon \mathbf{E}) &= \rho \\ \nabla \cdot (\sigma \mathbf{E}) &= 0. \end{aligned} \tag{4.20}$$

The field $\mathbf{E}$ should simultaneously obey two equations. Posed as a boundary-value problem for the scalar potential, V, we may determine V from the second equation and determine ρ as a "post-processing" result originating from the first equation.

4.5 Semiconductors

Intrinsic semiconductors are insulators at zero temperature. This is because the band structure of semiconductors consists of bands that are either filled or empty. At zero temperature, the chemical potential falls between the highest filled band which is called the valence band and the lowest empty band which is named the conduction band. The separation of the valance and conduction band is sufficiently small such that at some temperature, there is an appreciable amount of electrons that have an energy above the conduction band onset. As a consequence these electron are mobile and will contribute to the current if a voltage drop is put over the semiconducting material. The holes in the valance band act as positive charges with positive effective mass and therefore they also contribute to the net current. Intrinsic semiconductors are rather poor conductors but their resistance is very sensitive to the temperature ($\sim\exp(-A/T)$). By adding dopants to the intrinsic semiconductor, the chemical potential of the electrons and holes may by shifted up or down with respect to the band edges. Before going into further descriptions of dopant distributions, we would like to emphasize the following fact: *Each thermodynamic system in thermal equilibrium has constant intensive conjugated variables.* In particular, the temperature, T, conjugated to the internal energy of the system and the chemical potential, μ, conjugated to the number of

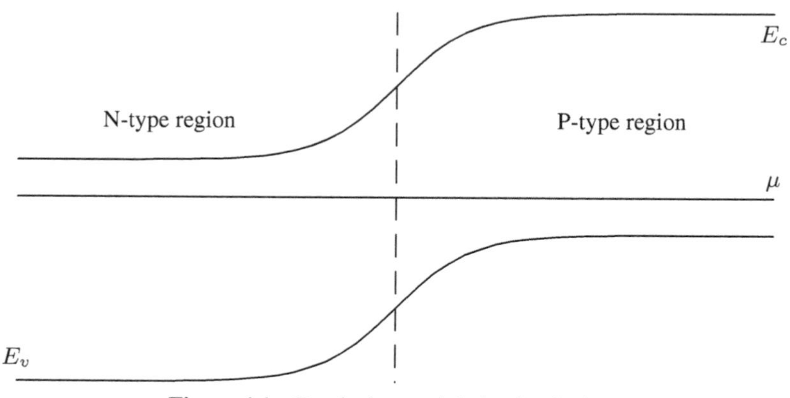

Figure 4.1 Band edge modulation by doping.

particles in the systems are constant for a system in equilibrium. Therefore, if the dopant distribution varies in the device and the distance between the chemical potential and the band edges is modulated, then for the device being in equilibrium, the band edges must vary in accordance with the dopant variations as is shown in Figure 4.1.

4.6 Currents in Semiconductors

Whereas in metals the high conductivity prevents local charge accumulation at an detectable time scale, the situation in semiconductors is quite different. In uniformly doped semiconductors, the decay of an excess charge spot occurs by a diffusion process, that takes place on much longer time scale. In non-uniformly doped semiconductors, there are depletion layers, or accumulation layers of charges that permanently exists even in thermal equilibrium.

The charge and current densities in semiconductors follow also from the general Boltzmann transport theory, but this theory needs to be complemented with specific details such as the band gap, the dopant distribution, and the properties related to the interfaces to other materials.

Starting from the Boltzmann transport equation, the *moment expansion* considers variables that are averaged quantities as far as the momentum dependence is concerned. The generic expression for the moment expansion is

$$\frac{1}{\Omega}\sum_{\mathbf{p}} Q(\mathbf{p})\left(\frac{\partial}{\partial t}+\frac{\mathbf{p}}{m}\cdot\boldsymbol{\nabla}_{\mathbf{q}}+\mathbf{F}\cdot\boldsymbol{\nabla}_{\mathbf{p}}\right) f(\mathbf{p},\mathbf{q},t)=\frac{1}{\Omega}\sum_{\mathbf{p}} Q(\mathbf{p})\left(\frac{\partial f}{\partial t}\right)_{\mathrm{c}} \tag{4.21}$$

where $Q(\mathbf{p})$ is an polynomial in the components of $\mathbf{p}$. and the normalization $1/\Omega$ allows for a smooth transition to integrate over all momentum states in the Brillouin zone.

$$\frac{1}{\Omega}\sum_{\mathbf{p}} \rightarrow \frac{1}{4\pi^3}\int_{\mathrm{BZ}} \mathrm{d}^3k \tag{4.22}$$

The zeroth order expansion gives (Lundstrom [1999])

$$\begin{aligned} \frac{\partial n}{\partial t}-\frac{1}{q}\boldsymbol{\nabla}\cdot\mathbf{J}_{\mathrm{n}} &= -U \\ \frac{\partial p}{\partial t}+\frac{1}{q}\boldsymbol{\nabla}\cdot\mathbf{J}_{\mathrm{p}} &= U \end{aligned} \tag{4.23}$$

and where the various variables are:

$$\begin{array}{ll} \underline{\text{electrons}} & \underline{\text{holes}} \\ n(\mathbf{r},t) = \dfrac{1}{\Omega}\sum_{\mathbf{p}} f_{\mathrm{n}}(\mathbf{p},\mathbf{r},t) & p(\mathbf{r},t) = \dfrac{1}{\Omega}\sum_{\mathbf{p}} f_{\mathrm{p}}(\mathbf{p},\mathbf{r},t) \\ \mathbf{J}_{\mathrm{n}}(\mathbf{r},t) = -q\, n(\mathbf{r},t)\mathbf{v}_{\mathrm{n}}(\mathbf{r},t) & \mathbf{J}_{\mathrm{p}}(\mathbf{r},t) = q\, p(\mathbf{r},t)\mathbf{v}_{\mathrm{p}}(\mathbf{r},t) \\ \mathbf{v}_{\mathrm{n}}(\mathbf{r},t) = \dfrac{1}{\Omega}\sum_{\mathbf{p}} \dfrac{\mathbf{p}}{m} f_{\mathrm{n}}(\mathbf{p},\mathbf{r},t) & \mathbf{v}_{\mathrm{p}}(\mathbf{r},t) = \dfrac{1}{\Omega}\sum_{\mathbf{p}} \dfrac{\mathbf{p}}{m} f_{\mathrm{p}}(\mathbf{p},\mathbf{r},t) \end{array} \tag{4.24}$$

and

$$U = \frac{1}{\Omega}\sum_{\mathbf{p}} \left(\frac{\partial f}{\partial t}\right)_{\mathrm{c}} = R - G \tag{4.25}$$

The particle velocities give an expression for the current densities but by choosing $Q(\mathbf{p}) = \mathbf{p}$, we obtain the first moment of the expansion that can be further approximated to give alternative expressions for the current densities. Defining the momentum relaxation time τ_{p} as a characteristic time for the momentum to reach thermal equilibrium from a non-equilibrium state and the electron and hole temperature tensors (Forghieri et al. [1988])

$$\begin{aligned} \frac{1}{2} n k_{\mathrm{B}}\, T_{\mathrm{n},ij}(\mathbf{r},t) &= \tfrac{1}{\Omega}\textstyle\sum_{\mathbf{p}} \tfrac{1}{2m}(p_i - m v_{\mathrm{n},i})(p_j - m v_{\mathrm{n},j}) f_{\mathrm{n}}(\mathbf{p},\mathbf{r},t) \\ &= \tfrac{1}{2} n k_{\mathrm{B}} T_{\mathrm{n}}(\mathbf{r},t)\,\delta_{ij} \\ \frac{1}{2} p k_{\mathrm{B}}\, T_{\mathrm{p},ij}(\mathbf{r},t) &= \tfrac{1}{\Omega}\textstyle\sum_{\mathbf{p}} \tfrac{1}{2m}(p_i - m v_{\mathrm{p},i})(p_j - m v_{\mathrm{p},j}) f_{\mathrm{p}}(\mathbf{p},\mathbf{r},t) \\ &= \tfrac{1}{2} p k_{\mathrm{B}} T_{\mathrm{p}}(\mathbf{r},t)\,\delta_{ij} \end{aligned} \tag{4.26}$$

where the last equality follows from assuming an isotropic behavior. then one arrives at the following constitutive equation for the currents in semiconducting materials

$$\begin{aligned} \mathbf{J}_{\mathrm{n}} + n\, \tau_{\mathrm{pn}} \frac{\mathrm{d}}{\mathrm{d}t}\left(\frac{\mathbf{J}_{\mathrm{n}}}{n}\right) &= q\, \mu_{\mathrm{n}}\, n \left(\mathbf{E} + \frac{k_{\mathrm{B}}}{q}\boldsymbol{\nabla} T_{\mathrm{n}}\right) + q D_{\mathrm{n}} \boldsymbol{\nabla} n \\ \mathbf{J}_{\mathrm{p}} + p\, \tau_{\mathrm{pp}} \frac{\mathrm{d}}{\mathrm{d}t}\left(\frac{\mathbf{J}_{\mathrm{p}}}{p}\right) &= q\, \mu_{\mathrm{p}}\, p \left(\mathbf{E} - \frac{k_{\mathrm{B}}}{q}\boldsymbol{\nabla} T_{\mathrm{p}}\right) - q D_{\mathrm{p}} \boldsymbol{\nabla} p. \end{aligned} \tag{4.27}$$

The momentum relaxation times, the electron and hole mobilities and the electron and hole diffusivities are related through the Einstein relations

$$D = \frac{k_{\mathrm{B}} T}{q}\mu = \frac{k_{\mathrm{B}} T}{m}\tau. \tag{4.28}$$

The second terms on the left-hand sides of Equation (4.27) are the *convective currents*. The procedure of taking moments of the Boltzmann transport equation always involves a truncation, i.e. the n-th order equation in the expansion demands information of the (n+1)-th order moment to be supplied. For the second-order moment, one thus needs to provide information on the third moment

$$\frac{1}{\Omega}\sum_{\mathbf{p}} p_i p_j p_k f(\mathbf{p}, \mathbf{r}, t). \tag{4.29}$$

In the above scheme the second-order expansion leads to the *hydrodynamic model* (Forghieri et al. [1988]). In this model the carrier temperatures are determined self-consistently with the carrier densities. The closure of the system of equations is achieved by assuming a model for the term (4.29) that only contains lower order variables. The thermal flux Q, being the energy that gets transported through thermal conductance can be expressed as

$$\mathbf{Q} = \frac{1}{\Omega}\sum_{\mathbf{p}} \frac{1}{2m}|\mathbf{p} - m\mathbf{v}|^2 \left(\frac{\mathbf{p}}{m} - \mathbf{v}\right) = -\kappa \boldsymbol{\nabla} T, \tag{4.30}$$

where $\kappa = \kappa_n,\ \kappa_p$ are the thermal conductivities.

Besides the momentum flux, a balance equation is obtained for the energy flux:

$$\begin{aligned}
\frac{\partial(nw_{\mathrm{n}})}{\partial t} + \boldsymbol{\nabla}\cdot\mathbf{S}_{\mathrm{n}} &= \mathbf{E}\cdot\mathbf{J}_{\mathrm{n}} + n\left(\frac{\partial w_{\mathrm{n}}}{\partial t}\right)_{\mathrm{c}} \\
\frac{\partial(pw_{\mathrm{p}})}{\partial t} + \boldsymbol{\nabla}\cdot\mathbf{S}_{\mathrm{p}} &= \mathbf{E}\cdot\mathbf{J}_{\mathrm{p}} + p\left(\frac{\partial w_{\mathrm{p}}}{\partial t}\right)_{\mathrm{c}}.
\end{aligned} \tag{4.31}$$

The energy flux is denoted as $\mathbf{S}$ and w is the energy density. In the isotropic approximation, the latter reads

$$\begin{aligned}
w_{\mathrm{n}} &= \frac{3}{2}k_{\mathrm{B}}T_{\mathrm{n}} + \frac{1}{2}m_{\mathrm{n}}v_{\mathrm{n}}^2 \\
w_{\mathrm{p}} &= \frac{3}{2}k_{\mathrm{B}}T_{\mathrm{p}} + \frac{1}{2}m_{\mathrm{p}}v_{\mathrm{p}}^2.
\end{aligned} \tag{4.32}$$

The energy flux can be further specified as

$$\begin{aligned}
\mathbf{S}_{\mathrm{n}} &= \kappa_{\mathrm{n}}\boldsymbol{\nabla}T_{\mathrm{n}} - (w_{\mathrm{n}} + k_{\mathrm{B}}T_{\mathrm{n}})\frac{\mathbf{J}_{\mathrm{n}}}{q} \\
\mathbf{S}_{\mathrm{p}} &= \kappa_{\mathrm{p}}\boldsymbol{\nabla}T_{\mathrm{p}} + (w_{\mathrm{p}} + k_{\mathrm{B}}T_{\mathrm{p}})\frac{\mathbf{J}_{\mathrm{p}}}{q}.
\end{aligned} \tag{4.33}$$

Just as for the momentum, one usually assumes a characteristic time, τ_{e}, for a non-equilibrium energy distribution to relax to equilibrium. Then the collision term in the energy balance equation becomes

$$n\left(\frac{\partial w_{\mathrm{n}}}{\partial t}\right)_{\mathrm{c}} = -n\frac{w_{\mathrm{n}} - w^*}{\tau_{\mathrm{en}}} - Uw_{\mathrm{n}}$$
$$p\left(\frac{\partial w_{\mathrm{p}}}{\partial t}\right)_{\mathrm{c}} = -p\frac{w_{\mathrm{p}} - w^*}{\tau_{\mathrm{ep}}} - Uw_{\mathrm{p}} \tag{4.34}$$

and w^* is the carrier mean energy at the lattice temperature. In order to complete the hydrodynamic model the thermal conductivities are given by the Wiedemann-Franz law for thermal conductivity

$$\kappa = \left(\frac{k_{\mathrm{B}}}{q}\right)^2 T\sigma(T)\Delta(T). \tag{4.35}$$

Herein is $\Delta(T)$ a value obtained from evaluating the steady-state Boltzmann transport equation for uniform electric fields and $\sigma(T) = q\mu c$ the electrical conductivity ($c = n, p$). If a power-law dependence for the energy relaxation times can be assumed, i.e.

$$\tau_{\mathrm{e}} = \tau_0\left(\frac{w}{k_{\mathrm{B}}T^*}\right)^{\nu} \tag{4.36}$$

then $\Delta(T) = 5/2 + \nu$. Occasionally, ν is considered to be a constant ($\nu = 0.5$). However, this results into too restrictive an expression for the $\tau_{\mathrm{e}}(w)$. Therefore $\Delta_(T)$ is often tuned towards Monte-Carlo data.

Comparing the present elaboration on deriving constitutive equations from the Boltzmann transport equation with the derivation of the currents in metals we note that we did not refer to a displaced Maxwellian distribution. Such a derivation is also possible for semiconductor currents. The method was used by (Stratton [1962]). A difference pops up in the diffusion term of the carrier current. For the above results we obtained

$$\mathbf{J}\text{ (diffusive part)} \propto \mu\boldsymbol{\nabla}T.$$

In Stratton's model one obtains

$$\mathbf{J}\text{ (diffusive part)} \propto \boldsymbol{\nabla}\left(\mu T\right),$$

the difference being a term

$$\xi = \frac{\partial \log \mu(T)}{\partial \log (T)}.$$

Stratton's model is usually referred to as the *energy transport* model.

For the semiconductor environment, the Scharfetter-Gummel scheme provides a means to discretize the current equations on a grid (Scharfetter and Gummel [1969]). In the case that no carrier heating effects are considered (T is constant) the diffusion equations are

$$\mathbf{J} = q\mu c\mathbf{E} \pm kT\mu\nabla c, \tag{4.37}$$

where the plus (minus) sign refers to negatively (positively) charged particles and c denotes the corresponding carrier density. It is assumed that both the current $\mathbf{J}$ and the electric field $\mathbf{E}$ are constant along a link and that the potential V varies linearly along the link. Adopting a local coordinate axis u with $u = 0$ corresponding to node i, and $u = h_{ij}$ corresponding to node j, we may integrate Equation (4.37) along the link ij to obtain

$$J_{ij} = q\mu_{ij}c\left(\frac{V_i - V_j}{h_{ij}}\right) \pm kT\mu_{ij}\frac{\mathrm{d}c}{\mathrm{d}u}, \tag{4.38}$$

which is a first-order differential equation in c. The latter is solved using the aforementioned boundary conditions and gives rise to a non-linear carrier profile. The current J_{ij} can then be rewritten as

$$\frac{J_{ij}}{\mu_{ij}} = -\frac{\alpha}{h_{ij}}B\left(\frac{-\beta_{ij}}{\alpha}\right)c_i + \frac{\alpha}{h_{ij}}B\left(\frac{\beta_{ij}}{\alpha}\right)c_j, \tag{4.39}$$

using the Bernoulli function

$$B(x) = \frac{x}{\mathrm{e}^x - 1}. \tag{4.40}$$

Furthermore, we used $\alpha = \pm kT$ and $\beta_{ij} = q(V_i - V_j)$.
Before turning to the consideration of insulating materials, we briefly discuss the influence of strong magnetic fields on the currents. These fields will bend the trajectories due to the Lorentz force. In the derivation of the macroscopic current densities from the Boltzmann transport equation, we should include this force. The result is that in the constitutive current expression we must

make the replacement: $\mathbf{E} \to \mathbf{E} + q\mathbf{v} \times \mathbf{B}$. Since $\mathbf{J} = qc\mathbf{v}$, we arrive at the following *implicit* relation for $\mathbf{J}$:

$$\mathbf{J} = \sigma\mathbf{E} + \mu\mathbf{J} \times \mathbf{B}, \tag{4.41}$$

where $\sigma = q\mu c$ is the conductivity and μ is the mobility. This relation can be made *explicit* by solving the following set of linear equations:

$$\begin{bmatrix} 1 & -\mu B_z & \mu B_y \\ \mu B_z & 1 & -\mu B_x \\ -\mu B_y & \mu B_x & 1 \end{bmatrix} \cdot \begin{bmatrix} J_x \\ J_y \\ J_z \end{bmatrix} = \begin{bmatrix} \sigma E_x \\ \sigma E_y \\ \sigma E_x \end{bmatrix} \tag{4.42}$$

of which the solution is:

$$\mathbf{J} = \left[\sigma\mathbf{E} + \mu\sigma\,\mathbf{E} \times \mathbf{B} + \mu^2\sigma\,(\mathbf{E} \cdot \mathbf{B})\,\mathbf{B}\right] / (1 + \mu^2 B^2). \tag{4.43}$$

Above considerations are required for the description of Hall sensors. Here we will not further elaborate on this extension, nor will we consider the consequences of an-isotropic conductivity properties.

4.7 Insulators

So far, we have been rather sloppy in classifying materials as being an insulator, semiconductor or metal. We have referred to the reader's qualitative awareness of the conduction quality of a material under consideration. For the time being we will sustain in this practice and define insulators as having a negligible conductivity. Therefore, in an insulating material there are no conduction currents. The constitutive equation for $\mathbf{J}$ becomes trivial.

$$\mathbf{J} = \mathbf{0} \tag{4.44}$$

Recently, there is an increased interest in currents in insulating materials. The gate dielectric material SiO_2 that has been used in mainstream CMOS technology has a band gap of 3.9 eV and therefore acts as a perfect insulator for normal voltage operation conditions around 3 Volts and using 60 Å thick oxides. However, the continuous down scaling of the transistor architecture requires that the oxides thicknesses are also reduced. For device generation below 100 nm gate length, the oxide thickness should be less than 20 Å. For these thin layers, direct tunneling through the layer barrier becomes a dominating current leakage in integrated CMOS devices.

4.8 Dielectric Media

A dielectric material increases the storage capacity of a condenser or a capacitor by neutralizing charges at the electrodes that would otherwise contribute to the external field. Faraday identified this phenomenon as dielectric polarization. The polarization is caused by a microscopic alignment of dipole charges with respect to the external field. Looking at the macroscopic scale, we may introduce a polarization vector field, $\mathbf{P}$.

In order to give an accurate formulation of dielectric polarization we first consider an arbitrary charge distribution localized around the origin. The electric potential in some point $\mathbf{r}$, is

$$V(\mathbf{r}) = \frac{1}{4\pi\epsilon_0}\int \frac{\rho(\mathbf{r}')}{|\mathbf{r}-\mathbf{r}'|}\mathrm{d}\tau' \tag{4.45}$$

Now let $\mathbf{r}$ be a point outside the localization region of the charge distribution, i.e. $|\mathbf{r}| > |\mathbf{r}'|$. From the completeness of the series of the spherical harmonics, $Y_{lm}(\theta,\phi)$, one obtains

$$\frac{1}{|\mathbf{r}-\mathbf{r}'|} = 4\pi\sum_{l=0}^{\infty}\sum_{m=-l}^{l}\frac{1}{2l+1}\frac{|\mathbf{r}'|^l}{|\mathbf{r}|^{l+1}}Y_{lm}^*(\theta',\phi')Y_{lm}(\theta,\phi), \tag{4.46}$$

where

$$Y_{lm}(\theta,\phi) = \sqrt{\frac{2l+1}{4\pi}\frac{(l-m)!}{(l+m)!}}P_l^m(\cos\theta)\mathrm{e}^{\mathrm{i}m\phi} \tag{4.47}$$

and

$$P_l^m(x) = (-1)^m(1-x^2)^{\frac{m}{2}}\frac{\mathrm{d}^m}{\mathrm{d}x^m}P_l(x) \tag{4.48}$$

are the associated Legendre polynomials. Using above expansion, the potential of the charge distribution can be written as:

$$V(\mathbf{r}) = \frac{1}{4\pi\epsilon_0}\sum_{l=0}^{\infty}\sum_{m=-l}^{l}\frac{4\pi}{2l+1}q_{lm}\frac{Y_{lm}(\theta,\phi)}{r^{l+1}} \tag{4.49}$$

and

$$q_{lm} = \int Y_{lm}^*(\theta',\phi')(r')^l\rho(\mathbf{r}')\,\mathrm{d}\tau' \tag{4.50}$$

are the *multipole moments* of the charge distribution. The zeroth-order expansion coefficient

$$q_{00} = \frac{1}{4\pi}\int \rho(\mathbf{r})\,\mathrm{d}\tau = \frac{Q}{4\pi} \tag{4.51}$$

corresponds to total charge of the localized charge distribution. The total charge can be referred to as the electric *monopole* moment. The electric dipole moment

$$\mathbf{p} = \int \mathbf{r}\rho(\mathbf{r})\, \mathrm{d}\tau \tag{4.52}$$

and the first order expansion coefficients are related according to

$$\begin{aligned} q_{1,1} &= -\sqrt{\frac{3}{8\pi}}(p_x - ip_y) \\ q_{1,-1} &= \sqrt{\frac{3}{8\pi}}(p_x + ip_y) \\ q_{1,0} &= \sqrt{\frac{3}{4\pi}}p_z. \end{aligned} \tag{4.53}$$

The higher-order moments depend on the precise choice of the origin inside the charge distribution and therefore their usage is mainly restricted to cases where a preferred choice of the origin is dictated by the physical systems[3]. The potential of the charge distribution, ignoring second and higher order terms is

$$V(\mathbf{r}) = \frac{1}{4\pi\epsilon_0}\left(\frac{q}{r} + \frac{\mathbf{p}\cdot\mathbf{r}}{r^3}\right) \tag{4.54}$$

and the electric field of a dipole $\mathbf{p}$ located at the origin is

$$\mathbf{E}(\mathbf{r}) = \frac{3\hat{\mathbf{n}}(\mathbf{p}\cdot\hat{\mathbf{n}}) - \mathbf{p}}{4\pi\epsilon_0 r^3} \tag{4.55}$$

and $\hat{\mathbf{n}} = \mathbf{r}/|\mathbf{r}|$. This formula is correct provided that $\mathbf{r} \neq \mathbf{0}$. An idealized dipole sheet at $x = 0$ is described by a charge distribution

$$\rho(\mathbf{r}) = \frac{\sigma}{4\pi\epsilon_0}\delta'(x), \tag{4.56}$$

where δ' is the derivative of the delta function. The corresponding electric field is

$$\mathbf{E}(\mathbf{r}) = -\frac{\sigma}{4\pi\epsilon_0}\delta(x). \tag{4.57}$$

We will now consider the polarization of dielectric media and derive the macroscopic version of Gauss' law. If an electric field is applied to a

[3]For example, the center of a nucleus provides a preferred choice of the origin. The quadrupole moment of a nucleus is an important quantity in describing the nuclear structure.

medium consisting of a large number of atoms and molecules, the molecular charge distribution will be distorted. In the medium an electric polarization is produced. The latter can be quantitatively described as a macroscopic variable or cell variable such as $\mathbf{P} = \Delta\mathbf{p}/\Delta V$, i.e. as the dipole moment per unit volume. On a macroscopic scale, we may consider the polarization as a vector field, i.e. $\mathbf{P}(\mathbf{r})$. The potential $V(\mathbf{r})$ can be constructed by linear superposition of the contributions from each volume element $\Delta\Omega$ located at $\mathbf{r}'$. Each volume element gives a contribution originating from the net charge and a contributions arising from the the dipole moment.

$$\Delta V(\mathbf{r}) = \frac{1}{4\pi\epsilon_0}\left(\frac{\rho(\mathbf{r}')}{|\mathbf{r}-\mathbf{r}'|}\Delta\Omega + \frac{\mathbf{P}(\mathbf{r}')\cdot(\mathbf{r}-\mathbf{r}')}{|\mathbf{r}-\mathbf{r}'|^3}\right). \tag{4.58}$$

Adding all contributions and using the fact that

$$\boldsymbol{\nabla}'\left(\frac{1}{|\mathbf{r}-\mathbf{r}'|}\right) = \frac{\mathbf{r}-\mathbf{r}'}{|\mathbf{r}-\mathbf{r}'|^3}, \tag{4.59}$$

we obtain

$$V(\mathbf{r}) = \frac{1}{4\pi\epsilon_0}\int \mathrm{d}\tau' \frac{1}{|\mathbf{r}-\mathbf{r}'|}\left(\rho(\mathbf{r}') - \boldsymbol{\nabla}'\cdot\mathbf{P}(\mathbf{r}')\right). \tag{4.60}$$

This corresponds to the potential of a charge distribution $\rho - \boldsymbol{\nabla}\cdot\mathbf{P}$. Since the microscopic equation $\boldsymbol{\nabla}\times\mathbf{E} = 0$ does apply also on the macroscopic scale, we conclude that $\mathbf{E}$ is still derivable from a potential field, $\mathbf{E} = -\boldsymbol{\nabla}V$, and

$$\boldsymbol{\nabla}\cdot\mathbf{E} = \frac{1}{\epsilon_0}(\rho - \boldsymbol{\nabla}\cdot\mathbf{P}). \tag{4.61}$$

This result can be easily confirmed by using

$$\nabla^2\left(\frac{1}{|\mathbf{r}-\mathbf{r}'|}\right) = -4\pi\delta(\mathbf{r}-\mathbf{r}'). \tag{4.62}$$

The electric displacement, $\mathbf{D}$, is defined as

$$\mathbf{D} = \epsilon_0\mathbf{E} + \mathbf{P} \tag{4.63}$$

and the first Maxwell equation becomes

$$\boldsymbol{\nabla}\cdot\mathbf{D} = \rho. \tag{4.64}$$

If the response of the medium to the electric field is linear and isotropic then the coefficient of proportionality is the electric susceptibility, χ_{e} and the polarization reads

$$\mathbf{P} = \epsilon_0 \chi_{\mathrm{e}} \mathbf{E}. \tag{4.65}$$

and consequently,

$$\mathbf{D} = \epsilon_0 (1 + \chi_{\mathrm{e}}) \mathbf{E} = \epsilon_0 \epsilon_{\mathrm{r}} \mathbf{E}. \tag{4.66}$$

This is a *constitutive* relation connecting $\mathbf{D}$ and $\mathbf{E}$, necessary to solve the field equations. Here we have limited ourselves to consider an elementary connection. However, in general the connection can be non-linear and anisotropic, such that $\mathbf{P} = \mathbf{P}(\mathbf{E})$ will involve a non-trivial expression.

It is instructive to apply above terminology to a parallel-plate capacitor. The storage capacity C of two electrodes with charges $\pm Q$ in vacuum is $C = Q/V$, where V is the voltage drop. Filling the volume between the plates with a dielectric material results into a voltage drop

$$V = \frac{Q/\epsilon_{\mathrm{r}}}{C}. \tag{4.67}$$

This equation may be interpreted as stating that of the total charge Q, the *free* charge Q/ϵ_{r} contributes to the voltage drop, whereas the *bound* charge $(1 - 1/\epsilon_{\mathrm{r}})Q$, is neutralized by the polarization of the dielectric material. The electric susceptibility, χ_{e} emerges as the ratio of the bound charge and the free charge:

$$\chi_{\mathrm{e}} = \frac{(1 - 1/\epsilon_{\mathrm{r}})Q}{Q/\epsilon_{\mathrm{r}}} = \epsilon_{\mathrm{r}} - 1. \tag{4.68}$$

The displacement and the polarization both have the dimension [charge/area]. These variables correspond to electric flux densities. Given an infinitesimal area element $\mathbf{dS}$ on an electrode, the normal component of $\mathbf{D}$ corresponds to the charge $\mathrm{d}Q = \mathbf{D} \cdot \mathbf{dS}$ on the area element and the normal component of $\mathbf{P}$ represents the bound charge $(1 - 1/\epsilon_{\mathrm{r}})dQ$ on the area element. Finally, the normal component of $\epsilon_0 \mathbf{E}$ corresponds to the free charge $\mathrm{d}Q/\epsilon_{\mathrm{r}}$ residing on the area element. The question arises how the displacement $\mathbf{D}$, the polarization $\mathbf{P}$ and $\epsilon_0 \mathbf{E}$ can be associated to flux densities while there is no flow. In fact, the terminology is justified by analogy or mathematical equivalence with real flows. Consider for instance a stationary flow of water in $\mathbb{R}^3$. There exists a one-parameter family of maps $\phi_t : \mathbb{R}^3 \rightarrow \mathbb{R}^3$ that takes the molecule located at the position $\mathbf{r}_0$ at t_0 to the position $\mathbf{r}_1$ at t_1. Associated to the flow there exists a flux field

$$\mathbf{J}(\mathbf{r}) = \frac{\mathrm{d}\mathbf{r}}{dt}. \tag{4.69}$$

The velocity field describes the streamlines of the flow. For an incompressible stationary flow we have that for any volume Ω

$$\oint_{\partial\Omega} \mathbf{J} \cdot \mathbf{dS} = 0 \qquad \text{or} \qquad \boldsymbol{\nabla} \cdot \mathbf{J} = 0. \tag{4.70}$$

The number of water molecules that enter a volume exactly balances the number of water molecules that leave the volume. Now suppose that it is possible that water molecules are created or annihilated, e.g. by a chemical reaction $2H_2O \leftrightarrow O_2 + 2H_2$ in some volume. This process corresponds to a source/sink, Σ in the balance equation.

$$\boldsymbol{\nabla} \cdot \mathbf{J}(\mathbf{r}) = \Sigma(\mathbf{r}). \tag{4.71}$$

Comparing this equation with the first Maxwell equation, we observe the mathematical equivalence. The charge density ρ acts as a source/sink for the flux field $\mathbf{D}$.

4.9 Magnetic Media

A stationary current density, $\mathbf{J}(\mathbf{r})$, generates a magnetic induction given by

$$\mathbf{B}(\mathbf{r}) = \frac{\mu_0}{4\pi} \int \mathrm{d}\tau' \, \mathbf{J}(\mathbf{r}') \times \frac{\mathbf{r} - \mathbf{r}'}{|\mathbf{r} - \mathbf{r}'|^3} \tag{4.72}$$

This result is essentially the finding of Biot, Savart and Ampère. With the help of Equation (4.59) we may write (4.72) as

$$\mathbf{B}(\mathbf{r}) = \frac{\mu_0}{4\pi} \, \boldsymbol{\nabla}\times \int \mathrm{d}\tau' \, \frac{\mathbf{J}(\mathbf{r}')}{|\mathbf{r} - \mathbf{r}'|} \tag{4.73}$$

An immediate consequence is $\boldsymbol{\nabla} \cdot \mathbf{B} = 0$. Using the identity $\boldsymbol{\nabla}\times \boldsymbol{\nabla}\times \mathbf{A} = \boldsymbol{\nabla}(\boldsymbol{\nabla} \cdot \mathbf{A}) - \nabla^2\mathbf{A} = 0$, and the fact that $\mathbf{J} = 0$, as well as Equation (4.62) one obtains that

$$\boldsymbol{\nabla}\times\mathbf{B} = \mu_0 \, \mathbf{J}. \tag{4.74}$$

Helmholtz' theorem implies that there will be a vector field $\mathbf{A}$ such that $\mathbf{B} = \boldsymbol{\nabla}\times\mathbf{A}$ and a comparison with Equation (4.73) shows that

$$\mathbf{A}(\mathbf{r}, t) = \frac{\mu_0}{4\pi} \int \mathrm{d}\tau' \, \frac{\mathbf{J}(\mathbf{r}')}{|\mathbf{r} - \mathbf{r}'|} + \boldsymbol{\nabla}\chi(\mathbf{r}, t), \tag{4.75}$$

where χ is an arbitrary scalar function. The arbitrariness in the solution (4.75) for $\mathbf{A}$ illustrates the freedom to perform gauge transformations. This freedom however is lifted by fixing a gauge condition, i.e. by inserting an additional constraint that the component of $\mathbf{A}$ should obey, such that not all components are independent anymore. A particular choice is the Coulomb gauge, $\boldsymbol{\nabla}\times\mathbf{A} = \mathbf{0}$. In that case, χ is a solution of Laplace's equation $\nabla^2\chi = 0$. Provided that there are no sources at infinity and space is unbounded, the unique solution for χ is a constant, such that

$$\mathbf{A}(\mathbf{r}, t) = \frac{\mu_0}{4\pi} \int \mathrm{d}\tau' \, \frac{\mathbf{J}(\mathbf{r}')}{|\mathbf{r} - \mathbf{r}'|}. \tag{4.76}$$

We will now consider a localized current distribution around some origin, $\mathbf{0}$. Then we may expand Equation (4.76) for $|\mathbf{r}| > |\mathbf{r}'|$ using

$$\frac{1}{|\mathbf{r} - \mathbf{r}'|} = \frac{1}{|\mathbf{r}|} + \frac{\mathbf{r} \cdot \mathbf{r}'}{|\mathbf{r}|^3} + \cdots \tag{4.77}$$

as

$$\mathbf{A}(\mathbf{r}) = \frac{\mu_0}{4\pi r} \int \mathrm{d}\tau' \, \mathbf{J}(\mathbf{r}') + \frac{\mu_0}{4\pi r^3} \int \mathrm{d}\tau' \, (\mathbf{r} \cdot \mathbf{r}') \, \mathbf{J}(\mathbf{r}'). \tag{4.78}$$

The first integral is zero, i.e. $\int \mathrm{d}\tau \, \mathbf{J}(\mathbf{r}) = 0$, whereas the second integral gives

$$\begin{aligned} \mathbf{A}(\mathbf{r}) &= \frac{\mu_0}{4\pi} \frac{\mathbf{m} \times \mathbf{r}}{r^3}, \\ \mathbf{m} &= \frac{1}{2} \int \mathrm{d}\tau \, \mathbf{r} \times \mathbf{J}(\mathbf{r}). \end{aligned} \tag{4.79}$$

The variable $\mathbf{m}$ is the *magnetic moment* of the current distribution. Following a similar reasoning as was done for the dielectric media, we consider the macroscopic effects of magnetic materials. Since $\boldsymbol{\nabla} \cdot \mathbf{B} = 0$ at the microscopic scale, this equation also is valid at macroscopic scale. Therefore, Helmholtz' theorem is still applicable. By dividing space into volume elements ΔV, we can assign to each volume element a magnetic moment

$$\Delta \mathbf{m} = \mathbf{M}(\mathbf{r}) \, \Delta V, \tag{4.80}$$

where $\mathbf{M}$ is the magnetization or magnetic moment density. For a substance consisting of k different atoms or molecules with partial densities ρ_i, $(i = 1, ..k)$ and with magnetic moment $\mathbf{m}_i$ for the i-th atom or molecule, the magnetization is

$$\mathbf{M}(\mathbf{r}) = \sum_{i=1}^{k} \rho_i(\mathbf{r}) \, \mathbf{m}_i. \tag{4.81}$$

The free-charge current density and the magnetization of the volume element ΔV at location $\mathbf{r}'$, give rise to a contribution to the the vector potential at location $\mathbf{r}$ being

$$\Delta \mathbf{A}(\mathbf{r}) = \frac{\mu_0}{4\pi} \frac{\mathbf{J}(\mathbf{r}')}{|\mathbf{r} - \mathbf{r}'|} \Delta V + \frac{\mu_0}{4\pi} \frac{\mathbf{M}(\mathbf{r}') \times (\mathbf{r} - \mathbf{r}')}{|\mathbf{r} - \mathbf{r}'|^3} \Delta V. \tag{4.82}$$

Adding all contributions

$$\mathbf{A}(\mathbf{r}) = \frac{\mu_0}{4\pi} \int \mathrm{d}\tau' \, \frac{\mathbf{J}(\mathbf{r}') + \boldsymbol{\nabla} \times' \mathbf{M}(\mathbf{r}')}{|\mathbf{r} - \mathbf{r}'|}. \tag{4.83}$$

This corresponds to the vector potential of a current distribution $\mathbf{J} + \boldsymbol{\nabla} \times \mathbf{M}$ and therefore

$$\boldsymbol{\nabla} \times \mathbf{B} = \mu_0 \left(\mathbf{J} + \boldsymbol{\nabla} \times \mathbf{M} \right). \tag{4.84}$$

The magnetic *field* is defined as

$$\mathbf{H} = \frac{1}{\mu_0} \mathbf{B} - \mathbf{M}. \tag{4.85}$$

Then the stationary macroscopic equations become

$$\begin{aligned} \boldsymbol{\nabla} \times \mathbf{H} &= \mathbf{J}, \\ \boldsymbol{\nabla} \cdot \mathbf{B} &= 0. \end{aligned} \tag{4.86}$$

If we follow a strict analogy with the discussion on electrical polarization we should adopt a linear relation between the magnetization $\mathbf{M}$ and the induction $\mathbf{B}$ in order to obtain a constitutive relation between $\mathbf{H}$ and $\mathbf{B}$. However, historically it has become customary to define the *magnetic susceptibility* χ_{m} as the ratio of the magnetization and the magnetic field

$$\mathbf{M} = \chi_{\mathrm{m}} \, \mathbf{H}. \tag{4.87}$$

Then we obtain

$$\mathbf{B} = \mu_0 \left(\mathbf{H} + \mathbf{M} \right) = \mu_0 \, \left(1 + \chi_{\mathrm{m}} \right) \, \mathbf{H} = \mu_0 \mu_{\mathrm{r}} \mathbf{H} = \mu \mathbf{H}. \tag{4.88}$$

In here, μ is the *permeability* and μ_{r} is the *relative* permeability.

Just as is the case for electrical polarization, the constitutive relation, $\mathbf{B} = \mathbf{B}(\mathbf{H})$, can be anisotropic and non-linear. In fact, the $\mathbf{B}(\mathbf{H})$ relation may be multiple-valued depending on the history of the preparation of the material or the history of the applied magnetic fields (hysteresis).

In deriving the macroscopic field equations, we have so far been concerned with stationary phenomena. Both the charge distributions and the current distributions were assumed to be time-independent. The resulting equations are

$$\nabla \times \mathbf{E} = 0, \tag{4.89}$$
$$\nabla \cdot \mathbf{B} = 0, \tag{4.90}$$
$$\nabla \cdot \mathbf{D} = \rho, \tag{4.91}$$
$$\nabla \times \mathbf{H} = \mathbf{J}. \tag{4.92}$$

Faraday's law that was obtained from experimental observation, relates the circulation of the electric field to the time variation of the magnetic flux

$$\oint \mathbf{E} \cdot \mathbf{dr} = -\frac{\mathrm{d}}{\mathrm{d}t} \int \mathbf{B} \cdot \mathbf{dS}, \tag{4.93}$$

or

$$\nabla \times \mathbf{E} + \frac{\partial \mathbf{B}}{\partial t} = 0. \tag{4.94}$$

Magnetic monopoles have never been observed nor mimiced by time-varying fields. Therefore, the equation $\nabla \cdot \mathbf{B} = 0$ holds in all circumstances. Maxwell observed that the simplest generalization of the Equations (4.91) and (4.92) that apply to time-dependent situations and that are consistent with charge conservation, are obtained by substituting $\mathbf{J}$ in Equation (4.92) by $\mathbf{J} + \partial\mathbf{D}/\partial t$, since using the charge conservation and Gauss' law gives

$$\nabla \cdot \left(\mathbf{J} + \frac{\partial \mathbf{D}}{\partial t} \right) = 0, \tag{4.95}$$

such that the left- and right-hand side of

$$\nabla \times \mathbf{H} = \mathbf{J} + \frac{\partial \mathbf{D}}{\partial t} \tag{4.96}$$

are both divergenceless. Equations (4.90), (4.94), (4.91) and (4.96) are referred to as the (macroscopic) *Maxwell equations*. From a theoretical point of view, the Maxwell Equations (4.90) and (4.94) found their proper meaning within the geometrical interpretation of electrodynamics, where they are identified as the Bianci identities for the curvature (see Chapter 9).

5

Wave Guides and Transmission Lines

An important application of the Maxwell theory concerns the engineering of physical devices that are capable of transporting electromagnetic energy. This transport takes place in a wave-like manner. The static limit does not take into account the wave behavior of the Maxwell equations. The easiest way to implement this feature is by confining the field in two dimensions, allowing it to move freely along the third dimension (i.e. longitudinal sections are much larger than transverse directions). In this way, guided waves are recovered. A particular case of this model is the transmission line.

The wave guide consists of boundary surfaces that are good conductors. In practical realizations these surfaces are metallic materials such that the ohmic losses will be low. In the description of wave guides one usually assumes that the surfaces are perfectly conducting in a first approximation and that for large but finite conductivity, the ohmic losses can be calculated by perturbative methods. Besides the (idealized) boundary surfaces, the wave guide consists of a dielectric medium with no internal charges ($\rho = 0$), no internal currents ($\mathbf{J} = \mathbf{0}$). Furthermore, for an idealized description it is assumed that the conductivity of the dielectric medium vanishes ($\sigma = 0$). Finally, a wave guide is translational invariant in one direction. It has become customary, to choose the z-axis parallel to this direction.

In order to solve the Maxwell equations for wave guides, one considers harmonic fields (modes). The generic solution may be obtained as a superposition of different modes. The physical fields $\mathbf{E}(\mathbf{r}, t)$ and $\mathbf{H}(\mathbf{r}, t)$ are obtained from

$$\mathbf{E}(\mathbf{r}, t) = \Re\left(\mathbf{E}(\mathbf{r})\, \mathrm{e}^{\mathrm{i}\omega t}\right) \qquad \mathbf{H}(\mathbf{r}, t) = \Re\left(\mathbf{H}(\mathbf{r})\, \mathrm{e}^{\mathrm{i}\omega t}\right), \tag{5.1}$$

where $\mathbf{E}(\mathbf{r})$ and $\mathbf{H}(\mathbf{r})$ are complex phasors. The Maxwell equations governing these phasors are

$$\begin{aligned} \nabla \cdot \mathbf{E} &= 0 & \nabla \cdot \mathbf{H} &= 0 \\ \nabla \times \mathbf{E} &= -\mathrm{i}\omega\mu\mathbf{H} & \nabla \times \mathbf{H} &= \mathrm{i}\omega\epsilon\mathbf{E}. \end{aligned} \tag{5.2}$$

Defining $\omega\mu = k\zeta$ and $\omega\epsilon = k/\zeta$ then $k = \omega\sqrt{\mu\epsilon}$ and $\zeta = \sqrt{\mu/\epsilon}$. From the Equation (5.2) it follows that the phasors satisfy the following equation:

$$(\nabla^2 + k^2)\left\{ \begin{array}{c} \mathbf{E} \\ \mathbf{H} \end{array} \right\} = 0. \tag{5.3}$$

The translational invariance implies that if $\mathbf{E}(\mathbf{r}), \mathbf{H}(\mathbf{r})$ is a solution of Equation (5.3), then $\mathbf{E}(\mathbf{r} + \boldsymbol{a})$, $\mathbf{H}(\mathbf{r} + \boldsymbol{a})$ with $\boldsymbol{a} = a\mathbf{e}_\mathrm{z}$, is also a solution of Equation (5.3). We may therefore introduce a shift operator, $\hat{S}(\boldsymbol{a})$ such that

$$\hat{S}(\boldsymbol{a})\left\{ \begin{array}{c} \mathbf{E}(\mathbf{r}) \\ \mathbf{H}(\mathbf{r}) \end{array} \right\} = \left\{ \begin{array}{c} \mathbf{E}(\mathbf{r} + \boldsymbol{a}) \\ \mathbf{H}(\mathbf{r} + \boldsymbol{a}) \end{array} \right\}. \tag{5.4}$$

Performing a Taylor series expansion gives

$$\mathbf{E}(\mathbf{r} + \boldsymbol{a}) = \sum_{n=0}^{\infty} \frac{a^n}{n!} \frac{\partial^n}{\partial z^n} \mathbf{E}(\mathbf{r}) = \exp\left(a\frac{\partial}{\partial z}\right)\mathbf{E}(\mathbf{r}) \tag{5.5}$$

and therefore $\hat{S}(\boldsymbol{a}) = \exp\left(a\frac{\partial}{\partial z}\right) = \exp\left(\mathrm{i}a\hat{k}\right)$ with $\hat{k} = -\mathrm{i}\frac{\partial}{\partial z}$. The Helmholtz operator $\hat{H} = \nabla^2 + k^2$ commutes with $\hat{k}$ i.e. $\left[\hat{H}, \hat{k}\right] = 0$. As a consequence we can write the solutions of Equation (5.3) in such a way that they are simultaneously eigenfunctions of $\hat{H}$ and $\hat{k}$. The eigenfunctions of $\hat{k}$ are easily found to be

$$f(z) = \mathrm{e}^{\mathrm{i}\kappa z},$$

since

$$-\mathrm{i}\frac{d}{dz}f(z) = \kappa f(z).$$

Thus from the translational invariance we may conclude that it suffices to consider solutions for $\mathbf{E}$ and $\mathbf{H}$ of the form $\mathbf{E}(x, y)\mathrm{e}^{\mathrm{i}\kappa z}$ and $\mathbf{H}(x, y)\mathrm{e}^{\mathrm{i}\kappa z}$. Defining explicitly the transversal and the longitudinal components of the fields

$$\begin{array}{ll} \mathbf{E}(x, y) = \mathbf{E}_\mathrm{T}(x, y) + \mathbf{E}_\mathrm{L}(x, y) & \mathbf{E}_\mathrm{L}(x, y) = E_z(x, y)\ \mathbf{e}_\mathrm{z} \\ \mathbf{H}(x, y) = \mathbf{H}_\mathrm{T}(x, y) + \mathbf{H}_\mathrm{L}(x, y) & \mathbf{H}_\mathrm{L}(x, y) = H_z(x, y)\ \mathbf{e}_\mathrm{z} \end{array} \tag{5.6}$$

and

$$\nabla^2 = {\nabla^2}_\mathrm{T} + \frac{\partial}{\partial z^2} = {\nabla^2}_\mathrm{T} - \kappa^2, \tag{5.7}$$

where the subscript T stands for a transverse field in the x-y-plane, while the subscript L denotes the longitudinal fields along the z-axis, we obtain

$$\begin{aligned} \left(\nabla^2{}_{\mathrm{T}} + k^2 - \kappa^2\right) \left\{ \begin{array}{c} \mathbf{E}_{\mathrm{T}}(x,y) \\ \mathbf{H}_{\mathrm{T}}(x,y) \end{array} \right\} &= 0 \\ \left(\nabla^2{}_{\mathrm{T}} + k^2 - \kappa^2\right) \left\{ \begin{array}{c} E_z(x,y) \\ H_z(x,y) \end{array} \right\} &= 0. \end{aligned} \tag{5.8}$$

The transverse equations correspond to an eigenvalue problem with fields vanishing at the boundaries in the transverse directions. The characteristic equations that need to be solved are the Helmholtz equations resulting into eigenvalue problems, where the eigenvalues are $p^2 = k^2 - \kappa^2$. The boundary conditions for the fields on the boundary surfaces are

$$\mathbf{n} \times \mathbf{E} = 0 \qquad\qquad \mathbf{n} \cdot \mathbf{H} = 0. \tag{5.9}$$

For the transverse components, going back to the full Maxwell equations, we get from Equation (5.2)

$$\boldsymbol{\nabla}_{\mathrm{T}} E_z - \frac{\partial}{\partial z}\mathbf{E}_{\mathrm{T}} = -\mathrm{i}\omega\mu \, \mathbf{e}_{\mathrm{z}} \times \mathbf{H}_{\mathrm{T}} \tag{5.10}$$

and

$$\boldsymbol{\nabla}_{\mathrm{T}} H_z - \frac{\partial}{\partial z}\mathbf{H}_{\mathrm{T}} = \mathrm{i}\omega\epsilon \, \mathbf{e}_{\mathrm{z}} \times \mathbf{E}_{\mathrm{T}}. \tag{5.11}$$

Combining (5.10) and (5.11), gives

$$\begin{aligned} p^2\mathbf{E}_{\mathrm{T}} &= \mathrm{i}\omega\mu \, \mathbf{e}_{\mathrm{z}} \times \boldsymbol{\nabla}_{\mathrm{T}} H_z + \mathrm{i}\kappa \boldsymbol{\nabla}_{\mathrm{T}} E_z \\ p^2\mathbf{H}_{\mathrm{T}} &= -\mathrm{i}\omega\epsilon \, \mathbf{e}_{\mathrm{z}} \times \boldsymbol{\nabla}_{\mathrm{T}} E_z + \mathrm{i}\kappa \boldsymbol{\nabla}_{\mathrm{T}} H_z. \end{aligned}$$

We may define the transversal fields as

$$\begin{aligned} \mathbf{E}_{\mathrm{T}} &\propto V(z)\, \mathbf{e}_{\mathrm{t}}^{(1)} \\ \mathbf{H}_{\mathrm{T}} &\propto I(z)\, \mathbf{e}_{\mathrm{t}}^{(2)}, \end{aligned} \tag{5.12}$$

where $\mathbf{e}_{\mathrm{t}}^{(1)}$ and $\mathbf{e}_{\mathrm{t}}^{(2)}$ are transversal vectors independent of z.

5.1 TEM Modes

Inspired by waves in free space, we might look for modes that have a transverse behavior for both electric as magnetic field component, i.e.

$E_z = H_z = 0$. These solutions are the transverse electromagnetic or TEM modes.

$$\left[\nabla^2{}_\mathrm{T} + p^2\right] \mathbf{E}_\mathrm{T} = \mathbf{0}$$
$$\left[\nabla^2{}_\mathrm{T} + p^2\right] \mathbf{H}_\mathrm{T} = \mathbf{0}. \tag{5.13}$$

For the TEM mode, the Maxwell equations result into $\kappa = k$. As a consequence Equations (5.13) are void. However, one also obtains from the Maxwell Equations (5.10) and (5.11) that

$$\boldsymbol{\nabla} \times \mathbf{E}_\mathrm{T} = 0, \qquad \boldsymbol{\nabla} \cdot \mathbf{E}_\mathrm{T} = 0, \qquad \mathbf{H}_\mathrm{T} = \frac{1}{\zeta} \mathbf{e}_\mathrm{z} \times \mathbf{E}_\mathrm{T}. \tag{5.14}$$

Therefore the TEM modes are as in an infinite medium. Since $E_z = 0$, the surfaces are equipotential boundaries and therefore at least two surfaces are needed to carry the wave. Since in any plane with constant z, we have a static potential, we can consider an arbitrary path going from one conductor to another. The voltage drop will be

$$V(z) = \int_{\Gamma^1} \mathbf{E}_\mathrm{T} \cdot \mathbf{dr}. \tag{5.15}$$

The current in one conductor can be evaluated by taking a closed çontour around the conductor and evaluate the field circulation. This is illustrated in Figure 5.1.

$$I(z) = \oint_{\Gamma^2} \mathbf{H}_\mathrm{T} \cdot \mathbf{dr}. \tag{5.16}$$

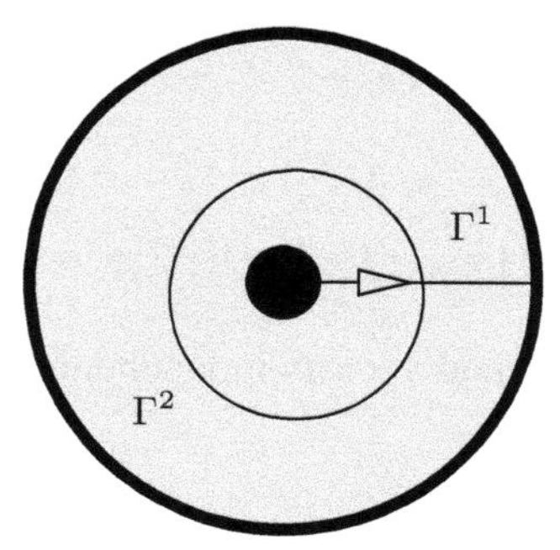

Figure 5.1 Contours for evaluating voltage drops and currents of a two-conductor system in a TEM mode.

5.2 TM Modes

When we look at solutions for which the longitudinal magnetic field vanishes ($H_z = 0$ everywhere), the magnetic field is always in the transverse direction. These solutions are the transverse magnetic or TM modes.

$$\left[\nabla^2{}_{\mathrm{T}} + p^2\right] E_z = \mathbf{0}, \tag{5.17}$$

$$p^2 \mathbf{E}_{\mathrm{T}} = \mathrm{i}\kappa \boldsymbol{\nabla}_{\mathrm{T}} E_z \tag{5.18}$$

$$p^2 \mathbf{H}_{\mathrm{T}} = -\mathrm{i}\omega\epsilon \, \mathbf{e}_{\mathrm{z}} \times \boldsymbol{\nabla}_{\mathrm{T}} E_z. \tag{5.19}$$

To find the solution of these equations, we need to solve a Helmholtz equation for E_z, and from Equations (5.18) and (5.19), the transverse field components are derived. Equation (5.18) implies that $\boldsymbol{\nabla}_{\mathrm{T}} \times \mathbf{E}_{\mathrm{T}} = 0$ and also that $\boldsymbol{\nabla}_{\mathrm{T}} \times \boldsymbol{e}_t^{(1)} = \mathbf{0}$. Therefore, we may introduce a (complex) transverse potential ϕ such that

$$\boldsymbol{e}_{\mathrm{t}}^{(1)} = -\boldsymbol{\nabla}_{\mathrm{T}} \phi. \tag{5.20}$$

This potential is proportional to E_z, i.e.

$$E_z = -\frac{p^2}{\mathrm{i}\kappa} V(z) \, \phi. \tag{5.21}$$

Substitution of the (5.12) and (5.12) into (5.11) gives that $\boldsymbol{e}_{\mathrm{t}}^{(2)} = \mathbf{e}_{\mathrm{z}} \times \boldsymbol{e}_{\mathrm{t}}^{(1)}$ and $V(z) = -(\kappa/\omega\epsilon) I(z)$.

5.3 TE Modes

Similarly, when we look at solutions for which the longitudinal electric field vanishes ($E_z = 0$ everywhere), the electric field is always in the transverse direction. These solutions are the transverse electric or TE modes.

$$\left[\nabla^2{}_{\mathrm{T}} + p^2\right] B_z = 0, \tag{5.22}$$

$$p^2 \mathbf{E}_{\mathrm{T}} = \mathrm{i}\omega\mu \, \mathbf{e}_{\mathrm{z}} \times \boldsymbol{\nabla}_{\mathrm{T}} H_z \tag{5.23}$$

$$p^2 \mathbf{H}_{\mathrm{T}} = \mathrm{i}\kappa \, \boldsymbol{\nabla}_{\mathrm{T}} H_z. \tag{5.24}$$

To find the solution of these equations, we need to solve a Helmholtz equation for B_z, and from Equations (5.23) and (5.24), the transverse field components are derived. Since in this case $\boldsymbol{\nabla}_{\mathrm{T}} \times \mathbf{H}_{\mathrm{T}} = \mathbf{0}$ there exists a scalar potential ψ such that

$$\boldsymbol{e}_{\mathrm{t}}^{(2)} = -\boldsymbol{\nabla}_{\mathrm{T}} \psi. \tag{5.25}$$

Following a similar reasoning as above we obtain that

$$H_z = \frac{p^2}{\mathrm{i}k\zeta} V(z)\, \psi. \tag{5.26}$$

Furthermore, we find that $\mathbf{e}_t^{(1)} = -\mathbf{e}_z \times \mathbf{e}_t^{(2)}$ and $V(z) = -(\omega\mu/\kappa)I(z)$.

5.4 Transmission Line Theory – S Parameters

The structure of the transverse components of the electric and magnetic fields gives rise to an equivalent-circuit description. In order to show this, we will study the TM mode, but the TE description follows the same reasoning. By returning to the generic transmission-line solutions

$$V(z) = V_+ \mathrm{e}^{-\mathrm{i}\kappa z} + V_- \mathrm{e}^{\mathrm{i}\kappa z}, \tag{5.27}$$

$$I(z) = \frac{1}{Z_\mathrm{c}} \left(V_+ \mathrm{e}^{-\mathrm{i}\kappa z} - V_- \mathrm{e}^{\mathrm{i}\kappa z}\right), \tag{5.28}$$

where Z_c is the characteristic impedance of the transmission line or the "telegraph" equations

$$\frac{\mathrm{d}V(z)}{\mathrm{d}z} = -ZI(z) \tag{5.29}$$

$$\frac{\mathrm{d}I(z)}{\mathrm{d}z} = -YV(z). \tag{5.30}$$

In these equations, the series impedance is denoted by Z and Y is the shunt admittance of the equivalent transmission line model. Each propagating mode corresponds to an eigenvalue p and we find that

$$Z = \frac{p^2 - k^2}{\mathrm{i}\omega\varepsilon}$$
$$Y = \mathrm{i}\omega\varepsilon. \tag{5.31}$$

From these expressions, the resulting equivalent circuit can be constructed.

A very important application of wave guides in electronics is the coax cable. A cross-section of the coax cable is shown in Figure 5.1. In order to perform clean measurements of devices it is required that the signals to and from the devices-under-test (DUTs) to the measurement equipment are not perturbed by external distortions. In particular, for high-frequency measurements it becomes a delicate matter to screen off the electro-magnetic noise that surrounds the device-under-test and measurement system. Coax cables play a crucial role to achieve this goal. If the transient measurement signals varies fast in time one must account for the wave character of the electromagnetic variations and the measurement data are most conveniently captured in variables corresponding to these waves. S-parameters are variables for measurement of RF characteristics of (linearized) network. The underlying idea is that the device under test (DUT) is connected to the measurement equipment with coax cables. The electromagnetic signals in such cables are transverse electric and magnetic fields propagating along the cable. The energy flow in the cable is given by I*V, where I is the current on the inner wire and V is the voltage difference between the inner wire and the outer shield. The cable is terminated by a ground-signal-ground or SGS configuration as is illustrated in Figure 5.2.

The propagation along the cable of the voltage and currents as function of time and position are given by transmission line theory or wave-guide modeling. Both descriptions contain a subtle difference in terms of used variables. For a TEM wave approach, one identifies the integral of the electric field from the inner wire to the outer wire as a potential difference. The key assumption is that at the end of the coax cable, this potential difference is assigned unmodified to the contact pads and each contact pad is an equipotential plate.

Let us revisit Equations (5.27) and (5.28) with insertion of the time dependence $exp(\mathrm{i}\omega t)$ It is seen that the combination:

$$\frac{V(z) + Z_0 I(z)}{2} = V_+ cos(-\kappa z + \omega t). \tag{5.32}$$

Figure 5.2 Ground-Signal-Ground (GSG) probe. The tip and the connector to the coax cable are shown. Photo of Infinity Probe ® courtesy of Cascade Microtech, a FormFactor company.

corresponds to a wave running into the positive z-direction. The combination

$$\frac{V(z) - Z_0 I(z)}{2} = V_{-} cos(\kappa z + \omega t). \tag{5.33}$$

corresponds to a wave running into the negative z-direction. The importance of these waves is found in the fact that these waves are directly associated with the observables that are addressed in a measurement set up. The DUT is equipped with a number of probe-landing places where the three-tip probes are placed. Actually a better way of looking at the landing of the probe tip is that the probe tips are skating into the contact areas in order to make a good connection.

Starting from port 1, an AC voltage is put over the contact pads of port 1. An impedance of Z_0 is placed over all other ports. For a two-port set up this is illustrated in Figure 5.3. The impedance is chosen 50 Ω because the measurement systems use cables with characteristic impedance of 50 $Omega$. The s-parameter definition follow from the idea that at each port there exists an incoming wave with amplitude a_i and an outgoing wave with amplitude b_i. The amplitudes at the source side and at the load side are:

$$\begin{aligned} a_1 &= \frac{V_S + Z_C I_S}{2} \quad a_2 = \frac{V_L + Z_C I_L}{2} \\ b_1 &= \frac{V_S - Z_C I_S}{2} \quad b_2 = \frac{V_L - Z_C I_L}{2} \end{aligned} \tag{5.34}$$

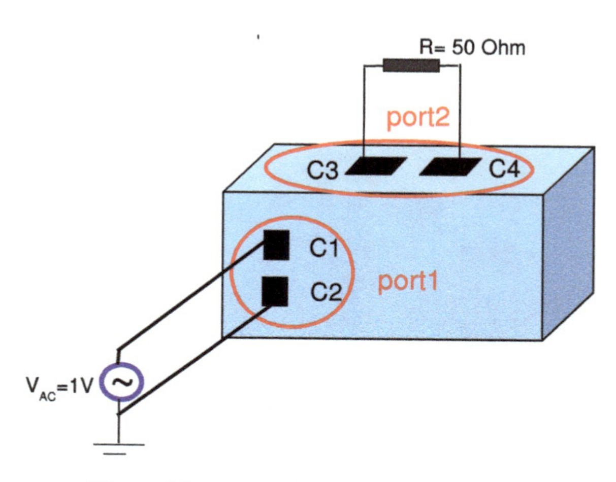

Figure 5.3 Set up of the s-matrix simulation.

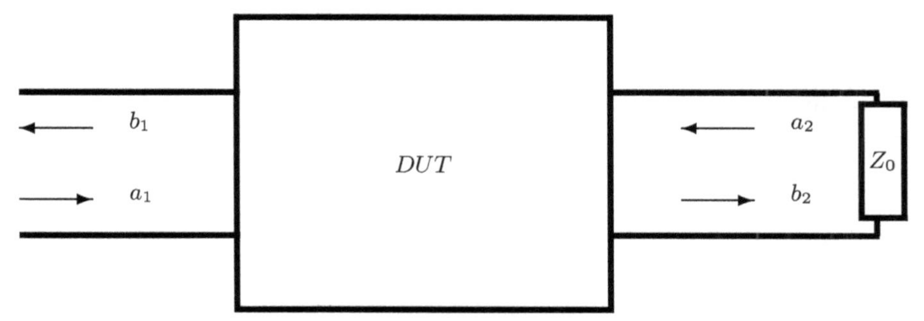

Figure 5.4 Two port s-parameter set up.

It is shown in a slightly more abstract way in Figure 5.4.

One often finds that the current is defined positive as it flows into the device at the source side and if it lows outwards at the load side.

$$b = \frac{V_S + Z_C I_S}{2} \tag{5.35}$$

We define a current positive if it flows outwards at any port. At the load side we have

$$a = \frac{V_{-} Z_L I_L}{2} \tag{5.36}$$

$$b = \frac{V_L + Z_L I_L}{2} \tag{5.37}$$

The scattering matrix is

$$\mathbf{b} = S\mathbf{a} \tag{5.38}$$

or

$$b_i = S_{ij} a_j \tag{5.39}$$

We can get all $a_i = 0$ except for the driver port by closing the loaded ports. As a consequence

$$S_{ij} = \frac{b_i}{a_j} \tag{5.40}$$

Thus the s-parameter module collects all voltages and currents for the configuration of Figure 5.3. With these voltages and currents we compose the a an b coefficients from which we next derive the S matrix. The s-parameters will depend (strongly) on the value of the characteristic impedance that is used for the conversion of the current voltages to s-parameters.

We must now relate above considerations with the construction of the boundary conditions for simulation using solvEM. The basic observation is that a simulator can measure using idealized conditions. In particular, contrary to measurements, the solvEM tool has direct access to the phase rotations of the voltages and currents. In order to compute the s parameters we return to the network as given in above figure. By including a voltage difference over the S-port that is composed from contacts C1 and C2 and a load over the L-port that is composed from contacts C3 and C4, a complete boundary value problem is defined. The boundary condition is limited to the two contacts of the source port.

With the applied voltage at contact C1, a current will be flow into C1 and a equal but opposite current will leave C2. This current is the outcome of the computation. In a similar way a current will flow through the load RL that is usually fixed at 50 Ω. Using Equation (1.2) we compose the amplitudes a1 and a2. We also compose the linear combinations b1 and b2. With these four variables we can now solve for the sparameters.

Once that the s-parameters are known, the Z, Y and H parameters can also be obtained. In above approach is should be noted that a ground can be chosen only once. We select one of the contacts of the source at 0 Volt. As a consequence, no contact at the load can be fixed upfront at 0 Volt but these contacts will have a voltage being both different from zero. The set up of Figure 5.3 illustrates this point.

5.5 Classical Ghosts Fields

We have already introduced the scalar potential V and the magnetic vector potential $\mathbf{A}$. The magnetic induction $\mathbf{B}$ is given by

$$\mathbf{B} = \nabla \times \mathbf{A} \tag{5.41}$$

The electric field is given by

$$\mathbf{E} = -\nabla V - \frac{\partial \mathbf{A}}{\partial t}$$

The Maxwell equations are expressed in terms of the potential formulation and an additional ghost field χ as follows:

$$-\nabla \cdot \left(\epsilon \nabla V + \epsilon \frac{\partial \mathbf{A}}{\partial t} \right) = \rho(V, \mathbf{A}) \tag{5.42}$$

$$\nabla \times \nabla \times \mathbf{A} - \gamma \nabla \chi = \mu_0 \mathbf{J}(V, \mathbf{A}) - \mu_0 \epsilon \frac{\partial}{\partial t} \left(\nabla V + \frac{\partial \mathbf{A}}{\partial t} \right) \tag{5.43}$$

where the gauge condition is given by

$$\nabla \cdot \mathbf{A} + \nabla^2 \chi = 0 \tag{5.44}$$

Acting with ∇ on (5.43), we obtain $\nabla^2 \chi = 0$. Putting $\chi = 0$ on the boundary implies that the Coulomb gauge condition $\nabla \cdot \mathbf{A} = 0$ is recovered. The sole purpose of the ghost field χ is used to make the operator $\nabla \times \nabla \times$ well defined. An alternative gauge condition is given by

$$\nabla \cdot (\epsilon \mathbf{A} + \epsilon \nabla \chi) = 0 \tag{5.45}$$

that leads to a simplified Poisson equation:

$$-\nabla \cdot \epsilon \nabla V = \rho(V, \mathbf{A}) \tag{5.46}$$

and the corresponding equation for the vector potential is:

$$\nabla \times \nabla \times \mathbf{A} - \gamma \epsilon \nabla \chi = \mu_0 \mathbf{J}(V, \mathbf{A}) - \mu_0 \epsilon \frac{\partial}{\partial t} \left(\nabla V + \frac{\partial \mathbf{A}}{\partial t} \right) \tag{5.47}$$

The constant γ guarantees matching of dimensions. On a discretization grid there is a vector potential variable $A_{ij} = \mathbf{A} \cdot \hat{\mathbf{e}}_{ij}$ associated to each link, where $\hat{\mathbf{e}}_{ij}$ is a unit vector in the direction of the link between nodes i and j.

The ghost field as used here, is an interesting tool to illustrate the subtleties that come with using an operator like $\nabla \times \nabla \times$ acting on a vector field $\mathbf{A}$. However, once one is aware of these subtleties one may use this knowledge and construct a formulation of the electromagnetic field problem.

A 'naive' way to proceed is by noting that

$$\nabla \times \nabla \times \mathbf{A} = -\nabla^2 \mathbf{A} + \nabla (\nabla \cdot \mathbf{A}) \tag{5.48}$$

and using the gauge condition $\nabla \cdot \mathbf{A} = 0$. However, this replacement must be used with care. In fact, the correct way to look at this equation is by saying that

$$\nabla_V^2 \mathbf{A} = -\nabla \times \nabla \times \mathbf{A} + \nabla (\nabla \cdot \mathbf{A}) \tag{5.49}$$

In other words, we can define a 'Laplacian' ∇_V^2 acting on vector field as the right-hand side of above equation. For unstructured grids this is the proper way to proceed. Thus an alternative form of the Equations (5.42), (5.43) and (5.44) is

$$-\nabla \cdot \left(\epsilon \nabla V + \epsilon \frac{\partial \mathbf{A}}{\partial t}\right) = \rho(V, \mathbf{A}) \tag{5.50}$$

$$\nabla \times \nabla \times \mathbf{A} - \nabla (\nabla \cdot \mathbf{A}) = -\nabla (\nabla \cdot \mathbf{A}) + \mu_0 \mathbf{J}(V, \mathbf{A}) - \mu_0 \epsilon \frac{\partial}{\partial t} \left(\nabla V + \frac{\partial \mathbf{A}}{\partial t}\right) \tag{5.51}$$

where the gauge condition is given by

$$\nabla \cdot \mathbf{A} + \frac{\xi}{c^2} \frac{\partial V}{\partial t} = 0 \tag{5.52}$$

In here, ξ is a slider that allows us to select different gauge conditions. For example for $\xi = 0$ we recover the Coulomb gauge and for $\xi = \epsilon_r$ we recover the Lorenz gauge.

Equation (5.51) can now be written as

$$\nabla_V^2 \mathbf{A} = \mu_0 \mathbf{J}(V, \mathbf{A}) - \mu_0 \epsilon \frac{\partial^2}{\partial t^2} \mathbf{A} + \nabla \left[\left(\frac{\xi}{c^2} - \mu_0 \epsilon\right) \frac{\partial V}{\partial t}\right] \tag{5.53}$$

The last term is equal zero $\xi = \epsilon_r$, so the result resembles the usual wave equation appearance Equation (5.51) can now be written as

$$\nabla_V^2 \mathbf{A} = \mu_0 \mathbf{J}(V, \mathbf{A}) - \mu_0 \epsilon \frac{\partial^2}{\partial t^2} \mathbf{A} \tag{5.54}$$

In the small-signal approximation, the time evolution is described in the Fourier domain. All variables now become dependent on the operation frequency, ω. In the Fourier domain the potential description becomes for the selected gauge

$$\nabla \cdot (\varepsilon \nabla V + \mathrm{i} \varepsilon \omega \mathbf{A}) = -\rho \tag{5.55}$$

$$\nabla \times \nabla \times \mathbf{A} - \gamma \nabla \chi = \mu_0 \mathbf{J} - \mathrm{i} \omega \mu_0 \varepsilon \nabla V + \mu_0 \varepsilon \omega^2 \mathbf{A} \tag{5.56}$$

$$\nabla \cdot \mathbf{A} + \nabla^2 \chi = 0 \tag{5.57}$$

5.6 The Static Approach and Dynamic Parts

The electrostatic field, V_0, is obtained by solving the Poisson equation

$$\nabla \cdot (\varepsilon \nabla V_0) = \rho(V_0), \tag{5.58}$$

and the corresponding charge distribution $\rho(V_0)$ must be calculated self-consistently for (a) bounded surface charges on the boundary surfaces of the dielectric regions taking into account the appropriate boundary conditions, (b) free surface charges on the boundaries of a conductor and (c) space charge in the doped semiconductor volume. The current density $\mathbf{J}_0$, gives rise to the vector potential $\mathbf{A_0}$, being the solution of

$$\nabla \times \nabla \times \mathbf{A}_0 - \gamma \nabla \chi_0 = \mu_0 \mathbf{J}_0 \left(V_0 \right) \quad (5.59)$$

and submitted to the gauge condition

$$\nabla \cdot \mathbf{A}_0 + \nabla^2 \chi_0 = 0 \quad (5.60)$$

Inside conducting media the latter equation is supplemented by

$$\begin{aligned} \rho &= 0 && (5.61) \\ \mathbf{J}_0 &= \sigma \mathbf{E}_0 && (5.62) \\ \mathbf{E}_0 &= -\nabla V_0 && (5.63) \\ \nabla \cdot \mathbf{J}_0 &= 0 && (5.64) \end{aligned}$$

whereas in the semiconducting regions the following equations apply

$$\begin{aligned} \rho_0 &= q(p_0 - n_0 + N_D - N_A) && (5.65) \\ \mathbf{J}_{n0} &= q\mu_n n_0 \mathbf{E_0} + kT\mu_n \nabla n_0 && (5.66) \\ \mathbf{J}_{p0} &= q\mu_p p_0 \mathbf{E_0} - kT\mu_p \nabla p_0 && (5.67) \\ \nabla \cdot \mathbf{J}_{n0} - U(n_0, p_0) &= 0 && (5.68) \\ \nabla \cdot \mathbf{J}_{p0} + U(n_0, p_0) &= 0 && (5.69) \end{aligned}$$

The equations that determine the amplitudes and phases of the harmonic perturbations are obtained as linear perturbations of the full system. Returning to Equations (5.55–5.57), one obtains

$$\begin{aligned} \nabla \cdot (\varepsilon \nabla V_R - \varepsilon \omega \mathbf{A}_I) + \rho_R &= 0 && (5.70) \\ \nabla \cdot (\varepsilon \nabla V_I + \varepsilon \omega \mathbf{A}_R) + \rho_I &= 0 && (5.71) \\ \nabla \times \nabla \times \mathbf{A}_R - \mu_0 \varepsilon \omega^2 \mathbf{A}_R - \mu_0 \mathbf{J}_R - \mu_0 \varepsilon \omega \nabla V_I - \gamma \nabla \chi_R &= 0 && (5.72) \\ \nabla \times \nabla \times \mathbf{A}_I - \mu_0 \varepsilon \omega^2 \mathbf{A}_I - \mu_0 \mathbf{J}_I + \mu_0 \varepsilon \omega \nabla V_R - \gamma \nabla \chi_I &= 0 && (5.73) \\ \nabla^2 \chi_R + \nabla \cdot \mathbf{A}_R &= 0 && (5.74) \\ \nabla^2 \chi_I + \nabla \cdot \mathbf{A}_I &= 0 && (5.75) \end{aligned}$$

where the sources $\mathbf{J}_R$, $\mathbf{J}_I$, ρ_R and ρ_I must be determined by the non-linear constitutive equations.

5.7 Interface Conditions

In general, the structures consist of insulating, semiconducting and metallic regions. As a consequence, there will be four types of interface nodes, i.e.

- insulator/metal interface nodes
- insulator/semiconductor interface nodes
- metal/semiconductor interface nodes
- insulator/semiconductor/metal 'triple' points

At the metal/semiconductor interface nodes, we implement the idealized interface Schottky contact condition, as for a boundary condition for a semiconductor region, by setting $\phi_p = \phi_n = V_{\text{metal}}$, where V_{metal} is the value of the Poisson potential at the metal side of the interface. The Poisson potential at the semiconductor side of the interface is $V_{\text{semi}} = V_{\text{metal}} - \delta V$, where δV represents the contact potential between the two materials. Using

$$\begin{aligned} n &= n_i \exp \frac{q}{kT}(V - \phi_p) \\ p &= n_i \exp \frac{q}{kT}(\phi_p - V) \end{aligned} \tag{5.76}$$

and applying the neutrality condition $p - n - N = 0$, where $N = N_D - N_A$, we obtain for p-type semiconductor regions ($N < 0$)

$$\delta V = \log \left(-\frac{N}{2n_i} \left(1 + \sqrt{1 + \frac{4n_i^2}{N^2}} \right) \right) \tag{5.77}$$

and for n-type semiconductor regions ($N > 0$)

$$\delta V = -\log \left(\frac{N}{2n_i} \left(1 + \sqrt{1 + \frac{4n_i^2}{N^2}} \right) \right) \tag{5.78}$$

At a metal/semiconductor interface node there is *one* variable (V_{metal}) that needs to be solved. The equation for this variable assigned to the node i is the current-continuity equation,

$$\sum_j J_{ij}\, S_{ij} = 0 \tag{5.79}$$

where J_{ij} is the current density in discretized form for the link (ij) from node i to node j, and S_{ij} the perpendicular cross section of the link (ij). Note that for an idealized Schottky contact the Poisson potential is double-valued.

At metal/insulator interface nodes we assume continuity of the Poisson potential. For these nodes there is, apart from the variables $\mathbf{A}$ and χ, one unknown V_i, and the corresponding equation is the current-continuity equation. The Poisson equation determines the interface charge, ρ_i and can be obtained by post-processing, once V is determined.

At insulator/semiconductor interface nodes there are three unknowns to be determined, V, n and p. These variables are treated in the usual way as is done in device simulation tools, i.e. the Poisson equation is solved self-consistently with the current-continuity equations for n and p, while V is continuous at the insulator/semiconductor interface.

At triple point nodes, the Poisson potential is triple-valued. One arrives at different values depending on the material in which one approaches the node. For computational convenience we take the value of the Poisson potential in the insulator at the midpoint between V_{metal} and V_{semi}, i.e.

$$\lim_{\mathbf{x}\to\mathbf{x}_{\text{tr}}} V_{\text{insul}}(\mathbf{x}) = V_{\text{metal}}(\mathbf{x}_{\text{tr}}) - \frac{1}{2}\delta V \tag{5.80}$$

The interface conditions for the vector field $\mathbf{A}$ and the ghost field χ are straightforward. The choice of the gauge condition, Equation (5.57), is independent of specific material parameters. During assembling of Equation (5.56), the current associated to each link is uniquely determined in an earlier iteration of the Gummel loop and therefore the vector potential (as well as χ) is single-valued.

5.8 Boundary Conditions

The boundary conditions represent a crucial ingredient for the solution that is ultimately obtained. Furthermore, the determination of the boundary conditions are an essential step in setting up a well-defined formulation of the problem under consideration. In fact, the selection of the boundary conditions is not only determined by the desire to obtain a well-defined mathematical problem: they also are the vehicle to select the physical restrictions of the problem under consideration. In particular, the boundary conditions for radiation applications will differ from energy-confining conditions.

The MAGWEL solver is tailored to the physical conditions that are encountered in TCAD device simulation. In particular, these conditions are extended in such way that one can deal with the electromagnetic fields in the high-frequency regime. As a consequence, the boundary conditions are a minimal extension of the boundary conditions of the static electromagnetic formulation.

Since the MAGWEL solver starts from the formulation of the Maxwell equations in terms of the potentials, the boundary conditions also should be given for the *potentials*, i.e. the scalar potential V and the vector potential $\mathbf{A}$. Furthermore, the appearance of the differential equations for these potentials depend on the choice of the gauge condition. The MAGWEL solver applies the Lorenz gauge condition:

$$\nabla \cdot \mathbf{A} + \mathrm{i}\frac{\omega}{c^2}V = 0 \tag{5.81}$$

The Coulomb gauge $\nabla \cdot \mathbf{A} = 0$ has the advantage that Poisson's equation remains unaltered in going form the static to the dynamic regime and inside the simulation domain. However, at the surface of the simulation domain one should retain Gauss' equation in its original form since $\mathbf{A}$ does not necessarily vanish outside the simulation domain.

The Lorenz gauge is well suited to address wave like solutions. This gauge also has the advantage that some couplings in the field equations vanish.

Inside the simulation domain, we simulate the *device under study*. The device is terminated with contacts at the borders of the simulation domain. The contacts are the only boundaries along which we can transfer currents through the edge of the simulation domain.

On those contacts, the applied voltage can be specified. A constant electric potential (used in the static part of the simulation) as well as a constant small-signal (high-frequency) electric potential can be applied. As the applied voltages at the borders of the device are already specified, there is no need to specify the currents. Those will be calculated self-consistently in the simulator. This is a generalization of the fact that one should either specify Dirichlet's boundary conditions or Neumann boundary conditions at each given position of the domain boundary.

To keep the simulation domain separated from the rest of the world, the simulation domain should not be able to 'see' electromagnetic things happening in the rest of the world. This imposes certain conditions at the boundaries of the simulation domain. Those conditions will influence the $\mathbf{E}$ and $\mathbf{B}$ field at the boundaries and inside the simulation domain. To ease the discussion of those fields at the boundaries, they will be split into a normal and a tangential component:

$$\begin{aligned}\mathbf{E} &= \mathbf{E}_{\mathrm{n}} + \mathbf{E}_{\mathrm{t}} \\ \mathbf{B} &= \mathbf{B}_{\mathrm{n}} + \mathbf{B}_{\mathrm{t}}\end{aligned}$$

In order to guarantee minimal coupling between the simulation domain and the rest of the world, we divide the boundary of the simulation domain into two parts: First there are the contacts through which currents can enter and leave the simulation domain. These contacts are characterized by assigning fixed voltage values to them. Secondly, the remainder of the boundary is characterized by demanding that the perpendicular component of the electric field vanishes:

$$\mathbf{E}_{\mathrm{n}} = 0$$

Voltages In order to have $\mathbf{E}_{\mathrm{n}} = 0$, the electric potential at the boundary surface in the normal direction of that surface should compensate the normal component of the vector potential:

$$\frac{\partial V}{\partial \mathbf{n}} = -\mathrm{i}\omega \mathbf{A}_{\mathrm{n}}$$

since the electric field is given by $\mathbf{E} = -\nabla V - \mathrm{i}\omega \mathbf{A}$.

Currents To hide all currents in the rest of the world from the simulation domain, no **B** field should be allowed to enter the simulation domain. This forces:

$$\mathbf{B}_{\mathrm{n}} = 0$$

at the boundaries. This forces all magnetic field lines to remain inside the simulation domain.

Another way of looking at this situation is to regards the currents that enter the simulation domain through the contact areas by perpendicular leads (that are not simulated). These current-carrying leads bring with them circulating B fields that are parallel to the surface that contains the contact. In the Coulomb gauge, the vector potential **A** that is induced by this lead is given by:

$$\mathbf{A}(\mathbf{x}) = -\frac{1}{4\pi} \int_{\mathrm{lead}} \mathrm{dv} \frac{\mathbf{J}(\mathbf{x}')}{|\mathbf{x} - \mathbf{x}'|}$$

As a consequence, the impinging vector potential is parallel to in/out going current and also perpendicular to the surface containing the contact area.

Writing $\mathbf{A} = \mathbf{A}_{\mathrm{n}} + \mathbf{A}_{\mathrm{t}}$, we set $\mathbf{A}_{\mathrm{t}} = 0$. This conditions is applied for the complete surface of the simulation domain, irrespective of being at a contact area.

The above boundary conditions correspond to a minimal modification of the TCAD boundary conditions. The physical motivation behind setting $\mathbf{E}_{\mathrm{n}} = 0$ implies that if one takes a surface integral of this normal field one finds that the enclosed charge sums up to zero. In other words: the total charge in the structure under consideration does not change. Another consequence of this approach is that contacts may be arbitrarily distributed over the surface of the structure. A typical unification of contacts to realize 'ports' is lacking in this view. In general, a 'port' is defined as an area of the surface through which energy can enter or leave the simulation domain. The above 'TCAD'-like boundary conditions imply that the Poynting vector can *only* point inwards to the simulation domain. Thus above setting of the boundary conditions mean that the total surface of the simulation domain (except for the metallic contacts segments) act as a single 'port' by which energy can only enter. (This energy is dissipated in the conductive regions and excellent balance is obtained).

It should be noted that in order to have a Poynting vector unequal to zero that $\mathbf{E}_{\mathrm{t}} \neq 0$. An artifact of a box-shaped simulation domain is that at the edge of the box the tangential electric field components make an abrupt change of direction when moving from one adjacent surface element to another. Simulations show that the field intensities are strongly suppressed in the edges and corners of the simulation domain.

Enhanced boundary conditions A major limitation of above setting of the boundary conditions is that radiation cannot leave the simulation domain. Of course, it is feasible that the correct physical situation corresponds to emission of energy, when the device is excited. Thus it should be possible to divide the surface of the simulation domain into several parts: At some parts energy enters to simulation domain whereas through other parts of the surface energy is leaving. Taking into account the conservation of energy, there should be more energy entering the system at some ports than that there should leave radiation at other parts of the surface, since some energy is consumed in dissipative current flows.

A detailed analysis of the high-frequency energy theorem shows that the surface integral of the Poynting vector can include radiation if $\mathbf{A}_{\mathrm{t}} \neq 0$. Starting from

$$\oint_{\partial\Omega} \mathbf{S}.\mathbf{da} = \frac{1}{2}\oint_{\partial\Omega} \mathbf{E} \times \mathbf{H}^{*}.\mathbf{da} \tag{5.82}$$

we may substitute $\mathbf{E} = \nabla V - \mathrm{i}\omega\mathbf{A}$.

By dividing $\partial\Omega$ into regions where $\mathbf{A}_{\mathrm{t}} = 0$ and regions where $\mathbf{A}_{\mathrm{t}} \neq 0$, the surface integral can be further elaborated on. It is important to realize that the statement $\mathbf{A}_{\mathrm{t}} = 0$ depends on the choice of the gauge condition and that it must be evaluated taking into account the scalar potential V. We will discuss in detail the meaning of the setting $\mathbf{A}_{\mathrm{t}} = 0$ at the surface of the simulation domain.

It should be noted that a two-port contact system with incoming or outgoing energy flux can be properly described by the 'TCAD'-like boundary condition. The two ports are at the same face of the simulation domain and a TEM wave can be used describing $V, \mathbf{A}$ where $\mathbf{A} = A_{\mathrm{n}}\mathbf{n}$. For the coaxial contact pair of Figure 5.1, the TEM wave has the following properties:

- $\mathbf{E}_{\mathrm{n}} = 0$.
- $\mathbf{B}_{\mathrm{n}} = 0$.

From the Maxwell equations one finds

- $\nabla_{\mathrm{t}} \times \mathbf{E}_{\mathrm{t}} = 0$
- $\nabla_{\mathrm{t}} \cdot \mathbf{E}_{\mathrm{t}} = 0$

From the equation $\mathbf{B} = \nabla \times \mathbf{A}$ follows for the TEM that

- $\nabla_{\mathrm{t}} \times \mathbf{A}_{\mathrm{t}} = 0$

It is interesting to look at the line integral over $\mathbf{E}_{\mathrm{t}}$ from contact no. 1 to contact no. 2.

$$\begin{aligned}\int_{\mathbf{x}_1}^{\mathbf{x}_2} \mathbf{E}_{\mathrm{t}} \cdot \mathbf{dl} &= -\int_{\mathbf{x}_1}^{\mathbf{x}_2} \nabla_{\mathrm{t}} V \cdot \mathbf{dl} - \mathrm{i}\,\omega \int_{\mathbf{x}_1}^{\mathbf{x}_2} \mathbf{A}_{\mathrm{t}} \cdot \mathbf{dl} \\ &= V_1 - V_2 - \mathrm{i}\,\omega \int_{\mathbf{x}_1}^{\mathbf{x}_2} \mathbf{A}_{\mathrm{t}} \cdot \mathbf{dl}\end{aligned}$$

We can also write this equation as:

$$V_1 - V_2 = \int_{\mathbf{x}_1}^{\mathbf{x}_2} \mathbf{E}_{\mathrm{t}} \cdot \mathbf{dl} + \mathrm{i}\,\omega \int_{\mathbf{x}_1}^{\mathbf{x}_2} \mathbf{A}_{\mathrm{t}} \cdot \mathbf{dl} \tag{5.83}$$

Since $\mathbf{E}_{\mathrm{t}}$ and $\mathbf{A}_{\mathrm{t}}$ satisfy $\nabla_{\mathrm{t}} \times \mathbf{E}_{\mathrm{t}} = 0$ and $\nabla_{\mathrm{t}} \times \mathbf{A}_{\mathrm{t}} = 0$ there exist scalar functions ψ and ϕ such that $\mathbf{E}_{t} = -\nabla_{\mathrm{t}}\psi$ and $\mathbf{A}_{\mathrm{t}} = -\nabla_{\mathrm{t}}\phi$. This leads to the result:

$$V_1 - V_2 = \psi_1 - \psi_2 + \mathrm{i}\omega(\phi_1 - \phi_2)$$

which can be viewed as a constraint between V, ψ and ϕ:

$$V(\mathbf{x}) - \psi(\mathbf{x}) - \mathrm{i}\omega\ \phi(\mathbf{x}) = \mathrm{const}$$

The constraint can be exploited to ease analytic computations. For the analysis of TEM in wave guides one chooses

$$\phi(\mathbf{x}) = \mathrm{const} \quad \Rightarrow \quad \mathbf{A}_{\mathrm{t}} = 0,$$

whereas for the analysis of free-field radiation it is common to set

$$V(\mathbf{x}) = \mathrm{const} \quad \Rightarrow \psi(\mathbf{x}) = -\mathrm{i}\omega\ \phi(\mathbf{x}) \quad \Rightarrow \quad \mathbf{E}_{\mathrm{t}} = -\mathrm{i}\omega\mathbf{A}_{\mathrm{t}} \tag{5.84}$$

In general, the fields in source-free regions are conveniently solved in the radiation or Coulomb gauge, $\nabla \cdot \mathbf{A} = 0$ and the absence of sources gives $V = 0$.

We will now consider in detail the solution of a coaxial port. Without loss of generality we may set the axial direction to be the z coordinate. With $\mathbf{A}_{\mathrm{t}} = 0$ we find that $\mathbf{E}_{\mathrm{t}} = -\nabla_{\mathrm{t}} V$. From $\nabla_{\mathrm{t}}^2 V = 0$ we can solve the 2D Poisson problem:

$$\frac{1}{r}\frac{\partial}{\partial r}\left(r\frac{\partial V}{\partial r}\right) + \frac{1}{r^2}\frac{\partial^2 V}{\partial \phi^2} = 0 \tag{5.85}$$

With $\frac{\partial V}{\partial \phi} = 0$, we find that $V(r) = \int_{r_1}^{r} c/r'dr'$ and with $V_1 - V_2 = \Delta V$ we obtain $V(r) = \Delta V\ \ln{(r_2/r_1)} \ln{(r/r_0)}$. The electric field is:

$$\mathbf{E}_{\mathrm{t}}(r) = -\mathbf{e}_{\mathrm{r}}\, \Delta V\ \ln(r_2/r_1)\ \frac{1}{r} \tag{5.86}$$

The magnetic field is:

$$\mathbf{H}_{\mathrm{t}} = \frac{1}{\sqrt{\frac{\mu}{\epsilon}}}\,\mathbf{e}_z \times \mathbf{E}_{\mathrm{t}} = \frac{1}{\sqrt{\frac{\mu}{\epsilon}}}\,\mathbf{e}_\phi\, \Delta V \ln(r_2/r_1)\frac{1}{r} \tag{5.87}$$

The injected power is $P = 2\pi\,\sqrt{\frac{\epsilon}{\mu}}\,\Delta V^2\,[\ln{(r_2/r_1)}]^3$.
The magnetic induction is:

$$\mathbf{B} = \mathbf{B}_{\mathrm{t}} = \sqrt{\mu\epsilon}\,\mathbf{e}_\phi\, \Delta V \ln{(r_2/r_1)}\ \frac{1}{r} \tag{5.88}$$

Furthermore, $\mathbf{B}_{\mathrm{t}} = \mathbf{e}_\phi\ \left(\frac{\partial A_{\mathrm{r}}}{\partial z} - \frac{\partial A_{\mathrm{z}}}{\partial r}\right)$. Since $A_{\mathrm{r}} = 0$, we obtain

$$\frac{\partial A_z}{\partial r} = -\sqrt{\mu\epsilon}\,\Delta V \ln{(r_2/r_1)}\ \frac{1}{r} \tag{5.89}$$

As a consequence,

$$A_{\mathrm{z}} = -\sqrt{\mu\epsilon}\,\Delta V \ln\left(r_2/r_1\right)\,\ln\left(r/r_0\right) \tag{5.90}$$

and $\mathbf{A}_{\mathrm{t}} = 0$. Since $\mathbf{E}_{\mathrm{z}} = 0 \Rightarrow \frac{\partial V}{\partial z} = -\mathrm{j}\omega A_{\mathrm{z}}$, we conclude that

$$V(r,z) = \mathrm{j}\omega\,\sqrt{\mu\epsilon}\,\Delta V \ln\left(r_2/r_1\right)\,\ln\left(r/r_0\right) z + \mathrm{const} \tag{5.91}$$

The voltage grows unlimited along the coax that produces the TEM boundary condition. At first sight this seems weird, but it should be noted that the coax itself is not part of the problem to be solved. All that matters is that the tangential fields $\mathbf{E}$ and $\mathbf{B}$ are produced at the surface. The Coulomb gauge condition is also satisfied:

$$\nabla\cdot\mathbf{A} = 0 \quad \text{and} \quad \mathbf{A}_{\mathrm{t}} = 0 \quad \Rightarrow \quad \frac{\partial A_z}{\partial z} = 0 \tag{5.92}$$

which is in agreement with Equation (5.90).

This specific example shows that with a two-port contact system one is able to inject energy in the structure and that such an injection according to a TEM mode can be described using the boundary condition as formulated with the TCAD-like approach.

The above analytic computation has a numerical implementation that is summarized by the following boundary conditions:

- $\mathbf{A}_{\mathrm{t}} = 0$
- $\mathbf{E}_{\mathrm{n}} = 0$
- $\frac{\partial A_z}{\partial z} = 0$, where z denotes the normal direction.
- $\frac{\partial V}{\partial z} = -\mathrm{i}\omega A_z$

The last equation is implemented in the box-integration method by putting $\mathbf{E}_{\mathrm{n}} = 0$ for nodes at the surface of the simulation domain. This corresponds to a modified Neumann boundary condition for V.

As can be seen from Equation (5.91), their is no wave character in the fields V and A_z in the normal direction. We can include a wave-like description by turning to the Lorenz gauge. Then

$$\nabla\cdot\mathbf{A} + \mathrm{i}\frac{\omega}{c^2}V = 0 \tag{5.93}$$

Insertion of the boundary condition $\frac{\partial V}{\partial z} = -\mathrm{i}\omega\, A_z$ into this equation gives

$$\frac{\partial^2 V}{\partial z^2} + \frac{\omega^2}{c^2}V = 0 \tag{5.94}$$

The solution (along the coax) is

$$V(r,z) = V(r)\ \exp\left(ikz\right) \tag{5.95}$$

with $k = \frac{\omega}{c}$.

Thus, in order to meet the following three conditions simultaneously:

- $\mathbf{A}_{\mathrm{t}} = 0$
- $\mathbf{E}_{\mathrm{n}} = 0$
- $V(r,z) = V(r)\ \exp\left(ikz\right)$ i.e. wave like behavior in the boundary region

then we must adopt the Lorenz gauge condition. On the other hand, if we want to meet the following three conditions simultaneously:

- $\mathbf{E}_{\mathrm{t}} = 0$
- $\nabla \cdot \mathbf{A} = 0$, stick to the Coulomb gauge
- $A(r,z) = A(r)\ \exp\left(ikz\right)$, i.e. wave-like behavior

then we must allow for tangential components of the vector potential, i.e. $\mathbf{A}_{\mathrm{t}} \neq 0$. This can be understood from the fact that for a plane wave, the wave vector is perpendicular to the surface (outgoing wave). Moreover, the gauge condition $\nabla \cdot \mathbf{A} = 0$ leads to the conclusion that $\mathbf{k} \perp \mathbf{A}$. In other words: the links in the surface of the simulation domain should allow for Neumann-type of boundary conditions contrary to the TCAD implementation that is of the Dirichlet type $\mathbf{A}_{\mathrm{t}} = 0$ for which the link variables $A_{ij} = 0$. A pure Neumann boundary condition for $\mathbf{A}$ takes the form: $\frac{\partial A_{\mathrm{t}}}{\partial z} = 0$. However, a mixed version of Neumann and Dirichlet comes closer to the incorporation of radiative boundary conditions. The condition

$$\alpha\ \frac{\partial A_{\mathrm{t}}}{\partial z} + \beta\ A_{\mathrm{t}} = 0 \tag{5.96}$$

is compatible with free field radiation, provided that α and β are chosen in agreement with the Mur conditions. These advanced boundary conditions are presently under development. In particular, the tag for selecting alternative boundary condition in the editor is not active yet. The tag "perfect-electric boundary" refers to the assumption that the boundary surface element is perfectly conducting and that no tangential electric field can exist on this surface element. We apply this assumption to the contact areas. However, outside contact areas the present implementation is based on the assumption that only a tangential component exists.

Semiconductor contacts Contacts that are attached to semiconducting regions need to impose boundary conditions for the electromagnetic potentials as well as for the carrier concentrations or quasi-fermi levels. We will assume ohmic boundary conditions. These are valid at highly-doped regions, which is usually the case for contacting semiconductors. For lowly-doped regions, Schottky contacts need to be modeled which is not possible in this release.

The ohmic boundary conditions correspond to a very high recombination time of electrons and holes, leading to the condition of space-charge neutrality at the contact: $pn = n_i^2$ and $p - n + N_\mathrm{D} - N_\mathrm{A} = 0$. The quasi-fermi levels at the contact are equal to the applied voltage: $\phi_p = \phi_n = V_\mathrm{app}$.

Circuits Often it is useful to connect the *device under study* to the circuit located outside the simulation area, e.g. for the calculation of S-parameters.

6

Energy Calculations and the Poynting Vector

For the whole simulation area we can formulate the Poynting theorem. We will show in this section that this theorem enables us to have a first guess on the accuracy of the results.

6.1 Static Case

In this section, the static Poynting theorem is explained. We start with the energy dissipated in the simulation domain. This can be written as:

$$E_{\text{joule}} = \int_V \mathbf{J} \cdot \mathbf{E} \, \mathrm{dv}, \tag{6.1}$$

integrated over the simulation domain.

In the static case, all time derivatives vanish. Hence $\frac{\partial \mathbf{B}}{\partial t} = 0$, and also $\nabla \times \mathbf{E} = 0$ due to Faraday's law. On the other hand, we can rewrite Ampère's equation as $\nabla \times \mathbf{H} = \mathbf{J}$. This is sufficient to rewrite (6.1) as:

$$\begin{aligned} E_{\text{joule}} &= \int_V (\nabla \times \mathbf{H}) \cdot \mathbf{E} \, \mathrm{dv} \\ &= \int_V \nabla \cdot (\mathbf{E} \times \mathbf{H}) \, \mathrm{dv} \\ &= \oint_{\partial V} \mathbf{S} \cdot \mathbf{da}, \end{aligned}$$

with $\mathbf{S}$ denoting the Poynting vector, defined by $\mathbf{S} = \mathbf{E} \times \mathbf{H}$. We can rewrite $\mathbf{E}$ in (6.1) using the definition of the electric field in function of the potential as $\mathbf{E} = -\nabla V$. The current continuity requires that $\nabla \cdot \mathbf{J} = 0$. This enables us to write (6.1) as:

$$
\begin{aligned}
E_{\text{joule}} &= -\int_V \mathbf{J} \cdot \nabla V \, \mathrm{dv} \\
&= -\int_V \nabla \cdot (\mathbf{J}V) \, \mathrm{dv} \\
&= -\oint_{\partial V} V \, \mathbf{J} \cdot \mathbf{da}
\end{aligned}
$$

The integral is evaluated over the surface of the simulation area. However, the only locations of non-zero current density across the surface are the contacts. At these contacts the static potential is constant, and therefore we can write:

$$
\int_V \mathbf{J} \cdot \mathbf{E} \, \mathrm{dv} = -\sum_{\text{contacts}} I_\alpha V_\alpha \tag{6.2}
$$

Because we can calculate each of the terms of (6.2) in different ways, the equality of these terms is a good control for the consistency of the calculation. We expect that the terms n Equation (6.2) are equal.

6.2 High-Frequency Case

Along the same lines, we write the dissipated energy in phasor notation:

$$
E_{\text{joule}} = \frac{1}{2} \int_{\mathrm{V}} \mathbf{J}^* \cdot \mathbf{E} \; \mathrm{dv}, \tag{6.3}
$$

whose real part gives the time average rate of work done by the fields in the volume V. Using Maxwell's equations this becomes:

$$
\begin{aligned}
E_{\text{joule}} &= \frac{1}{2} \int_{\mathrm{V}} (\nabla \times \mathbf{H}^* + \mathrm{i}\omega \mathbf{D}^*) \cdot \mathbf{E} \; \mathrm{dv} \\
&= \frac{1}{2} \int_{\mathrm{V}} [-\nabla \cdot (\mathbf{E} \times \mathbf{H}^*) + \mathbf{H}^* \cdot (\nabla \times \mathbf{E}) + \mathrm{i}\omega \mathbf{D}^* \cdot \mathbf{E}] \; \mathrm{dv} \\
&= \frac{1}{2} \int_{\mathrm{V}} [-\nabla \cdot (\mathbf{E} \times \mathbf{H}^*) + \mathrm{i}\omega \, (\mathbf{D}^* \cdot \mathbf{E} - \mathbf{H}^* \cdot \mathbf{B})] \; \mathrm{dv} \\
&= -\frac{1}{2} \oint_{\partial \mathrm{V}} \mathbf{E} \times \mathbf{H}^* \cdot \mathbf{da} + \mathrm{i}\omega \int_V (\mathbf{D}^* \cdot \mathbf{E} - \mathbf{H}^* \cdot \mathbf{B}) \; \mathrm{dv} \\
&= -\oint_{\partial \mathrm{V}} \mathbf{S} \cdot \mathbf{da} + 2\mathrm{i}\omega \int_V (w_{\mathrm{e}} - w_{\mathrm{m}}) \, \mathrm{dv}
\end{aligned}
$$

with the following definitions of electric and magnetic energy density and Poynting vector:

$$
\begin{aligned}
w_{\rm e} &= \frac{1}{4}\mathbf{D}^* \cdot \mathbf{E} \\
w_{\rm m} &= \frac{1}{4}\mathbf{H}^* \cdot \mathbf{B} \\
\mathbf{S} &= \frac{1}{2}\mathbf{E} \times \mathbf{H}^*
\end{aligned}
$$

The Poynting term is integrated over the boundary of the full simulation domain. The potential expression for the electric field can be substituted:

$$
-\oint_{\partial \mathrm{V}} \mathbf{S} \cdot \mathbf{da} = -\frac{1}{2}\oint_{\partial \mathrm{V}} (-\nabla V - \mathrm{i}\omega \mathbf{A}) \times \mathbf{H}^* \cdot \mathbf{da}
$$

Since the boundary conditions are such that $\mathbf{H}^*$ is parallel to the boundary and $\mathbf{A}$ is perpendicular to the boundary, $\mathbf{A} \times \mathbf{H}^*$ is parallel to the boundary. As a consequence, the second term does not contribute. The first term can be written as:

$$
\begin{aligned}
\frac{1}{2}\oint_{\partial \mathrm{V}} \mathbf{da} \cdot\ \nabla V \times \mathbf{H}^* = &\frac{1}{2}\oint_{\partial \mathrm{V}} \mathbf{da} \cdot\ \nabla \times (V\, \mathbf{H}^*) \\
&-\frac{1}{2}\oint_{\partial \mathrm{V}} \mathbf{da} \cdot\ (\nabla \times \mathbf{H}^*)\ V
\end{aligned}
$$

Since the boundary of the simulation domain is closed, i.e. it has no edge, the first integral does not contribute. Therefore, the Poynting integral becomes:

$$
\begin{aligned}
-\oint_{\partial \mathrm{V}} \mathbf{S} \cdot \mathbf{da} &= -\frac{1}{2}\oint_{\partial \mathrm{V}} \mathbf{da} \cdot (\nabla \times \mathbf{H}^*)\ V \\
&= -\frac{1}{2}\oint_{\partial \mathrm{V}} \mathbf{da} \cdot (\mathbf{J}^* - \mathrm{i}\omega \mathbf{D}^*)\ V \\
&= -\frac{1}{2}\oint_{\partial \mathrm{V}} \mathbf{da} \cdot \mathbf{J}^*\ V + \frac{\mathrm{i}\omega}{2}\oint_{\partial \mathrm{V}} \mathbf{da} \cdot \mathbf{D}^*\ V
\end{aligned}
$$

The first integral contributes only over the boundary contacts ($A_{\rm cont}$). The second integral also only contributes at the contact since away from the contact areas and on the surface of the simulation domain $\mathbf{E}_{\rm n} = 0$. As consequence $\mathbf{D}_{\rm n} = 0$. Since V is also constant on the contact area one obtains:

$$
-\frac{1}{2}\int_{A_{\rm cont}} \mathbf{da} \cdot \mathbf{J}^*\, V = -\sum_{\rm contacts} \frac{1}{2} I_i^* V_i
$$

where

$$I_i = \int_{A_i} \mathbf{da} \cdot (\mathbf{J} + \mathrm{i}\omega\mathbf{D})$$

is the total current of the i-th contact. When all this is combined we get:

$$\sum_{\text{contacts}} \frac{1}{2} I_i^* V_i = -E_{\text{joule}} + 2\mathrm{i}\omega \int_V (w_\mathrm{e} - w_\mathrm{m}) \, \mathrm{dv}$$

In the case when there are no dielectric losses (ε real), and assuming no polarization effects in metals (σ real) we note that the joule energy will always be positive and real, while the electric/magnetic energy term will be imaginary. Hence we can write:

$$\sum_{\text{contacts}} \Re \left[\frac{1}{2} I_i^* V_i \right] = -E_{\text{joule}} \tag{6.4}$$

$$\sum_{\text{contacts}} \Im \left[\frac{1}{2} I_i^* V_i \right] = 2\omega \int_V (w_\mathrm{e} - w_\mathrm{m}) \, \mathrm{dv} \tag{6.5}$$

These equations can be extended to include dielectric losses. For lossy dielectrics the permittivity becomes:

$$\epsilon = \epsilon_\mathrm{R} - \mathrm{i}\epsilon_\mathrm{I}$$

In that case the energy balance equations get modified as follows:

$$\sum_{\text{contacts}} \Re \left[\frac{1}{2} I_i^* V_i \right] = -E_{\text{joule}} - 2\omega \int_V \frac{1}{4} \epsilon_\mathrm{R} \omega \mathbf{E} \cdot \mathbf{E}^* \mathrm{dv} \tag{6.6}$$

$$\sum_{\text{contacts}} \Im \left[\frac{1}{2} I_i^* V_i \right] = 2\omega \int_V \left(\frac{1}{4} \epsilon_\mathrm{R} \omega \mathbf{E} \cdot \mathbf{E}^* - w_\mathrm{m} \right) \mathrm{dv} \tag{6.7}$$

This impedance is calculated, based upon energy considerations for the whole simulation area. The static energies (total dissipated energy E_{joule}, electric energy E_{elec} and magnetic energy E_{magn}) define the static values for resistance, inductance and capacitance as follows:

$$R_{\text{DC}} = E_{\text{joule}} / I^2 \tag{6.8}$$

$$C_{\text{DC}} = 2 \cdot E_{\text{elec}} / V^2 \tag{6.9}$$

$$L_{\text{DC}} = 2 \cdot E_{\text{magn}} / I^2 \tag{6.10}$$

7

From Macroscopic Field Theory to Electric Circuits

7.1 Kirchhoff's Laws

Electronic circuits consist of electronic components or devices integrated in a network. The number of components may range form a few to several billion. In the latter case the network is usually subdivided in functional blocks and each block has a unique functional description. The hierarchical approach is vital to the progress of electronic design and reuse of functional blocks (sometimes referred to as intellectual property) determines the time-to-market of new electronic products. Besides the commercial value of the hierarchical approach, there is also a scientific benefit. It is not possible to design advanced electronic circuits by solving the Maxwell equations using the boundary conditions that are imposed by the circuit. The complexity of the problem simply does not allow such an approach taking into account the available compute power and the constraints that are imposed on the design time. Moreover, a full solution of the Maxwell equations is often not very instructive in obtaining insight into the operation of the circuit. In order to understand the operation or input/output response of a circuit, it is beneficial to describe the circuit in effective variables. These coarse-grained variables (in the introduction we referred to these variables as "baskets") should be detailed enough such that a physical meaning can be given to them, whereas on the other hand they should be sufficiently "coarse" so as to mask details that are not relevant for understanding the circuit properties. The delicate balancing between these two requirements has resulted into "electronic circuit theory". The latter is based on the physical laws that are expressed by Maxwell's equations, and the laws of energy and charge conservation. The purpose of this section is to analyze how the circuit equations may be extracted from these microscopic physical laws. It should be emphasized that the extraction is not a rigorous derivation in the mathematical sense but relies on the validity of a number of approximations and assumptions reflecting the ideal behavior of electric circuits. These assumptions should be critically revised if one wishes to apply the circuit equations in areas that are outside the original scope of circuit theory. A simple example is a capacitor consisting of two large, conducting parallel plates separated by a relatively thin insulating layer: its capacity may be a suitable, characteristic variable for describing its impact in a circuit at low and moderately high frequencies. However, at extremely large frequencies the same device may act as a wave guide or an antenna, partly radiating the stored electromagnetic energy.

Being aware of such pitfalls, we continue our search for effective formulations of the circuit equations. In fact, the underlying prescriptions are given by the following (plausible) statements:

- A circuit can be represented by a topological network that consists of branches and nodes.
- **Kirchhoff's voltage law** (KVL) – The algebraic sum of all voltages along any arbitrary loop of the network equals zero at every instant of time.
- **Kirchhoff's current law** (KCL) – The algebraic sum of all currents entering or leaving any particular network node equals zero at every instant of time.

In order to make sense out of these statements we first need to have a clear understanding of the various words that were encountered ; in particular, we must explain what is meant by a node, a branch, a voltage and a current. For that purpose we consider the most elementary circuit: a battery and a resistor that connects the poles of the battery. The circuit is depicted in Figure 7.1. We have explicitly taken into account the finite resistance of the leads. In fact, a more realistic drawing is presented in Figure 7.2, where we account for the fact that the leads have a finite volume. In particular, we have divided the full circuit volume into four different regions: (1) the battery region Ω_{B}, (2) the left lead region $\Omega_{1\mathrm{L}}$, (3) the right lead region $\Omega_{2\mathrm{L}}$, and (4) the resistor region Ω_{A}.

We will now consider the power supplied by the battery to the circuit volume. The work done by the electromagnetic field on all charges in the circuit volume per unit time is given by

$$\frac{\mathrm{d}E_{\mathrm{MECH}}}{\mathrm{d}t} = \int_{\Omega} \mathbf{J} \cdot \mathbf{E} \, \mathrm{d}\tau. \tag{7.1}$$

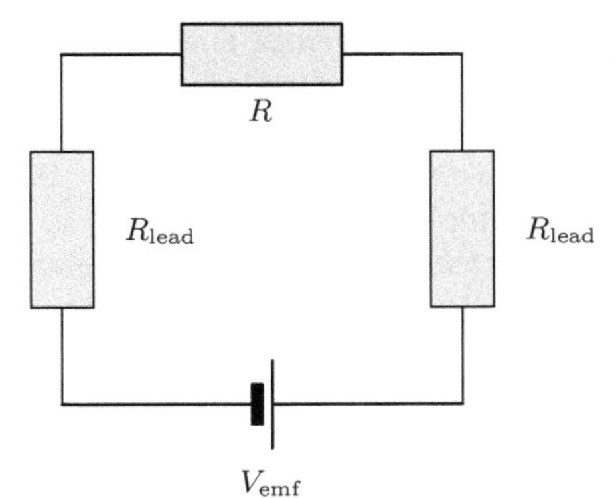

Figure 7.1 Closed electric circuit containing a resistor connected to a DC power supply through two resistive leads.

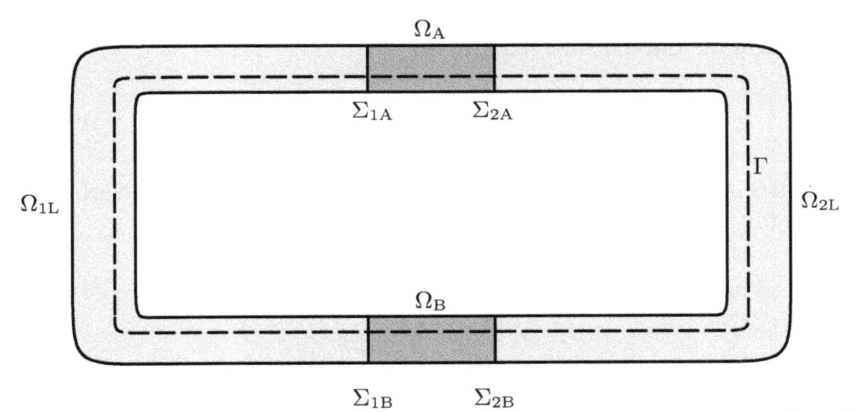

Figure 7.2 The electric circuit of Figure 7.1, taking into account the spatial extension of the leads. Γ is a circuit loop, i.e. an internal, closed loop encircling the "hole" of the circuit.

This corresponds to the dissipated power in steady-state conditions for which $(\partial\rho/\partial t) = 0$. As a consequence, $\nabla \cdot \mathbf{J} = 0$ and therefore we may apply the $\mathbf{J} \cdot \mathbf{E}$ theorem (see Appendix). We obtain:

$$\int_{\Omega} \mathbf{J} \cdot \mathbf{E} \, d\tau = \left(\oint_{\Sigma} \mathbf{J} \cdot \mathbf{dS} \right) \left(\oint_{\Gamma} \mathbf{E} \cdot \mathbf{dr} \right), \tag{7.2}$$

where Σ is an arbitrary cross section of the circuit and Γ is a circuit loop, i.e. an arbitrary closed path inside the circuit region. We identify the first integral of the right-hand side of Equation (7.2) as the *current* in the circuit. The second integral of the right-hand side of Equation (7.2) is identified as the *electromotive force* (EMF) or the *voltage* that is supplied by the battery, V_ϵ. The latter is nothing but the work done per unit charge by the electric field when the charge has made one full revolution around the circuit. Note the integral $\oint_\Gamma \mathbf{E}\cdot d\mathbf{r}$ is *non-zero*, although $\nabla\times\mathbf{E} = \mathbf{0}$. This is possible because the circuit is not a simply connected region in $\mathbb{R}^3$. More precisely, the topology of the circuit is that of a manifold of genus one, say a torus or a toroidal region with one "hole". We may now consider the left-hand side of Equation (7.2) and consider the contributions to Equation (7.2). For region (2) we obtain:

$$\int_{\Omega_{1L}} \mathbf{J}\cdot\mathbf{E}\,d\tau = -\int_{\Omega_{1L}} (\nabla V)\cdot\mathbf{J}\,d\tau = -\int_{\Omega_{1L}} \nabla\cdot(V\mathbf{J})\,d\tau + \int_{\Omega_{1L}} V\nabla\cdot\mathbf{J}\,d\tau. \tag{7.3}$$

The first equality is valid since $\mathbf{E} = -\nabla V$, in a simply-connected region such as Ω_{1L}, Ω_{2L}, Ω_A or Ω_B. The last integral is equal to zero, since $\nabla\cdot\mathbf{J} = 0$ and the one-but-last integral is

$$-\int_{\Omega_{1L}} \nabla\cdot(V\mathbf{J})\,d\tau = -\oint_{\partial\Omega_{1L}} V\mathbf{J}\cdot d\mathbf{S}. \tag{7.4}$$

If we now *assume* that the potential is constant on a cross-section of the circuit, then this integral has two contributions:

$$-\oint_{\partial\Omega_{1L}} V\mathbf{J}\cdot d\mathbf{S} = -V_{\Sigma_{1B}}\int_{\Sigma_{1B}} \mathbf{J}\cdot d\mathbf{S} - V_{\Sigma_{1A}}\int_{\Sigma_{1A}} \mathbf{J}\cdot d\mathbf{S}. \tag{7.5}$$

Using Gauss' theorem we may identify the two remaining surface integrals can be identified as the total current I. Indeed, in the steady state regime ($\partial\rho/\partial t = 0$) the divergence of $\mathbf{J}$ vanishes while $\mathbf{J}$ is assumed to be tangential to the circuit boundary $\partial\Omega$. Therefore, the vanishing volume integral of $\nabla\cdot\mathbf{J}$ over Ω_{1L} reduces to

$$0 = \int_{\partial\Omega_{1L}} \mathbf{J}\cdot d\mathbf{S} = \int_{\Sigma_{1A}} \mathbf{J}\cdot d\mathbf{S} - \int_{\Sigma_{1B}} \mathbf{J}\cdot d\mathbf{S}, \tag{7.6}$$

which justifies the identification

$$\int_{\Sigma_{1A}} \mathbf{J}\cdot d\mathbf{S} = \int_{\Sigma_{1B}} \mathbf{J}\cdot d\mathbf{S} \equiv I \tag{7.7}$$

whence

$$-\oint_{\partial\Omega_{1L}} V\mathbf{J}\cdot d\mathbf{S} = I\,(\,V_{\Sigma_{1A}} - V_{\Sigma_{1B}}). \tag{7.8}$$

The regions (3) and (4) can be evaluated in a similar manner. As a consequence we obtain:

$$I\ (V_{\Sigma_{2B}} - V_{\Sigma_{2A}}) + I\ (V_{\Sigma_{2A}} - V_{\Sigma_{1A}}) + I\ (V_{\Sigma_{1A}} - V_{\Sigma_{1B}}) + \int_{\Omega_B} \mathbf{J} \cdot \mathbf{E}\, \mathrm{d}\tau = I\, V_\epsilon \tag{7.9}$$

The final integral that applies to the battery region, is also equal to zero. This is because the electric field consists of two components: a conservative component and a non-conservative component, i.e. $\mathbf{E} = \mathbf{E}_\mathrm{C} + \mathbf{E}_\mathrm{NC}$. The purpose of the ideal[4] battery is to cancel the conservative field, such that after a full revolution around the circuit a net energy supply is obtained from the electric field. Then we finally arrive at the following result:

$$V_\epsilon = V_{\Sigma_{2B}} - V_{\Sigma_{1B}}. \tag{7.10}$$

Equation (7.10) is not a trivial result: having been derived from energy considerations, it relates the EMF of the battery, arising from a non-conservative field, to the potential difference at its terminals, i.e. a quantity characterizing a conservative field. Physically, it reflects the concept that an ideal battery is capable of maintaining a constant potential difference at its terminals even if a current is flowing through the circuit. This example illustrates how Kirchhoff's laws can be extracted from the underlying physical laws. It should be emphasized that we achieved more than what is provided by Kirchhoff's laws. Often Kirchhoff's voltage law is presented as a *trivial* identity, i.e. by putting N nodes on a closed path, as we have done by selecting a series of cross sections, it is always true that

$$(V_1 - V_2) + (V_2 - V_3) + \cdots + (V_{N-1} - V_N) + (V_N - V_1) = 0. \tag{7.11}$$

Physics enters this identity (turning it into a useful equation) by relating the potential differences to their physical origin. In the example above, the potential difference, $V_{\Sigma_{2B}} - V_{\Sigma_{1B}}$, is the result of a power supply.

By the in-depth discussion of the simple circuit, we have implicitly provided a detailed understanding of what is understood to be a voltage, a current, a node and a branch in a Kirchhoff network. The nodes are geometrically idealized regions of the circuit to which network branches can be attached. The nodes can be electrically described by a single voltage value. A branch is also a geometrical idealization. Knowledge of the current *density* inside the branch is not required. All that counts is the total current

[4] The internal resistance of a real battery is neglected here.

in the branch. We also have seen that at some stages only progress could be made by making simplifying assumptions and finally that all variables are time independent. The last condition is a severe limitation. In the next section, we will discuss the consequences of eliminating this restriction. We can insert more physics in the network description. So far, we have not exploited Ohm's law, $\mathbf{J} = \sigma\mathbf{E}$. For a resistor with length L, cross sectional area A and constant resistivity σ, we find that

$$\int_{\Omega_A} \mathbf{J} \cdot \mathbf{E}\,\mathrm{d}\tau = \sigma \int (\boldsymbol{\nabla} V)^2\,\mathrm{d}\tau = \sigma L \cdot A \left(\frac{V_{\Sigma_{2A}} - V_{\Sigma_{1A}}}{L}\right)^2$$
$$= I\,(V_{\Sigma_{2A}} - V_{\Sigma_{1A}})\,. \tag{7.12}$$

As a result, we can "define" the resistance as the ratio of the potential difference and the current:

$$V_{\Sigma_{2A}} - V_{\Sigma_{1A}} = RI, \quad R = \frac{L}{\sigma A}. \tag{7.13}$$

7.2 Circuit Rules

In the foregoing section, we have considered DC steady-state currents, for which $\boldsymbol{\nabla} \cdot \mathbf{J} = 0$ and $\partial\mathbf{B}/\partial t = \mathbf{0}$, such that the $\mathbf{J} \cdot \mathbf{E}$ – theorem could be applied. In general, these conditions are not valid and the justification of using the Kirchhoff's laws becomes more difficult. Nevertheless, the guiding principles remain unaltered, i.e. the conservation of charge and energy will help us in formulating the circuit equations. On the other hand, as was already mentioned in the previous section, the idealization of a real circuit involves a number of approximations and assumptions that are summarized below in a – non-exhaustive – list of circuit rules:

1. An electric circuit, or more generally, a circuit network, is a manifold of genus $N \geq 1$, i.e. a multiply connected region with N holes. The branches of this manifold consist of distinct circuit segments or devices, mainly active and passive components, interconnecting conductors and seats of EMF.
2. The active components typically include devices that are actively processing signals, such as transistors, vacuum tubes, operational amplifiers, A/D converters.
3. Passive components refer to ohmic resistors, capacitors and inductors or coils, diodes, tunneling junctions, Coulomb blockade islands etc.

They are representing energy dissipation, induction effects, quantum mechanical tunneling processes and many other phenomena.

4. The seats of EMF include both DC and AC power supplies, i.e. chemical batteries, EMFs induced by externally applied magnetic fields, all different kinds of current and voltages sources and generators, etc. The electromagnetic power supplied by the EMF sources is dissipated entirely in the circuit. No energy is released to the environment of the circuit through radiation or any other mechanism.
5. In compliance with the previous rule, all circuit devices are assumed to behave in an ideal manner. First, all conductors are taken to be perfect conductors. Considering perfect conduction as the infinite conductivity limit of realistic conduction ($\mathbf{J} = \sigma\mathbf{E}$), it is clear that no electric fields can survive inside a perfect conductor which therefore can be considered an equipotential volume. Clearly, from $\boldsymbol{\nabla} \cdot \mathbf{E} = \rho/\epsilon$ it follows that the charge density also vanishes inside the conductor. Furthermore, a perfect conductor is perfectly shielded from any magnetic field. Strictly speaking, this is not a direct consequence of Maxwell's third equation, since $\boldsymbol{\nabla}\times\mathbf{E} = \mathbf{0}$ would only imply $\partial\mathbf{B}/\partial t = \mathbf{0}$ but the effect of static magnetic fields on the circuit behavior will not be considered here. It should also be noted that a perfect conductor is not the same as a superconductor. Although for both devices the penetration of magnetic fields is restricted to a very narrow boundary layer, called penetration depth, only the superconductor hosts a number of "normal" electrons (subjected to dissipative transport) and will even switch entirely to the normal state when the supercurrent attains its critical value. Furthermore, a supercurrent can be seen as a coherent, collective motion of so-called Cooper pairs of electrons, i.e. *bosons* while perfect conduction is carried by unpaired electrons or holes, i.e. *fermions*. Next, all energy dissipation exclusively takes place inside the circuit resistors. This implies that all capacitors and inductors are assumed to be made of perfect conductors. Inside the windings of an inductor and the plates of a capacitor, no electric or magnetic fields are present. The latter exist only in the cores of the inductors[5] while the corresponding vector potential and induced electric field are localized in the inductor. Similarly, the electric charge on the plates of a capacitor are residing in a surface layer and the corresponding, conservative electric field is strictly localized between the plates while all stray fields are ignored. Finally, the ideal

[5]Topologically, the cores are not part of the circuit region Ω

behavior of the seats of EMF is reflected in the absence of internal resistances and the strict localization of the non-conservative electric fields that are causing the EMFs.

6. The current density vector $\mathbf{J}$ defines a positive orientation of the circuit loop Γ. It corresponds the motion of a positive charge moving from the anode to the cathode outside the EMF seat and from cathode to anode inside the EMF seat.

7.3 Inclusion of Time Dependence

The previous set of rules will guide us towards the derivation of the final circuit equations. However, before turning to the latter, it is worth to have a second look at Equation (7.11). In the continuum, this identity can be given in the following way:

$$\oint_{\Gamma} \mathbf{dr} \cdot \nabla V(\mathbf{r}, t) = 0, \tag{7.14}$$

where Γ is an arbitrary closed loop. Note that above equation includes time-dependent fields $V(\mathbf{r}, t)$. In order to validate the first Kirchhoff law (KVL), we insert into Equation (7.14) the potential that corresponds to

$$\nabla V = -\mathbf{E} - \frac{\partial \mathbf{A}}{\partial t}. \tag{7.15}$$

Of course, if we were to plug this expression into Equation (7.14), we would just arrive at Faraday's law. The transition to the circuit equations is realized by cutting the loop into discrete segments (rule 1) and assigning to each segment appropriate lumped variables. To illustrate this approach we revisit the circuit of Figure 7.1, where we have now folded the resistor of the left lead into a helix and, according to the circuit rules, its resistance is taken to be zero whereas the top resistor is replaced by a capacitor. The resulting, idealized circuit depicted in Figure 7.3 has four segments.

The battery region, that now may produce a time-dependent EMF, and the right-lead region can be handled as was done in the foregoing section. According to the circuit rules, it is assumed that all resistance is concentrated in the resistor located between node 3 and node 4, while both the inductor and the capacitor are made of perfect conductors and no leakage current is flowing between the capacitor plates. Starting from the identities

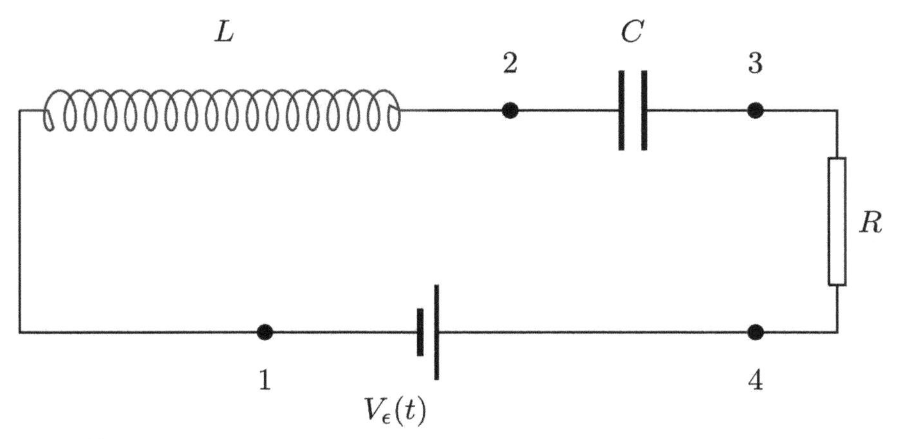

Figure 7.3 The electric circuit of Figure 7.2 with a helix-shaped "resistor".

$$V_1 - V_2 + V_2 - V_3 + V_3 - V_4 + V_4 - V_1 = 0, \tag{7.16}$$

$$\oint_\Gamma \mathbf{E} \cdot \mathbf{dr} + \frac{\partial}{\partial t} \oint_\Gamma \mathbf{A} \cdot \mathbf{dr} = 0,, \tag{7.17}$$

we decompose the electric field into a conservative, an external and induced component:

$$\mathbf{E} = \mathbf{E}_{\mathrm{C}} + \mathbf{E}_{\mathrm{EX}} + \mathbf{E}_{\mathrm{IN}} \tag{7.18}$$

where

$$\begin{aligned} \mathbf{A} &= \mathbf{A}_{\mathrm{EX}} + \mathbf{A}_{\mathrm{IN}} \\ \mathbf{E}_{\mathrm{C}} &= -\boldsymbol{\nabla} V \\ \mathbf{E}_{\mathrm{EX}} &= -\frac{\partial}{\partial t}\mathbf{A}_{\mathrm{EX}} \\ \mathbf{E}_{\mathrm{IN}} &= -\frac{\partial}{\partial t}\mathbf{A}_{\mathrm{IN}}. \end{aligned} \tag{7.19}$$

Since the battery and the inductor are perfect conductors, the total electric field in these devices is identically zero:

$$\int_1^4 \mathbf{dr} \cdot \mathbf{E} = 0 \tag{7.20}$$

$$\int_2^1 \mathbf{dr} \cdot \mathbf{E} = 0 \tag{7.21}$$

Following the circuit rules, we assume that the induced electric field and the external field are only present in the inductor region and the battery region respectively. Then Equation (7.21) can be evaluated as

$$\int_2^1 \mathbf{dr} \cdot \mathbf{E} = \int_2^1 \mathbf{dr} \cdot (\mathbf{E}_{\mathrm{C}} + \mathbf{E}_{\mathrm{IN}}) = V_2 - V_1 + \int_2^1 \mathbf{dr} \cdot \mathbf{E}_{\mathrm{IN}} = 0 \quad (7.22)$$

and therefore

$$V_1 - V_2 = \int_2^1 \mathbf{dr} \cdot \mathbf{E}_{\mathrm{IN}}. \quad (7.23)$$

For the battery region we obtain:

$$\int_1^4 \mathbf{dr} \cdot \mathbf{E} = \int_1^4 \mathbf{dr} \cdot (\mathbf{E}_{\mathrm{C}} + \mathbf{E}_{\mathrm{EX}}) = V_1 - V_4 + \int_1^4 \mathbf{dr} \cdot \mathbf{E}_{\mathrm{EX}} = 0 \quad (7.24)$$

and therefore

$$V_4 - V_1 = \int_1^4 \mathbf{dr} \cdot \mathbf{E}_{\mathrm{EX}} = \oint \mathbf{dr} \cdot \mathbf{E}_{\mathrm{EX}} = V_\epsilon. \quad (7.25)$$

Inside the capacitor, the induced and external fields are zero, and therefore we obtain

$$\int_3^2 \mathbf{dr} \cdot \mathbf{E} = \int_3^2 \mathbf{dr} \cdot \mathbf{E}_{\mathrm{C}} = V_3 - V_2. \quad (7.26)$$

On the other hand, the potential difference between the capacitor is assumed to be proportional to charge Q stored on one of the plates, i.e. $Q = CV$, where C is the *capacitance*. The resistor is treated in an analogous manner:

$$V_4 - V_3 = \int_4^3 \mathbf{dr} \cdot \mathbf{E}_{\mathrm{C}} = IR \quad (7.27)$$

Insertion of all these results into Equation (7.16) gives:

$$\int_1^2 \mathbf{E}_{\mathrm{IN}} \cdot \mathbf{dr} = -V_\epsilon + IR + \frac{Q}{C}, \quad (7.28)$$

where we anticipated that the electric field between the capacitor plates is given by $Q/(Cd)$ and d is the thickness of the dielectric. The integral at the left-hand side of Equation (7.28) can be obtained by using Faraday's law once again:

$$\int_2^1 \mathbf{E}_{\mathrm{IN}} \cdot \mathbf{dr} \simeq \oint_\Gamma \mathbf{E}_{\mathrm{IN}} \cdot \mathbf{dr} = -\frac{\partial}{\partial t} \oint_\Gamma \mathbf{A}_{\mathrm{IN}} \cdot \mathbf{dr} = -\frac{\partial}{\partial t} \int_{S(\Gamma)} \mathbf{B}_{\mathrm{IN}} \cdot \mathbf{dS}, \quad (7.29)$$

where $S(\Gamma)$ is the area enclosed by the loop Γ. Since the magnetic field $\mathbf{B}$ is only appreciably different from zero inside the core of the inductor, the integral may be identified as the magnetic self-flux Φ_{M} of the inductor. This flux is proportional to the circuit current I that also flows through the windings of the coil. Hence, $\Phi_{\mathrm{M}} = LI$, where L is the *inductance* of the inductor and therefore Equation (7.23) becomes:

$$V_1 - V_2 = -L\frac{\mathrm{d}I}{\mathrm{d}t}. \tag{7.30}$$

We are now in the position to write down the circuit equation for the simple circuit of Figure 7.3. Starting from the identity of Equation (7.16), we find

$$-L\frac{\mathrm{d}I}{\mathrm{d}t} + V_\epsilon - IR - \frac{Q}{C} = 0 \tag{7.31}$$

So far, we have not considered energy conservation for the time-dependent circuit equations. However, this conservation law is important for determining explicit expressions for the inductances and capacitances. Integrating the electromagnetic energy density u_{EM} over an arbitrarily large volume Ω_∞ with a boundary surface $\partial\Omega_\infty$, we obtain the total energy content of the electromagetic field:

$$U_{\mathrm{EM}} = \frac{1}{2}\int_{\Omega_\infty} \mathrm{d}\tau \big(\epsilon E^2 + \frac{B^2}{\mu}\big) = \frac{1}{2}\int_{\Omega_\infty} \mathrm{d}\tau \big(\mathbf{E}\cdot\mathbf{D} + \mathbf{B}\cdot\mathbf{H}\big). \tag{7.32}$$

Replacing $\mathbf{E}$ and $\mathbf{B}$ by $-\boldsymbol{\nabla}V - \partial\mathbf{A}/\partial t$ and $\boldsymbol{\nabla}\times\mathbf{A}$ respectively, we may rewrite Equation (7.32) as

$$U_{\mathrm{EM}} = \frac{1}{2}\int_{\Omega_\infty} \mathrm{d}\tau \left[-\left(\boldsymbol{\nabla}V + \frac{\partial\mathbf{A}}{\partial t}\right)\cdot\mathbf{D} + \mathbf{H}\cdot\boldsymbol{\nabla}\times\mathbf{A}\right] \tag{7.33}$$

Next, exploiting the vector identity (10.39), we applying Gauss' theorem to the volume Ω_∞ thereby neglecting all fields at the outer surface $\partial\Omega_\infty$, i.e.

$$\int_{\partial\Omega_\infty} \mathbf{dS}\cdot(V\mathbf{D}) = 0, \tag{7.34}$$

we obtain:

$$-\int_{\Omega_\infty} \mathrm{d}\tau \boldsymbol{\nabla}V\cdot\mathbf{D} = \int_{\Omega_\infty} \mathrm{d}\tau\, V\,\boldsymbol{\nabla}\cdot\mathbf{D} = \int_{\Omega_\infty} \mathrm{d}\tau\, \rho V \tag{7.35}$$

where the last equality follows from the first Maxwell equation $\nabla \cdot \mathbf{D} = \rho$. Similarly, using the identity (10.38) and inserting the fourth Maxwell equation, we may convert the volume integral of $\mathbf{H} \cdot \nabla\times\mathbf{A}$ appearing in Equation (7.33):

$$\int_{\Omega_\infty} \mathrm{d}\tau \mathbf{H} \cdot \nabla\times\mathbf{A} = \int_{\Omega_\infty} \mathrm{d}\tau \mathbf{A} \cdot \left(\mathbf{J} + \frac{\partial \mathbf{D}}{\partial t}\right) \tag{7.36}$$

Putting everything together, we may express the total electromagnetic energy as follows:

$$U_{\mathrm{EM}} = \frac{1}{2}\int_{\Omega_\infty} \mathrm{d}\tau \left[\rho V - \frac{\partial \mathbf{A}}{\partial t} \cdot \mathbf{D} + \mathbf{A} \cdot \left(\mathbf{J} + \frac{\partial \mathbf{D}}{\partial t}\right)\right] \tag{7.37}$$

$$= \frac{1}{2}\int_{\Omega} \mathrm{d}\tau \left[\rho V - \frac{\partial \mathbf{A}}{\partial t} \cdot \mathbf{D} + \mathbf{A} \cdot \left(\mathbf{J} + \frac{\partial \mathbf{D}}{\partial t}\right)\right]. \tag{7.38}$$

where the last integral is restricted to the circuit region Ω in view of the circuit rules stating that all electromagnetic fields are vanishing outside the circuit region. It is easy to identify in Equation (7.38) the "electric" and "magnetic" contributions respectively referring to E^2 and B^2 in Equation (7.32):

$$U_{\mathrm{EM}} = U_{\mathrm{E}} + U_{\mathrm{M}} \tag{7.39}$$

$$U_{\mathrm{E}} = \frac{1}{2}\int_{\Omega} \mathrm{d}\tau \left(\rho V - \frac{\partial \mathbf{A}}{\partial t} \cdot \mathbf{D}\right) \tag{7.40}$$

$$U_{\mathrm{M}} = \frac{1}{2}\int_{\Omega} \mathrm{d}\tau \, \mathbf{A} \cdot \left(\mathbf{J} + \frac{\partial \mathbf{D}}{\partial t}\right). \tag{7.41}$$

Neglecting the magnetic field inside the ideal circuit conductors according to the circuit rules, we take $\nabla\times\mathbf{A}$ to be zero inside the circuit. Moreover, bearing in mind that the identity

$$\nabla \cdot \left(\mathbf{J} + \frac{\partial \mathbf{D}}{\partial t}\right) = 0 \tag{7.42}$$

is generally valid, we may now apply the $\mathbf{J} \cdot \mathbf{E}$ – theorem to the combination $\mathbf{A} \cdot (\mathbf{J} + \partial \mathbf{D}/\partial t)$:

$$U_{\mathrm{M}} = \frac{1}{2}\left(\oint_{\Gamma} \mathbf{dr} \cdot \mathbf{A}\right)\left(\int_{\Sigma} \mathbf{dS} \cdot \left(\mathbf{J} + \frac{\partial \mathbf{D}}{\partial t}\right)\right) \tag{7.43}$$

The loop integral clearly reduces to the total magnetic flux, which consists of the self-flux Φ_{M} and the external flux Φ_{ex}. Furthermore, due to

Equation (7.42), the surface integral of Equation (7.41) can be calculated for any cross-section Σ that does not contain accumulated charge. Taking Σ in a perfectly conducting lead, we have $\mathbf{D} = \mathbf{0}$ and the integral reduces to the total current $I = \int_\Sigma \mathrm{d}\mathbf{S} \cdot \mathbf{J}$. On the other hand, if we were choosing Σ to cross the capacitor dielectric, the current density would vanish and the integral would be equal to $\mathrm{d}\Phi_\mathrm{D}(t)/\mathrm{d}t$ where

$$\Phi_\mathrm{D}(t) = \int_\Sigma \mathrm{d}\mathbf{S} \cdot \mathbf{D}(\mathbf{r}, t) \tag{7.44}$$

is the flux of the displacement vector. Since both choices of Σ should give rise to identical results, we conclude that

$$I(t) = \frac{\mathrm{d}\Phi_\mathrm{D}(t)}{\mathrm{d}t} \tag{7.45}$$

which confirms the observation that the circuit of Figure 7.3 where the capacitor is in series with the other components, can only carry charging and discharging currents. In any case, we are left with

$$U_\mathrm{M} = \frac{1}{2}(\Phi_\mathrm{M} + \Phi_\mathrm{ex})I \tag{7.46}$$

or, reusing the "definition" of inductance, i.e. $\Phi_\mathrm{M} = LI$,

$$U_\mathrm{M} = \frac{1}{2}LI^2 + \frac{1}{2}\Phi_\mathrm{ex}I \tag{7.47}$$

where $(1/2)LI^2$ is the familiar expression for the magnetic energy stored in the core of the inductor.

The electric energy may be rewritten in terms of capacitances in an analogous manner. The contribution of $\partial\mathbf{A}/\partial t \cdot \mathbf{D}$ in Equation (7.40) vanishes because $\partial\mathbf{A}/\partial t$, representing the non-conservative electric field, is non-zero only inside the inductor and the generator regions, where the total electric field reduces to zero. On the other hand, for perfectly conducting leads that are also equipotential domains, the first term gives:

$$U_\mathrm{E} = \frac{1}{2}\sum_n Q_n V_n. \tag{7.48}$$

where V_n generally denotes the potential of the n-th (ideal) conductor, containing a charge Q_n. Being expressed in terms of bare potentials, the result of Equation (7.48) seems to be gauge dependent at a first glimpse. It should be noted however, that Equation (7.48) has been derived within the circuit approximation, which implies that the charged conductors are not arbitrarily distributed in space, but are all part of the – localized – circuit. In particular, the charges Q_n are assumed to be stored on the plates of the capacitors of the circuit, and as such the entire set $\{Q_n\}$ can be divided into pairs of opposite charges $\{(Q_j, -Q_j)\}$. Hence, equation (7.48) should be read

$$U_{\mathrm{E}} = \frac{1}{2}\sum_j Q_j(V_{1j} - V_{2j}) = \frac{1}{2}\sum_j C_j(V_{1j} - V_{2j})^2, \tag{7.49}$$

where $V_{1j} - V_{2j}$ is the gauge-invariant potential difference between the plates of the j-th capacitor.

The second Kirchhoff law (KCL), follows from charge conservation. The branches in the network can not store charge, unless capacitors are included. The integrated charge is denoted by Q_n and

$$\frac{\mathrm{d}Q_j}{\mathrm{d}t} = -\int \mathbf{J}\cdot\mathbf{dS} = \sum_k I_{jk}, \tag{7.50}$$

where the surface integral is over a surface around charge-storage domain and I_{jk} is the current flowing from the charge-storage region j into the j-th circuit branch. As in the steady-state case, the Kirchhoff laws, in particular the expressions for the various voltage differences could only be obtained if some simplifying assumptions are made. For the inductor it was assumed that the induced magnetic field is only different from zero inside the core. For the capacitor, in a similar way it was assumed that the energy of storing the charge is localized completely between the plates. These assumptions need to be carefully checked before applying the network equations. As an illustration of this remark we emphasize that we ignored the volume integrals that are not parts of the circuits. In particular, the integral of the electric energy outside the circuit is the kinetic part of the radiation energy:

$$U_{\mathrm{E}}^{\mathrm{rad}} = -\frac{1}{2}\int_{\Omega_\infty\backslash\Omega} \mathrm{d}\tau\, \frac{\partial\mathbf{A}}{\partial t}\cdot\mathbf{D} = \frac{1}{2}\epsilon\int_{\Omega_\infty\backslash\Omega} \mathrm{d}\tau\, \frac{\partial\mathbf{A}}{\partial t}\cdot\frac{\partial\mathbf{A}}{\partial t}, \tag{7.51}$$

and the potential energy of the radiation field:

$$U_{\mathrm{M}}^{\mathrm{rad}} = -\frac{1}{2\mu}\int_{\Omega_\infty\backslash\Omega} \mathrm{d}\tau\; (\boldsymbol{\nabla}\times\mathbf{A})\cdot(\boldsymbol{\nabla}\times\mathbf{A}) \tag{7.52}$$

are not considered at the level of circuit modeling.

8

Gauge Conditions

The Maxwell theory of electrodynamics describes the interaction between radiation and charged particles. The electromagnetic fields are described by six quantities, the vector components of $\mathbf{E}$ and $\mathbf{B}$. The sources of the radiation fields are represented by the charge density ρ and the current density $\mathbf{J}$. If the sources are prescribed functions $\rho(\mathbf{r}, t)$ and $\mathbf{J}(\mathbf{r}, t)$, then the evolution of $\mathbf{E}(\mathbf{r}, t)$ and $\mathbf{B}(\mathbf{r}, t)$ is completely determined. The fields $\mathbf{E}$ and $\mathbf{B}$ may be obtained from a scalar potential V and a vector potential $\mathbf{A}$ such that

$$\mathbf{E} = -\nabla V - \frac{\partial \mathbf{A}}{\partial t}, \qquad \mathbf{B} = \nabla \times \mathbf{A}. \tag{8.1}$$

As was mentioned already in Section 3, the potentials $(V, \mathbf{A})$ are not unique. The choice

$$V \to V' = V - \frac{\partial \Lambda}{\partial t} \qquad \mathbf{A} \to \mathbf{A}' = \mathbf{A} + \nabla \Lambda \tag{8.2}$$

gives rise two the same fields $\mathbf{E}$ and $\mathbf{B}$. A change in potential according to Equation (8.2) is a gauge transformation. The Lagrangian density

$$\mathcal{L} = \frac{1}{2}\epsilon_0 \left(\nabla V + \frac{\partial \mathbf{A}}{\partial t} \right)^2 - \frac{1}{2\mu_0} (\nabla \times \mathbf{A})^2 + \mathbf{J} \cdot \mathbf{A} - \rho V \tag{8.3}$$

gives rise to an action integral

$$S = \int \mathrm{d}t \int \mathrm{d}^3 r \, \mathcal{L}(\mathbf{r}, t) \tag{8.4}$$

that is gauge invariant under the transformation (8.2). The gauge invariance of the Maxwell equations has been found a posteriori. It was the outcome of a consistent theory for numerous experimental facts. In modern physics invariance principles play a key role in order to classify experimental results.

One often postulates some symmetry or some gauge invariance and evaluates the consequences such that one can decide whether the supposed symmetry is capable of correctly ordering the experimental data.

The equations of motion that follow from the variation of the action S are

$$-\epsilon_0\left(\nabla^2 V + \nabla \cdot \frac{\partial \mathbf{A}}{\partial t}\right) = \rho \tag{8.5}$$

$$\frac{1}{\mu_0}\nabla\times\nabla\times\mathbf{A} = \mathbf{J} - \epsilon_0\frac{\partial}{\partial t}\left(\nabla V + \frac{\partial \mathbf{A}}{\partial t}\right). \tag{8.6}$$

These equations may be written as

$$M * \begin{bmatrix} V \\ \mathbf{A} \end{bmatrix} = \begin{bmatrix} \rho \\ \mathbf{J} \end{bmatrix} \tag{8.7}$$

where the matrix operator M is defined as

$$M = \begin{bmatrix} -\epsilon_0 \nabla^2 & -\epsilon_0\, \nabla\cdot\frac{\partial}{\partial t} \\ \epsilon_0\, \nabla\cdot\frac{\partial}{\partial t} & \epsilon_0\frac{\partial^2}{\partial t^2} + \frac{1}{\mu_0}\nabla\times\nabla\times \end{bmatrix} \tag{8.8}$$

This operator is *singular*, i.e. there exist non-zero fields $(X, \mathbf{Y})$ such that

$$M * \begin{bmatrix} X \\ \mathbf{Y} \end{bmatrix} = \begin{bmatrix} 0 \\ \mathbf{0} \end{bmatrix}. \tag{8.9}$$

An example is the pair $(X, \mathbf{Y}) = (-\partial\Lambda/\partial t, \nabla\Lambda)$, where $\Lambda(\mathbf{r}, t)$ is an arbitrary scalar field.

The matrix M corresponds to the second variation of the action integral and therefore $\mathcal{L}$ corresponds to a singular Lagrangian density. The singularity of M implies that there does not exist an unique inverse matrix M^{-1} and therefore, Equation (8.7) cannot be solved for the fields $(V, \mathbf{A})$ for given sources $(\rho, \mathbf{J})$. The singularity of the Lagrangian density also implies that not all the fields $(V, \mathbf{A})$ are independent. In particular, the canonical momentum conjugated to the generalized coordinate $V(\mathbf{r}, t)$ vanishes

$$\frac{\partial \mathcal{L}}{\partial\left(\frac{\partial V}{\partial t}\right)} = 0.$$

In fact, Gauss' law can be seen as a constraint for the field degrees of freedom and we are forced to restrict the set of field configurations by a gauge condition.

A gauge condition breaks the gauge invariance but it should not effect the theory such that the physical outcome is sensitive to it. In different words : the gauge condition should not influence the results of the calculation of the fields $\mathbf{E}$ and $\mathbf{B}$ and, furthermore, it must not make any field configurations of $\mathbf{E}$ and $\mathbf{B}$ "unreachable". Finally, the gauge condition should result into a non-singular Lagrangian density such that the potentials can be uniquely determined from the source distributions. We will now discuss a selection of gauge conditions that can be found in the physics literature.

8.1 The Coulomb Gauge

The Coulomb gauge is a constraint on the components of the vector potential such

$$C[\mathbf{A}] \equiv \boldsymbol{\nabla} \cdot \mathbf{A} = 0. \tag{8.10}$$

The constraint can be included in the action, S, by adding a term to the Lagrangian that explicitly breaks the gauge invariance of the action. The new action becomes "gauge-conditioned". We set:

$$S \rightarrow S_{g.c.} = S_0 + \frac{\lambda}{\mu_0} \int \mathrm{d}t\ \mathrm{d}\tau\ C^2[\mathbf{A}], \tag{8.11}$$

where $S_{g.c.}$ it the gauge-conditioned action, S_0 is the gauge-invariant action and λ is a dimensionless parameter. Then the equations for the potentials are

$$-\epsilon_0 \left(\nabla^2 V + \boldsymbol{\nabla} \cdot \frac{\partial \mathbf{A}}{\partial t} \right) = \rho \tag{8.12}$$

$$\frac{1}{\mu_0} \boldsymbol{\nabla}\times \boldsymbol{\nabla}\times\mathbf{A} - 2\frac{\lambda}{\mu_0}\boldsymbol{\nabla}\left(\boldsymbol{\nabla}\cdot\mathbf{A}\right) = \mathbf{J} - \epsilon_0 \frac{\partial}{\partial t}\left(\boldsymbol{\nabla} V + \frac{\partial \mathbf{A}}{\partial t} \right). \tag{8.13}$$

The parameter λ, can be chosen freely. Exploiting the constraint in Equations (8.10) and (8.12), we obtain

$$-\epsilon_0 \nabla^2 V = \rho \tag{8.14}$$

$$\left(\epsilon_0 \frac{\partial^2}{\partial t} - \frac{1}{\mu_0}\nabla^2 \right) \mathbf{A} = \mathbf{J} - \epsilon_0 \frac{\partial}{\partial t}\left(\boldsymbol{\nabla} V\right) \tag{8.15}$$

$$\boldsymbol{\nabla}\cdot\mathbf{A} = 0. \tag{8.16}$$

Equation (8.14) justifies the name of this gauge: the scalar potential is the instantaneous Coulomb potential of the charge distribution.

Equations (8.14) and (8.15) can be formally solved by Green functions. In general, a Green function function corresponding to a differential operator Δ is the solution of the following equation:

$$\Delta * G(\mathbf{r}, \mathbf{r}') = \delta(\mathbf{r} - \mathbf{r}'). \tag{8.17}$$

We have already seen that the Coulomb problem can be solved by the Green function $G(\mathbf{r}, \mathbf{r}') = -(1/4\pi)\ \delta(\mathbf{r} - \mathbf{r}')$. It should be emphasized that the Green function is not only determined by the structure of the differential operator but also by the boundary conditions. The wave Equation (8.15) can also be formally solved by a Green function obeying

$$\left(\frac{1}{c^2}\frac{\partial^2}{\partial t^2} - \nabla^2\right) G(\mathbf{r}, t, \mathbf{r}', t') = \delta(\mathbf{r} - \mathbf{r}')\ \delta(t - t'), \tag{8.18}$$

such that

$$\mathbf{A}(\mathbf{r}, t) = \int_{-\infty}^{\infty} \mathrm{d}t' \int \mathrm{d}\tau'\ G(\mathbf{r}, t, \mathbf{r}', t') \left(\mathbf{J}(\mathbf{r}', t') - \epsilon \frac{\partial}{\partial t} \boldsymbol{\nabla} V\right). \tag{8.19}$$

In free space the Green function is easily found by carrying out a Fourier expansion

$$G(\mathbf{r}, t, \mathbf{r}', t') = \frac{1}{(2\pi)^4} \int_{-\infty}^{\infty} \mathrm{d}\omega \int \mathrm{d}^3\mathbf{k}\ G(\omega, \mathbf{k}) \cdot \exp\left[\, \mathrm{i}\,(\omega(t - t') - \mathbf{k} \cdot (\mathbf{r} - \mathbf{r}'))\,\right]. \tag{8.20}$$

Defining $k^2 = (\omega/c)^2 - |\mathbf{k}|^2$, the Green function is $G(\omega, \mathbf{k}) = k^{-2}$. In order to respect physical causality the $(\omega, \mathbf{k})$ - integration should be done in such a way that the *retarded* Green function is obtained. This can be done by adding an infinitesimal positive shift to the poles of the Green function or propagator in the momentum representation, i.e. $G(\omega, \mathbf{k}) = 1/(k^2 - \mathrm{i}\epsilon)$. The ω-integral then generates a step function in the difference of the time arguments

$$\frac{1}{2\pi} \int_{-\infty}^{\infty} \mathrm{d}\omega \frac{\mathrm{e}^{\mathrm{i}\omega(t - t')}}{\omega - \omega_0 - \mathrm{i}\epsilon} = \mathrm{i}\theta(t - t')\mathrm{e}^{\mathrm{i}\omega_0(t - t')}. \tag{8.21}$$

8.2 The Lorenz Gauge

The next most commonly used gauge condition is the Lorenz gauge. In this gauge the scalar potential and vector potential are treated on an equal footing. The condition reads

$$C[\mathbf{A}, V] \equiv \boldsymbol{\nabla} \cdot \mathbf{A} + \frac{1}{c^2}\frac{\partial V}{\partial t} = 0; \tag{8.22}$$

where $c^{-1} = \sqrt{\mu_0 \epsilon_0}$ is the (vacuum) speed of light. The generic equations of motion (8.5) and (8.6) then lead to

$$\left(\frac{1}{c^2}\frac{\partial^2}{\partial t^2} - \nabla^2\right) V = \frac{\rho}{\epsilon_0} \tag{8.23}$$

$$\left(\frac{1}{c^2}\frac{\partial^2}{\partial t^2} - \nabla^2\right) \mathbf{A} = \mu_0 \mathbf{J}. \tag{8.24}$$

The Lorenz gauge is very suitable for performing calculations in the radiation regime. First of all, the similar treatment of all potentials simplifies the calculations and next, the travelling time intervals of the waves are not obscured by the "instantaneous" adaption of the fields to the sources as is done in the Coulomb gauge. This point is not manifest for free-field radiation, since for sourceless field solutions the absence of charges leads to $\nabla \cdot \mathbf{E} = 0$ which is solved by $V(\mathbf{r}, t) = 0$. Therefore the Coulomb gauge is suitable to handle plane electromagnetic waves. These waves have two transverse polarization modes. In the case of extended charge distributions, Gauss' law gets modified and as a consequence the scalar potential can not be taken identically equal to zero anymore. In the Lorenz gauge, there are four fields participating in the free-field solution. Definitely two of these fields are fictitious and, as such, they are called "ghost" fields. The longitudinal polarization of an electromagnetic wave corresponds to a ghost field. Care must be taken that these unphysical fields do not have an impact on the calculation of the physical quantities $\mathbf{E}$ and $\mathbf{B}$.

8.3 The Landau Gauge

Various derivations of the integer quantum Hall effect (IQHE) are based on the Landau gauge. The IQHE that was discovered by (Von Klitzing et al. [1980]) may generally occur in two-dimensional conductors with a finite width, such as the conduction channel in the inversion layer of a metal-oxide-semiconductor field-effect transistor (MOSFET) or the potential well of a semiconductor heterojunction.

Consider a two-dimensional electron gas (2DEG) confined to a ribbon $0 \leq x \leq L$, $|y| \leq W/2$, $z = 0$ carrying an electron current I in the x-direction. When a homogeneous magnetic field $\mathbf{B}$ is applied perpendicularly to the strip, the electrons are deflected by the Lorentz force $-e\mathbf{v} \times \mathbf{B}$ and start piling up at one side of the strip leaving a positive charge at the other side. As a result, a transverse Hall voltage V_H arises and prevents any further lateral transfer of deflected electrons. This phenomenon is of course nothing but the classical Hall effect for which the Hall field is probed by the Hall resistance being defined as the ratio of the Hall voltage and the longitudinal current I:

$$R_{\mathrm{H}} = \frac{V_{\mathrm{H}}}{I}. \tag{8.25}$$

However, if the ribbon is cooled down to cryogenic temperatures and the density of the 2DEG is systematically increased by changing the gate voltage, one may observe subsequent plateaus in the Hall resistance, corresponding to a series of quantized values

$$R_{\mathrm{H}} = \frac{h}{2e^2\nu} = \frac{R_{\mathrm{K}}}{\nu}, \tag{8.26}$$

where $R_{\mathrm{K}} = h/2e^2 = 25812.8\ \Omega$ is the von Klitzing resistance and ν is a positive integer.

Moreover, each time the Hall resistance attains a plateau, the longitudinal resistance of the ribbon drops to zero, which is a clear indication of ballistic, scattering free transport. For extensive discussions on the theory of the quantum Hall effect, we refer to (Butcher et al. [1993] Datta [1995] Dittrich et al. [1997] Ezawa [2000]) and all references therein. Here we would merely like to sketch how the choice of a particular gauge may facilitate the description of electron transport in terms of spatially separated, current carrying states (edge states).

The one-electron Hamiltonian reads

$$H = \frac{(\mathbf{p} + e\mathbf{A})^2}{2m} + U(y), \tag{8.27}$$

where $\mathbf{A}$ is the vector potential incorporating the external magnetic field and $U(y)$ describes the confining potential in the lateral direction. In view of the longitudinal, macroscopic current, it is quite natural to inquire whether the eigensolutions of $H\psi(x, y, z) = E\psi(x, y, z)$ are modulated by plane waves propagating along the x-direction, i.e.

$$\psi(x, y) = \frac{1}{\sqrt{L}} \mathrm{e}^{ikx} \chi_k(y), \tag{8.28}$$

where the wave number k would be an integer multiple of $2\pi/L$ to comply with periodic boundary conditions. Clearly, the establishment of full translational invariance for the Hamiltonian proposed in Equation (8.27) is a prerequisite and so we need to construct a suitable gauge such that the non-zero components of $\mathbf{A}$ do not depend on x. The simplest gauge meeting this requirement is the Landau gauge, which presently takes the form

$$\mathbf{A} = (-By, 0, 0), \tag{8.29}$$

thereby giving rise to the correct magnetic field $\nabla \times \mathbf{A} = B\mathbf{e}_z$. Combining Equations (8.27), (8.28) and (8.29), we obtain an effective Schrödinger equation for the "transverse" wave functions $\chi_k(y)$:

$$-\frac{\hbar^2}{2m}\frac{\mathrm{d}^2\chi_k(y)}{\mathrm{d}y^2} + \left[\tilde{U}_k(y) - E\right]\chi_k(y) = 0 \tag{8.30}$$

with

$$\tilde{U}_k(y) = U(y) + \frac{1}{2}m\omega_c^2(y - y_k)^2. \tag{8.31}$$

$\tilde{U}_k(y)$ acts as an effective confinement potential, centered around its minimum at $y = y_k$ (see Figure 8.1) where

$$y_k = \frac{\hbar k}{eB} \tag{8.32}$$

and $\omega_c = eB/m$ is the cyclotron frequency. For strong magnetic fields, the eigenfunctions of Equation (8.30) corresponding to a given wave number k are strongly peaked around $y = y_k$ where the probability of finding an electron outside the effective potential well falls off very rapidly. In particular, when $|k|$ increases, y_k will become of the same order of magnitude as the ribbon half-width or get even larger, so that the corresponding eigenstates – the so-called "edge states" – are strongly localized near the edges of the Hall bar while states with positive momenta $\hbar k$ have no significant lateral overlap with states having negative momenta. The spatial separation of edge states with different propagation directions and the resulting reduction of scattering matrix elements is crucial for the occurrence of the quantized Hall plateaus and can obviously be investigated most conveniently by adopting the Landau gauge since the latter ensures translational invariance in the direction of the current. It should be noted however that a full analytic solution cannot be given in terms of the familiar harmonic oscillator functions (Hermite functions) because of the edge-related boundary condition

$$\chi_k\left(\pm\frac{W}{2}\right) = 0. \tag{8.33}$$

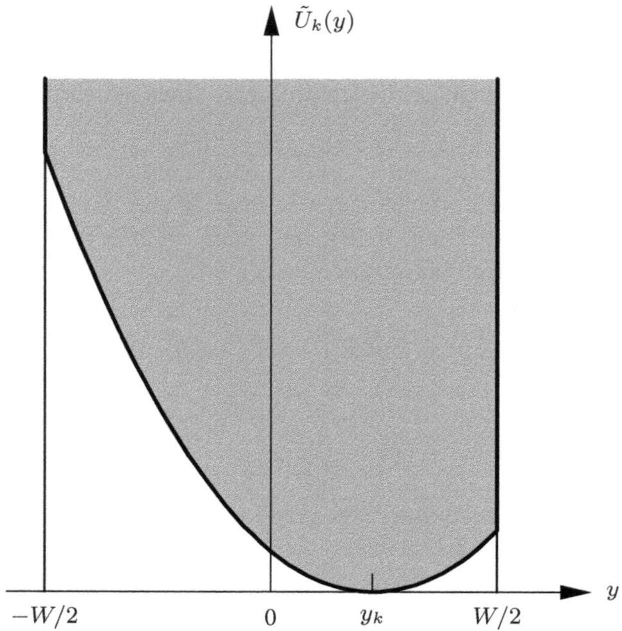

Figure 8.1 Effective confinement potential in a Hall bar (shaded area). The bare confinement is invoked by a "hard wall" restricting the lateral motion to the interval $|y| < W/2$.

8.4 The Temporal Gauge

The temporal gauge is given by the condition that the scalar field V vanish identically.

$$V(\mathbf{r}, t) = 0. \tag{8.34}$$

The electric field is then solely represented by the time derivative of the vector potential.

$$\mathbf{E}(\mathbf{r}, t) = -\frac{\partial \mathbf{A}(\mathbf{r}, t)}{\partial t}. \tag{8.35}$$

In particular, this implies that for a static field the vector potential grows unboundedly in time.

This gauge has the nice property that from a Lagrangian point of view the electric field is just the canonical momentum conjugated to the vector field variables, i.e.

$$\mathcal{L} = \frac{1}{2}\epsilon_0 \left(\frac{\partial \mathbf{A}}{\partial t}\right)^2 - \frac{1}{2\mu_0} (\boldsymbol{\nabla} \times \mathbf{A})^2 . \tag{8.36}$$

8.5 The Axial Gauge

The axial gauge is a variation of the theme above. In this gauge one component of the vector potential, e.g. A_z is set identically equal to zero.

$$A_z = 0. \tag{8.37}$$

This gauge may be exploited if a cylindrical symmetry is present. This symmetry can be inserted by setting

$$\mathbf{A}(\rho, \phi, z) = (A_\rho(\rho, \phi), A_\phi(\rho, \phi), 0) \tag{8.38}$$

in cylindrical coordinates (ρ, ϕ, z). Then

$$\mathbf{B} = \boldsymbol{\nabla}\times\mathbf{A} = \mathbf{e}_z \frac{1}{\rho}\left(\frac{\partial}{\partial \rho}(\rho A_\phi) - \frac{\partial A_\rho}{\partial \phi}\right). \tag{8.39}$$

An infinitely thin solenoid along the z-axis corresponds to a magnetic field distribution like a "needle", i.e. $\mathbf{B} = \Phi\delta(x)\ \delta(y)\mathbf{e}_z$. Such a field can be represented by the following vector potential:

$$\mathbf{A} = \frac{\Phi}{2\pi\rho}\,\mathbf{e}_\phi, \tag{8.40}$$

where Φ denotes the magnetic flux generated by the solenoid.

8.6 The 't Hooft Gauge

The selection of a gauge should be done by first identifying the problem that one wants to solve. Experience has shown that a proper selection of the gauge condition is essential to handle a particular issue. At all times it should be avoided that in the process of constructing the solution one should jump ad-hoc from one gauge condition to another. There can be found examples in the literature, where this is done, e.g. a sudden jump is taken from the Coulomb gauge to the temporal gauge, without defining the transition function that accompanies such a gauge transformation. Moreover, the demonstration that the physical results are insensitive to such transitions is often neither given. The gauge fixing method due to ('t Hooft [1971]) carefully takes the above considerations into account. It illustrates the freedom in choosing a gauge condition as well as the sliding in going from one gauge condition to another. Whereas 't Hooft's original work deals with the theory of weak interactions, the ideas can also be applied to condensed matter physics. Suppose that the physical system consists of the electromagnetic fields $(V, \mathbf{A})$ and some charged scalar field ϕ. For the latter, there is a Lagrangian density

$$\mathcal{L}_{\text{scalar}} = \frac{1}{2}\mathrm{i}\hbar\left(\phi^* \frac{\partial \phi}{\partial t} - \phi \frac{\partial \phi^*}{\partial t}\right) - \frac{\hbar^2}{2m}(\boldsymbol{\nabla}\phi^*) \cdot (\boldsymbol{\nabla}\phi) - W(\phi^*\phi). \quad (8.41)$$

The potential W describes the (massive) mode of this scalar field and possible self-interactions. If this potential has the form

$$W(\phi^*\phi) = c_2|\phi|^2 + c_3|\phi|^3 + c_4|\phi|^4 \quad (8.42)$$

with c_2 a positive number the field ϕ then this Lagrangian density describes massive scalar particles and the vacuum corresponds to $\phi = 0$. On the other hand, if $c_2 < 0$ then the minimum of W occurs at $|\phi| \equiv \phi_0 \neq 0$. In condensed matter physics, the ground state of a superconductor has non-zero expectation value for the presence of Cooper pairs. These Cooper pairs can be considered as a new particle having zero spin, i.e. it is a boson and its charge is $2e$. The corresponding field for these bosons can be given by ϕ as above, and the ground state is characterized by some non-zero value of ϕ. This can be realized by setting $c_2 < 0$. The interaction of this scalar field with the electromagnetic field is provided by the minimal substitution procedure and leads to the following Lagrangian

$$\begin{aligned}
\mathcal{L} &= \mathcal{L}_{\text{EM}} + \mathcal{L}_{\text{scalar}} + \mathcal{L}_{\text{int}}, \\
\mathcal{L}_{\text{int}} &= \mathbf{J} \cdot \mathbf{A} - \rho V - \frac{e}{m}\rho A^2, \\
\rho &= -e\phi^*\phi, \\
\mathbf{J} &= \frac{\mathrm{i}e\hbar}{2m}\left[\phi^*\boldsymbol{\nabla}\phi - \left(\boldsymbol{\nabla}\phi^*\right)\phi\right] + \frac{e}{m}\rho\mathbf{A}.
\end{aligned} \quad (8.43)$$

The complex field $\phi = \phi_1 + \mathrm{i}\phi_2$ can now be expanded around the vacuum expectation value $\phi = \phi_0 + \chi + \mathrm{i}\phi_2$. The interaction Lagrangian will contain terms being quadratic in the fields that mix the electromagnetic potentials with the scalar fields. Such terms can be eliminated by choosing the gauge condition in such way that these terms cancel, i.e. by properly selecting the constants α_1 and α_2 in

$$C[\mathbf{A}, V, \chi, \phi_2] \equiv \boldsymbol{\nabla} \cdot \mathbf{A} + \frac{1}{c}\frac{\partial V}{\partial t} + \alpha_1\,\chi + \alpha_2\,\phi_2 = 0. \quad (8.44)$$

9

The Geometry of Electrodynamics

Electrodynamics was discovered as a phenomenological theory. Starting from early experiments with amber, permanent magnets and conducting wires, one finally arrived after much effort at Gauss' law. Biot-Savart's law and Faraday's law of induction. Only Maxwell's laws were obtained by theoretical reasoning being confirmed experimentally later on by Herz. Maxwell's great achievement was later equalized by Einstein who proposed in the general theory of relativity that

gravity = curvature

Ever since Einsteins's achievement of describing gravity in terms of non-Euclidean geometry, theoretical physics has witnessed a stunning development based on geometrical reasoning. Nowadays it is generally accepted that the standard model of matter, based on gauge theories, is the correct description (within present-day experimental accessibility) of matter and its interaction. These gauge theories have a geometrical interpretation very analogous to Einstein's theory of gravity. In fact, we may widen our definition of "geometry" such that gravity (coordinate covariance) and the standard theory (gauge covariance) are two realizations of the same mechanism. Electrodynamics is the low-energy part of the standard model. Being a major aspect of this book, it deserves special attention and in this interpretation. Besides the aesthetic beauty that results from these insights, there is also pragmatic benefit. Solving electrodynamic problems on the computer, guided by the geometrical meaning of the variables has been a decisive factor for the success of the calculation. This was already realized by (Wilson [1974]) when he performed computer calculations of the quantum aspects of gauge theories. In order to perform computer calculations of the classical fields, geometry plays an important role as is discussed in chapter II. However, the classical fields $\mathbf{E}$ and $\mathbf{B}$ as well as the sources ρ and $\mathbf{J}$ are invariant under gauge transformations and therefore their deeper geometrical meaning is hidden. In fact, we can identify the proper geometric character for these variables, such as scalars (zero-forms), force fields (one-forms), fluxes (two-forms) or volume densities (three-forms) as can be done for any other fluid dynamic system, but this can be done without making any reference to the geometric nature of electrodynamics in the sense that $\mathbf{E}$ and $\mathbf{B}$ represent the *curvature* in the geometrical interpretation of electrodynamics. Therefore, in this section we will consider the scalar potential and vector potential fields that do depend on gauge transformations and as such will give access to the geometry of electrodynamics.

9.1 Gravity as a Gauge Theory

The history of the principle of gauge invariance begins with the discovery of the principle of general covariance in general relativity. According to this principle the physical laws should maintain their form for all coordinate systems. In 1918, Hermann Weyl made an attempt to unify electrodynamics with gravity in (Weyl [1918]). According to the general theory of relativity, the gravitational field corresponds to curvature of space-time, and therefore, if a vector is parallel transported along a closed loop, the angle between the starting vector and the final vector will differ from zero. Furthermore, this angle is a measure for the curvature in space. Weyl extended the Riemann geometry in such a way that not only the angle changes but also the *length* of the vector. The relative change in length is described by an anti-symmetric tensor and this tensor is invariant under changing the "unit of length". This invariance is closely related to charge conservation. Weyl called this "Maszstab Invarianz". The theory turned out to be contradictory and was abandoned, but the term "Maszstab Invarianz" survived (Maszstab = measure = gauge). With the arrival of quantum mechanics the principle of gauge invariance obtained its final interpretation: gauge invariance should refer to the phase transformations that may be applied on the wave functions. In particular, the phase transformation may be applied with different angles for different points in space and time.

$$\psi(\mathbf{r},t) \rightarrow \psi'(\mathbf{r},t) = \exp\left(\frac{ie}{\hbar}\chi(\mathbf{r},t)\right)\psi(\mathbf{r},t). \tag{9.1}$$

At first sight it looks as if we have lost the geometrical connection and the link is only historical. However, a closer look at gravity shows that the link is still present.

Starting from the idea that all coordinate systems are equivalent, we may consider a general coordinate transformation

$$x^\mu \to x'^\mu = x'^\mu(x^\nu). \tag{9.2}$$

The transformation rule for coordinate differentials is

$$\mathrm{d}x'^\mu = \frac{\partial x'^\mu}{\partial x^\nu}\mathrm{d}x^\nu. \tag{9.3}$$

An ordered set of functions transforming under a change of coordinates in the same way as the coordinate differentials is defined to be a *contravariant vector*

$$V'^\mu = \frac{\partial x'^\mu}{\partial x^\nu}V^\nu. \tag{9.4}$$

A *scalar* transforms in an invariant way, i.e.

$$\phi(x) \to \phi'(x') = \phi(x). \tag{9.5}$$

The derivatives of a scalar transform as

$$V'_\mu = \frac{\partial x^\nu}{\partial x'^\mu}V_\nu. \tag{9.6}$$

Any ordered set of functions transforming under a change of coordinates as the derivatives of a scalar function is a *covariant vector*. In general, *tensors* transform according to a multiple set of pre-factors, i.e.

$$V'^{\alpha_1\alpha_2\dots}_{\mu_1\mu_2\dots} = \frac{\partial x'^{\alpha_1}}{\partial x^{\beta_1}}\frac{\partial x'^{\alpha_2}}{\partial x^{\beta_2}}\frac{\partial x^{\nu_1}}{\partial x'^{\mu_1}}\frac{\partial x^{\nu_2}}{\partial x'^{\mu_2}}\cdots V^{\beta_1\beta_2\dots}_{\nu_1\nu_2,\dots} \quad . \tag{9.7}$$

The principle of general coordinate covariance can be implemented by claiming that all physical laws should be expressed as tensor equations. Since left and right hand sides will transform with equal sets of pre-factors, the form invariance is guaranteed.

So far, we have only been concerned with the change from one arbitrary coordinate system to another. One might argue that this will just hide well-known results in a thick shell of notation complexity. In order to peal off these shells and to find the physical implications one must refer to the *intrinsic* properties of the geometric structure. Occasionally, the intrinsic structure is simple, e.g. flat space time, and the familiar relations are recovered. It was Einstein's discovery that space-time is *not* flat in the presence of matter and therefore the physical laws are more involved.

Riemann geometry is a generalization of Euclidean geometry in the sense that locally one can still find coordinate systems $\xi^{\mu} = (ict, \mathbf{x})$, such that the distance between two near-by points is given by Pythagoras' theorem, i.e.

$$\mathrm{d}s^2 = \delta_{\mu\nu}\mathrm{d}\xi^{\mu}\mathrm{d}\xi^{\nu}. \tag{9.8}$$

In an arbitrary coordinate system the distance is given by

$$\mathrm{d}s^2 = g_{\mu\nu}(x)\mathrm{d}x^{\mu}\mathrm{d}x^{\nu}, \tag{9.9}$$

where

$$g_{\mu\nu}(x) = \frac{\partial\xi^{\alpha}}{\partial x^{\mu}}\frac{\partial\xi^{\beta}}{\partial x^{\nu}}\delta_{\alpha\beta} \tag{9.10}$$

is the metric tensor of the coordinate system.

In the local coordinate system, ξ, the equation of motion of a freely falling particle is given by

$$\frac{\mathrm{d}^2\xi^{\mu}}{\mathrm{d}s^2} = 0. \tag{9.11}$$

In an arbitrary coordinate system, this equation becomes

$$\frac{\mathrm{d}}{\mathrm{d}s}\left(\frac{\partial\xi^{\mu}}{\partial x^{\alpha}}\frac{\mathrm{d}x^{\alpha}}{\mathrm{d}s}\right) = 0. \tag{9.12}$$

This can be evaluated to

$$\frac{\mathrm{d}^2x^{\alpha}}{\mathrm{d}s^2} + \Gamma^{\alpha}_{\mu\nu}\frac{\mathrm{d}x^{\mu}}{\mathrm{d}s}\frac{\mathrm{d}x^{\nu}}{\mathrm{d}s} = 0, \tag{9.13}$$

where $\Gamma^{\alpha}_{\mu\nu}$ is the *affine connection*

$$\Gamma^{\alpha}_{\mu\nu} = \frac{\partial x^{\alpha}}{\partial\xi^{\beta}}\frac{\partial^2\xi^{\beta}}{\partial x^{\mu}\partial x^{\nu}}. \tag{9.14}$$

The affine connection transform under general coordinate transformations as

$$\Gamma'^{\alpha}_{\mu\nu} = \frac{\partial x'^{\alpha}}{\partial x^{\rho}}\frac{\partial x^{\tau}}{\partial x'^{\mu}}\frac{\partial x^{\sigma}}{\partial x'^{\nu}}\Gamma^{\rho}_{\tau\sigma} + \frac{\partial x'^{\alpha}}{\partial x^{\rho}}\frac{\partial^2 x^{\rho}}{\partial x'^{\mu}\partial x'^{\nu}}. \tag{9.15}$$

The second term destroys the covariance of the affine connection, i.e. the affine connection is *not* a tensor.

The metric tensor $g_{\mu\nu}(x)$ contains information on the local curvature of the Riemann geometry. Now consider a vector $V^\mu(\tau)$ along a curve $x^\mu(\tau)$. In the locally Euclidean coordinate system (ξ), the change of the vector along the curve is $\mathrm{d}V^\mu/\mathrm{d}\tau$.

In another coordinate system (x'), we find from the transformation rule (9.4)

$$\frac{\mathrm{d}V'^\mu}{\mathrm{d}\tau} = \frac{\partial x'^\mu}{\partial x^\nu}\frac{\mathrm{d}V^\nu}{\mathrm{d}\tau} + \frac{\partial^2 x'^\mu}{\partial x^\nu \partial x^\lambda}\frac{\partial x^\lambda}{\partial \tau}V^\nu(\tau). \tag{9.16}$$

The second derivative in the second term is an inhomogeneous term in the transformation rule that prevents $\mathrm{d}V^\mu/\mathrm{d}\tau$ from being a vector and contains the key to curvature. This term is directly related to the affine connection. The combination

$$\frac{DV^\mu}{D\tau} = \frac{\mathrm{d}V^\mu}{\mathrm{d}\tau} + \Gamma^\mu_{\nu\lambda}\frac{\mathrm{d}x^\lambda}{\mathrm{d}\tau}V^\nu \tag{9.17}$$

transforms as a vector and is called the *covariant* derivative along the curve. In the restricted region where we can use the Euclidean coordinates, ξ, we may apply Euclidean geometrical methods, and in particular we can shift a vector over an infinitesimal distance from one base point to another and keep the initial and final vector parallel. This is depicted in Figure 9.1. The component of the vector do not alter by the shift operation: $\delta V^\mu = 0$. Furthermore, in the local frame $x^\mu = \xi^\mu_{x(\tau)}$, the affine connection vanishes, i.e. $\Gamma^\alpha_{\mu\nu} = 0$. Therefore, the conventional operation of parallel shifting a vector in the locally Euclidean coordinate system can be expressed by the equation $DV^\mu/D\tau = 0$. Being a tensor equation, this it true in all coordinate systems. A vector, whose covariant derivative along a curve vanishes

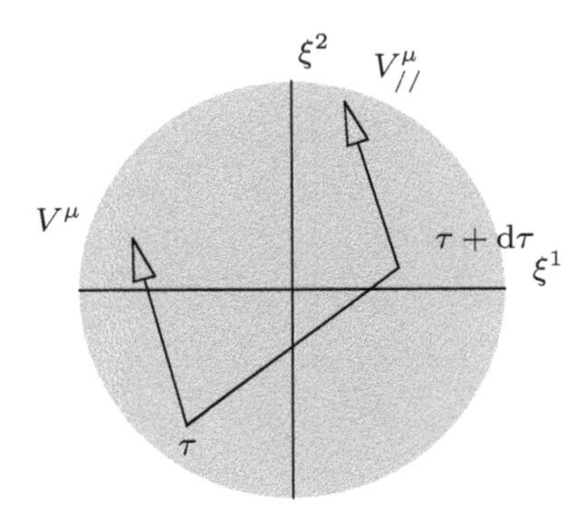

Figure 9.1 Parallel displacement in the locally Euclidean coordinate system.

is said to be *parallel* transported along the curve. The coordinates satisfy the following first-order differential equations:

$$\frac{\mathrm{d}V^{\mu}}{\mathrm{d}\tau} = -\Gamma^{\mu}_{\nu\lambda}\frac{\mathrm{d}x^{\lambda}}{\mathrm{d}\tau}V^{\nu}. \tag{9.18}$$

The parallel transport of a vector V^{μ} over a small distance $\mathrm{d}x^{\nu}$ changes the components of the vector by amounts

$$\delta V^{\mu} = -\Gamma^{\mu}_{\nu\lambda}V^{\nu}\delta x^{\lambda}. \tag{9.19}$$

In general, if we want to perform the differentiation of a tensor field with respect to the coordinates, we must compare tensors in two nearby points. In fact, the comparison corresponds to subtraction, but a subtraction is only defined if the tensors are anchored to the same point. (In different points, we have different local coordinate systems.) Therefore we must first parallel transport the initial tensor to the nearby point before the subtraction can be performed. This is illustrated in Figure 9.2. For example, the covariant differential of a vector field is

$$DV^{\mu} = \mathrm{d}V^{\mu} - \delta V^{\mu} = \left(\frac{\partial V^{\mu}}{\partial x^{\lambda}} + \Gamma^{\mu}_{\lambda\kappa}V^{\kappa}\right)\mathrm{d}x^{\lambda} = D_{\lambda}V^{\mu}\mathrm{d}x^{\lambda}. \tag{9.20}$$

So far, the general coordinate systems include both accelerations originating from non-uniform boosts of the coordinate systems as well as acceleration that may be caused by gravitational field due to the presence of matter.

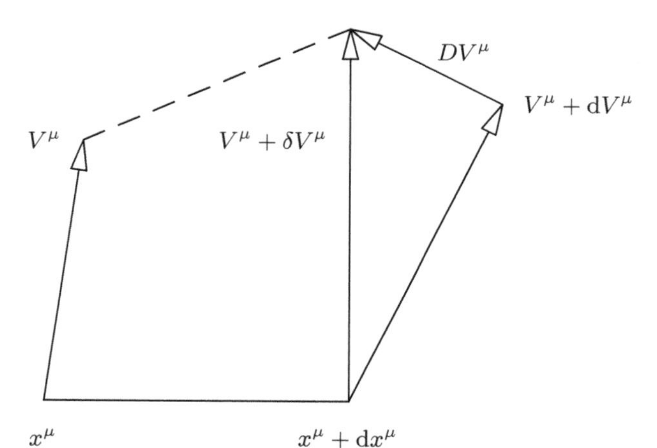

Figure 9.2 The covariant derivative of a vector field.

In the first case, space-time is not really curved. In the second case space-time is curved. In order to find out whether gravitation is present one must extract information about the intrinsic properties of space-time. This can be done by the parallel transport of a vector field along a closed loop. If the initial and final vector differ, one can conclude that gravity is present. The difference that a closed loop (see Figure 9.3) transport generates is given by

$$\Delta V^{\mu} = V^{\mu}_{\text{via B}} - V^{\mu}_{\text{via D}} = R^{\mu}_{\rho\lambda\sigma} V^{\rho} \delta x^{\lambda} \delta x^{\sigma}, \tag{9.21}$$

where

$$R^{\mu}_{\rho\lambda\sigma} = \frac{\partial \Gamma^{\mu}_{\rho\lambda}}{\partial x^{\sigma}} - \frac{\partial \Gamma^{\mu}_{\rho\sigma}}{\partial x^{\lambda}} + \Gamma^{\eta}_{\rho\lambda}\Gamma^{\mu}_{\sigma\eta} - \Gamma^{\eta}_{\rho\sigma}\Gamma^{\mu}_{\lambda\eta} \tag{9.22}$$

is the *curvature* tensor or Riemann tensor. This tensor describes the intrinsic curvature in a point.

We are now prepared to consider the geometrical basis of electrodynamics and other gauge theories but we will first summarize a few important facts:

- in each space-time point a local frame may be erected,
- the affine connection is a path-dependent quantity,
- the affine connection does not transform as a tensor,
- the field strength (curvature) may be obtained by performing a parallel transport along a closed loop.

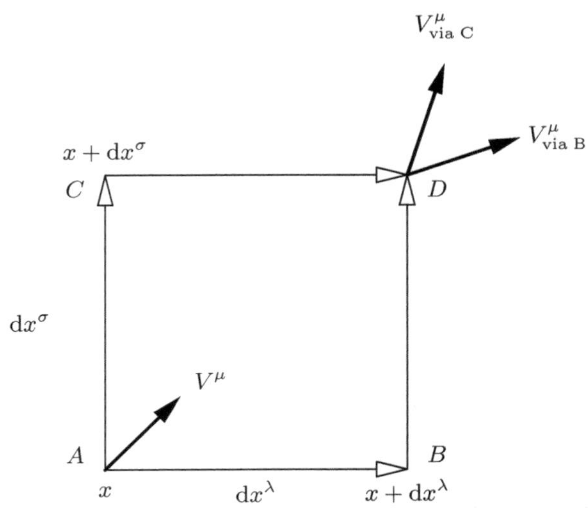

Figure 9.3 Determination of the curvature from a round trip along a closed loop.

9.2 The Geometrical Interpretation of Electrodynamics

As for the local Euclidean coordinate systems, we will consider the possibility of setting up in each space-time point a local frame for fixing the phase of the complex wave function $\psi(\mathbf{r}, t)$. Since the choice of such a local frame (gauge) is not unique we may rotate the frame without altering the physical content of a frame fixing.

We can guarantee the latter by demanding appropriate transformation properties (see the above section about tensors) of the variables. Changing the local frame for the phase of a wave function amounts to

$$\begin{aligned} \psi'(\mathbf{r}, t) &= \exp\left(\frac{\mathrm{ie}}{\hbar}\chi(\mathbf{r}, t)\right)\psi(\mathbf{r}, t), \\ \psi'^{*}(\mathbf{r}, t) &= \exp\left(-\frac{\mathrm{ie}}{\hbar}\chi(\mathbf{r}, t)\right)\psi^{*}(\mathbf{r}, t). \end{aligned} \tag{9.23}$$

These transformation rules are similar to the contravariant and covariant transformation rules for vectors in the foregoing section. We can similarly construct a "scalar" by taking $\psi^{*}\psi$. The derivative of the wave function transforms as

$$\frac{\partial \psi'}{\partial x^{\mu}} = \exp\left(\frac{\mathrm{ie}}{\hbar}\chi\right)\frac{\partial \psi}{\partial x^{\mu}} + \frac{\mathrm{ie}}{\hbar}\frac{\partial \chi}{\partial x^{\mu}}\exp\left(\frac{\mathrm{ie}}{\hbar}\chi\right)\psi. \tag{9.24}$$

The second term prevents the derivative of ψ from transforming as a "vector" under the change of gauge. However, geometry will now be of help to construct gauge covariant variables from derivatives. We must therefore postulate

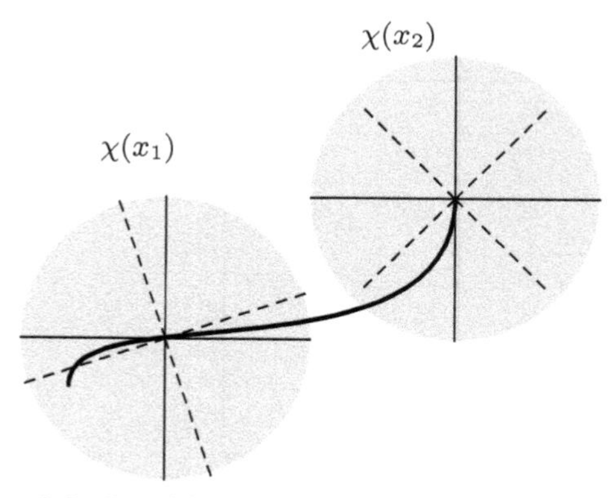

Figure 9.4 Local frames for the phase of a wave function.

an "affine connection", such that a covariant derivative can be defined. For that purpose a connection, A_μ, is proposed that transforms as

$$A_\mu = A_\mu + \frac{\partial \chi}{\partial x^\mu}. \tag{9.25}$$

The covariant derivative is

$$D_\mu = \frac{\partial}{\partial x^\mu} + \frac{\mathrm{ie}}{\hbar} A_\mu. \tag{9.26}$$

Similar to the gravitational affine connection, the field A_μ can be used to construct "parallel" transport. Therefore, the field A_μ must be assigned to the *paths* along which the transport takes place. The curvature of the connection can also be constructed by making a complete turn around a closed loop. The result is

$$F_{\mu\nu} \delta x^\mu \delta x^\nu = \oint \mathrm{d}x^\mu A_\mu, \tag{9.27}$$

where

$$F_{\mu\nu} = \frac{\partial A_\mu}{\partial x^\nu} - \frac{\partial A_\nu}{\partial x^\mu} \tag{9.28}$$

is the electromagnetic field tensor.

10

Integral Theorems

Integral theorems borrowed from the differential geometry of curves, surfaces and connected regions (Morse and Feshbach [1953] Magnus and Schoenmaker [1998]) turn out to be useful and perhaps even indispensable for a thorough understanding of elementary electromagnetic theory. Not only are they quite helpful in converting the differential form of Maxwell's equations into their equivalent integral form, but they also offer a convenient tool to define a discretized version of the field variables in the framework of numerical simulation. Moreover, they naturally bridge the gap between the microscopic interaction of the electromagnetic fields and charges in a solid-state conductor and the global circuit models envisaged on the macroscopic level.

The first three integral theorems that are are summarized below, are extensively referred to in Section 2. The fourth one is the Helmholtz theorem, which allows one to decompose any well-behaved vector field into a longitudinal and a transverse part.

Theorem 10.1 (Stokes' theorem) *Let Σ be an open, orientable, multiply connected surface in $\mathbb{R}^3$ bounded by an outer, closed curve $\partial\Sigma_0$ and n inner, closed curves $\partial\Sigma_1, \ldots, \partial\Sigma_n$ defining n holes. If Σ is oriented by a surface element $\mathbf{dS}$ and if $\mathbf{A}$ is a differentiable vector field defined on Σ, then*

$$\int_{\Sigma} \nabla\times\mathbf{A}\cdot\mathbf{dS} = \oint_{\partial\Sigma_0} \mathbf{A}\cdot\mathbf{dr} - \sum_{j=1}^{n} \oint_{\partial\Sigma_j} \mathbf{A}\cdot\mathbf{dr} \tag{10.1}$$

where the orientation of all boundary curves is uniquely determined by the orientation of $\mathbf{dS}$.

Theorem 10.2 (Gauss' theorem) *Let Ω be a closed, orientable, multiply connected subset of $\mathbb{R}^3$ bounded by an outer, closed surface $\partial\Omega_0$ and n inner,* *closed surfaces defining n holes. If* $\mathbf{E}$ *is a differentiable vector field defined on* Ω *then*

$$\int_{\Omega} \boldsymbol{\nabla}\cdot\mathbf{E}\,\mathrm{d}\tau \;=\; \int_{\partial\Omega_0} \mathbf{E}\cdot\mathbf{dS} \;-\; \sum_{j=1}^{n}\int_{\partial\Omega_j} \mathbf{E}\cdot\mathbf{dS} \tag{10.2}$$

and

$$\int_{\Omega} \boldsymbol{\nabla}\times\mathbf{E}\,\mathrm{d}\tau \;=\; \int_{\partial\Omega_0} \mathbf{dS}\times\mathbf{E} \;-\; \sum_{j=1}^{n}\int_{\partial\Omega_j} \mathbf{dS}\times\mathbf{E} \tag{10.3}$$

where all boundary surfaces have the same orientation as the outward pointing surface element of the outer surface.

The scalar Gauss theorem (10.2) reduces to **Green's Theorem** when the vector field takes the form $\mathbf{E} = f\boldsymbol{\nabla}g - g\boldsymbol{\nabla}f$

$$\begin{aligned}\int_{\Omega}(f\nabla^2 g - g\nabla^2 f)\,\mathrm{d}\tau \;&=\; \int_{\partial\Omega_0}(f\boldsymbol{\nabla}g - g\boldsymbol{\nabla}f)\cdot\mathbf{dS}\\ &\quad-\sum_{j=1}^{n}\int_{\partial\Omega_j}(f\boldsymbol{\nabla}g - g\boldsymbol{\nabla}f)\cdot\mathbf{dS}\end{aligned} \tag{10.4}$$

where the scalar fields f and g are differentiable on Ω.

Theorem 10.3 ($\mathbf{J}\cdot\mathbf{E}$ theorem) *Let* Ω *be a closed, multiply connected, bounded subset of* $\mathbb{R}^3$ *with one hole and boundary surface* $\partial\Omega$*. If* $\mathbf{J}$ *and* $\mathbf{E}$ *are two differentiable vector fields on* Ω*, circulating around the hole and satisfying the conditions*

$$\boldsymbol{\nabla}\cdot\mathbf{J} \;=\; 0 \tag{10.5}$$

$$\boldsymbol{\nabla}\times\mathbf{E} \;=\; \mathbf{0} \tag{10.6}$$

$$\mathbf{J}\parallel\partial\Omega \quad or \quad \mathbf{J} \;=\; \mathbf{0}\ \textit{in each point point of}\ \partial\Omega \tag{10.7}$$

then

$$\int_{\Omega}\mathbf{J}\cdot\mathbf{E}\,\mathrm{d}\tau \;=\; \left(\int_{\Sigma}\mathbf{J}\cdot\mathbf{dS}\right)\left(\oint_{\Gamma}\mathbf{E}\cdot\mathbf{dr}\right) \tag{10.8}$$

where Σ *is an arbitrary cross section, intersecting* Ω *only once and* Γ *is a simple closed curve, encircling the hole and lying within* Ω *but not intersecting* $\partial\Omega$*. The orientation of* Σ *is uniquely determined by the positive orientation of* Γ*.*

Proof. Without any loss of generality one may define curvilinear coordinates (x^1, x^2, x^3) and a corresponding set of covariant basis vectors $(\boldsymbol{a_1}, \boldsymbol{a_2}, \boldsymbol{a_3})$ and its contravariant counterpart, which are compatible with the topology of the toroidal (torus-like) region Ω. More precisely, x^1, x^2 and x^3 may be chosen such that the boundary surface $\partial\Omega$ coincides with one of the coordinate surfaces $\mathrm{d}x^1 = 0$ while the curves $\mathrm{d}x^1 = \mathrm{d}x^2 = 0$ are closed paths encircling the hole only once and x^3 is a cyclic coordinate. Then the inner volume contained within Ω may be conveniently parametrized by restricting the range of (x^1, x^2, x^3) to some rectangular interval $[c^1, d^1] \times [c^2, d^2] \times [c^3, d^3]$.

Since Ω is multiply connected, the irrotational vector field $\mathbf{E}$ cannot generally be derived from a scalar potential for the whole region Ω. However, for the given topology of Ω, it is always possible to assign such a potential to the "transverse" components of $\mathbf{E}$ only:

$$E_1(x^1, x^2, x^3) = -\frac{\partial V(x^1, x^2, x^3)}{\partial x^1} \tag{10.9}$$

$$E_2(x^1, x^2, x^3) = -\frac{\partial V(x^1, x^2, x^3)}{\partial x^2} \tag{10.10}$$

but

$$E_3(x^1, x^2, x^3) \neq -\frac{\partial V(x^1, x^2, x^3)}{\partial x^3} \tag{10.11}$$

where $V(x^1, x^2, x^3)$ can be constructed straightaway by invoking the first two components of $\boldsymbol{\nabla}\times\mathbf{E} = \mathbf{0}$:

$$\begin{aligned} V(x^1, x^2, x^3) &= V(c^1, c^2, x^3) - \int_{c^1}^{x^1} \mathrm{d}s\, E_1(s, x^2, x^3) \\ &\quad - \int_{c^2}^{x^2} \mathrm{d}t\, E_2(c^1, t, x^3) \end{aligned} \tag{10.12}$$

The potential term $V(c^1, c^2, x^3)$ naturally arises as an integration constant which, depending on x^3 only, may be absorbed in the definition of $V(x^1, x^2, x^3)$ and will therefore be omitted. Equations (10.9) and (10.10) are now easily recovered by taking the derivative of (10.12) with respect to x^1 and x^2, and inserting the third component of $\boldsymbol{\nabla}\times\mathbf{E} = \mathbf{0}$.
Finally, taking also the derivative with respect to x^3, one obtains:

$$E_3(x^1, x^2, x^3) = -\frac{\partial V(x^1, x^2, x^3)}{\partial x^3} + E_3(c^1, c^2, x^3) \tag{10.13}$$

From Equations (10.9), (10.10) and (10.13) arises a natural decomposition of $\mathbf{E}$ into a conservative vector field $\mathbf{E}_\mathrm{C}$ and a non-conservative field $\mathbf{E}_\mathrm{NC}$ that is oriented along $\boldsymbol{a}_3$, thereby depending only on the cyclic coordinate x^3:

$$\mathbf{E} = \mathbf{E}_\mathrm{C} + \mathbf{E}_\mathrm{NC} \tag{10.14}$$

with

$$\mathbf{E}_\mathrm{C}(x^1, x^2, x^3) = -\ \boldsymbol{\nabla} V(x^1, x^2, x^3) \tag{10.15}$$

$$\mathbf{E}_\mathrm{NC}(x^1, x^2, x^3) = E_3(c^1, c^2, x^3)\ \boldsymbol{a}^3 \tag{10.16}$$

The conservative part of $\mathbf{E}$ does not contribute to the volume integral of $\mathbf{J} \cdot \mathbf{E}$. Indeed, from (10.15) it follows

$$\begin{aligned} \int_\Omega \mathbf{J} \cdot \mathbf{E}_\mathrm{C}\ \mathrm{d}\tau &= \int_\Omega \mathbf{J} \cdot \boldsymbol{\nabla} V\ \mathrm{d}\tau \\ &= \int_\Omega \boldsymbol{\nabla} \cdot (V\mathbf{J})\ \mathrm{d}\tau\ -\ \int_\Omega V\ \boldsymbol{\nabla} \cdot \mathbf{J}\ \mathrm{d}\tau \end{aligned} \tag{10.17}$$

With the help of Gauss' theorem – which is also valid for multiply connected regions – the first term of the right-hand side of Equation (10.17) can be rewritten as a surface integral of $V\mathbf{J}$ which is seen to vanish as $\mathbf{J}$ is assumed to be tangential to the surface $\partial\Omega$ in all of its points. Clearly, the second integral in the right-hand side of (10.17) is identically zero due to $\boldsymbol{\nabla} \cdot \mathbf{J} = 0$ and one is therefore lead to the conclusion

$$\int_\Omega \mathbf{J} \cdot \mathbf{E}_\mathrm{C}\ \mathrm{d}\tau\ =\ 0 \tag{10.18}$$

On the other hand, the contribution of $\mathbf{E}_\mathrm{NC}$ can readily be evaluated in terms of the curvilinear coordinates. Denoting the Jacobian determinant by $g(x^1, x^2, x^3)$ one may express the volume integral as a threefold integral over the basic interval $[c^1, d^1] \times [c^2, d^2] \times [c^3, d^3]$, thereby exploiting the fact that the non-conservative contribution merely depends on x^3:

$$\begin{aligned} \int_\Omega \mathbf{J} \cdot \mathbf{E}\ \mathrm{d}\tau\ &=\ \int_\Omega \mathbf{J} \cdot \mathbf{E}_\mathrm{NC}\ \mathrm{d}\tau \\ &= \int_{c^3}^{d^3} \mathrm{d}x^3\ E_3(c^1, c^2, x^3) \int_{c^1}^{d^1} \mathrm{d}x^1 \\ &\quad \cdot \int_{c^2}^{d^2} \mathrm{d}x^2\ g(x^1, x^2, x^3)\ J^3(x^1, x^2, x^3) \end{aligned} \tag{10.19}$$

The last integral can conveniently be interpreted as the flux of $\mathbf{J}$ through the single cross section $\Sigma(x^3)$ defined by

$$\Sigma(x^3) = \{(x^1, x^2, x^3) \mid c^1 \leq x^1 \leq d^1;\ c^2 \leq x^2 \leq d^2; x^3 \text{ fixed}\} \quad (10.20)$$

Indeed, expanding the Jacobian determinant as a mixed product of the three basis vectors, i.e.

$$g = \boldsymbol{a_1} \times \boldsymbol{a_2} \cdot \boldsymbol{a_3} \quad (10.21)$$

and identifying the two-form $\boldsymbol{a_1} \times \boldsymbol{a_2}\ \mathrm{d}x^1 \mathrm{d}x^2$ as a generic surface element $\mathbf{dS}$ perpendicular to $\Sigma(x^3)$, one easily arrives at

$$\begin{aligned}
\int_{c^1}^{d^1} \mathrm{d}x^1 \int_{c^2}^{d^2} & \mathrm{d}x^2 g(x^1, x^2, x^3) J^3(x^1, x^2, x^3) \\
&= \int_{c^1}^{d^1} \mathrm{d}x^1 \int_{c^2}^{d^2} \mathrm{d}x^2\ \boldsymbol{a_1} \times \boldsymbol{a_2} \cdot \mathbf{J}(x^1, x^2, x^3) \\
&= \int_{\Sigma(x^3)} \mathbf{dS} \cdot \mathbf{J} \\
&\equiv I(x^3)
\end{aligned} \quad (10.22)$$

and

$$\int_{\Omega} \mathbf{J} \cdot \mathbf{E}\ \mathrm{d}\tau = \int_{c^3}^{d^3} \mathrm{d}x^3\ E_3(c^1, c^2, x^3)\ I(x^3) \quad (10.23)$$

The sign of the flux $I(x^3)$ obviously depends on the orientation of $\Sigma(x^3)$, which is unequivocally determined by the surface element $\mathbf{dS} = \boldsymbol{a_1} \times \boldsymbol{a_2}\ \mathrm{d}x^1 \mathrm{d}x^2$. As long as only positive body volumes are concerned, one may equally require that each infinitesimal volume element $\mathrm{d}\tau = g\ \mathrm{d}x^1 \mathrm{d}x^2 \mathrm{d}x^3$ be positive for positive incremental values $\mathrm{d}x^1$, $\mathrm{d}x^2$ and $\mathrm{d}x^3$. Moreover, since $\mathbf{dr} = \mathrm{d}x^3 \boldsymbol{a_3}$ is the elementary tangent vector of the coordinate curve $\Gamma(x^1, x^2) = \{(x^1, x^2, x^3) \mid x^1, x^2 \text{ fixed};\ c^3 \leq x^3 \leq d^3\}$ orienting $\Gamma(x^1, x^2)$ in a positive traversal sense through increasing x^3, one easily arrives at

$$\mathrm{d}\tau = \mathbf{dS} \cdot \mathbf{dr} > 0 \quad (10.24)$$

In other words, the orientation of $\Sigma(x^3)$ is completely fixed by the positive traversal sense of $\Gamma(x^1, x^2)$. However, since $\mathbf{J}$ is solenoidal within Ω as well as tangential to $\partial\Omega$, one may conclude from Gauss' theorem that the value of the flux $I(x^3)$ does not depend on the particular choice of the cross section $\Sigma(x^3)$ which may thus be replaced by any other single cross section

Σ provided that the orientation is preserved. Consequently, $I(x^3)$ reduces to a constant value I and may be taken out of the integral of Equation (10.23) which now simplifies to:

$$\int_{\Omega} \mathbf{J} \cdot \mathbf{E} \, \mathrm{d}\tau \;=\; I \int_{c^3}^{d^3} \mathrm{d}x^3 \; E_3(c^1, c^2, x^3)$$

The remaining integral turns out to be the line integral of $\mathbf{E}$ along the coordinate curve $\Gamma(c^1, c^2)$:

$$\int_{\Omega} \mathbf{J} \cdot \mathbf{E} \, \mathrm{d}\tau \;=\; I V_{\epsilon}(c^1, c^2) \tag{10.25}$$

with

$$V_{\epsilon}(c^1, c^2) \;=\; \oint_{\Gamma(c^1, c^2)} \mathbf{E} \cdot \mathbf{dr} \tag{10.26}$$

Since $\mathbf{E}$ is irrotational, according to Stokes' theorem its circulation does not depend on the particular choice of the circulation curve as was already discussed in more detail in the previous section. Consequently, $\Gamma(c^1, c^2)$ may be replaced by any other interior closed curve Γ encircling the hole region and sharing the traversal sense with $\Gamma(c^1, c^2)$:

$$V_{\epsilon}(c^1, c^2) \;=\; V_{\epsilon} \equiv \oint_{\Gamma} \mathbf{E} \cdot \mathbf{dr} \tag{10.27}$$

Hence,

$$\int_{\Omega} \mathbf{J} \cdot \mathbf{E} \, \mathrm{d}\tau \;=\; I V_{\epsilon} \tag{10.28}$$

This completes the proof.

Theorem 10.4 (Helmholtz' theorem) *Let Ω be a simply connected, bounded subset of $\mathbb{R}^3$. Then, any finite, continuous vector field $\boldsymbol{F}$ defined on Ω can be derived from a differentiable vector potential $\mathbf{A}$ and a differentiable scalar potential χ such that*

$$\boldsymbol{F} = \boldsymbol{F}_{\mathrm{L}} + \boldsymbol{F}_{\mathrm{T}} \tag{10.29}$$

$$\boldsymbol{F}_{\mathrm{L}} = \boldsymbol{\nabla} \chi \tag{10.30}$$

$$\boldsymbol{F}_{\mathrm{T}} = \boldsymbol{\nabla} \times \mathbf{A} \tag{10.31}$$

Due to the obvious properties

$$\begin{aligned} \boldsymbol{\nabla}\times\boldsymbol{F}_{\mathrm{L}} &= \mathbf{0} \\ \boldsymbol{\nabla}\cdot\boldsymbol{F}_{\mathrm{T}} &= 0 \end{aligned} \tag{10.32}$$

$\boldsymbol{F}_{\mathrm{L}}$ and $\boldsymbol{F}_{\mathrm{T}}$ are respectively called the longitudinal and transverse components of $\boldsymbol{F}$.

10.1 Vector Identities

Let f, $\mathbf{A}$ and $\mathbf{B}$ represent a scalar field and two vector fields defined on a connected subset Ω of $\mathbb{R}^3$, all being differentiable on Ω. Then the following (non-exhaustive) list of identities may be derived using familiar vector calculus :

$$\boldsymbol{\nabla}\cdot(\boldsymbol{\nabla}\times\mathbf{A}) \equiv 0 \tag{10.33}$$

$$\boldsymbol{\nabla}\times(\boldsymbol{\nabla}f) \equiv \mathbf{0} \tag{10.34}$$

$$\begin{aligned} \boldsymbol{\nabla}(\mathbf{A}\cdot\mathbf{B}) = {} & \mathbf{A}(\boldsymbol{\nabla}\cdot\mathbf{B}) + \mathbf{B}(\boldsymbol{\nabla}\cdot\mathbf{A}) + (\mathbf{A}\cdot\boldsymbol{\nabla})\mathbf{B} + \\ & (\mathbf{B}\cdot\boldsymbol{\nabla})\mathbf{A} + \mathbf{A}\times(\boldsymbol{\nabla}\times\mathbf{B}) + \mathbf{A}\times(\boldsymbol{\nabla}\times\mathbf{B}) \end{aligned} \tag{10.35}$$

$$\begin{aligned} \boldsymbol{\nabla}\times(\mathbf{A}\times\mathbf{B}) = {} & -\mathbf{A}(\boldsymbol{\nabla}\cdot\mathbf{B}) + \mathbf{B}(\boldsymbol{\nabla}\cdot\mathbf{A}) - (\mathbf{A}\cdot\boldsymbol{\nabla})\mathbf{B} + \\ & (\mathbf{B}\cdot\boldsymbol{\nabla})\mathbf{A} - \mathbf{A}\times(\boldsymbol{\nabla}\times\mathbf{B}) + \mathbf{A}\times(\boldsymbol{\nabla}\times\mathbf{B}) \end{aligned} \tag{10.36}$$

$$\boldsymbol{\nabla}\times(f\mathbf{A}) = f\,\boldsymbol{\nabla}\times\mathbf{A} + \boldsymbol{\nabla}f\times\mathbf{A} \tag{10.37}$$

$$\boldsymbol{\nabla}\cdot(\mathbf{A}\times\mathbf{B}) = \mathbf{B}\cdot\boldsymbol{\nabla}\times\mathbf{A} - \mathbf{A}\cdot\boldsymbol{\nabla}\times\mathbf{B} \tag{10.38}$$

$$\boldsymbol{\nabla}\cdot(f\mathbf{A}) = f\,\boldsymbol{\nabla}\cdot\mathbf{A} + \boldsymbol{\nabla}f\cdot\mathbf{A} \tag{10.39}$$

$$\boldsymbol{\nabla}\times(\boldsymbol{\nabla}\times\mathbf{A}) = \boldsymbol{\nabla}(\boldsymbol{\nabla}\cdot\mathbf{A}) - \boldsymbol{\nabla}^{\mathbf{2}}\mathbf{A} \tag{10.40}$$

It should be noted that Equation (10.40) should be considered as a definition of the vectorial Laplace operator ("Laplacian"), rather than a vector identity. Clearly, if one expands the left-hand side of Equation (10.40) in Cartesian coordinates, one may straightforwardly obtain

$$[\boldsymbol{\nabla}\times(\boldsymbol{\nabla}\times\mathbf{A})]_x = \frac{\partial}{\partial x}\boldsymbol{\nabla}\cdot\mathbf{A} - \nabla^2 A_x \tag{10.41}$$

etc., which does indeed justify the identification $\boldsymbol{\nabla}^{\mathbf{2}}\mathbf{A} = (\nabla^2 A_x, \nabla^2 A_y, \nabla^2 A_z)$ for Cartesian coordinates, but not for an arbitrary system of curvilinear coordinates.

PART II

Discretization Methods for Sources and Fields

Of the plethora of discretization methods that have been formulated in due time, their derivation is often taken for granted since one addresses the Maxwell equations as 'just-another' set of partial differential equations and next refers to the available toolbox collection that was developed for the solution of the latter. However, this approach ignores the geometric structure of the Maxwell equations and therefore lacks robustness for difficult applications. In Part II we will focus on the discretization aspects of the Maxwell equations that accounts for the geometrical structure of the Maxwell equations of the fields and their sources. There are two mainstream approaches to do so. The first approach is a generalization of the finite-element method and is based on the construction of base functions in the elements of the computational grid. The base functions are specified according to the mesh entity under consideration. The other approach is based on the finite-integration technique. Here the variables are assigned to mesh entities without reference to interpolation where the mesh element is not located. Besides these approaches there are also techniques that start from the integral formulation of the differential equations using Green functions such as the method of moments (MoM) or the boundary element method (BEM). We will not elaborate on these methods. Not because they are not useful, but they are not inherently depending on the use of the underlying gauge principle. Moreover, these techniques are complementary to the methods that are discussed in this book. Finally, we also do not elaborate on the finite-difference time domain. This method method does formulate the Maxwell equations on a computational grid. From this point-of-view it deserves to be discussed in detail. However, the time evolution is computed explicitly. As a consequence, the solving of large linear systems is avoided. The transient method that we discussed does implicit or Eulerian time integration.

As was said earlier the solution of the Maxwell equations can be obtained only for very limited situations that are usually marked by a high level of symmetry of the boundary conditions and structural set up as well as simple material parameters. As soon as one of these conditions is not met it becomes very awkward to still obtain the solution by analytic means. Therefore, over time several computational techniques have been designed to obtain the solutions numerically. In fact, numerous techniques have been developed and can be recognized by there acronyms. We list here the methods that have found wide acceptance.

- Finite Difference Scheme – FD
- Finite Element Method – FEM
- Finite Integration Technique – FIT
- Finite Volume Method – FVM
- Method of Moments – MoM
- Boundary Element Method – BEM
- Partial Element Electrical Circuit method – PEEC
- Fast Multipole Method – FMM
- Finite Difference Time Domain – FDTD

The list of methods can be categorized in differential equation solvers and integral equation solvers. For example the method of moments starts from formulating the Maxwell equations in integral form. Often BEM is consider a synonym for MoM. The method that is the subject of this volume can be seen as a variant of the FIT and FVM completed with a finite-surface method (FSM). We will discuss only a small subset of the method in somewhat more depth except FVM and FSM which are the essential tools of this work.

11

The Finite Difference Method

Finite-Difference (FD) methods are rather straightforward discretizations of differential operators. The method goes back to the original Newton-Leibniz definition of a differentiation as the limit of a division of two near-by function evaluation at two near-by input variables

$$\frac{df}{dx} = \lim_{h \to 0} \frac{f(x+h) - f(x)}{h} \tag{11.1}$$

Therefore if we approximate the real axis by a finite set of points a function will be given by its value in the various points and an approximate derivative can be given as

$$\frac{df}{dx} \simeq \frac{f(x_{i+1}) - f(x_i)}{x_{i+1} - x_i} \tag{11.2}$$

Here we have taken the *forward* finite difference. An equally valid choice is to take the backward difference

$$\frac{df}{dx} \simeq \frac{f(x_{i-1}) - f(x_i)}{x_{i-1} - x_i} \tag{11.3}$$

In order to get some insight over the induced errors in the finite-difference scheme we analyze the discretization process in more detail. For that purpose we define the shift operator E as follows

$$E_{\Delta} f(x) = f(x + \Delta) \tag{11.4}$$

The shift operator maps a function $f(x)$ onto another function $f(x) \rightarrow g(x) = E * f(x)$. From the shift operator we can construct the difference operator,

$$\Delta f(x) = (E - 1) * f(x) = f(x + \Delta) - f(x) \tag{11.5}$$

We may view Δ also as an operator that assign to a function another function. We may repeat the use of the operator and find:

$$\Delta^2 f(x) = \Delta(\Delta f(x)) = \Delta(f(x+\Delta) - f(x)) = \Delta f(x+\Delta) - \Delta f(x) \tag{11.6}$$

Now consider

$$f(x+\Delta) = f(x) + \Delta f(x) = (1+\Delta) f(x) = Ef(x) \tag{11.7}$$

We may repeatedly apply the operator E and find

$$E^n f(x) = (1+\Delta_+)^n f(x) = f(x+n\Delta) \tag{11.8}$$

By applying the binomial theorem we find an expansion of $f(x+n\Delta)$ as

$$\begin{aligned} f(x+n\Delta) &= \sum_{k=0}^{k=n} \binom{n}{k} \Delta^k f(x) \\ &= f(x) + n\Delta f(x) + \frac{n(n-1)}{2!} \Delta^2 f(x) \\ &+ \frac{n(n-1)(n-2)}{3!} \Delta^3 f(x) + ... \end{aligned} \tag{11.9}$$

This equation suggests that there is a close correspondence between the operator Δ_+ and the regular differential operation $df/dx = d_x f$. This is indeed the situation. As is known from a Taylor expansion we may write

$$Ef(x) = f(x+\Delta) = \sum_{k=0}^{\infty} \frac{\Delta^k}{k!} d_x^k f(x). \tag{11.10}$$

In other words:

$$E = \mathrm{e}^{\Delta d_x} \tag{11.11}$$

Using $\Delta = E - 1$ we find

$$\Delta^n = \left(\mathrm{e}^{\Delta d_x} - 1\right)^n \tag{11.12}$$

So

$$\begin{aligned} \Delta f(x) &= \Delta d_x f + \frac{\Delta^2}{2!} d_x^2 f + ... \\ \Delta^2 f(x) &= \Delta^2 d_x^2 f + \Delta^3 d_x^3 f + \frac{7}{12} \Delta^4 d_x^4 f + ... \end{aligned} \tag{11.13}$$

We also may use the connection (11.12) in the opposite direction. From

$$\Delta d_x = ln(1+\Delta) = \sum_{k=0}^{\infty}(-1)^k\frac{1}{k}\Delta^k \tag{11.14}$$

and

$$(\Delta d_x)^n = \left(\Delta - \frac{1}{2}\Delta^2 + -\frac{1}{3}\Delta^3 + ...\right)^n \tag{11.15}$$

we find

$$\begin{aligned} d_x f &= \frac{1}{\Delta}\Delta f - \frac{1}{2}\Delta^2 f + ... \\ d_x^2 f &= \frac{1}{\Delta^2}\Delta_+^2 f - \frac{1}{\Delta^2}\Delta^3 f + \frac{1}{\Delta^2}\frac{11}{12}\Delta^4 f + ... \end{aligned} \tag{11.16}$$

So any order derivative may be approximated by a a series expansion of finite difference operators. It may be applied to any appearance of an operator "d_x". Moreover, it is an exclusive mathematical recipe. This is also the reason that success may be limited. In electromagnetic simulations it is essential that charge is strictly conserved. By blind application of above for mules the solving process may spontaneously create charge from truncation errors and numerical inaccuracy. Therefore, the above for mules have found limited application and we will not pursue the use of these equations any further. Finite difference methods have been abandoned as a starting point for spatial discretization except when a problem with some other method needs to be analyzed in-depth and there is serious suspicious that there is a flaw in the discretization method itself. The equations presented above *do* play an important role in putting numerical methods on solid footing. The finite-difference scheme is very valuable if parts of the problem characteristics can be addressed by analytic means. In this way some critical conservation laws are often implicitly addressed due to exploitation of symmetry and the remaining parts of the problem may be solved by straight-forward schemes as presented above.

12

The Finite Element Method

The finite-element method is based on the following key idea: Since the (Maxwell) equations are formulated for piece-wise continuous functions over space and time, the ultimate goal is to find a valuable approximation of these continuous functions. In other words: the aim is to replace the error-free or exact continuous solution of the equations by an approximate piece-wise continuous functions. The most essential concept in the finite-element method is the element. The element corresponds to a specific function defined on a restricted interval and the approximation of the piece-wise continuous function is obtained by constructing a weighted sum of all element contributions. In order to make this statement more transparent we will describe the finite-element method using a one-dimensional example. The extension to two and three dimensions is rather straightforward but in order to applied the finite-element method successfully for electromagnetic problems some serious changes must be adapted.

12.1 Trial Solutions

Consider a function of one variable $U(x)$. The function satisfies some one-dimensional differential equation

$$\frac{d}{dx}\left(a(x)\frac{dU(x)}{dx}\right) = b(x). \tag{12.1}$$

We assume that the functions $a(x)$ and $b(x)$ are known expressions. For example $a(x) = 2x$ and $b(x) = \frac{1}{x}$. Furthermore we must select boundary conditions if we want to determine the function $U(x)$ as the solution for Equation (12.1). Since we are dealing with a second-order differential equation we must provide two boundary conditions. Here we assume that for $x = 1$ we have $U(1) = 5$ and for $x = 10$

we have $U(10) = 0$. We can now proceed by constructing a trial solution $\tilde{U}(x)$ as a polynomial of x, e.g.

$$\tilde{U}(x) = w_0 + w_1 x + w_2 x + w_3 x + ... + w_n x^n \tag{12.2}$$

and determine all parameters w_k from the boundary conditions and inserting the trial solution into the Equation (12.1). If we select a limited set of weights w_k then in general a substitution of the trial function into the equation will result into an approximation. Equation (12.1) can be written as

$$\frac{d}{dx}\left(a(x)\frac{dU(x)}{dx}\right) - b(x) = 0. \tag{12.3}$$

and when using $\tilde{U}(x)$ we obtain

$$\frac{d}{dx}\left(a(x)\frac{d\tilde{U}(x)}{dx}\right) - b(x) = R(x; w_k). \tag{12.4}$$

$R(x; w_k)$ is the residual of the equation. Now a good trial solution would of course result into a small residual. The question arises of what is a small residue. A possible identification of smallness can be that for each entry in the trial function we have

$$\begin{aligned}
\int_1^2 dx R(x, w_k) w_0 &= 0 \\
\int_1^2 dx R(x, w_k) w_1 x &= 0 \\
\int_1^2 dx R(x, w_k) w_2 x^2 &= 0 \\
&\cdots \\
\int_1^2 dx R(x, w_k) w_n x^n &= 0
\end{aligned} \tag{12.5}$$

This method is known as the Galerkin method. There exists different criteria for identifying suitable conditions for constructing a small residue but the Galerkin method turned out to be very generally applicable.

12.2 The Element Concept

So far we considered a single polynomial that should result into a good approximation over the entire interval. For one-dimensional problems this may be a good choice but as said in the introduction, we may be confronted with functions that possess discontinuities or jumps. For example the potential field will show a jump in its value when crossing a dipole layer. Therefore it is interesting to consider a collection of trial functions where each trial function is valid for a restricted interval. It implies that the solution domain must be divided in fragments. Each fragment is called an element and the collection of all elements is a mesh. Mesh creation is an important aspect in finding good approximate solutions. It it evident that if a function is very smooth over a certain region then a few element suffice to capture its values. However, if the function has much variation over a region it is required to use many elements in order to a get a faithful approximation. The algorithms that construct the meshes are known as mesh generators. For the one-dimensional case the elements are simple line segments. For two-dimensional problems the elements are conveniently chosen to be triangles and for three-dimensional problems, the elements are chosen as tetrahedrons. The motivation for these choices is that the trial functions are constructed on the mesh elements as piece-wise linear and each trial function is equal to one at one element corner and zero at the remaining corner. In Figure 12.1 we show some function over an interval and split the domain into segments (elements) Over each element we now use two base function $u_1(x) = \frac{x-x_1}{x_2-x_1}$ and $u_2 = \frac{x-x_2}{x_2-x_1}$. The functions are shown in Figure 12.2. We can now consider an approximation of the function over the full domain as a linear combination of all all base functions over all elements.

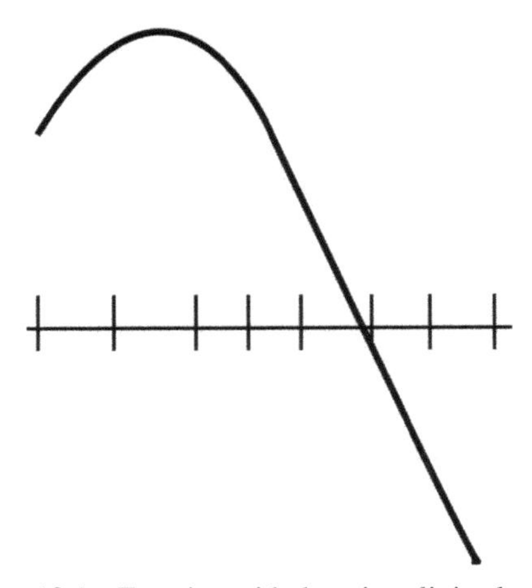

Figure 12.1 Function with domain split in elements.

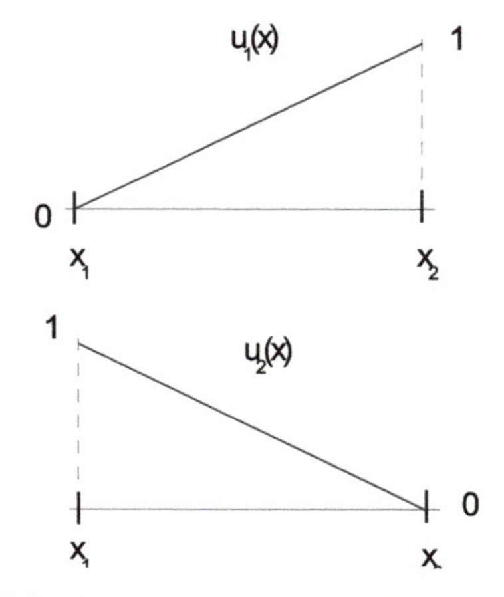

Figure 12.2 Base functions for one element of the domain.

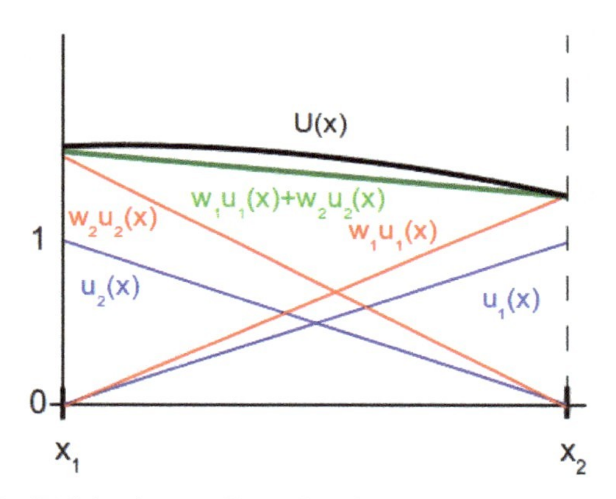

Figure 12.3 Weighted sum of base functions for one element of the domain.

$$\tilde{f}(x) = \sum_{elements\ e} (w_1^e u_1^e + w_2^e u_2^e) \tag{12.6}$$

The weighted sum is shown in Figure 12.3 In order to find the weights w_1^e we next apply the Galerkin method for each element. By insertion of the trial function $\tilde{f}(x)$ into the differential equation and obtain the Galerkin residual

$$R(x; w) = \frac{d}{dx}\left(a(x) \sum_{e;k=1,2} w_k^e \frac{du^k}{dx} - b(x)\right) \tag{12.7}$$

This residual should now be minimized in the Galerkin meaning. That means that if we select for each element a test function $t(x)$ such that

$$\int_{x_i}^{x_{i+1}} dx\ t(x)R(x;w) = 0 \tag{12.8}$$

It is common to choose the test functions equal to the base functions. This results into

$$\int_{x_i}^{x_{i+1}} dx\ u_l^{e'}(x) \left\{ \frac{d}{dx}\left[a(x)\left(\sum_{e;k=1,2} w_k^e \frac{du_k^e}{dx}(x)\right)\right] - b(x)\right\} = 0 \tag{12.9}$$

For the first contribution we perform an integration by parts such that we obtain

$$-\int_{x_i}^{x_{i+1}} dx\ \frac{du_l^{e'}}{dx}(x)\left[a(x)\sum_{e;k=1,2} w_k^e \frac{du_k^e}{dx}(x)\right] - \int_{x_i}^{x_{i+1}} dx\ u_l^{e'}(x)b(x)$$
$$+\left[u_l^{e'}(x)a(x)\sum_{e;k=1,2} w_k^e \frac{du_k^e}{dx}(x)\right]_{x_i}^{x_{i+1}} = 0 \tag{12.10}$$

Now the equation is empty if the element e' is not equal to the element e: each test function provides an equation for one element only. Each term in Equation (12.10) will be discussed separately. Let us start with the first term. It can be rewritten as:

$$-\int_{x_i}^{x_{i+1}} dx\ \frac{du_l^e}{dx}(x)\left[a(x)\sum_{e;k=1,2} w_k^e \frac{du_k^e}{dx}(x)\right] = -a(x)\sum_{e;k=1,2} K_{lk}^e w_k^e \tag{12.11}$$

and

$$K_{lk}^e = \int_{x_i}^{x_{i+1}} dx\ \frac{du_l^e}{dx}(x)\left[a(x)\frac{du_k^e}{dx}(x)\right] \tag{12.12}$$

The matrix K is called the *stiffness* matrix. The second term can be written as

$$-\int_{x_i}^{x_{i+1}} dx\ u_l^e(x)b(x) = F_l^e \tag{12.13}$$

This term is independent of the weights w_k^e. The vector F is called the *load* vector. Finally the boundary terms can be evaluated by using the fact that $u_1^e(x_i) = 0$ and $u_1^e(x_{i+1}) = 1$ and furthermore that $u_2^e(x_i) = 1$ and $u_2^e(x_{i+1}) = 0$. Therefore

$$\left[u_l^e(x) a(x) \sum_{e;k=1,2} w_k^e \frac{du_k^e}{dx}(x) \right]_{x_i}^{x_{i+1}} = \begin{cases} \left(a(x) \frac{d\tilde{U}(x)}{dx} \right)_{x_{i+1}} & l = 1 \\ -\left(a(x) \frac{\tilde{U}(x)}{dx} \right)_{xi} & l = 2 \end{cases} \tag{12.14}$$

We can summarize the two equations for the weights w_1 and w_2 as follows:

$$\begin{bmatrix} K_{11}^e & K_{12}^e \\ K_{21}^e & K_{22}^e \end{bmatrix} * \begin{bmatrix} w_1^e \\ w_2^e \end{bmatrix} + \begin{bmatrix} F_1^e \\ F_2^e \end{bmatrix} = \begin{bmatrix} \left(a(x) \frac{\tilde{U}(x)}{dx} \right)_{x_{i+1}} \\ -\left(a(x) \frac{\tilde{U}(x)}{dx} \right)_{x_i} \end{bmatrix} \tag{12.15}$$

We have arrived at a linear system of equations for the two weight factors where the right-hand side is determined by the boundary conditions at the end points of the interval.

The next step in the development of the finite element method is to consider two sequential elements e_1 and e_2. Following the same reasoning as above we will have now four weight factors and the corresponding linear system of equations will take the following form:

$$\begin{bmatrix} K_{11}^{e_1} & K_{12}^{e_1} & 0 & 0 \\ K_{21}^{e_1} & K_{22}^{e_1} & 0 & 0 \\ 0 & 0 & K_{21}^{e_2} & K_{22}^{e_2} \\ 0 & 0 & K_{21}^{e_2} & K_{22}^{e_2} \end{bmatrix} * \begin{bmatrix} w_1^{e_1} \\ w_2^{e_1} \\ w_1^{e_2} \\ w_2^{e_2} \end{bmatrix} + \begin{bmatrix} F_1^{e_1} \\ F_2^{e_1} \\ F_1^{e_2} \\ F_2^{e_2} \end{bmatrix}$$

$$= \begin{bmatrix} \left(a(x) \frac{\tilde{U}(x)}{dx} \right)_{x_{i+1}} \\ -\left(a(x) \frac{\tilde{U}(x)}{dx} \right)_{x_i} \\ \left(a(x) \frac{\tilde{U}(x)}{dx} \right)_{x_{i+2}} \\ -\left(a(x) \frac{\tilde{U}(x)}{dx} \right)_{x_{i+1}} \end{bmatrix} \tag{12.16}$$

From the continuity of the function $U(x)$ at the point x_{i+1} we obtain that $w_1^{e1} = w_2^{e2}$. Therefore, using this information we can add the first and last row of the system of equations and obtain:

$$\begin{bmatrix} K_{11}^{e_1} + K_{22}^{e_2} & K_{12}^{e_1} & 0 \\ K_{21}^{e_1} & K_{22}^{e_1} & 0 \\ K_{22}^{e_2} & 0 & K_{21}^{e_2} \end{bmatrix} * \begin{bmatrix} w_1^{e_1} \\ w_2^{e_1} \\ w_1^{e_2} \end{bmatrix} + \begin{bmatrix} F_1^{e_1} + F_2^{e_2} \\ F_2^{e_1} \\ F_1^{e_2} \end{bmatrix}$$

$$= \begin{bmatrix} 0 \\ -\left(a(x)\frac{\tilde{U}(x)}{dx}\right)_{x_i} \\ \left(a(x)\frac{\tilde{U}(x)}{dx}\right)_{x_{i+2}} \end{bmatrix} \tag{12.17}$$

From this calculation we can deduce some important observations. First of all the internal "boundary conditions" cancel at the point x_{i+1}. This is not a coincidence but a very general property. The boundary conditions are at the edge of the full simulation domain. Another observation is that the stiffness matrix and load vector entries can be assembled element after element and that there contributions are inserted additivily.

Note that with the linear trial functions we have two weight factors in each element but the weight factors are not independent since the resulting function is supposed to be continuous. To make this point more explicit we may work with the function values at the mesh points directly instead of the weight factors, The collection of the unknowns is $\mathbf{U} = \{U(x_1), U)(x_2), \ldots, U(x_n)\}$. We can redo above assembling for this set of unknowns as above. Again we will end up with a stiffness matrix, a load vector and a right-hand side that contains the boundary conditions. The method can be also extended to higher dimensions. In two dimensions it is common to deal with triangular-shaped elements and the base functions are chosen to be linear over the element. Moreover, the base functions are zero in two nodes of the triangle and equal to the unknown function value at the third node. Since a triangle has three corners there are three base functions in each element in two dimensions. For three dimensions the elements have the shape of a tetrahedron, which has four corners and therefore there are four linear base functions (Burnett [1987]).

The finite-element method is very useful for the simulation of scalar-valued fields. It the fields are vector valued as is the case for electromagnetic fields it is important to realize that the base function respect the geometric character of the field. A finite-element approach that just does that exploits the so-called Whitney elements: first the geometrical entity is identified e.g. node, edge, surface, volume and next a specific base function is constructed. We will not consider the finite-element method further here. The key property is that the approximate solution is defined in every space point of the simulation domain via the element by element provided base functions. Applications of the finite-element method in electromagnetic computing can be found (Cendes [1991] Lee et al. [1991]).

13

The Finite Volume Method and Finite Surface Method

Contrary to the finite-element method, the guiding principle of the finite-volume method, is not to arrive at piece-wise continuous approximations of the exact solutions but one is satisfied with a discretized representation of the exact solution at some discrete set of space-time points.

In order to discuss the numerical consequences of the fact that the vector potential needs to be assigned to the links of the grid we consider the following example. We will solve the steady-state equation

$$
\begin{aligned}
\boldsymbol{\nabla}\times\mathbf{B} &= \mu_0\mathbf{J}, \\
\mathbf{B} &= \boldsymbol{\nabla}\times\mathbf{A}, \\
\mathbf{J} &= \sigma\mathbf{E}, \\
\mathbf{E} &= -\boldsymbol{\nabla}V,
\end{aligned}
\tag{13.1}
$$

by discretizing the set of equations on a regular Cartesian grid having N nodes in each direction. The total number of nodes in D dimensions is $M_{\text{nodes}} = N^D$. To each node we may associate D links along the positive directions, and therefore the grid has approximately DN^D links. There are $2D$ sides with each a number of $N^{(D-1)}$ nodes. Half the fraction of side nodes will not contribute a link in the positive direction. Therefore, the precise number of links in the lattice is $M_{\text{links}} = DN^D(1 - \frac{1}{N})$.

As far as the description of the electromagnetic field is concerned, the counting of unknowns for the full lattice results into M_{links} variables (A_{ij}) for the links, and M_{nodes} variables (V_i) for the nodes. Since each link (node) gives rise to one equation, the naive counting is consistent. However, we have not yet implemented the gauge condition. The conventional Coulomb gauge $\boldsymbol{\nabla}\cdot\mathbf{A} = 0$, constraints the link degrees of freedom and therefore not all link fields are independent. There are $3N^3(1 - \frac{1}{N})$ link variables and $3N^3(1 - \frac{1}{N}) + N^3$ equations, including the constraints. As a consequence, at first sight it seems that we are confronted with an overdetermined system of equations, since each node provides an extra equation for $\mathbf{A}$. However, the translation of the Maxwell-Ampère equation on the lattice leads to a singular matrix, i.e. not all rows are independent. The rank of the corresponding matrix is $3N^3(1 - \frac{1}{N})$, whereas there are $3N^3(1 - \frac{1}{N}) + N^3$ rows and $3N^3(1-\frac{1}{N})$ columns. Such a situation is highly inconvenient for solving non-linear systems of equations, where the non-linearity stems from the source terms being explicitly dependent on the fields. The application of the Newton-Raphson method requires that the matrices in the related Newton equation be non-singular and square. In fact, the non-singular and square form of the Newton-Raphson matrix can be recovered by introducing the more general gauge $\boldsymbol{\nabla} \cdot \mathbf{A} + \nabla^2 \chi = 0$, where an additional field χ, i.e. one unknown per node, is introduced. This is illustrated in Figure 13.1. In this way the number of unknowns and the number of equations match again. In the continuum limit ($N \to \infty$), the field χ and one component of $\mathbf{A}$ can be eliminated. Though being irrelevant for theoretical understanding, the auxiliary field χ is essential for obtaining numerical stability on a discrete, finite lattice. In other words, our specific gauge solely serves as a tool to obtain a discretization scheme that generates a regular Newton-Raphson matrix, as explained in (Meuris et al. [2001a]).

It should be emphasized that the inclusion of the gauge-fixing field χ should not lead to unphysical currents. As a consequence, the χ-field should be a solution of $\boldsymbol{\nabla}\chi = 0$. To summarize, instead of solving the problem

$$\begin{aligned} \boldsymbol{\nabla}\times \boldsymbol{\nabla}\times \mathbf{A} &= \mu_0 \, \mathbf{J}(\mathbf{A}), \\ \boldsymbol{\nabla} \cdot \mathbf{A} &= 0, \end{aligned} \tag{13.2}$$

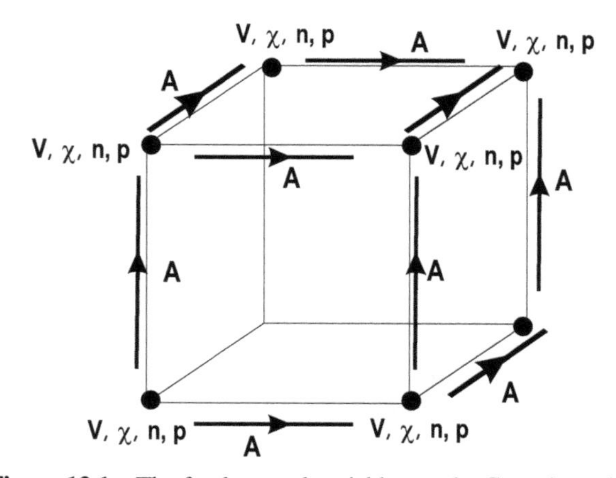

Figure 13.1 The fundamental variables on the Cartesian grid.

we solve the equivalent system of equations

$$\begin{aligned}\boldsymbol{\nabla}\times\boldsymbol{\nabla}\times\mathbf{A}-\gamma\boldsymbol{\nabla}\chi &= \mu_0\,\mathbf{J}(\mathbf{A}),\\ \boldsymbol{\nabla}\cdot\mathbf{A}+\nabla^2\chi &= 0.\end{aligned} \tag{13.3}$$

The equivalence of both sets of Equations (13.2) and (13.3) can be demonstrated by considering the action integral

$$S=-\frac{1}{2\mu_0}\int \mathrm{d}\tau\ |\boldsymbol{\nabla}\times\mathbf{A}|^2+\int \mathrm{d}\tau\ \mathbf{J}\cdot\mathbf{A}. \tag{13.4}$$

Functional differentiation with respect to $\mathbf{A}$ yields the field equations

$$\frac{\delta S}{\delta\mathbf{A}}=-\frac{1}{\mu_0}\boldsymbol{\nabla}\times\boldsymbol{\nabla}\times\mathbf{A}+\mathbf{J}=0. \tag{13.5}$$

The constraint corresponding to the Coulomb gauge can be taken into account by adding a Lagrange multiplier term to the action integral

$$S=-\frac{1}{2\mu_0}\int \mathrm{d}\tau\ |\boldsymbol{\nabla}\times\mathbf{A}|^2+\int \mathrm{d}\tau\ \mathbf{J}\cdot\mathbf{A}+\gamma\int \mathrm{d}\tau\ \chi\,\boldsymbol{\nabla}\cdot\mathbf{A} \tag{13.6}$$

and perform the functional differentiation with respect to χ

$$\frac{\delta S}{\delta\chi}=\boldsymbol{\nabla}\cdot\mathbf{A}=0. \tag{13.7}$$

Finally, the Lagrange multiplier field χ becomes a dynamical variable by adding a free-field part to the action integral

$$S=-\frac{1}{2\mu_0}\int \mathrm{d}\tau\,|\boldsymbol{\nabla}\times\mathbf{A}|^2+\int \mathrm{d}\tau\ \mathbf{J}\cdot\mathbf{A}+\gamma\int \mathrm{d}\tau\ \chi\boldsymbol{\nabla}\cdot\mathbf{A}-\frac{1}{2}\gamma\int \mathrm{d}\tau\ |\boldsymbol{\nabla}\chi|^2 \tag{13.8}$$

and functional differentiation with respect to $\mathbf{A}$ and χ results into the new system of equations. Physical equivalence is guaranteed provided that $\boldsymbol{\nabla}\chi$ does not lead to an additional current source. Therefore, it is required that $\boldsymbol{\nabla}\chi = 0$. In fact, acting with the divergence operator on the first equation of (13.3) gives Laplace's equation for χ. The solution of the Laplace equation is identically zero if the solution vanishes at the boundary.

We achieved to implement the gauge condition resulting into a unique solution and simultaneously to arrive at a system containing the same number of equations and unknowns. Hence a square Newton-Raphson matrix is guaranteed while solving the full set of non-linear equations.

13.1 Differential Operators in Cartesian Grids

Integrated over a test volume ΔV_i surrounding a node i, the divergence operator, acting on vector potential $\mathbf{A}$, can be discretized as a combination of 6 neighboring links

$$\int_{\Delta V_i} \nabla \cdot \mathbf{A} \, \mathrm{d}\tau = \int_{\partial(\Delta V_i)} \mathbf{A} \cdot \mathbf{dS} \sim \sum_{k}^{6} S_{ik} A_{ik} \tag{13.9}$$

The symbol $\sim$ represents the conversion to the grid formulation and $\partial(\Delta V_i)$ denotes the boundary of ΔV_i.

Similarly, the gradient operator acting on the ghost field χ or any scalar field V, can be discretized for a link ij using the nodes i and j. Integration over a surface S_{ij} perpendicular to the link ij gives

$$\int_{\Delta S_{ij}} \nabla \chi \cdot \mathbf{dS} \sim \frac{\chi_j - \chi_i}{h_{ij}} S_{ij} \tag{13.10}$$

where h_{ij} denotes the length of the link between the nodes i and j.

The gradient operator for a link ij, integrated along the link ij, is given by

$$\int_{\Delta L_{ij}} \nabla \chi \cdot \mathbf{dr} \sim \chi_j - \chi_i \tag{13.11}$$

The *curl-curl* operator can be discretized for a link ij using a combination of 12 neighboring links and the link ij itself. As indicated in Figure 13.2, the field $\mathbf{B}_i$ in the center of the "wing" i, can be constructed by taking the circulation of the vector potential $\mathbf{A}$ around the wing i $(i = 1, 4)$

$$\mathbf{B}_i S_i = \sum_{j=1}^{3} \mathbf{A}_{ij} h_{ij} + \mathbf{A}_0 h_0 \tag{13.12}$$

where h_α is the length of the corresponding link α. Integration over a surface S_{ij} perpendicular to the link ij yields a linear combination of different $\mathbf{A}_{ij}$'s, the coefficients of which are denoted by Λ_{ij}.

$$\begin{aligned} \int_{\Delta S_{ij}} \nabla \times \nabla \times \mathbf{A} \cdot \mathbf{dS} &= \int_{\partial(\Delta S_{ij})} \nabla \times \mathbf{A} \cdot \mathbf{dr} \\ &= \int_{\partial(\Delta S_{ij})} \mathbf{B} \cdot \mathbf{dr} \\ &\sim \Lambda_{ij} \, A_{ij} + \sum_{kl}^{12} \Lambda_{ij}^{kl} \, A_{kl} \end{aligned} \tag{13.13}$$

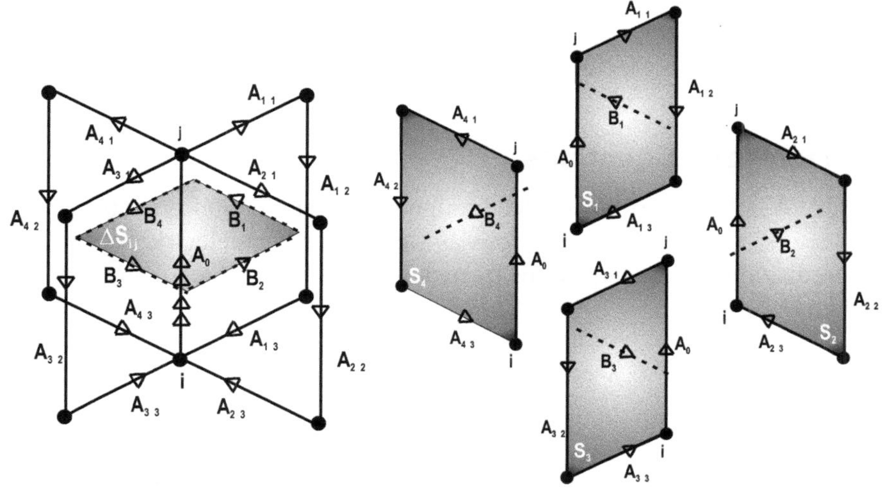

Figure 13.2 The assembly of the $\nabla\times\nabla\times$-operator using 12 contributions of neighboring links.

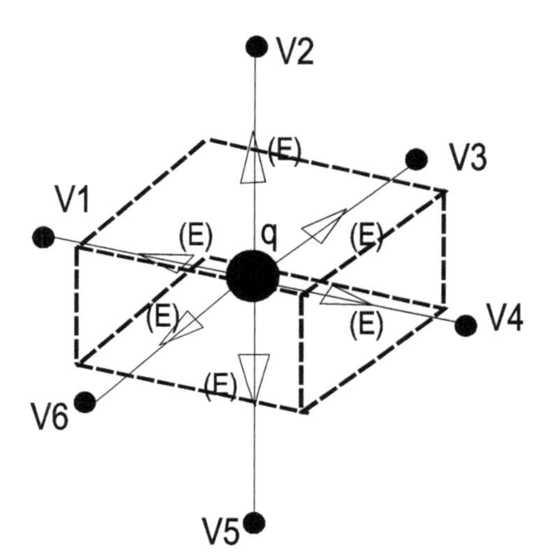

Figure 13.3 The assembly of the $\nabla\cdot\nabla$-operator using 6 contributions of neighboring nodes.

The div-grad (Laplacian) operator can be discretized (see Figure 13.3) being integrated over a test volume ΔV_i surrounding a node i as a combination of 6 neighboring nodes and the node i itself.

$$\int_{\Delta V_i} \boldsymbol{\nabla} \cdot (\boldsymbol{\nabla}\chi)\, \mathrm{d}\tau = \int_{\partial(\Delta V_i)} \boldsymbol{\nabla}\chi \cdot \mathbf{dS} \sim \sum_k^6 S_{ik} \frac{\chi_k - \chi_i}{h_{ik}} \tag{13.14}$$

13.2 Discretized Equations

The fields $(\mathbf{A}, \chi)$ need to be solved throughout the simulation domain, i.e. for conductors, semiconducting regions as well as for the dielectric regions. The discretization of these equations by means of the box/surface-integration method gives

$$\int_{\Delta S} (\boldsymbol{\nabla}\times \boldsymbol{\nabla}\times \mathbf{A} - \gamma\boldsymbol{\nabla}\chi - \mu_0\mathbf{J}) \cdot \mathbf{dS} = 0 \tag{13.15}$$

$$\int_{\Delta V} \boldsymbol{\nabla} \cdot \mathbf{J}\, \mathrm{d}\tau = 0 \tag{13.16}$$

$$\int_{\Delta V} (\boldsymbol{\nabla} \cdot \mathbf{A} + \nabla^2\chi)\, \mathrm{d}\tau = 0 \tag{13.17}$$

leading for the independent variables $\mathbf{A}, \chi$ to

$$\Lambda_{ij}\, A_{ij} + \sum_{kl}^{12} \Lambda_{ij}^{kl}\, A_{kl} - \mu_0\, S_{ij} J_{ij} - \gamma S_{ij} \frac{\chi_j - \chi_i}{h_{ij}} = 0 \tag{13.18}$$

$$\sum_k^6 S_{ik} J_{ik} = 0 \tag{13.19}$$

$$\sum_k^6 S_{ik} \left(A_{ik} + \frac{\chi_k - \chi_i}{h_{ik}} \right) = 0 \tag{13.20}$$

Depending on the region under consideration, the source terms $(Q_i, \mathbf{J}_{ij})$ differ. In a conductor we implement Ohm's law, $\mathbf{J} = \sigma\mathbf{E}$ on a link ij:

$$J_{ij} = -\sigma_{ij} \left(\frac{V_j - V_i}{h_{ij}} \right) \tag{13.21}$$

and Q_i is determined by charge conservation.

13.3 The No-Ghost Approach

In the foregoing section we have introduced a ghost field that needs to be added in order to avoid singular and/or non-square systems of equations. On the other hand, when arriving at the solution we find that this field is identical zero. Therefore, the question arises if we can avoid the calculation of the ghost field at all. This is indeed the case as is seen below. Let us return to the Maxwell's equations:

$$\begin{aligned} \nabla.\mathbf{D} &= \rho, \\ \mathbf{E} &= -\nabla V - \frac{\partial \mathrm{A}}{\partial t}, \\ \mathbf{B} &= \nabla \times \mathbf{A}, \\ \nabla \times \mathbf{H} &= \mathbf{J}_c + \frac{\partial \mathbf{D}}{\partial t}. \end{aligned} \tag{13.22}$$

We need to do gauge fixing in order to have to unique scalar and vector potentials. and use the gauge condition:

$$\nabla.\mathbf{A} + f = 0 \tag{13.23}$$

We replace the fields $\mathbf{E} = -\nabla V - \partial_t \mathbf{A}$ and $\mathbf{B} = \nabla \times \mathbf{A}$ in the Maxwell-Ampere system and add the gauge condition (13.23) as an additional equation. The modified Maxwell's equations are;

$$\begin{aligned} \nabla.\mathbf{D} - \rho &= 0 \quad \text{Gauss's law} \\ \nabla \times \mathbf{H} - \mathbf{J}_c - \frac{\partial \mathbf{D}}{\partial t} &= 0 \quad \text{Maxwell-Ampere} \\ \nabla.\mathbf{A} + f &= 0. \quad \text{Gauge condition} \end{aligned} \tag{13.24}$$

In terms of V and $\mathbf{A}$ we obtain:

$$\mathbf{H} = \frac{1}{\mu}\mathbf{B} = \frac{1}{\mu}\nabla \times \mathbf{A} = \mathbf{H}(\mathbf{A}), \tag{13.25}$$

$$\mathbf{D} = \epsilon \mathbf{E} = -\epsilon\left(\nabla V + \frac{\partial \mathbf{A}}{\partial t}\right) = \mathbf{D}(\mathbf{A}, V). \tag{13.26}$$

We can add the gauge condition to the Maxwell-Ampere equation since it is zero anyway. The modified Maxwell-Ampere equation with some weight factor γ is:

$$\nabla \times \mathbf{H}(\mathbf{A}) - \mathbf{J}_c(\mathbf{A}, V) - \mathbf{J}_d(\mathbf{A}, V) + \gamma \nabla (\nabla.\mathbf{A} + f) = 0, \tag{13.27}$$

where $\mathbf{J}_d(\mathbf{A}, V) = \frac{\partial}{\partial t}\mathbf{D}(\mathbf{A}, V)$.

We can get the divergence of Equation (13.27), so that we can simplify the problem to;

$$\begin{aligned} \nabla.\left[\nabla \times \mathbf{H}(\mathbf{A}) - \mathbf{J}_c(\mathbf{A}, V) - \frac{\partial}{\partial t}\mathbf{D}(\mathbf{A}, V) + \gamma\nabla(\nabla.\mathbf{A} + f)\right] &= 0, \\ \nabla.\nabla \times \mathbf{H}(\mathbf{A}) - \nabla.(\mathbf{J}_c(\mathbf{A}, V) + \mathbf{J}_d(\mathbf{A}, V)) + \gamma\nabla^2(\nabla.\mathbf{A} + f) &= 0, \\ -\nabla.(\mathbf{J}_c(\mathbf{A}, V) + \mathbf{J}_d(\mathbf{A}, V)) + \gamma\nabla^2(\nabla.\mathbf{A} + f) &= 0, \end{aligned}$$

since $\nabla.(\nabla \times \mathbf{H}(\mathbf{A})) = 0$. We observe that *if* we solve the current continuity

$$\nabla.(\mathbf{J}_c(\mathbf{A}, V) + \mathbf{J}_d(\mathbf{A}, V)) = 0$$

then,

$$\nabla^2(\nabla.\mathbf{A} + f) = 0.$$

This is the reason, why we need to solve the current-continuity equation. It is clear if we solve the above current continuity equation we do not need to solve the Gauss's law equation in order to eliminate the equation redundancy. It is demanded to make sure that the solutions of the modified Maxwell-Ampere equation, will coincide with the usual Maxwell-Ampere equation. However, although solving the current continuity equation is demanded, it is not sufficient. We want to make sure that that gauge condition is recovered and not only the Laplacian acting on the gauge condition. There is a nice mathematical theorem that claims the following: if $\nabla^2\chi = 0$ and $\chi|_{\partial\Omega} = 0$, where $\partial\Omega$ is the edge of some domain Ω and $\chi = 0$ everywhere in Ω. Thus the second requirement is that of the gauge condition which must be set equal zero at the edge of the simulation domain. The constant γ and the function f where left undetermined. A convenient choice is to take

$$\begin{aligned} \gamma &= -\frac{1}{\mu_0} \\ f &= \xi\mu_0\epsilon\frac{\partial V}{\partial t} \end{aligned}$$

In order to solve the AV-formulation we just need to solve the equations below

$$\begin{aligned} \nabla \times \mathbf{H}(\mathbf{A}) - \mathbf{J}_c(\mathbf{A}, V) - \mathbf{J}_d(\mathbf{A}, V) + \gamma\nabla(\nabla.\mathbf{A} + f) &= 0 \\ \nabla.(\mathbf{J}_c(\mathbf{A}, V) + \mathbf{J}_d(\mathbf{A}, V)) &= 0 \end{aligned} \qquad (13.28)$$

and impose $\nabla.\mathbf{A}+f=0$ on the edges of the domain. If we restrict ourselves to metals and insulators:

$$\mathbf{J}_c(\mathbf{A},V)=\begin{cases}\sigma\mathbf{E}(\mathbf{A},V), & \text{for metals}\\ 0, & \text{for insulators}.\end{cases} \tag{13.29}$$

Thus for the case of the insulators, we need to have a special treatment, we identify a material as an insulator as having the property that the conductance $\sigma=0$, as a consequence, $\mathbf{J}=0$ and the continuity equation implies that $\frac{\partial\rho}{\partial t}=0$. Thus ρ in independent of time then, applying the small signal analysis $\rho=\rho_0+\rho_1 e^{\mathrm{i}\omega t}=\rho_0$ and $\rho_1=0$. From Gauss' law:

$$\nabla.\mathbf{D}(\mathbf{A},V)=\rho, \tag{13.30}$$

we obtain

$$\nabla.\mathbf{D}_0(\mathbf{A},V)=\rho_0$$
$$\nabla.\mathbf{D}_1(\mathbf{A},V)=0$$

Since the setup of the small signal analysis is rooted in expanding every term as:

$$\mathbf{X}(\mathbf{r},t)=\mathbf{X}(\mathbf{r},0)+\mathbf{X}(\mathbf{r})\,e^{\mathrm{i}\omega t},$$

where $\mathbf{X}(\mathbf{r},t)$ is any field and let $\mathbf{X}_1=\mathbf{X}(\mathbf{r})\,e^{\mathrm{i}\omega t}$ is the harmonic state and is a complex -valued function, and $\mathbf{X}_0=\mathbf{X}(\mathbf{r},0)$ is the static state. Thus $\mathbf{E}(\mathbf{A},V)=-\nabla V_1-\mathrm{i}\omega\mathbf{A}_1$. For metals we obtain $\mathbf{J}_c(\mathbf{A},V)=-\sigma(\nabla V_1+\mathrm{i}\omega\mathbf{A}_1)$ and the displacement current

$$\mathbf{J}_d(\mathbf{A},V)=\frac{\partial}{\partial t}\epsilon\mathbf{E}(\mathbf{A},V) \tag{13.31}$$
$$=\epsilon\frac{\partial}{\partial t}\mathbf{E}(\mathbf{A},V),\ \text{Assume }\epsilon\text{ is a constant.} \tag{13.32}$$
$$=\epsilon\mathrm{i}\omega\left[-\nabla V_1-\mathrm{i}\omega\mathbf{A}_1\right], \tag{13.33}$$
$$=-\mathrm{i}\omega\epsilon\left(\nabla V_1+\mathrm{i}\omega\mathbf{A}_1\right). \tag{13.34}$$

Furthermore, $\mathbf{H}_1(\mathbf{A})=\frac{1}{\mu}\nabla\times\mathbf{A}_1$ and $f=\xi\mu_0\epsilon\mathrm{i}\omega V_1$. If we make the above substitution in the Maxwell-Ampere equation and let $\mu=\mu_0\mu_r$ we obtain,

$$\nabla\times\frac{1}{\mu_r}(\nabla\times\mathbf{A}_1)+\mu_0(\sigma+\mathrm{i}\omega\epsilon)(\nabla V_1+\mathrm{i}\omega\mathbf{A}_1)$$
$$-\nabla(\nabla.\mathbf{A}_1+\xi\mu_0\epsilon\mathrm{i}\omega V_1)=0. \tag{13.35}$$

The Poisson/current-continuity equation becomes

$$\begin{aligned} -\nabla.\left[(\sigma + \mathrm{i}\omega\epsilon)\left(\nabla V_1 + \mathrm{i}\omega \mathbf{A}_1\right)\right] &= 0, \quad \text{for case of metals,} \\ -\nabla.\left[\epsilon\left(\nabla V_1 + \mathrm{i}\omega \mathbf{A}_1\right)\right] &= 0, \quad \text{for case of insulators.} \end{aligned} \tag{13.36}$$

with $\nabla.\mathbf{A}_1 + \xi\mu_0\epsilon\mathrm{i}\omega V_1 = 0$ at the edge of the domain. If we consider the Poisson equation for the insulator case;

$$\begin{aligned} \nabla.\left[\,(\nabla V_1 + \mathrm{i}\omega \mathbf{A}_1)\right] &= 0, \quad \text{in regions of constant } \epsilon \text{ or:} \\ \nabla.\nabla V_1 + \mathrm{i}\omega\nabla.\mathbf{A}_1 &= 0, \text{ then} \\ \nabla^2 V_1 + \mathrm{i}\omega\left(-\xi\mu_0\epsilon\mathrm{i}\omega V_1\right) &= 0, \quad \text{by using the gauge condition} \\ & \qquad \nabla.\mathbf{A}_1 = -\xi\mu_0\epsilon\mathrm{i}\omega V_1. \end{aligned}$$

Therefore,

$$\nabla^2 V_1 + \xi\mu_0\epsilon\omega^2 V_1 = 0, \tag{13.37}$$

It should be emphasized that we did rewrite Gauss's law using the gauge condition. Therefore, (13.37) is still representing Gauss' law.

Equations (13.35), (13.36) or Equation (13.37) for the case of constant σ and ϵ is what we call the AV-formulation.

For the semiconductor environment we follow the Scharfetter-Gummel scheme (Scharfetter and Gummel [1969]). In this approach, the diffusion equations

$$\mathbf{J} = q\mu c\mathbf{E} \pm kT\mu\boldsymbol{\nabla} c \tag{13.38}$$

where the plus (minus) sign refers to negatively (positively) charged particles and c denotes the corresponding carrier density. It is assumed that both the current $\mathbf{J}$ and vector potential $\mathbf{A}$ are constant along a link and that the potential V and the gauge field χ vary linearly along the link. Adopting a local coordinate axis u with $u = 0$ corresponding to node i, and $u = h_{ij}$ corresponding to node j, we may integrate Equation (13.38) along the link ij to obtain

$$J_{ij} = q\mu_{ij}c\left(\frac{V_i - V_j}{h_{ij}}\right) \pm k_{\mathrm{B}}T\mu_{ij}\frac{dc}{du} \tag{13.39}$$

which is a first-order differential equation in c. The latter is solved using the aforementioned boundary conditions and gives rise to a non-linear carrier profile. The current J_{ij} can then be rewritten as

$$\frac{J_{ij}}{\mu_{ij}} = -\frac{\alpha}{h_{ij}}B\left(\frac{-\beta_{ij}}{\alpha}\right)c_i + \frac{\alpha}{h_{ij}}B\left(\frac{\beta_{ij}}{\alpha}\right)c_j, \tag{13.40}$$

where $B(x)$ is the Bernoulli function

$$B(x) = \frac{x}{\mathrm{e}^x - 1} \tag{13.41}$$

and

$$\alpha = \pm k_{\mathrm{B}} T, \tag{13.42}$$
$$\beta_{ij} = q\,(V_i - V_j) \tag{13.43}$$

13.4 Current Continuity Equation

The current continuity equations, can be written for holes and electrons as follows:

$$\begin{aligned} \nabla \cdot \mathbf{J}_p + \mathrm{i}\, q\, \omega\, p + q(R - G) &= 0 \\ -\nabla \cdot \mathbf{J}_n + \mathrm{i}\, q\, \omega\, n + q(R - G) &= 0 \end{aligned}$$

Where $\mathbf{J}_p$ and $\mathbf{J}_n$ stand for the hole and electron current densities, q denotes the positive unit of charge, p, n, R and G are the carrier densities and the recombination and generation rate.

When we subtract the second equation from the first we get the charge continuity:

$$\nabla \cdot \mathbf{J}_p + \nabla \cdot \mathbf{J}_n + \mathrm{i}\, q\, \omega\, (p - n) = 0$$

It should be noted that the convention is that i is the starting node and j is the end node of the link $\langle i, j \rangle$ as is illustrated in Figure 13.4. We follow the Sharfetter-Gummmel approach for which the drift-diffusion equation becomes:

$$\begin{aligned} \mathbf{J}_p &= q\, \mu_p\, p\mathbf{E} - \mathrm{k_B} T\, \mu_p\, \nabla p \\ \mathbf{J}_n &= q\, \mu_n\, n\mathbf{E} + \mathrm{k_B} T\, \mu_n\, \nabla n \end{aligned}$$

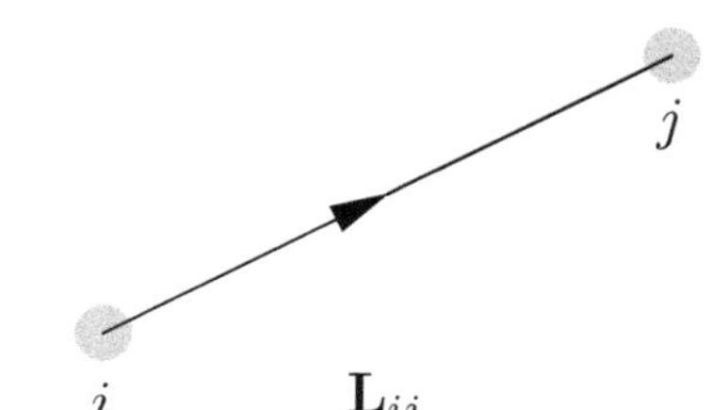

Figure 13.4 Identification of link direction.

This can be rewritten as:

$$\mathbf{J}_p = -\,\mu_p\,\mathrm{k_B}T\,\left(-\frac{q}{\mathrm{k_B}T}p\mathbf{E} + \nabla p\right)$$
$$\mathbf{J}_n = -\,\mu_n\,\mathrm{k_B}T\,\left(-\frac{q}{\mathrm{k_B}T}n\mathbf{E} - \nabla n\right)$$

Defining the variable

$$X_{ij} = -\frac{q}{\mathrm{k_B}T}\int_i^j \mathbf{E}\cdot \mathrm{d}\mathbf{l} \tag{13.44}$$

we obtain after projection on the link direction and integration along the link that has a length h_{ij}:

$$E_{ij} = -\frac{\mathrm{k_B}T}{q}\frac{X_{ij}}{h_{ij}}$$

where E_{ij} is the constant electric field along the link.

$$J^p_{ij} = -\mu_p \mathrm{k_B}T\left(p\frac{X_{ij}}{h_{ij}} + \frac{dp}{ds}\right)$$
$$J^n_{ij} = -\mu_n \mathrm{k_B}T\left(n\frac{X_{ij}}{h_{ij}} - \frac{dn}{ds}\right)$$

and s is a parameter along the link with direction $\hat{\mathbf{e}}$, i.e.

$$\hat{\mathbf{e}}\cdot\nabla p = dp/ds$$
$$\hat{\mathbf{e}}\cdot\nabla n = dn/ds.$$

The perpendicular surface is d_{ij}. Integrated over a link, the classical Sharfetter-Gummel approach results in:

$$\mathbf{J}^p_{ij} = q\frac{\mu_{p,i,j}}{h_{ij}}\left(p_i B[X_{ij}] - p_j B[-X_{ij}]\right) \tag{13.45}$$
$$\mathbf{J}^n_{ij} = -q\frac{\mu_{n,i,j}}{h_{ij}}\left(n_i B[-X_{ij}] - n_j B[X_{ij}]\right) \tag{13.46}$$

with:

$$X_{ij} = \frac{q}{\mathrm{k_B}T}\left(V_j - V_i + \mathrm{j}\,\omega\,\mathrm{sgn}_{ij}\,A_{ij}h_{ij}\right)$$

where $\text{sgn}_{ij} = \pm 1$ depending if the direction $\hat{\mathbf{e}}$ is in the positive or negative direction with respect to the coordinate-frame axis $(\hat{\mathbf{e}}_x, \hat{\mathbf{e}}_y, \hat{\mathbf{e}}_z)$. The Bernoulli function $B[X]$ given by:

$$B[X] = \frac{X}{\exp(X) - 1}$$

13.5 Computational Details of the Hole Transport Equation

The solution of hole current-continuity equation is done by assembling the equation around each node [6]. Each primary volume element, i.e. cube, contributes to the assembling. A loop over cubes and within each cube, another loop over the links provides the final assembling. At the inner loop the following contribution to the assembling is made at node i for the holes and for the link $\langle i, j\rangle$

$$I_{ij} = \frac{\mu_{ij} d_{ij}}{h_{ij}} \left(p_i B[X_{ij}] - p_j B[-X_{ij}]\right) + \mathrm{i}\,\omega\, p_i v_{ij} + (R - G)_i v_{ij} = 0 \tag{13.47}$$

with v_{ij} being the element's contribution the volume around the node i associated to the direction of the link with other node j.

The symbol I_{ij} is also used to describe the current flowing from node i to node i, through an area associated to a volume element and link $\langle ij\rangle$.

$$I_{ij} = \frac{\mu_{ij} d_{ij}}{h_{ij}} \left(p_i B[X_{ij}] - p_j B[-X_{ij}]\right) \tag{13.48}$$

The use of the symbols will be clear from the context. For later linearization we define:

$$\begin{aligned}
X_{ij} &= \frac{q}{\mathrm{k_B} T} \left(V_j - V_i + \mathrm{i}\,\omega\, \text{sgn}_{i,j}\, A_{ij}\, h_{ij}\right) \\
X_{ij}^0 &= \frac{q}{\mathrm{k_B} T} \left(V_j^0 - V_i^0\right) \\
X_{ij}^1 &= \frac{q}{\mathrm{k_B} T} \left(V_j^1 - V_i^1 + \mathrm{i}\,\omega\, \text{sgn}_{i,j}\, A_{ij}^1\, h_{ij}\right) \\
n &= n_{\text{intr}} \exp\left(\frac{V - \phi_n}{\mathrm{k_B} T/q}\right)
\end{aligned}$$

[6]This section may be skipped at first reading.

$$n^0 = n_{\text{intr}} \exp\left(\frac{V^0 - \phi_n^0}{\mathrm{k_B} T/q}\right)$$

$$n^1 = n^0 \left(\frac{V^1 - \phi_n^1}{\mathrm{k_B} T/q}\right)$$

$$p = n_{\text{intr}} \exp\left(\frac{\phi_p - V}{\mathrm{k_B} T/q}\right)$$

$$p^0 = n_{\text{intr}} \exp\left(\frac{\phi_p^0 - V^0}{\mathrm{k_B} T/q}\right)$$

$$p^1 = p^0 \left(\frac{\phi_p^1 - V^1}{\mathrm{k_B} T/q}\right)$$

where $\text{sgn}_{ij} = \pm 1$ dependent on the orientation of the link, A_{ij}, $\mathrm{k_B}$, T, q denote the vector potential along the link, Boltzmann's constant, the temperature and the charge unit respectively. The intrinsic carrier density, and the local carrier densities for holes and electrons in the semiconductor is written as n_i, p and n.

The equations are usually evaluated after scaling. Then the voltages and Fermi potentials are scaled with respect to the 'thermal' voltage $V_{\mathrm{T}} = \frac{\mathrm{k_B} T}{q}$. This simplifies above expressions as follows:

$$\begin{aligned}
X_{ij} &= \left(V_j - V_i + \mathrm{i}\,\omega\, \text{sgn}_{ij}\, A_{ij}\, h_{ij}\right) \\
X_{ij}^0 &= \left(V_j^0 - V_i^0\right) \\
X_{ij}^1 &= \left(V_j^1 - V_i^1 + \mathrm{i}\,\omega\, \text{sgn}_{ij}\, A_{ij}^1\, h_{ij}\right) \\
n &= n_{\text{intr}} \exp\left(V - \phi_n\right) \\
n^0 &= \mathrm{n}_{\text{intr}}^0 \exp\left(V^0 - \phi_n^0\right) \rightarrow \frac{\partial n^0}{\partial \phi_n^0} = -n^0 \\
n^1 &= n^0 \left(V^1 - \phi_n^1\right) \\
p &= n_{\text{intr}} \exp\left(\phi_p - V\right) \\
p^0 &= \mathrm{n}_{\text{intr}}^0 \exp\left(\phi_p^0 - V^0\right) \rightarrow \frac{\partial p^0}{\partial \phi_p^0} = p^0 \\
p^1 &= p^0 \left(\phi_p^1 - V^1\right)
\end{aligned}$$

and $\mathrm{n}_{\text{intr}}^0$ is the scaled value of the intrinsic concentration n_{intr}.

We linearise (13.47) as follows:

$$I_{ij} = I_{ij}^0 + I_{ij}^1 + O(2) = 0 \tag{13.49}$$

with

$$\begin{aligned} I_{ij}^0 &= \frac{\mu_{ij} d_{ij}}{h_{ij}} \left(p_i^0 B[X_{ij}^0] - p_j^0 B[-X_{ij}^0] \right) + (R^0 - G^0)_i \, v_{ij} \\ I_{ij}^1 &= \frac{\mu_{ij} d_{ij}}{h_{ij}} \left(p_i^1 \nu_i B[X_{ij}^0] - p_j^1 \nu_j B[-X_{ij}^0] \right) \\ &+ \frac{\mu_{ij} d_{ij}}{h_{ij}} \left(p_i^0 \frac{\partial B}{\partial X}[X_{ij}^0] + p_j^0 \frac{\partial B}{\partial X}[-X_{ij}^0] \right) X_{ij}^1 \\ &+ j \, \omega \, p_i^1 \, \nu_i \, v_{ij} \\ &+ \left(\left. \frac{\partial (R-G)}{dp_i} \right|_0 p_i^1 + \left. \frac{\partial (R-G)}{dn_i} \right|_0 n_i^1 \right) \, \nu_i \, v_{ij} \end{aligned} \tag{13.50}$$

R^0 and G^0 are the recombination and generation contribution for the static equations, i.e. these are the formulas for which the concentrations n^0 and p^0 are substituted. The used models are:

- No recombination: $R = 0$
- Shockley-Reed-Hall (SRH) recombination:

$$R^{SRH} = \frac{n \cdot p - n_i^2}{\tau_p (n + n_i) + \tau_n (p + n_i)} \tag{13.51}$$

- SRH + Auger recombination:

$$R^{AU} = (C_{cn} n + C_{cp} p) \cdot (np - n_i^2) \tag{13.52}$$

From above equations we can obtain the following expressions for the currents in a volume element along the link $\langle ij \rangle$:

$$\begin{aligned} I_{ij}^0 &= \frac{\mu_{ij} d_{ij}}{h_{ij}} \left(p_i^0 B[X_{ij}^0] - p_j^0 B[-X_{ij}^0] \right) \\ I_{ij}^1 &= \frac{\mu_{ij} d_{ij}}{h_{ij}} \left(p_i^1 \nu_i B[X_{ij}^0] - p_j^1 \nu_j B[-X_{ij}^0] \right) \\ &+ \frac{\mu_{ij} d_{ij}}{h_{ij}} \left(p_i^0 \frac{\partial B}{\partial X}[X_{ij}^0] + p_j^0 \frac{\partial B}{\partial X}[-X_{ij}^0] \right) X_{ij}^1 \end{aligned}$$

Since the derivatives of the Bernoulli are always less than or equal zero, we may define a semiconductor link hole conductance as follows:

$$\sigma_{ij}^{\text{hole}} = -\mu_{ij}^{\text{hole}} \left(p_i^0 \frac{\partial B}{\partial X}[X_{ij}^0] + p_j^0 \frac{\partial B}{\partial X}[-X_{ij}^0] \right) \tag{13.53}$$

Then the small-signal current reads

$$I_{ij}^1 = -\sigma_{ij}^{\text{hole}} \frac{d_{ij}}{h_{ij}} X_{ij}^1 + \frac{\mu_{ij} d_{ij}}{h_{ij}} \left(p_i^1 \nu_i B[X_{ij}^0] - p_j^1 \nu_j B[-X_{ij}^0] \right) \tag{13.54}$$

13.5.1 Scaling

In what follows we scale the Fermi levels and electric potential by the thermal voltage $k_{\text{B}}T/q$, and the densities by the intrinsic carrier density. Therefore:

$$\begin{aligned}
n^0 &= \mathrm{n}_{\text{intr}}^0 \exp\left(V^0 - \phi_n^0\right) \\
p^0 &= \mathrm{n}_{\text{intr}}^0 \exp\left(\phi_p^0 - V^0\right) \\
n^1 &= n^0 \left(V^1 - \phi_n^1\right) \\
p^1 &= p^0 \left(\phi_p^1 - V^1\right) \\
X_{ij}^1 &= V_j^1 - V_i^1 + \mathrm{i}\,\omega\, \mathrm{sgn}_{ij}\, A_{ij}^1\, h_{ij} \\
X_{ij}^0 &= V_j^0 - V_i^0
\end{aligned}$$

and we can evaluate (13.50) for the coupling between I and the electric potential and the Fermi levels to construct the Newton-Raphson matrices, starting from:

$$I_{ij}^0 = \frac{\mu_{ij} d_{ij}}{h_{ij}} \left(p_i^0 B[X_{ij}^0] - p_j^0 B[-X_{ij}^0] \right) + (R^0 - G^0)_i\, v_{ij}$$

and

$$\begin{aligned}
I_{ij}^1 = &\frac{\mu_{ij} d_{ij}}{h_{ij}} \left(p_i^0 \left(\phi_{p,i}^1 - V_i^1\right) B[X_{ij}^0]\, \nu_i - p_j^0 \left(\phi_{p,j}^1 - V_j^1\right) B[-X_{ij}^0]\, \nu_j \right) \\
&+ \frac{\mu_{ij} d_{ij}}{h_{ij}} \left(p_i^0 \frac{\partial B}{\partial X}[X_{ij}^0] + p_j^0 \frac{\partial B}{\partial X}[-X_{ij}^0] \right) \left(V_j^1 - V_i^1 + \mathrm{i}\,\omega\, \mathrm{sgn}_{ij}\, A_{ij}^1\, h_{ij} \right) \\
&+ \mathrm{i}\,\omega\, p_i^0 \left(\phi_{p,i}^1 - V_i^1\right) \nu_i\, v_{ij} \\
&+ \left(\left.\frac{\partial(R-G)}{dp_i}\right|_0 p_i^0 \left(\phi_{p,i}^1 - V_i^1\right) + \left.\frac{\partial(R-G)}{dn_i}\right|_0 n_i^0 \left(V_i^1 - \phi_{n,i}^1\right) \right) \nu_i\, v_{ij}
\end{aligned} \tag{13.55}$$

or

$$\frac{I_{ij}^0}{p_i^0} = \frac{\mu_{ij} d_{ij}}{h_{ij}} \left(B[X_{ij}^0] - \frac{p_j^0}{p_i^0} B[-X_{ij}^0] \right) + \frac{(R^0 - G^0)_i}{p_i^0}\, v_{ij}$$

$$\frac{I_{ij}^1}{p_i^0} = \frac{\mu_{ij} d_{ij}}{h_{ij}} \left(\left(\phi_{p,i}^1 - V_i^1\right) B[X_{ij}^0]\, \nu_i - \frac{p_j^0}{p_i^0} \left(\phi_{p,j}^1 - V_j^1\right) B[-X_{ij}^0]\, \nu_j \right)$$
$$+ \frac{\mu_{ij} d_{ij}}{h_{ij}} \left(\frac{\partial B}{\partial X}[X_{ij}^0] + \frac{p_j^0}{p_i^0} \frac{\partial B}{\partial X}[-X_{ij}^0] \right) \left(V_j^1 - V_i^1 + \mathrm{j}\, \omega\, \mathrm{sgn}_{ij}\, A_{ij}^1\, h_{ij}\right)$$
$$+ \mathrm{i}\, \omega\, \left(\phi_{p,i}^1 - V_i^1\right) \nu_i v_{ij}$$
$$+ \left(\left.\frac{\partial (R-G)}{dp_i}\right|_0 \left(\phi_{p,i}^1 - V_i^1\right) + \left.\frac{\partial (R-G)}{dn_i}\right|_0 \frac{n_i^0}{p_i^0} \left(V_i^1 - \phi_{n,i}^1\right) \right) \nu_i v_{ij} \tag{13.56}$$

Static coupling to V^0 The static coupling deals with derivatives with respect to V^0 and ϕ_p^0. Special attention should be given to Semi-Metal interface nodes. At these nodes the concentrations are frozen and given by the neutrality condition. In other words:

$$n^0 p^0 = \left[\mathrm{n}_{\mathrm{intr}}^0\right]^2$$

and

$$p^0 - n^0 + N_D^0 - N_A^0 = 0$$

Here, N_D^0 and N_A^0 are the scaled concentrations. Furthermore, we set

$$\phi_p = \phi_n = V_{\mathrm{metal}}$$

and

$$V_{\mathrm{metal}} = V_{\mathrm{semi}} + \delta\psi$$

After scaling these equations become: $\phi_p^0 = \phi_n^0 = V_{\mathrm{metal}}^0$ and $V_{\mathrm{metal}}^0 = V_{\mathrm{semi}}^0 + \delta\psi^0$. For later use we note that at the metal-semiconductor interfaces, due to the frozen concentrations, $n^1 = p^1 = 0$. As a consequence one has that the concentrations at interface nodes are given as follows: For N-type material: $N = N_D > 0$:

$$n^0 = \mathrm{n}_{\mathrm{intr}}^0 \exp\left(V_{\mathrm{metal}}^0 - V_{\mathrm{semi}}^0\right) = \mathrm{n}_{\mathrm{intr}}^0 \exp\left(-\delta\psi^0\right) \tag{13.57}$$

$$
\begin{aligned}
n^0 &= \frac{1}{2} N_D^0 \left(1 + \sqrt{1 + 4 \left(\frac{\mathrm{n}_{\mathrm{intr}}^0}{N_D} \right)^2} \right) \\
\delta\psi^0 &= -\log \left[\frac{1}{2} \frac{N_D^0}{\mathrm{n}_{\mathrm{intr}}^0} \left(1 + \sqrt{1 + 4 \left(\frac{\mathrm{n}_{\mathrm{intr}}^0}{N_D} \right)^2} \right) \right] \\
p^0 &= \mathrm{n}_{\mathrm{intr}}^0 \exp\left(+\delta\psi^0\right)
\end{aligned}
$$

For P-type material: $N = -N_A < 0$:

$$
p^0 = \mathrm{n}_{\mathrm{intr}}^0 \exp\left(V_{\mathrm{semi}}^0 - V_{\mathrm{metal}}^0\right) = \mathrm{n}_{\mathrm{intr}}^0 \exp\left(\delta\psi^0\right) \tag{13.58}
$$

$$
\begin{aligned}
p^0 &= -\frac{1}{2} N_A^0 \left(1 + \sqrt{1 + 4 \left(\frac{\mathrm{n}_{\mathrm{intr}}^0}{N_A} \right)^2} \right) \\
\delta\psi^0 &= \log \left[-\frac{1}{2} \frac{N_A^0}{\mathrm{n}_{\mathrm{intr}}^0} \left(1 + \sqrt{1 + 4 \left(\frac{\mathrm{n}_{\mathrm{intr}}^0}{N_A} \right)^2} \right) \right] \\
n^0 &= \mathrm{n}_{\mathrm{intr}}^0 \exp\left(-\delta\psi^0\right)
\end{aligned}
$$

With this information in mind the derivatives must be evaluated taking into account the rigidity of the concentrations at the metal/semiconductor interface.

$$
\begin{aligned}
\partial I_{ij}^0 / \partial V_k^0 = {} & \frac{\mu_{ij} d_{ij}}{h_{ij}} \left(p_i^0 \frac{\partial B}{\partial X}[X_{ij}^0] + p_j^0 \frac{\partial B}{\partial X}[-X_{ij}^0] \right) (\delta_{jk} - \delta_{ik}) \\
& + \frac{\mu_{ij} d_{ij}}{h_{ij}} \left(-p_i^0 \, B[X_{ij}^0] \delta_{ik} \nu_k + p_j^0 \, B[-X_{ij}^0] \delta_{jk} \nu_k \right) \\
& + \left. \frac{\partial(R-G)}{\partial n_i} \right|_0 \delta_{ik} \nu_k n_i^0 \, v_{ij} - \left. \frac{\partial(R-G)}{\partial p_i} \right|_0 \delta_{ik} \nu_k p_i^0 \, v_{ij}
\end{aligned}
$$

where the variable ν_k is defined as

$$
\nu_k = \begin{cases} 0, & k \quad \text{is interface node} \\ 1, & k \quad \text{is NOT interface node} \end{cases} \tag{13.59}
$$

We obtain for the input in the

$$
\begin{aligned}
\partial\left[\frac{I_{ij}^0}{p_i^0}\right] \Big/ \partial V_k^0 = & -\frac{\mu_{ij}d_{ij}}{h_{ij}}(\delta_{ik}-\delta_{jk})\,\frac{1}{p_i^0}\,\left(p_i^0\frac{\partial B}{\partial X}[X_{ij}^0]+p_j^0\,\frac{\partial B}{\partial X}[-X_{ij}^0]\right) \\
& -\frac{\mu_{ij}d_{ij}}{h_{ij}}\frac{1}{p_i^0}\left(\,p_i^0\,B[X_{ij}^0]\,\delta_{ik}\nu_k - p_j^0\,B[-X_{ij}^0]\,\delta_{jk}\nu_k\,\right) \\
& +\left.\frac{\partial(R-G)}{\partial n_i}\right|_0 \delta_{ik}\nu_k\frac{n_i^0}{p_i^0}v_{ij} - \left.\frac{\partial(R-G)}{\partial p_i}\right|_0 \delta_{ik}\nu_k\,v_{ij} \\
& +\left(\frac{I_{ij}^0}{p_i^0}\right)\,\delta_{ik}\nu_k
\end{aligned} \tag{13.60}
$$

and with insertion of the last term

$$
\begin{aligned}
\partial\left[\frac{I_{ij}^0}{p_i^0}\right] \Big/ \partial V_k^0 = & -\frac{\mu_{ij}d_{ij}}{h_{ij}}(\delta_{ik}-\delta_{jk})\frac{1}{p_i^0}\,\left(p_i^0\frac{\partial B}{\partial X}[X_{ij}^0]+p_j^0\,\frac{\partial B}{\partial X}[-X_{ij}^0]\right) \\
& -\frac{\mu_{ij}d_{ij}}{h_{ij}}\frac{1}{p_i^0}\left(\,p_i^0\,B[X_{ij}^0]\,\delta_{ik}\nu_k - p_j^0\,B[-X_{ij}^0]\,\delta_{jk}\nu_k\,\right) \\
& +\left.\frac{\partial(R-G)}{\partial n_i}\right|_0 \delta_{ik}\nu_k\frac{n_i^0}{p_i^0}v_{ij} - \left.\frac{\partial(R-G)}{\partial p_i}\right|_0 \delta_{ik}\nu_k\,v_{ij} \\
& +\frac{\mu_{ij}d_{ij}}{h_{ij}}\frac{1}{p_i^0}\left(p_i^0B[X_{ij}^0]-p_j^0B[-X_{ij}^0]\right)\,\delta_{ik}\nu_k \\
& +\left(R^0-G^0\right)_i\frac{1}{p_i^0}\delta_{ik}\nu_k
\end{aligned} \tag{13.61}
$$

Combining the second and the fourth line, we obtain

$$
\begin{aligned}
\partial\left[\frac{I_{ij}^0}{p_i^0}\right] \Big/ \partial V_k^0 = & -\frac{\mu_{ij}d_{ij}}{h_{ij}}(\delta_{ik}-\delta_{jk})\frac{1}{p_i^0}\,\left(p_i^0\frac{\partial B}{\partial X}[X_{ij}^0]+p_j^0\frac{\partial B}{\partial X}[-X_{ij}^0]\right) \\
& -\frac{\mu_{ij}d_{ij}}{h_{ij}}\frac{p_j^0}{p_i^0}\,B[-X_{ij}^0]\,(\delta_{ik}-\delta_{jk})\,\nu_k \\
& +\left.\frac{\partial(R-G)}{\partial n_i}\right|_0 \delta_{ik}\nu_k\frac{n_i^0}{p_i^0}v_{ij} - \left.\frac{\partial(R-G)}{\partial p_i}\right|_0 \delta_{ik}\nu_k\,v_{ij} \\
& +\left(R^0-G^0\right)_i\frac{1}{p_i^0}\delta_{ik}\nu_k
\end{aligned} \tag{13.62}
$$

This can also be written as:

$$\partial\left[\frac{I_{ij}^0}{p_i^0}\right]\Big/\partial V_k^0 = -\frac{\mu_{ij}d_{ij}}{h_{ij}}(\delta_{ik}-\delta_{jk})$$

$$*\left(\frac{\partial B}{\partial X}[X_{ij}^0]+\frac{p_j^0}{p_i^0}\frac{\partial B}{\partial X}[-X_{ij}^0]+\frac{p_j^0}{p_i^0}B[-X_{ij}^0]\nu_k\right)$$

$$+\left.\frac{\partial(R-G)}{\partial n_i}\right|_0\delta_{ik}\nu_k\frac{n_i^0}{p_i^0}v_{ij}-\left.\frac{\partial(R-G)}{\partial p_i}\right|_0\delta_{ik}\nu_k\,v_{ij}$$

$$+\left(R^0-G^0\right)_i\frac{1}{p_i^0}\delta_{ik}\nu_k \qquad (13.63)$$

Static coupling to ϕ_p^0 In order to construct the solutions we also need the derivatives with respect to the hole densities and the Fermi levels for the construction of the Newton-Raphson Jacobian:

$$\partial I_{ij}^0/\,\partial p_k^0 = \frac{\mu_{ij}d_{ij}}{h_{ij}}\left(\delta_{ik}\nu_k B[X_{ij}^0]-\delta_{jk}\nu_k B[-X_{ij}^0]\right)$$

$$+\left.\frac{\partial(R-G)}{\partial p_i}\right|_0\delta_{ik}p_i^0v_{ij}$$

and for the Fermi level dependence:

$$\partial\left[\frac{I_{ij}^0}{p_i^0}\right]\Big/\partial\phi_{p,k}^0 = \frac{\mu_{ij}d_{ij}}{h_{ij}}\frac{1}{p_i^0}\left(p_i^0B[X_{ij}^0]\delta_{ik}-p_j^0B[-X_{ij}^0]\delta_{jk}\right)\nu_k$$

$$+\left.\frac{\partial(R-G)}{\partial p_i}\right|_0\delta_{ik}\nu_k v_{ij}-\frac{I_{ij}^0}{p_i^0}\delta_{ik}\nu_k\,v_{ij} \qquad (13.64)$$

An alternative formulation is given as:

$$\partial\left[\frac{I_{ij}^0}{p_i^0}\right] \Big/ \partial\phi_{p,k}^0 = \frac{\mu_{ij}d_{ij}}{h_{ij}}\left(\delta_{ik}-\delta_{jk}\right)\nu_k\frac{p_j^0}{p_i^0}B[-X_{ij}^0] + \left.\frac{\partial(R-G)}{\partial p_i}\right|_0 \delta_{ik}\nu_k v_{ij} - \left(R^0-G^0\right)_i\frac{1}{p_i^0}\delta_{ik}\nu_k\, v_{ij} \tag{13.65}$$

Static coupling to ϕ_n^0 The recombination term introduces a cross coupling between the two carrier types:

$$\partial I_{ij}^0 / \partial\phi_{n,k}^0 = -\left.\frac{\partial(R-G)}{\partial n_i}\right|_0 \delta_{ik}\nu_k\, n_i^0 v_{ij}$$

and

$$\partial\left[\frac{I_{ij}^0}{p_i^0}\right] \Big/ \partial\phi_{n,k}^0 = -\left.\frac{\partial(R-G)}{\partial n_i}\right|_0 \frac{n_i^0}{p_i^0}\delta_{ik}\nu_k\, v_{ij}$$

High-frequency coupling of the real part to $\phi_{p,Re}$ The high-frequency Jacobians can be constructed by considering the real and imaginary parts independently. These entries are discussed here:

$$\partial\left[\frac{I_{Re,ij}^1}{p_i^0}\right] \Big/ \partial\phi_{p,Re,k}^1 = \frac{\mu_{ij}d_{ij}}{h_{ij}}\left(\delta_{ik}B[X_{ij}^0] - \frac{p_j^0}{p_i^0}\delta_{jk}B[-X_{ij}^0]\right)\nu_k + \left.\frac{\partial(R-G)}{\partial p_i}\right|_0 \delta_{ik}\nu_k\, v_{ij}$$

High-frequency coupling of the real part to $\phi_{n,Re}$

$$\partial\left[\frac{I_{Re,ij}^1}{p_i^0}\right] \Big/ \partial\phi_{n,Re,k}^1 = -\left.\frac{\partial(R-G)}{dn_i}\right|_0 \frac{n_i^0}{p_i^0}\delta_{ik}\nu_k\, v_{ij}$$

High-frequency coupling of the real part to V_{Re}

$$\partial\left[\frac{I^1_{Re,ij}}{p^0_i}\right]\Big/\partial V^1_{Re,k} = \frac{\mu_{ij}d_{ij}}{h_{ij}}\left(-\delta_{ik}B[X^0_{ij}] + \frac{p^0_j}{p^0_i}\delta_{jk}B[-X^0_{ij}]\right)\nu_k + \frac{\mu_{ij}d_{ij}}{h_{ij}}\left(\frac{\partial B}{\partial X}[X^0_{ij}] + \frac{p^0_j}{p^0_i}\frac{\partial B}{\partial X}[-X^0_{ij}]\right)(\delta_{jk} - \delta_{ik}) + \left.\frac{\partial(R-G)}{\partial n_i}\right|_0 \frac{n^0_i}{p^0_i}\delta_{ik}\nu_k\, v_{ij} - \left.\frac{\partial(R-G)}{dp_i}\right|_0 \delta_{ik}\nu_k\, v_{ij}$$

High-frequency coupling of the real part to $\phi_{p,Im}$

$$\partial\left[\frac{I^1_{Re,ij}}{p^0_i}\right]\Big/\partial\phi^1_{p,Im,k} = -\omega\delta_{ik}\nu_k v_{ij}$$

High-frequency coupling of the real part to $\phi_{n,Im}$

$$\partial\left[\frac{I^1_{Re,ij}}{p^0_i}\right]\Big/\partial\phi^1_{n,Im,k} = 0$$

High-frequency coupling of the real part to V_{Im}

$$\partial\left[\frac{I^1_{Re,ij}}{p^0_i}\right]\Big/\partial V^1_{Im,k} = \omega\delta_{ik}\nu_k v_{ij}$$

High-frequency coupling of the real part to $\mathbf{A}_{Re}$

$$\partial\left[\frac{I^1_{Re,ij}}{p^0_i}\right]\Big/\partial\mathbf{A}^1_{Re,ij} = 0$$

High-frequency coupling of the real part to $\mathbf{A}_{Im}$

$$\partial\left[\frac{I^1_{Re,ij}}{p^0_i}\right]\Big/\partial\mathbf{A}^1_{Im,ij} = -\frac{\mu_{ij}d_{ij}}{h_{ij}}\left(\frac{\partial B}{\partial X}[X^0_{ij}] + \frac{p^0_j}{p^0_i}\frac{\partial B}{\partial X}[-X^0_{ij}]\right)\omega\ \mathrm{sgn}_{ij}\ h_{ij} \qquad (13.66)$$

High-frequency coupling of the imaginary part The coupling of equation $f(x_{Re}, x_{Im})$ to the imaginary part of the variable x can be calculated by using the following rules:

$$\frac{\partial f_{Im}}{\partial x_{Im}} = \frac{\partial f_{Re}}{\partial x_{Re}}$$

$$\frac{\partial f_{Im}}{\partial x_{Re}} = -\frac{\partial f_{Re}}{\partial x_{Im}}$$

13.6 Computational Details of the Electron Transport Equation

The calculation for the electrons is completely similar[7]. Equation (13.46) can be written for the electron on the link $\langle i, j \rangle$ and linearized as:

$$-I_{ij} = -I_{ij}^0 - I_{ij}^1 + O(2) = 0 \tag{13.67}$$

with

$$-I_{ij}^0 = \frac{\mu_{ij} d_{ij}}{h_{ij}} \left(n_i^0 B[-X_{ij}^0] - n_j^0 B[X_{ij}^0] \right)$$

and

$$\begin{aligned}
-I_{ij}^1 = & \frac{\mu_{ij} d_{ij}}{h_{ij}} \left(n_i^1 B[-X_{ij}^0]\, \nu_i - n_j^1 B[X_{ij}^0]\, \nu_j \right) \\
& - \frac{\mu_{ij} d_{ij}}{h_{ij}} \left(n_i^0 \frac{\partial B}{\partial X}[-X_{ij}^0] + n_j^0 \frac{\partial B}{\partial X}[X_{ij}^0] \right) \\
& \cdot \left(V_j^1 - V_i^1 + \mathrm{i}\,\omega \;\; \mathrm{sgn}_{ij}\; A_{ij}^1 h_{ij} \right) \\
& + \mathrm{i}\,\omega\, n_i^0 \left(V_i^1 - \phi_{n,i}^1 \right) \nu_i\, v_{ij} \\
& + \left(\left. \frac{\partial (R-G)}{dp_i} \right|_0 p_i^1 + \left. \frac{\partial (R-G)}{dn_i} \right|_0 n_i^1 \right) \nu_i\, v_{ij}
\end{aligned} \tag{13.68}$$

[7]This section may be skipped at first reading.

As was the case for holes, we can define the current in a volume element along the link $\langle ij \rangle$ as follows:

$$\begin{aligned} I_{ij}^1 &= -\frac{\mu_{ij} d_{ij}}{h_{ij}} \left(n_i^1 B[-X_{ij}^0]\, \nu_i - n_j^1 B[X_{ij}^0]\, \nu_j \right) \\ &\quad + \frac{\mu_{ij} d_{ij}}{h_{ij}} \left(n_i^0 \frac{\partial B}{\partial X}[-X_{ij}^0] + n_j^0 \frac{\partial B}{\partial X}[X_{ij}^0] \right) \\ &\quad \times (V_j^1 - V_i^1 + \mathrm{i}\,\omega\ \mathrm{sgn}_{ij}\ A_{ij}^1 h_{ij}) \end{aligned} \tag{13.69}$$

We can identify the electron conductance along the link ij as

$$\sigma_{ij}^{\mathrm{elec}} = -\mu_{ij}^{\mathrm{elec}} \left(n_i^0 \frac{\partial B}{\partial X}[-X_{ij}^0] + n_j^0 \frac{\partial B}{\partial X}[X_{ij}^0] \right) \quad \geq 0 \tag{13.70}$$

The small-signal current then becomes

$$\begin{aligned} I_{ij}^1 &= -\sigma_{ij}^{\mathrm{elec}} \frac{d_{ij}}{h_{ij}} (V_j^1 - V_i^1 + \mathrm{i}\,\omega\ \mathrm{sgn}_{ij}\ A_{ij}^1 h_{ij}\ ) \\ &\quad - \frac{\mu_{ij} d_{ij}}{h_{ij}} \left(n_i^1 B[-X_{ij}^0]\, \nu_i - n_j^1 B[X_{ij}^0]\, \nu_j \right) \end{aligned} \tag{13.71}$$

13.6.1 Couplings

In what follows we scale the fermi levels and electric potential by the thermal voltage kT/q and evaluate (13.68) for the coupling between I and the electric potential and the Fermi levels to construct the Newton-Raphson matrices, starting from the link contribution in each cube:

$$-I_{ij}^0 = \frac{\mu_{ij} d_{ij}}{h_{ij}} \left(n_i^0 B[-X_{ij}^0] - n_j^0 B[X_{ij}^0] \right) + (R^0 - G^0)_i v_{ij}$$

and

$$\begin{aligned} -I_{ij}^1 = &\frac{\mu_{ij} d_{ij}}{h_{ij}} \left(n_i^0\, (V_i^1 - \phi_{n,i}^1)\, B[-X_{ij}^0]\, \nu_i - n_j^0\, (V_j^1 - \phi_{n,j}^1)\, B[X_{ij}^0]\, \nu_j \right) \\ &- \frac{\mu_{ij} d_{ij}}{h_{ij}} \left(n_i^0 \frac{\partial B}{\partial X}[-X_{ij}^0] + n_j^0 \frac{\partial B}{\partial X}[X_{ij}^0] \right) \\ &\cdot (V_j^1 - V_i^1 + \mathrm{i}\,\omega\, \mathrm{sgn}_{ij} A_{ij}^1 h_{ij}) + \mathrm{i}\,\omega\, n_i^0\, (V_i^1 - \phi_{n,i}^1)\ \nu_i\, v_{ij} \\ &+ \left(\frac{\partial (R-G)}{dp_i}\bigg|_0 p_i^0\, (\phi_{p,i}^1 - V_i^1) + \frac{\partial (R-G)}{\partial n_i}\bigg|_0 n_i^0\, (V_i^1 - \phi_{n,i}^1) \right) \nu_i v_{ij} \end{aligned}$$

or

$$
-\frac{I_{ij}^0}{n_i^0} = \frac{\mu_{ij} d_{ij}}{h_{ij}} \left(B[-X_{ij}^0] - \frac{n_j^0}{n_i^0} B[X_{ij}^0] \right) + \frac{(R^0 - G^0)_i}{n_i^0} v_{ij} \tag{13.72}
$$

and

$$
\begin{aligned}
-\frac{I_{ij}^1}{n_i^0} = & \frac{\mu_{ij} d_{ij}}{h_{ij}} \left(\left(V_i^1 - \phi_{n,i}^1\right) B[-X_{ij}^0]\, \nu_i - \frac{n_j^0}{n_i^0} \left(V_j^1 - \phi_{n,j}^1\right) B[X_{ij}^0]\, \nu_j \right) \\
& - \frac{\mu_{ij} d_{ij}}{h_{ij}} \left(\frac{\partial B}{\partial X}[-X_{ij}^0] + \frac{n_j^0}{n_i^0} \frac{\partial B}{\partial X}[X_{ij}^0] \right) \\
& \cdot \left(V_j^1 - V_i^1 + \mathrm{i}\, \omega \, \mathrm{sgn}_{ij}\, A_{ij}^1 h_{ij}\right) + \mathrm{i}\, \omega \, \left(V_i^1 - \phi_{n,i}^1\right) \, \nu_i v_{ij} \\
& + \left(\left.\frac{\partial (R-G)}{\partial p_i}\right|_0 \frac{p_i^0}{n_i^0} \left(\phi_{p,i}^1 - V_i^1\right) + \left.\frac{\partial (R-G)}{\partial n_i}\right|_0 \left(V_i^1 - \phi_{n,i}^1\right) \right) \nu_i \, v_{ij}
\end{aligned}
$$

Static coupling to V^0

$$
\begin{aligned}
-\, \partial I_{ij}^0 / \, \partial V_k^0 = & \frac{\mu_{ij} d_{ij}}{h_{ij}} \left(n_i^0 \, \frac{\partial B}{\partial X}[-X_{ij}^0] \; + \; n_j^0 \, \frac{\partial B}{\partial X}[X_{ij}^0] \right) (\delta_{ik} - \delta_{jk}) \\
& + \frac{\mu_{ij} d_{ij}}{h_{ij}} \left(\, n_i^0 \; B[-X_{ij}^0] \; \nu_k \delta_{ik} - n_j^0 \; B[X_{ij}^0] \; \nu_k \delta_{jk} \right) \\
& + \left.\frac{\partial (R-G)}{\partial n_i}\right|_0 \nu_k \; \delta_{ik} n_i^0 v_{ij} - \left.\frac{\partial (R-G)}{\partial p_i}\right|_0 \; \nu_k \delta_{ik} p_i^0 v_{ij}
\end{aligned}
$$

$$
\begin{aligned}
-\, \partial \left[\frac{I_{ij}^0}{n_k^0} \right] \bigg/ \, \partial V_k^0 = & \frac{\mu_{ij} d_{ij}}{h_{ij}} (\delta_{ik} - \delta_{jk}) \\
& \times \left(\frac{\partial B}{\partial X}[-X_{ij}^0] + \frac{n_j^0}{n_i^0} \frac{\partial B}{\partial X}[X_{ij}^0] + \frac{n_j^0}{n_i^0} B[X_{ij}^0] \; \nu_k \right) \\
& - \left.\frac{\partial (R-G)}{\partial p_i}\right|_0 \frac{p_i^0}{n_i^0} \; \nu_k \; \delta_{ik} \; v_{ij} + \left.\frac{\partial (R-G)}{\partial n_i}\right|_0 \; \nu_k \; \delta_{ik} v_{ij} \\
& - (R^0 - G^0)_i \frac{1}{n_i^0} \; \nu_k \; \delta_{ik} \; v_{ij}
\end{aligned} \tag{13.73}
$$

Static coupling to ϕ_p^0

$$- \,\partial I_{ij}^0 / \,\partial \phi_p^0 \;=\; \left.\frac{\partial (R-G)}{\partial p_i}\right|_0 \,\nu_k \;\delta_{ik} p_i^0 v_{ij}$$

$$- \,\partial \left[\frac{I_{ij}^0}{n_i^0}\right] \Big/ \;\partial \phi_{p,k}^0 \;=\; \left.\frac{\partial (R-G)}{\partial p_i}\right|_0 \frac{p_i^0}{n_i^0} \,\nu_k \delta_{ik} v_{ij}$$

Static coupling to ϕ_n^0

$$\begin{aligned} - \,\partial I_{ij}^0 / \,\partial n_k^0 \;=\;& \frac{\mu_{ij} d_{ij}}{h_{ij}} \left(\,\nu_k \;\delta_{ik} B[-X_{ij}^0] - \;\nu_k \;\delta_{jk} B[X_{ij}^0] \,\right) \\ &+ \left.\frac{\partial (R-G)}{\partial p_i}\right|_0 n_i^0 \,\nu_k \;\delta_{ik} \;v_{ij} \end{aligned}$$

$$\begin{aligned} - \,\partial \left[\frac{I_{ij}^0}{n_i^0}\right] \Big/ \;\partial \phi_{n,k}^0 \;=\;& -\frac{\mu_{ij} d_{ij}}{h_{ij}} \,\nu_k \;\left(\delta_{ik} - \delta_{jk}\right) \;\frac{n_j^0}{n_i^0} \;B[X_{ij}^0] \\ &- \left.\frac{\partial (R-G)}{\partial n_i}\right|_0 \;\nu_k \;\delta_{ik} \;v_{ij} \\ &+ (R-G)_i \;\frac{1}{n_i^0} \,\nu_k \;\delta_{ik} \;v_{ij} \end{aligned} \tag{13.74}$$

High-frequency coupling of the real part to $\phi_{p,Re}$

$$- \,\partial \left[\frac{I_{Re,ij}^1}{n_i^0}\right] \Big/ \;\partial \phi_{p,Re,k}^1 \;=\; +\frac{p_i^0}{n_i^0} \left.\frac{\partial (R-G)}{\partial p_i}\right|_0 \;\nu_k \;\delta_{ik} v_{ij}$$

High-frequency coupling of the real part to $\phi_{n,Re}$

$$\begin{aligned} - \,\partial \left[\frac{I_{Re,ij}^1}{n_i^0}\right] \Big/ \;\partial \phi_{n,Re,k}^1 \;=\;& \frac{\mu_{ij} d_{ij}}{h_{ij}} \left(-\delta_{ik} B[-X_{ij}^0] + \frac{n_j^0}{n_i^0} \delta_{jk} B[X_{ij}^0] \right) \nu_k \\ &- \left.\frac{\partial (R-G)}{\partial n_i}\right|_0 \;\nu_k \;\delta_{ik} v_{ij} \end{aligned}$$

High-frequency coupling of the real part to V_{Re}

$$
\begin{aligned}
-\,\partial\left[\frac{I^1_{Re,ij}}{n^0_i}\right]\Big/\,\partial V^1_{Re,k} =\; & \frac{\mu_{ij}d_{ij}}{h_{ij}}\left(\delta_{ik}B[-X^0_{ij}]-\frac{n^0_j}{n^0_i}\delta_{jk}B[X^0_{ij}]\right)\;\nu_k \\
& -\frac{\mu_{ij}d_{ij}}{h_{ij}}\left(\frac{\partial B}{\partial X}[-X^0_{ij}]+\frac{n^0_j}{n^0_i}\frac{\partial B}{\partial X}[X^0_{ij}]\right)(\delta_{jk}-\delta_{ik}) \\
& +\left.\frac{\partial(R-G)}{dn_i}\right|_0 \nu_k\;\delta_{ik}v_{ij}-\frac{p^0_i}{n^0_i}\left.\frac{\partial(R-G)}{\partial p_i}\right|_0 \nu_k\;\delta_{ik}v_{ij}
\end{aligned}
$$

High-frequency coupling of the real part to $\phi_{p,Im}$

$$
-\,\partial\left[\frac{I^1_{Re,ij}}{n^0_i}\right]\Big/\,\partial\phi^1_{p,Im}\;=\;0 \tag{13.75}
$$

High-frequency coupling of the real part to $\phi_{n,Im}$

$$
-\,\partial\left[\frac{I^1_{Re,ij}}{n^0_i}\right]\Big/\,\partial\phi^1_{n,Im}=\omega\;\nu_k\;\delta_{ik}v_{ij} \tag{13.76}
$$

High-frequency coupling of the real part to V_{Im}

$$
-\,\partial\left[\frac{I^1_{Re,ij}}{n^0_i}\right]\Big/\,\partial V^1_{Im,k}=-\omega\;\nu_k\;\delta_{ik}v_{ij} \tag{13.77}
$$

High-frequency coupling of the real part to $\mathbf{A}_{Re}$

$$
-\,\partial\left[\frac{I^1_{Re,ij}}{n^0_i}\right]\Big/\,\partial\mathbf{A}^1_{Re,ij}=0
$$

High-frequency coupling of the real part to $\mathbf{A}_{Im}$

$$
\begin{aligned}
-\,\partial\left[\frac{I^1_{Re,ij}}{n^0_i}\right]\Big/\,\partial\mathbf{A}^1_{Im,ij}=\; & \frac{\mu_{ij}d_{ij}}{h_{ij}}\left(\frac{\partial B}{\partial X}[-X^0_{ij}]+\frac{n^0_j}{n^0_i}\frac{\partial B}{\partial X}[X^0_{ij}]\right) \\
& \omega\;\mathrm{sgn}_{ij}h_{ij}
\end{aligned} \tag{13.78}
$$

13.7 The Poisson Equation

The electric part of the Maxwell's equations is solved by the following Poisson equation:

$$-\nabla \cdot \varepsilon \left(\nabla V + \frac{\partial \mathbf{A}}{\partial t} \right) = \rho \quad (13.79)$$

We integrate (13.79) over a volume element V, and obtain:

$$\varepsilon_{ij} \frac{d_{ij}}{h_{ij}} (V_i - V_j - \mathrm{i}\, \omega \, \mathrm{sgn}_{ij} \; A_{ij} h_{ij}) - (p - n + D) \; v_{ij} = 0 \quad (13.80)$$

When we linearise the equation we get for the static solution:

$$\varepsilon_{ij} \frac{d_{ij}}{h_{ij}} (V_i^0 - V_j^0) - \left(p_i^0 - n_i^0 + D \right) \; v_{ij} = 0 \quad (13.81)$$

For the high-frequency component the RHS becomes:

$$\varepsilon_{ij} \frac{d_{ij}}{h_{ij}} \left(V_i^1 - V_j^1 - \mathrm{i}\, \omega \, \mathrm{sgn}_{ij} A_{ij}^1 h_{ij} \right) - \; \left(p_i^1 - n_i^1 \right) \; v_{ij} = 0$$

$$\varepsilon_{ij} \frac{d_{ij}}{h_{ij}} \left(V_i^1 - V_j^1 - \mathrm{i}\, \omega \, \mathrm{sgn}_{ij} A_{ij}^1 h_{ij} \right) - p^0 \left(\phi_p^1 - V^1 \right) \; v_{ij} + n^0$$

$$\times \left(V^1 - \phi_n^1 \right) \; v_{ij} = 0$$

We refer, when coding the system, to the 'Poisson equation' as the equation that serves as the equation that determines the electric potential. This means that 'Poisson' should be seen in a more general context. In particular in bulk metal, the 'Poisson' equation is actually the current-continuity equation:

$$\nabla \cdot \mathbf{J} = 0 \quad (13.82)$$

This will read in discretized form for a bulk node i:

$$\sum_{j=1}^{N} \sigma_{ij} \; \frac{d_{ij}}{h_{ij}} \; \left(\; V_i - V_j - \mathrm{i}\omega \; \mathrm{sgn}_{ij} A_{ij} h_{ij} \; \right) = 0 \quad (13.83)$$

where σ_{ij} is the conductivity of the link $\langle ij \rangle$ seen from the volume element under consideration.

At metal-insulator interface nodes, we use the current-continuity equation to obtain V. The Gauss' equation is used to determine the surface charge at the metal-interface interface. The surface charge provides a second independent variable that allows us to simultaneously obey the current-continuity:

$$\nabla \cdot (\sigma \, \mathbf{E}) = 0 \tag{13.84}$$

and

$$\nabla \cdot (\epsilon \, \mathbf{E}) = \rho \tag{13.85}$$

The ρ is obtained as a *post-processed* result.

Special care must also be taken at metal-semiconductor interface nodes. The 'Poisson' equation again actually is the current-continuity equation. Now the continuity of the current for an interface node i is:

$$\sum_{j=1}^{N_{\mathrm{metal}}} \sigma_{ij} \, \frac{d_{ij}}{h_{ij}} \, \left(\, V_i - V_j - \mathrm{i}\omega \, \mathrm{sgn}_{ij} A_{ij} h_{ij} \, \right) + \sum_{j=1}^{N_{\mathrm{semi}}} \, \left(\, I_{ij}^{\mathrm{holes}} \, + \, I_{ij}^{\mathrm{elec}} \, \right) = 0 \tag{13.86}$$

The first sum is to nodes in bulk metal and the second sum is in bulk semiconductor. If j is also an interface node, its contribution is not vanishing since there can be a lateral electron and hole current parallel to the interface because the nodes i and j are at different electrical potential. There is no contribution originating from a different carrier concentration in that case since for metal-semiconductor interface nodes, the carrier concentration are fixed at intrinsic concentration values. For the static equations we then obtain

$$\sum_{j=1}^{N_{\mathrm{metal}}} \sigma_{ij} \, \frac{d_{ij}}{h_{ij}} \, \left(\, V_i^0 - V_j^0 \, \right) \, + \, \sum_{j=1}^{N_{\mathrm{semi}}} \, \left(\, I_{ij}^{0,\mathrm{holes}} \, + \, I_{ij}^{0,\mathrm{elec}} \, \right) = 0 \tag{13.87}$$

and for the dynamic part

$$\sum_{j=1}^{N_{\mathrm{metal}}} \sigma_{ij} \, \frac{d_{ij}}{h_{ij}} \, \left(\, V_i^1 - V_j^1 - \mathrm{i}\omega \, \mathrm{sgn}_{ij} A_{ij}^1 h_{ij} \, \right)$$

$$+ \sum_{j=1}^{N_{\mathrm{semi}}} \, \left(\, I_{ij}^{1,\mathrm{holes}} \, + \, I_{ij}^{1,\mathrm{elec}} \, \right) = 0 \tag{13.88}$$

where the fluxes are given by:

$$
\begin{aligned}
I_{ij}^{0,\text{holes}} &= \frac{\mu_{ij}^{p} d_{ij}}{h_{ij}} \left(p_i^0 B[X_{ij}^0] - p_j^0 B[-X_{ij}^0]\right) \\
&= \frac{\mu_{ij}^{p} d_{ij}}{h_{ij}} \, \mathrm{n}_{\text{intr}}^0 (\exp\left(\phi_{p,i}^0 - V_i^0\right) B[X_{ij}^0] \\
&\quad - \exp\left(\phi_{p,j}^0 - V_j^0\right) B[-X_{ij}^0]) \\
I_{ij}^{1,\text{holes}} &= \frac{\mu_{ij}^{p} d_{ij}}{h_{ij}} \left(p_i^0 \left(\phi_{p,i}^1 - V_i^1\right) \, B[X_{ij}^0] \, \nu_i - p_j^0 \, \left(\phi_{p,j}^1 - V_j^1\right) \, B[-X_{ij}^0] \, \nu_j \right) \\
&\quad + \frac{\mu_{ij}^{p} d_{ij}}{h_{ij}} \left(p_i^0 \frac{\partial B}{\partial X}[X_{ij}^0] + p_j^0 \frac{\partial B}{\partial X}[-X_{ij}^0]\right) \left(V_j^1 - V_i^1 + \mathrm{i}\, \omega \, \mathrm{sgn}_{ij} \, A_{ij}^1 \, h_{ij}\right)
\end{aligned}
\tag{13.89}
$$

and for the electron contribution we have

$$
\begin{aligned}
& I_{ij}^{0,\text{elec}} \\
&= -\frac{\mu_{ij}^{n} d_{ij}}{h_{ij}} \left(n_i^0 B[-X_{ij}^0] - n_j^0 B[X_{ij}^0]\right) \\
&= -\frac{\mu_{ij}^{n} d_{ij}}{h_{ij}} \, \mathrm{n}_{\text{intr}}^0 \, \left(\exp\left(V_i^0 - \phi_{n,i}^0\right) B[-X_{ij}^0] - \exp\left(V_j^0 - \phi_{n,j}^0\right) B[X_{ij}^0]\right)
\end{aligned}
$$

$$
\begin{aligned}
I_{ij}^{1,\text{elec}} &= -\frac{\mu_{ij}^{n} d_{ij}}{h_{ij}} \left(n_i^0 (V_i^1 - \phi_{n,i}^1) \, B[-X_{ij}^0] \, \nu_i - n_j^0 \, \left(V_j^1 - \phi_{n,j}^1\right) B[X_{ij}^0] \, \nu_j \right) \\
&\quad + \frac{\mu_{ij}^{n} d_{ij}}{h_{ij}} \left(n_i^0 \frac{\partial B}{\partial X}[-X_{ij}^0] + n_j^0 \frac{\partial B}{\partial X}[X_{ij}^0]\right) \left(V_j^1 - V_i^1 + \mathrm{i}\, \omega \; \mathrm{sgn}_{ij} \, A_{ij}^1 h_{ij}\right)
\end{aligned}
\tag{13.90}
$$

Now the coupling to V is obtained as follows. Let P_{ij} be the contribution to the 'Poisson' equation at node i originating from the connection to node j. Then we find that

- Bulk metal nodes

$$\frac{\partial P_{ij}^0}{\partial V_k^0} = \sigma_{ij}\frac{d_{ij}}{h_{ij}}\left(\delta_{ik} - \delta_{jk}\right) \tag{13.91}$$

$$\frac{\partial P_{ij}^1}{\partial V_k^1} = \sigma_{ij}\frac{d_{ij}}{h_{ij}}\left(\delta_{ik} - \delta_{jk}\right) \tag{13.92}$$

- Bulk semiconductor nodes:

$$\frac{\partial P_{ij}^0}{\partial V_k^0} = \epsilon_{ij}\frac{d_{ij}}{h_{ij}}\left(\delta_{ik} - \delta_{jk}\right) + \left(p_i^0 + n_i^0\right)\ v_{ij} \tag{13.93}$$

$$\frac{\partial P_{ij}^1}{\partial V_k^1} = \epsilon_{ij}\frac{d_{ij}}{h_{ij}}\left(\delta_{ik} - \delta_{jk}\right) + \left(p_i^0 + n_i^0\right) v_{ij} \tag{13.94}$$

- Bulk insulator nodes:

$$\frac{\partial P_{ij}^0}{\partial V_k^0} = \epsilon_{ij}\frac{d_{ij}}{h_{ij}}\left(\delta_{ik} - \delta_{jk}\right) \tag{13.95}$$

$$\frac{\partial P_{ij}^1}{\partial V_k^1} = \epsilon_{ij}\frac{d_{ij}}{h_{ij}}\left(\delta_{ik} - \delta_{jk}\right) \tag{13.96}$$

- Semi-insulator interface nodes:

$$\frac{\partial P_{ij}^0}{\partial V_k^0} = \epsilon_{ij}\frac{d_{ij}}{h_{ij}}\left(\delta_{ik} - \delta_{jk}\right) + \left(p_i^0 + n_i^0\right)\ \bar{v}_{ij} \tag{13.97}$$

$$\frac{\partial P_{ij}^1}{\partial V_k^1} = \epsilon_{ij}\frac{d_{ij}}{h_{ij}}\left(\delta_{ik} - \delta_{jk}\right) + \left(p_i^0 + n_i^0\right) \bar{v}_{ij} \tag{13.98}$$

where $\bar{v}_{ij}$ is the semiconductor part of the nodal volume.

- Metal-Semiconductor nodes or Metal-Semiconductor-Insulator nodes (triple points)

$$
\begin{aligned}
\frac{\partial P_{ij}^0}{\partial V_k^0} &= \sigma_{ij}\frac{d_{ij}}{h_{ij}}\left(\delta_{ik}-\delta_{jk}\right) \\
&+ \frac{\mu_{ij}^p d_{ij}}{h_{ij}}\,\mathrm{n}_{\mathrm{intr}}^0\,\left(-\exp\left(\phi_{p,i}^0 - V_i^0\right) B[X_{ij}^0]\;\delta_{ik}\nu_k\right. \\
&+ \left.\exp\left(\phi_{p,j}^0 - V_j^0\right) B[-X_{ij}^0]\;\delta_{jk}\nu_k\right) \\
&+ \frac{\mu_{ij}^p d_{ij}}{h_{ij}}\,\mathrm{n}_{\mathrm{intr}}^0\,\left(\exp\left(\phi_{p,i}^0 - V_i^0\right)\frac{\partial B}{\partial X}[X_{ij}^0]\right. \\
&+ \left.\exp\left(\phi_{p,j}^0 - V_j^0\right)\frac{\partial B}{\partial X}[-X_{ij}^0]\right)\;\left(\delta_{jk}-\delta_{ik}\right) \\
&- \frac{\mu_{ij}^n d_{ij}}{h_{ij}}\,\mathrm{n}_{\mathrm{intr}}^0\,\left(\exp\left(V_i^0 - \phi_{n,i}^0\right) B[-X_{ij}^0]\;\delta_{ik}\nu_k\right. \\
&+ \left.\exp\left(V_j^0 - \phi_{n,j}^0\right) B[X_{ij}^0]\;\delta_{jk}\nu_k\right) \\
&+ \frac{\mu_{ij}^n d_{ij}}{h_{ij}}\,\mathrm{n}_{\mathrm{intr}}^0\,\left(\exp\left(V_i^0 - \phi_{n,i}^0\right)\frac{\partial B}{\partial X}[-X_{ij}^0]\right. \\
&+ \left.\exp\left(V_j^0 - \phi_{n,j}^0\right)\frac{\partial B}{\partial X}[X_{ij}^0]\right)\;\left(\delta_{jk}-\delta_{ik}\right) \qquad (13.99)
\end{aligned}
$$

$$
\begin{aligned}
\frac{\partial P_{ij}^1}{\partial V_k^1} &= \sigma_{ij}\frac{d_{ij}}{h_{ij}}\left(\delta_{ij}-\delta_{jk}\right) \\
&+ \frac{\mu_{ij}^p d_{ij}}{h_{ij}}\,\mathrm{n}_{\mathrm{intr}}^0\,\left(-\exp\left(\phi_{p,i}^0 - V_i^0\right)\; B[X_{ij}^0]\delta_{ik}\nu_k\right. \\
&+ \left.\exp\left(\phi_{p,j}^0 - V_j^0\right)\; B[-X_{ij}^0]\delta_{jk}\nu_k\right) \\
&+ \frac{\mu_{ij}^p d_{ij}}{h_{ij}}\,\mathrm{n}_{\mathrm{intr}}^0\,\left(\exp\left(\phi_{p,i}^0 - V_i^0\right)\frac{\partial B}{\partial X}[X_{ij}^0]\right. \\
&+ \left.\exp\left(\phi_{p,j}^0 - V_j^0\right)\frac{\partial B}{\partial X}[-X_{ij}^0]\right)\;\left(\delta_{jk}-\delta_{ik}\right) \\
&- \frac{\mu_{ij}^n d_{ij}}{h_{ij}}\,\mathrm{n}_{\mathrm{intr}}^0\,\left(\exp\left(V_i^0 - \phi_{n,i}^0\right)\; B[X_{ij}^0]\delta_{ik}\nu_k\right. \\
&- \left.\exp\left(V_j^0 - \phi_{p,j}^0\right)\; B[-X_{ij}^0]\delta_{jk}\nu_k\right)
\end{aligned}
$$

$$+\frac{\mu_{ij}^n d_{ij}}{h_{ij}}\,\mathrm{n}_{\mathrm{intr}}^0\left(\exp\left(V_i^0-\phi_{n,i}^0\right)\frac{\partial B}{\partial X}[-X_{ij}^0]\right.$$
$$\left.+\exp\left(V_j^0-\phi_{n,j}^0\right)\frac{\partial B}{\partial X}[X_{ij}^0]\right)\ (\delta_{jk}-\delta_{ik}) \tag{13.100}$$

Reminder: the variable ν_k is defined as

$$\nu_k=\begin{cases}0, & k \quad \text{is interface node}\\ 1, & k \quad \text{is NOT interface node}\end{cases} \tag{13.101}$$

For the derivatives with respect to Fermi level variables ϕ_p and ϕ_n, we have the following situation.

- Bulk metal and Bulk insulator: There are no Fermi levels, so derivatives are absent.
- Bulk semiconductor (v_{ij}) and Semiconductor-Insulator interface nodes ($v_{ij}\to\bar{v}_{ij}$)

$$\frac{\partial P_{ij}^0}{\partial\phi_{p,k}^0}=-\bar{v}_{ij}\mathrm{n}_{\mathrm{intr}}^0\ \exp\left(\phi_{p,i}^0-V_i^0\right)\nu_k\delta_{ik} \tag{13.102}$$

$$\frac{\partial P_{ij}^0}{\partial\phi_{n,k}^0}=-\bar{v}_{ij}\mathrm{n}_{\mathrm{intr}}^0\ \exp\left(V_i^0-\phi_{n,i}^0\right)\nu_k\delta_{ik} \tag{13.103}$$

  Note that no contributions are found for interface nodes.
  For the high-frequency derivatives, we obtain

$$\frac{\partial P_{ij}^1}{\partial\phi_{p,k}^1}=-\bar{v}_{ij}\mathrm{n}_{\mathrm{intr}}^0\ \exp\left(\phi_{p,i}^0-V_i^0\right)\nu_k\delta_{ik} \tag{13.104}$$

$$\frac{\partial P_{ij}^1}{\partial\phi_{n,k}^1}=-\bar{v}_{ij}\mathrm{n}_{\mathrm{intr}}^0\ \exp\left(V_i^0-\phi_{n,i}^0\right)\nu_k\delta_{ik} \tag{13.105}$$

- Metal-Semiconductor interface nodes
  Remember that i is the interface node and j is the connected node.

$$\frac{\partial P_{ij}^0}{\partial\phi_{p,k}^0}=-\frac{\mu_{ij}^p d_{ij}}{h_{ij}}\,\mathrm{n}_{\mathrm{intr}}^0\ \exp\left(\phi_{p,j}^0-V_j^0\right)\,B[-X_{ij}^0]\delta_{jk}\nu_k \tag{13.106}$$

$$\frac{\partial P_{ij}^0}{\partial\phi_{n,k}^0}=-\frac{\mu_{ij}^n d_{ij}}{h_{ij}}\,\mathrm{n}_{\mathrm{intr}}^0\ \exp\left(V_j^0-\phi_{n,j}^0\right)\,B[X_{ij}^0]\delta_{jk}\nu_k \tag{13.107}$$

For the high-frequency couplings we obtain

$$\frac{\partial P^1_{ij}}{\partial \phi^1_{p,k}} = -\frac{\mu^p_{ij} d_{ij}}{h_{ij}} \mathrm{n}^0_{\mathrm{intr}} \; \exp\left(\phi^0_{p,j} - V^0_j\right) B[-X^0_{ij}]\delta_{jk}\nu_k \tag{13.108}$$

$$\frac{\partial P^1_{ij}}{\partial \phi^1_{n,k}} = -\frac{\mu^n_{ij} d_{ij}}{h_{ij}} \mathrm{n}^0_{\mathrm{intr}} \; \exp\left(V^0_j - \phi^0_{n,j}\right) B[X^0_{ij}]\delta_{jk}\nu_k \tag{13.109}$$

We observe that:

$$\frac{\partial P^1_{ij}}{\partial V^1_k} = \frac{\partial P^0_{ij}}{\partial V^0_k}$$
$$\frac{\partial P^1_{ij}}{\partial \phi^1_{p,k}} = \frac{\partial P^0_{ij}}{\partial \phi^0_{p,k}}$$
$$\frac{\partial P^1_{ij}}{\partial \phi^1_{n,k}} = \frac{\partial P^0_{ij}}{\partial \phi^0_{n,k}}$$

13.8 Maxwell-Ampere Equation

The Maxwell-Ampere equation is:

$$\nabla \times \nabla \times \mathbf{A} + \nabla\chi - \mu_0 \mathbf{J} + \mu_0 \varepsilon \frac{\partial}{\partial t}\left(\nabla V + \frac{\partial \mathbf{A}}{\partial t}\right) = 0$$

With assigning the short-hand notation MAE to the left-hand side we can write it as $MAE = 0$.

When solving simultaneously in the small signal case, the couplings to V,ϕ_p and ϕ_n from the last two terms should be taken into account. We will next discuss the various couplings.

High-frequency coupling of the real part to V_{Re}

$$\begin{aligned}
\frac{\partial MAE_{Re}}{\partial V^1_{Re,k}} &= \frac{K\mu^p_{ij} d_{ij}}{h_{ij}} \left(\delta_{ik} p^0_i B[X^0_{ij}] - \delta_{jk} p^0_j B[-X^0_{ij}]\right) \nu_k \\
&\quad - \frac{K\mu^p_{ij} d_{ij}}{h_{ij}} \left(p^0_i \frac{\partial B}{\partial X}[X^0_{ij}] + p^0_j \frac{\partial B}{\partial X}[-X^0_{ij}]\right) (\delta_{jk} - \delta_{ik}) \\
&\quad + \frac{K\mu^n_{ij} d_{ij}}{h_{ij}} \left(n^0_i \delta_{ik} B[-X^0_{ij}] - n^0_j \delta_{jk} B[X^0_{ij}]\right) \nu_k
\end{aligned}$$

$$-\frac{K\mu_{ij}^{n}d_{ij}}{h_{ij}}\left(n_i^0\frac{\partial B}{\partial X}[-X_{ij}^0]+n_j^0\frac{\partial B}{\partial X}[X_{ij}^0]\right)(\delta_{jk}-\delta_{ik})$$
$$-\frac{Kd_{ij}}{h_{ij}}\omega\varepsilon_{Im}(\delta_{jk}-\delta_{ik})$$

High-frequency coupling of the real part to V_{Im}

$$\frac{\partial MAE_{Re}}{\partial V_{Im,k}^{1}} = -\frac{Kd_{ij}}{h_{ij}}\omega\varepsilon_{Re}(\delta_{jk}-\delta_{ik})$$

High-frequency coupling of the real part to $\phi_{p,Re}$

$$\frac{\partial MAE_{Re}}{\partial \phi_{p,Re,k}^{1}} = -\frac{K\mu_{ij}^{p}d_{ij}}{h_{ij}}\left(p_i^0\delta_{ik}B[X_{ij}^0]-p_j^0\delta_{jk}B[-X_{ij}^0]\right)\ \nu_k$$

High-frequency coupling of the real part to $\phi_{p,Im}$

$$\frac{\partial MAE_{Re}}{\partial \phi_{p,Im,k}^{1}} = 0$$

High-frequency coupling of the real part to $\phi_{n,Re}$

$$\frac{\partial MAE_{Re}}{\partial \phi_{n,Re,k}^{1}} = \frac{K\mu_{ij}^{n}d_{ij}}{h_{ij}}\left(-n_i^0\delta_{ik}B[-X_{ij}^0]+n_j^0\delta_{jk}B[X_{ij}^0]\right)\ \nu_k$$

High-frequency coupling of the real part to $\phi_{n,Im}$

$$\frac{\partial MAE_{Re}}{\partial \phi_{n,Im,k}^{1}} = 0$$

It should be noted that derivatives for $\nu_k = 0$ are not encountered if no variables for interface-Fermi levels are introduced.

So far, we have elaborated the calculations of the partial derivatives in detail for the real and imaginary parts. As said, these derivatives are required to construct the Newton-Raphson matrices that enter the calcalation of the updates towards finding the solution of the full system of equations. The subtleties are found in the treatment of the interfaces. It should be noted that one can also start with representing all unknowns in the frequency regime as complex variables. For example, when coding these equations in C++

the variables can be chosen as complex$<$ double $>$. This approach surely makes the effort to calculation of the partial derivatives more efficient and the MAGWEL solver does exploit this approach.

13.9 Using Gauge Conditions to Decrease Matrix Fill-In

In this section we explain how one can exploit the gauge equations to reduce matrix fill-in in solving the full-wave Maxwell system in which the currents are self-consistently obtained from the field solutions.

The gauge condition can be chosen in many different ways: There are two gauge conditions ("gauges") that have become quite popular. The Lorenz gauge is very convenient to solve wave equations since it minimizes the coupling of V and $\mathbf{A}$ for source-free Maxwell equations, i.e. free-space propagation. The Lorenz gauge reads after scaling of the Maxwell system:

$$\nabla.\mathbf{A} + k\mathrm{i}\omega\varepsilon_{\mathrm{r}}\, V = 0, \tag{13.110}$$

Another popular gauge condition is the Coulomb gauge, in which the Poisson equation takes a very simple form.

$$\nabla.\mathbf{A} = 0 \tag{13.111}$$

It should be realized that the Lorenz gauge and the Coulomb gauge were chosen based on analytic considerations, i.e. pencil and paper calculations are at the genesis of these gauges. These choices are not necessarily the best selection for numerical calculations and alternative options must be considered, keeping in mind the numerical context in which these gauge choices will be used. For example, the Lorenz gauge may be generalized to include the conductive character at some region in space.

$$\nabla.\mathbf{A} + k(\lambda\sigma + \mathrm{i}\omega\varepsilon_{\mathrm{r}})V = 0, \tag{13.112}$$

In the Maxwell-Ampere system, the use of $\lambda = 1$, will minimize the coupling to the Poisson field thereby decreasing the number of couplings in the full matrix. Note that λ is a free parameter, defining the gauge. This parameter is an arbitrary function of the position coordinate.

$$\nabla.\mathbf{A} + k\,\xi\,(\lambda\sigma + \mathrm{i}\omega\varepsilon_{\mathrm{r}})V = 0, \tag{13.113}$$

For $\xi = 0$ we obtain the Coulomb gauge and for $\xi = 1$ and $\lambda = 0$ we obtain the Lorenz gauge.

We start from the following set of equations:

$$\mathbf{J}_{\mathrm{c}} = \sigma\mathbf{E} \tag{13.114}$$

$$\mathbf{D} = \varepsilon_0\varepsilon_r\mathbf{E} \tag{13.115}$$

$$\mathbf{H} = \frac{1}{\mu_0\mu_r}\mathbf{B} \tag{13.116}$$

$$\mathbf{B} = \nabla\times\mathbf{A} \tag{13.117}$$

$$\mathbf{E} = -\nabla V - \frac{\partial\mathbf{A}}{\partial t} \tag{13.118}$$

For semiconductors,we will use

$$\mathbf{J}_{\mathrm{c}} = \sigma\mathbf{E} + \mathbf{J}_{\mathrm{diff}} \tag{13.119}$$

where $\mathbf{J}_{\mathrm{diff}} = \mathbf{J}_{\mathrm{diff}}^{\mathrm{hole}} + \mathbf{J}_{\mathrm{diff}}^{\mathrm{elec}}$ is identified as

$$\mathbf{J}_{\mathrm{diff}}^{\mathrm{hole}} \rightarrow \frac{\mu_{ij}^{\mathrm{hole}}d_{ij}}{h_{ij}}\left(p_i^1\nu_i B[X_{ij}^0] - p_j^1\nu_j B[-X_{ij}^0]\right) \tag{13.120}$$

$$\mathbf{J}_{\mathrm{diff}}^{\mathrm{elec}} \rightarrow -\frac{\mu_{ij}^{\mathrm{elec}}d_{ij}}{h_{ij}}\left(n_i^1 B[-X_{ij}^0]\,\nu_i - n_j^1 B[X_{ij}^0]\,\nu_j\right) \tag{13.121}$$

after discretization as can be seen from (13.54) and (13.71). The Ampere equation then becomes

$$\nabla\times\mathbf{H} - \frac{\partial\mathbf{D}}{\partial t} = \mathbf{J}_{\mathrm{c}} \tag{13.122}$$

We get after scaling:

$$\nabla\times\left(\frac{1}{\mu_{\mathrm{r}}}\nabla\times\mathbf{A}\right) - k\,(\sigma + \mathrm{i}\omega\varepsilon_{\mathrm{r}})\,(-\nabla V - \mathrm{i}\omega\mathbf{A}) - k\,\mathbf{J}_{\mathrm{diff}} = 0 \tag{13.123}$$

13.9.1 Poisson System

We will now analyze the Poisson system (electric system) and see how a minimal coupling the $\mathbf{A}$ can be obtained. Acting with the $\nabla.$ operator on Equation (13.123) we get:

$$\nabla.\left[(\sigma + \mathrm{i}\omega\varepsilon_r)\,(\nabla V + \mathrm{i}\omega\mathbf{A})\right] + \nabla.\mathbf{J}_{\mathrm{diff}} = 0 \tag{13.124}$$

This is essentially the current-continuity equation that is used for metals. In the remainder we will focus on the trade-off between V and $\mathbf{A}$ and put aside the diffusion current temporary: i.e. we assume that $\mathbf{J}_{\mathrm{diff}} = 0$.

13.9.2 Metals

Integrating over the dual volume ΔV over a node of the computational grid we get:

$$\int_{\Delta V} \nabla. \left[(\sigma + \mathrm{i}\omega\varepsilon_r) \left(\nabla V + \mathrm{i}\omega \mathbf{A} \right) \right] d^3x = 0$$

which is equivalent to:

$$\int_{\partial(\Delta V)} \left[(\sigma + \mathrm{i}\omega\varepsilon_{\mathrm{r}}) \nabla V \right] .d\mathbf{S} + \mathrm{i}\omega \int_{\partial(\Delta V)} \left[(\sigma + \mathrm{i}\omega\varepsilon_r) \mathbf{A} \right] .d\mathbf{S} = 0$$

The last term will now be split into two contributions:

$$\begin{aligned} \int_{\partial(\Delta V)} \left[(\sigma + \mathrm{i}\omega\varepsilon_r) \nabla V \right] .d\mathbf{S} + \mathrm{i}\alpha\omega \int_{\partial(\Delta V)} \left[(\sigma + \mathrm{i}\omega\varepsilon_r) \mathbf{A} \right] .d\mathbf{S} \\ + \mathrm{i}(1-\alpha)\omega \int_{\partial(\Delta V)} \left[(\sigma + \mathrm{i}\omega\varepsilon_r) \mathbf{A} \right] .d\mathbf{S} = 0 \end{aligned} \tag{13.125}$$

Equation (13.125) is fulfilled for all values of α. We can rewrite the gauge condition (13.113) by multiplying with $(\sigma + \mathrm{i}\omega\varepsilon_r)$ and integrating over the dual volume of a node to get:

$$\begin{aligned} \int_{\Delta V} \left[(\sigma + \mathrm{i}\omega\varepsilon_{\mathrm{r}}) \nabla.\mathbf{A} + k\ \xi\ (\sigma + \mathrm{i}\omega\varepsilon_r)(\lambda\sigma + \mathrm{i}\omega\varepsilon_r) V \right] d^3x \ &= \ 0 \\ &\Downarrow \\ \int_{\Delta V} \nabla. \left[(\sigma + \mathrm{i}\omega\varepsilon_r) \mathbf{A} \right] d^3x \ - \int_{\Delta V} \mathbf{A}.\nabla \left[\sigma + \mathrm{i}\omega\varepsilon_{\mathrm{r}} \right] d^3x & \\ + \int_{\Delta V} k\ \xi\ (\sigma + \mathrm{i}\omega\varepsilon_r)(\lambda\sigma + \mathrm{i}\omega\varepsilon_r) V d^3x \ &= \ 0 \end{aligned}$$

or equivalently,

$$\begin{aligned} \int_{\partial(\Delta V)} \left[(\sigma + \mathrm{i}\omega\varepsilon_{\mathrm{r}}) \mathbf{A} \right] .d\mathbf{S} \ &= \ \int_{\Delta V} \mathbf{A}.\nabla \left[\sigma + \mathrm{i}\omega\varepsilon_{\mathrm{r}} \right] d^3x \\ &\quad - \int_{\Delta V} k\ \xi\ (\sigma + \mathrm{i}\omega\varepsilon_{\mathrm{r}})(\lambda\sigma + \mathrm{i}\omega\varepsilon_{\mathrm{r}}) V d^3x \end{aligned}$$

When we substitute (13.126) in to (13.125) we get finally the following equation:

$$\int_{\partial(\Delta V)} (\sigma + \mathrm{i}\omega\varepsilon_{\mathrm{r}})\nabla V.d\mathbf{S} + \mathrm{i}\alpha\omega \int_{\partial(\Delta V)} (\sigma + \mathrm{i}\omega\varepsilon_{\mathrm{r}})\mathbf{A}.d\mathbf{S}$$
$$+\mathrm{i}(1-\alpha)\omega \int_{\Delta V} \mathbf{A}.\nabla(\sigma + \mathrm{i}\omega\varepsilon_r)d^3x$$
$$-\mathrm{i}\omega\,(1-\alpha)\,k\,\xi \int_{\Delta V} (\sigma + \mathrm{i}\omega\varepsilon_r)(\lambda\sigma + \mathrm{i}\omega\varepsilon_r)V d^3x \;=\; 0 \qquad (13.126)$$

This is the standard form in which the equation is programmed. We focus on the last term here:

$$I = \int_{\Delta V} \mathbf{A}.\nabla(\sigma + \mathrm{i}\omega\varepsilon_r)d^3x \qquad (13.127)$$

and assume that

$$(\sigma + \mathrm{i}\omega\varepsilon_{\mathrm{r}}) = (\sigma + \mathrm{i}\omega\varepsilon_{\mathrm{r}})_L + [(\sigma + \mathrm{i}\omega\varepsilon_{\mathrm{r}})_R - (\sigma + \mathrm{i}\omega\varepsilon_{\mathrm{r}})_L]\,\theta(x) \qquad (13.128)$$

with $\theta(x)$ the step function ($\theta(x) = 0$ if $x < 0$ and $\theta(x) = 1$ if $x > 0$). This means that we suppose that at $x = 0$ a material interface exists. This implies that:

$$\nabla(\sigma + \mathrm{i}\omega\varepsilon_{\mathrm{r}}) = [(\sigma + \mathrm{i}\omega\varepsilon_{\mathrm{r}})_R - (\sigma + \mathrm{i}\omega\varepsilon_{\mathrm{r}})_L]\,\delta(x)\mathbf{e_x} \qquad (13.129)$$

with δ denoting the delta function. Hence I becomes:

$$I = \int_{\Delta V} \mathbf{A}.\nabla(\sigma + \mathrm{i}\omega\varepsilon_{\mathrm{r}})d^3x \qquad (13.130)$$
$$= \int_{\Delta V} \mathbf{A}.\mathbf{e_x}\;\delta(x)dxdydz \qquad (13.131)$$
$$= \int_{S_{yz}} \mathbf{A}.\mathbf{e_x}\,[(\sigma + \mathrm{i}\omega\varepsilon_{\mathrm{r}})_R - (\sigma + \mathrm{i}\omega\varepsilon_{\mathrm{r}})_L]\,dydz \qquad (13.132)$$

The latter is a *surface* integral on the simulation domain. In order the perform the programming of this contribution, we must use the link sign to identify the label "R" and "L". For a volume element under consideration, the direction is given by going from the front (tail of the link arrow) node to the back (head of the link arrow) node of the link. So "R" is assigned to the back (head) and "L" is assigned to the front (tail). The approach is illustrated in Figure 13.5.

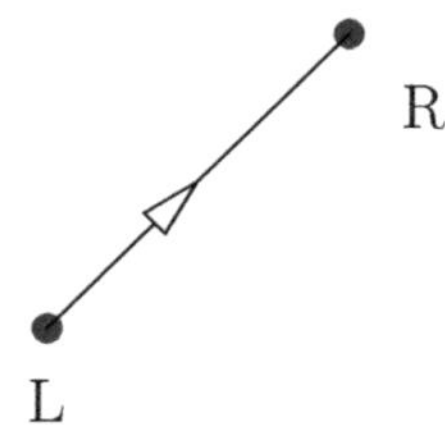

Figure 13.5 Labeling of the front (tail) and back (head) nodes of a link.

We can now exploit the parameters α and ξ to eliminate $\mathbf{A}$ coupling into the Poisson equation. As a consequence, the fill-in of the Newton-Raphson (Jacobian) can be substantially reduced. For example, by setting:

$$\xi = 0, \qquad \alpha = 0 \tag{13.133}$$

the Poisson equation reduces to

$$\int_{\partial(\Delta V)} (\sigma + \mathrm{i}\omega\varepsilon_{\mathrm{r}})\nabla V.d\mathbf{S} + \mathrm{i}\omega \int_{\Delta V} \mathbf{A}.\nabla(\sigma + \mathrm{i}\omega\varepsilon_r)d^3x = 0 \tag{13.134}$$

This equation has only a coupling to $\mathbf{A}$ at material interfaces.

13.9.3 Dielectrics

When $\sigma = 0$, Equation (13.126) becomes:

$$\begin{aligned} &\mathrm{i}\int_{\partial(\Delta V)} d\mathbf{S}\cdot(\varepsilon_r\nabla V) - \alpha\omega\int_{\partial(\Delta V)} .d\mathbf{S}\cdot(\varepsilon_r\mathbf{A}) \\ &+\mathrm{i}(1-\alpha)\omega\left\{\mathrm{i}\int_{\Delta V} d^3x\ \mathbf{A}.\nabla\varepsilon_r + k\omega\int_{\Delta V} d^3x\ \varepsilon_r^2 V d^3x\right\} = 0 \end{aligned}$$

However, it is needed for the matrix assembling to have positive close-to-real entries on the diagonal. Multiplying with i gives

$$\begin{aligned} &-\int_{\partial(\Delta V)} d\mathbf{S}\cdot[\varepsilon_r\nabla V] - \mathrm{i}\alpha\omega\int_{\partial(\Delta V)} d\mathbf{S}\cdot[\varepsilon_r\mathbf{A}] \\ &-(1-\alpha)\omega\left\{\mathrm{i}\int_{\Delta V} d^3x\ \mathbf{A}.\nabla\varepsilon_r + k\omega\int_{\Delta V} d^3x\ \varepsilon_r^2 V\right\} = 0 \end{aligned}$$

Furthermore, it is delicate to derive an expression from a variable which is known to be zero from start (i.e. the current density in insulators). Therefore, we prefer to start from Gauss' equation in insulators and semiconductors.

$$\nabla \cdot \mathbf{D} = \rho \tag{13.135}$$

Using the relation $\mathbf{D} = \varepsilon \mathbf{E}$ and $\mathbf{E} = -\nabla V - \mathrm{i}\,\omega \mathbf{A}$, we find

$$\nabla \cdot [\varepsilon\,(-\nabla V - \mathrm{i}\,\omega \mathbf{A})] - \rho = 0 \tag{13.136}$$

Note that the charge is now present in the complete equation for insulators and semiconductors. The Poisson equation is:

$$-\int_{\partial(\Delta V)} d\mathbf{S} \cdot [\varepsilon_{\mathrm{r}} \nabla V] - j\omega \int_{\Delta V} d^3x\ \nabla \cdot [\varepsilon_{\mathrm{r}} \mathbf{A}] - \int_{\Delta V} d^3x\ \rho = 0 \tag{13.137}$$

Now consider the gauge condition: since $\sigma = 0$ is for insulators or σ is small for semiconductors we use it in the following form:

$$\varepsilon_{\mathrm{r}} \nabla \cdot \mathbf{A} + k\mathrm{i}\omega\ \varepsilon_{\mathrm{r}}^2 V = 0 \tag{13.138}$$

A multiplication with ε_{r} is done. Integration over a cell gives:

$$\int_{\Delta V} d^3x\ \varepsilon_{\mathrm{r}} \nabla \cdot \mathbf{A} + \int_{\Delta V} d^3x\ k\mathrm{i}\omega\varepsilon_{\mathrm{r}}^2 V = 0 \tag{13.139}$$

or

$$\int_{\Delta V} d^3x \nabla \cdot [\varepsilon_{\mathrm{r}} \mathbf{A}] = \int_{\Delta V} d^3x \mathbf{A} \cdot \nabla \varepsilon_{\mathrm{r}} - \mathrm{i}\omega k \int_{\Delta V} d^3x\ \varepsilon_{\mathrm{r}}^2 V \tag{13.140}$$

We write the Poisson equation as follows:

$$\begin{aligned} &-\int_{\partial(\Delta V)} d\mathbf{S} \cdot [\varepsilon_{\mathrm{r}} \nabla V] - \mathrm{i}\omega\alpha \int_{\Delta V} d^3x \nabla \cdot [\varepsilon_{\mathrm{r}} \mathbf{A}] \\ &-\mathrm{i}\omega(1-\alpha) \int_{\Delta V} d^3x\ \nabla \cdot [\varepsilon_{\mathrm{r}} \mathbf{A}] - \int_{\Delta V} d^3x\ \rho = 0 \end{aligned} \tag{13.141}$$

Using the gauge condition we obtain

$$\begin{aligned} &-\int_{\partial(\Delta V)} d\mathbf{S} \cdot [\varepsilon_{\mathrm{r}} \nabla V] - \mathrm{i}\omega\alpha \int_{\partial(\Delta V)} d\mathbf{S}\ \cdot [\varepsilon_{\mathrm{r}} \mathbf{A}] \\ &- (1-\alpha)\mathrm{i}\omega \int_{\Delta V} d^3x\ \mathbf{A} \cdot \nabla \varepsilon_{\mathrm{r}} + (1-\alpha)\mathrm{i}\omega\mathrm{i}\omega\ k \int_{\Delta V} d^3x \varepsilon_{\mathrm{r}}^2 V \\ &- \int_{\Delta V} d^3x\ \rho = 0 \end{aligned} \tag{13.142}$$

This equation is coded as follows:

$$
\begin{aligned}
&- \int_{\partial(\Delta V)} d\mathbf{S} \cdot [\varepsilon_{\rm r} \nabla V] - \mathrm{i}\omega\alpha \int_{\partial(\Delta V)} d\mathbf{S} \cdot [\varepsilon_{\rm r} \mathbf{A}] \\
&- (1-\alpha)\omega^2 \, k \int_{\Delta V} d^3x \varepsilon_{\rm r}^2 V - (1-\alpha)\mathrm{i}\omega \int_{\Delta V} d^3x \, \mathbf{A} \cdot \nabla\varepsilon_{\rm r} \\
&- \int_{\Delta V} d^3x \, \rho = 0
\end{aligned}
\tag{13.143}
$$

It should be noted that the integral

$$
I_\delta = -(1-\alpha)\mathrm{i}\omega \int_{\Delta V} d^3x \, \mathbf{A} \cdot \nabla\varepsilon_{\rm r} \tag{13.144}
$$

is evaluated by visting all volumes. This will introduce double counting of the step function by coming once from the left and once from the right of the interface. This double counting must be corrected.

13.9.4 Maxwell-Ampere System

The starting point of the Maxwell-Ampere system will be Equation (13.123), i.e.

$$
\nabla \times \left(\frac{1}{\mu_{\rm r}} \nabla \times \mathbf{A} \right) + k \, (\sigma + \mathrm{i}\omega\varepsilon_{\rm r}) \, (\nabla V + \mathrm{i}\omega \mathbf{A}) - k \, \mathbf{J}_{\rm diff} = 0 \tag{13.145}
$$

The gauge condition that will be used is again:

$$
\nabla \cdot \mathbf{A} + k\xi(\lambda\sigma + \mathrm{i}\omega\varepsilon_{\rm r}) \, V = 0 \tag{13.146}
$$

For metallic links, σ is very large. Moreover, for metals it is often assumed that conduction currents over-flush the displacement currents, therefore the approximation $\varepsilon_{\rm r} = 0$ in metal should be allowed. Furthermore, $\lambda = 0$ should also be allowed. Therefore, in metals we will use the "standard" implementation. In general for any region is space, we allow:

- Conductors: $\sigma \neq 0$ and $\varepsilon_{\rm r} \geq 0$
- Insulators: $\sigma = 0$ and $\varepsilon_{\rm r} \neq 0$
- Seminconductors: $\sigma \neq 0$ and $\varepsilon_{\rm r} \neq 0$

Furthermore it should be realized that use of the gauge condition to eliminate V coupling is not possible in the Coulomb gauge, $\xi = 0$. Therefore, The "standard" implementation is used for the following cases.

- 'Metallic' links: The link has an volume attached that is metallic
- Coulomb gauge: $\xi = 0$.

13.9.5 "Standard" Implementation

The 'standard' method consists of lifting the singular character of the operator $\nabla \times \nabla\times$ by subtracting the divergence of the gauge condition from Equation (13.145). Then we obtain

$$\nabla \times \left(\frac{1}{\mu_{\mathrm{r}}} \nabla \times \mathbf{A} \right) - \nabla \left(\nabla \cdot \mathbf{A} \right) - \nabla \left[k\xi(\lambda\sigma + \mathrm{i}\omega\varepsilon_{\mathrm{r}})V \right]$$

$$+k \left(\sigma + \mathrm{i}\omega\varepsilon_{\mathrm{r}} \right) \left(\nabla V + j\omega \mathbf{A} \right) - k\, \mathbf{J}_{\mathrm{diff}} = 0 \quad (13.147)$$

This can be rewritten as:

$$\nabla \times \left(\frac{1}{\mu_{\mathrm{r}}} \nabla \times \mathbf{A} \right) - \nabla \left(\nabla \cdot \mathbf{A} \right) + k\mathrm{i}\omega \left(\sigma + \mathrm{i}\omega\varepsilon_{\mathrm{r}} \right) \mathbf{A}$$

$$+k \, \left\{ \left(1 - \lambda\xi \right) \sigma + (1 - \xi)\mathrm{i}\omega\varepsilon_{\mathrm{r}} \right\} \nabla V - k\mathbf{J}_{\mathrm{diff}} = 0 \quad (13.148)$$

13.9.6 Decoupling Implementation

For non-metallic links (defined as the complementary of the above use) and/or for $\xi \neq 0$, we can use the gauge condition the eliminate V terms in favor of $\mathbf{A}$ terms. Again we start with Equation (13.123)

$$\nabla\times \left(\frac{1}{\mu_{\mathrm{r}}} \nabla \times \mathbf{A} \right) + k \, (\sigma + \mathrm{i}\omega\varepsilon_{\mathrm{r}}) \left(\nabla V + \mathrm{i}\omega \mathbf{A} \right) - k \, \mathbf{J}_{\mathrm{diff}} = 0 \quad (13.149)$$

The gauge condition that will be used is again:

$$\nabla \cdot \mathbf{A} + k\xi(\lambda\sigma + \mathrm{i}\omega\varepsilon_{\mathrm{r}}) \, V = 0 \quad (13.150)$$

but it is now rewritten as:

$$V = -\frac{1}{k\xi(\lambda\sigma + \mathrm{i}\omega\varepsilon_{\mathrm{r}})} \nabla \cdot \mathbf{A} \quad (13.151)$$

We write

$$\nabla V = -\frac{1}{k\xi(\lambda\sigma + \mathrm{i}\omega\varepsilon_{\mathrm{r}})} \nabla \left(\nabla \cdot \mathbf{A} \right) \quad (13.152)$$

because we integrate the Maxwell-Ampere system over the area of a link. Then the change in material constant is perpendicular to the vector corresponding to this surface integration. Insertion into (13.149) gives

$$\nabla \times \left(\frac{1}{\mu_{\mathrm{r}}}\nabla \times \mathbf{A}\right) + k\ \mathrm{i}\omega\ (\sigma + \mathrm{i}\omega\varepsilon_{\mathrm{r}})\ \mathbf{A} - \frac{\sigma + \mathrm{i}\omega\varepsilon_{\mathrm{r}}}{\xi\ (\lambda\sigma + \mathrm{i}\omega\varepsilon_{\mathrm{r}})}\nabla\left(\nabla \cdot \mathbf{A}\right) - k\ \mathbf{J}_{\mathrm{diff}} = 0 \tag{13.153}$$

It is interesting to look at a few special cases: Suppose $\lambda = 0$ and $\xi = 1$. Then for insulators and semiconductors (metals are excluded) we find

$$\nabla \times \left(\frac{1}{\mu_{\mathrm{r}}}\nabla \times \mathbf{A}\right) + k\ \mathrm{i}\omega\ (\sigma + \mathrm{i}\omega\varepsilon_{\mathrm{r}})\mathbf{A} - (1 + \frac{\sigma}{\mathrm{i}\omega\varepsilon_{\mathrm{r}}})\nabla\left(\nabla \cdot \mathbf{A}\right) - k\ \mathbf{J}_{\mathrm{diff}} = 0 \tag{13.154}$$

As can be seen in Equation (13.154) a regularization of the singular operator is achieved.

13.10 The Generalized Coulomb Gauge

In this chapter we will introduce a new gauge condition which is a variation of the Coulomb gauge condition. Whereas the latter is a a direct divergence on $\mathbf{A}$ expression, the gauge is as follows:

$$\nabla \cdot \{(\sigma + \mathrm{i}\omega\varepsilon_{\mathrm{r}})\ \mathbf{A}\} = 0 \tag{13.155}$$

Note that the conductance pops up explicitly in the gauge condition. In this gauge the Poisson system simplifies substantially. The current-continuity and the Poisson equation read respectively

$$\nabla \cdot\ [(\sigma + \mathrm{i}\omega\varepsilon_{\mathrm{r}})\ (-\nabla V - \mathrm{i}\omega\mathbf{A})] = 0 \tag{13.156}$$

and

$$\nabla \cdot\ [\varepsilon_{\mathrm{r}}\ (-\nabla V - \mathrm{i}\omega\mathbf{A})] - \rho = 0 \tag{13.157}$$

Using the generalized gauge condition, these equations become

$$\nabla \cdot\ [(\sigma + \mathrm{i}\omega\varepsilon_{\mathrm{r}})\ (-\nabla V)] = 0 \tag{13.158}$$

and

$$\nabla \cdot\ [\varepsilon_{\mathrm{r}}\ (-\nabla V)] - \rho = 0 \tag{13.159}$$

By a proper choice of the gauge condition the coupling to $\mathbf{A}$ is completely disappeared! Let us now consider the Maxwell-Ampere system. The standard method can be applied.

The starting point of the Maxwell-Ampere system will be Equation (13.123), i.e.

$$\nabla \times \left(\frac{1}{\mu_{\mathrm{r}}}\nabla \times \mathbf{A}\right) + k\,(\sigma + \mathrm{i}\omega\varepsilon_{\mathrm{r}})\,(\nabla V + \mathrm{i}\omega\mathbf{A}) - k\,\mathbf{J}_{\text{diff}} = 0 \tag{13.160}$$

All we have to do is to subtract the grad of the gauge condition as is done for the standard implementation, to regularize the singular character of the curl-curl operation. We will introduce a slider ζ for the subtraction

$$\nabla \times \left(\frac{1}{\mu_{\mathrm{r}}}\nabla \times \mathbf{A}\right) - \zeta\nabla\left\{\nabla\cdot(\sigma + \mathrm{i}\omega\varepsilon_{\mathrm{r}})\,\mathbf{A}\right\} + k\,(\sigma + \mathrm{i}\omega\varepsilon_{\mathrm{r}})$$

$$\cdot\,(\nabla V + \mathrm{i}\omega\mathbf{A}) - k\,\mathbf{J}_{\text{diff}} = 0 \tag{13.161}$$

The role of ζ is to have a weight such that the curl-curl expression is of similar size as the subtracted term. Note that the ∇V term survives, but this is harmless since the Gummel cycle stops after the second cycle. For the implementation is suffices to take the standard implementation in the Coulomb gauge and to modify the evaluation of the $\nabla \cdot \mathbf{A}$ term.

13.10.1 Implementation Details of the Ampere-Maxwell System

For each link we consider the dual area and perform an area integral of the Ampere-Maxwell equation. In particular for Equation (13.161) this leads to

$$\int_{\Delta S} d\mathbf{S}\cdot\left[\nabla \times \left(\frac{1}{\mu_{\mathrm{r}}}\nabla \times \mathbf{A}\right)\right] - \zeta\int_{\Delta S} d\mathbf{S}\cdot\nabla\left\{\nabla\cdot(\sigma + \mathrm{i}\omega\varepsilon_{\mathrm{r}})\,\mathbf{A}\right\}$$

$$+k\int_{\Delta S} d\mathbf{S}\cdot\,(\sigma + \mathrm{i}\omega\varepsilon_{\mathrm{r}})\,(\nabla V + \mathrm{i}\omega\mathbf{A}) - k\int_{\Delta S} d\mathbf{S}\cdot\,\mathbf{J}_{\text{diff}} = 0 \tag{13.162}$$

Furthermore, we multiply this equation with the length of the link under consideration, i.e. ΔL. Then the gradient of the second term is replaced by the difference of the integrand in the end node and begin node of the link.

$$\Delta L\int_{\Delta S} d\mathbf{S}\cdot\left[\nabla \times \left(\frac{1}{\mu_{\mathrm{r}}}\nabla \times \mathbf{A}\right)\right]$$

$$-\zeta \int_{\Delta S} d\mathbf{S} \cdot \hat{\mathbf{e}} \left[\{\nabla \cdot (\sigma + \mathrm{i}\omega\varepsilon_{\mathrm{r}})\, \mathbf{A}\}_{back} - \{\nabla \cdot (\sigma + \mathrm{i}\omega\varepsilon_{\mathrm{r}})\, \mathbf{A}\}_{front} \right]$$

$$+ k\, \Delta L \int_{\Delta S} d\mathbf{S} \cdot\; (\sigma + \mathrm{i}\omega\varepsilon_{\mathrm{r}})\, (\nabla V + \mathrm{i}\omega \mathbf{A})$$

$$- k\, \Delta L \int_{\Delta S} d\mathbf{S} \cdot\; \mathbf{J}_{\mathrm{diff}} = 0 \qquad (13.163)$$

where "back" stands for the head of the link and "front" stands for the tail of the link. The area integral are made explicit, i.e. it is now assumed that the integrand is constant over this area. The object $\nabla \times \mathbf{A}$ is computed as

$$\hat{\mathbf{n}} \cdot \nabla \times \mathbf{A} = \frac{1}{\Delta S} \oint_{\partial(\Delta S)} \mathbf{A} \cdot\, d\mathbf{l} \qquad (13.164)$$

where $\hat{\mathbf{n}}$ is a unit vector normal to a surface element. The objects $\nabla \cdot (\sigma + \mathrm{i}\omega\varepsilon_{\mathrm{r}})\, \mathbf{A}$ are computed as

$$\nabla \cdot (\sigma + \mathrm{i}\omega\varepsilon_{\mathrm{r}})\, \mathbf{A} = \frac{1}{\Delta V} \int_{\Delta V} dv\; \nabla \cdot (\sigma + \mathrm{i}\omega\varepsilon_{\mathrm{r}})\, \mathbf{A}$$

$$= \frac{1}{\Delta V} \oint_{\partial(\Delta V)} d\mathbf{S} \cdot\; (\sigma + \mathrm{i}\omega\varepsilon_{\mathrm{r}})\, \mathbf{A}.$$

13.11 The *EV* Solver

In order to reduce the cross coupling between the V and $\mathbf{A}$ system, we will now consider another method to deal with the Ampere-Maxwell system. Consider the starting equation:

$$\nabla \times \left(\frac{1}{\mu_{\mathrm{r}}} \nabla \times \mathbf{A} \right) + k\, (\sigma + \mathrm{i}\omega\varepsilon_{\mathrm{r}})\, (\nabla V + \mathrm{i}\omega \mathbf{A}) - k\, \mathbf{J}_{\mathrm{diff}} = 0 \quad (13.165)$$

For notation convenience we will introduce the notation: $\phi = \sigma + \mathrm{i}\omega\varepsilon_{\mathrm{r}}$. Then we can write (13.165) as

$$\nabla \times \left(\frac{1}{\mu_{\mathrm{r}}} \nabla \times \mathbf{A} \right) + k\, \phi\, (\nabla V + \mathrm{i}\omega \mathbf{A}) - k\, \mathbf{J}_{\mathrm{diff}} = 0 \qquad (13.166)$$

Furthermore, we will need the gauge condition:

$$\nabla \cdot \mathbf{A} + \mathrm{i}\, \omega\xi k\; \varepsilon_{\mathrm{r}} V = 0 \qquad (13.167)$$

where ξ is the slider between zero (Coulomb gauge) and one (Lorenz gauge). The key observation is that for any scalar field, the following equation holds: $\nabla \times \nabla V = 0$.

As a consequence we may write:

$$\nabla \times \left(\frac{1}{\mu_{\mathrm{r}}} \nabla \times \left[\mathbf{A} + \frac{1}{\mathrm{i}\omega} \nabla V \right] \right) + k\, \phi\, (\nabla V + \mathrm{i}\omega \mathbf{A}) - k\, \mathbf{J}_{\mathrm{diff}} = 0 \tag{13.168}$$

This leads to

$$\nabla \times \left(\frac{1}{\mu_{\mathrm{r}}} \frac{1}{\mathrm{i}\omega}\, \nabla \times\, [\mathrm{i}\omega\, \mathbf{A} + \nabla V] \right) + k\, \phi\, (\nabla V + \mathrm{i}\omega \mathbf{A}) - k\, \mathbf{J}_{\mathrm{diff}} = 0 \tag{13.169}$$

We now recognize $\mathrm{i}\omega \mathbf{A} + \nabla V$ as $-\mathbf{E}$, i.e.

$$\mathbf{E} = -\nabla V - \mathrm{i}\, \omega \mathbf{A} \tag{13.170}$$

This then leads to:

$$\nabla \times \left(\frac{1}{\mu_{\mathrm{r}}} \frac{1}{\mathrm{i}\omega}\, \nabla \times\, \mathbf{E} \right) + k\, \phi\, \mathbf{E} + k\, \mathbf{J}_{\mathrm{diff}} = 0 \tag{13.171}$$

In a slightly different version it reads

$$\nabla \times \left(\frac{1}{\mu_{\mathrm{r}}}\, \nabla \times\, \mathbf{E} \right) + k\, \mathrm{i}\, \omega\, \phi\, \mathbf{E} + k\, \mathrm{i}\, \omega\, \mathbf{J}_{\mathrm{diff}} = 0 \tag{13.172}$$

Just as for the **A** system, we must regularize the operator $\nabla \times \nabla \times \mathbf{E}$. This is achieved by subtracting the gauge condition. Using

$$\mathbf{A} = \frac{\mathrm{i}}{\omega} [\mathbf{E} + \nabla V] \tag{13.173}$$

we obtain

$$\nabla \cdot \left\{ \frac{\mathrm{i}}{\omega} [\, \mathbf{E} + \nabla V] \right\} + \mathrm{i}\, \omega\, k\xi\, \varepsilon_{\mathrm{r}}\, V = 0 \tag{13.174}$$

This is equivalent to the following expression:

$$\nabla \cdot \mathbf{E} + \nabla^2\, V + \omega^2\, k\xi\, \varepsilon_{\mathrm{r}}\, V = 0 \tag{13.175}$$

The regularization is now achieved by subtraction the gradient of this equation from Equation (13.172).

$$\begin{aligned} \nabla \times \left(\frac{1}{\mu_{\mathrm{r}}}\, \nabla \times\, \mathbf{E} \right) &- \nabla (\nabla \cdot \mathbf{E}) \,+\, k\, \mathrm{i}\, \omega\, \phi\, \mathbf{E} \\ - \nabla \left(\nabla^2\, V \right) &- \,\omega^2\, k\xi\, \nabla\, (\varepsilon_{\mathrm{r}}\, V) \,+\, k\, \mathrm{i}\, \omega\, \mathbf{J}_{\mathrm{diff}} = 0 \end{aligned} \tag{13.176}$$

As is seen from this equation the coupling to the variables V has strength of order one and is not growing with σ anymore. Furthermore it should be noticed that the Poisson equation can not be used anymore. It is an implicit consequence of the Ampere-Maxwell system. Therefore, the equation that will be used to determine V, is the gauge condition:

$$\nabla^2 V + \nabla \cdot \mathbf{E} + k\, \xi\, \omega^2\, \varepsilon_{\mathrm{r}}\, V = 0 \tag{13.177}$$

With Equation (13.176) to find the solution for $\mathbf{E}$ and Equation (13.177) to find the solution for V, we can compute the full $\mathbf{E}V$ system. The cross couplings will not explode anymore for large σ in the bulk of metal.

13.11.1 Boundary Conditions for the *EV* System

Although no strong coupling exists in the bulk of the material, the boundary conditions introduce again this coupling in some circumstances. The boundary conditions for the $\mathbf{E}$ vector-field Equation (13.176) can be deduced from the boundary conditions for the vector potential $\mathbf{A}$. Since for each link in the surface of the simulation domain we have put the boundary condition $\mathbf{A} \cdot \hat{\mathbf{t}} = 0$, and $\hat{\mathbf{t}}$ is a tangential unit vector, we obtain

$$\mathbf{E} \cdot \hat{\mathbf{t}} = -\hat{\mathbf{t}} \cdot \nabla V \tag{13.178}$$

The boundary conditions for the scalar Equation (16.17), can be deduced from the condition that for surface regions outside the contacts, the outward pointing electric field component is taken equal to zero, i.e. $\mathbf{E} \cdot \hat{\mathbf{n}} = 0$ and $\hat{\mathbf{n}}$ is a tangential unit vector. However, this will not be sufficient to determine the boundary condition for V, since an additional unknown, $\partial V / \partial n$ needs to be given outside the contact regions. Fortunately, there is still room for further restriction. The boundary condition for $\mathbf{A}$ was only provided for the tangential components of $\mathbf{A}$. We will now include also a boundary condition for the normal component of $\mathbf{A}$ that consists of stating that the normal component of $\mathbf{A}$ will have be continuous when crossing the simulation surface:

$$\hat{\mathbf{n}} \cdot \mathbf{A}_{\mathrm{inside}} = \hat{\mathbf{n}} \cdot \mathbf{A}_{\mathrm{outside}} \tag{13.179}$$

This can also be written as $\partial A_{\perp} / \partial\, n = 0$, or in other words: a Neumann boundary condition is used for the perpendicular component of $\mathbf{A}$. However, the surface nodes of the simulation domain can also be determined by applying the Poisson equation and/or current continuity equation for these nodes.

$$\nabla \cdot \{\phi\, \mathbf{E}\} = 0 \tag{13.180}$$

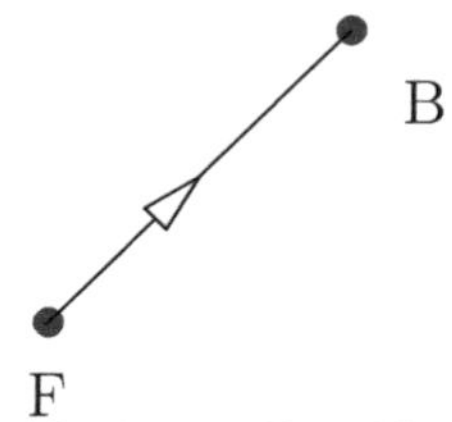

Figure 13.6 Labeling of the front (tail) and back (head) nodes of a link.

For internal nodes, this equation is a consequence of the Ampere-Maxwell system. However, at the surface it must be explicitly enforced by the boundary condition. Thus for the boundary nodes, we apply the usually Poisson and current-continuity equations, using the inwards pointing link variables E_{ij}. This enables us to get boundary conditions for the V variables on the simulation boundary.

13.11.2 Implementation Details of the *EV* System

A subtlety is present with the programming of the transformed equations. We must be beware that the curl-curl operator is calculated for A_{ij} and not for $\mathrm{sgn}_{ij}\ A_{ij}$. In order to re-use this code, we must transform the discretization of $\mathbf{E}$:

$$E_{ij} = -\frac{V_j - V_i}{h_{ij}} - \mathrm{i}\,\omega\ \mathrm{sgn}_{ij}\ A_{ij} \tag{13.181}$$

However, we should think here of $\mathbf{E}$ as a transformation of variables designed to eliminate intense cross coupling of the $\mathbf{A}V$ system. A link commonly has a front (F) node and a back (B) node. The terminology corresponds to arrow running from tail (front) to head (back) (see Figure 13.6). The orientation of the link direction is not related to some preferred direction of the coordinate system. In Figure (13.7) a collection of links is shown, including their orientation. As is seen the orientation of a link is just an convention of which node is used as the front node and which node is seen as the back node. We may now take the unit vector $\hat{\mathbf{n}}$ which points from the front node to the back node and project the electric field on this vector. This gives:

$$\mathbf{E}\cdot\hat{\mathbf{n}} = -\hat{\mathbf{n}}\cdot\nabla V - \mathrm{i}\omega\ \hat{\mathbf{n}}\cdot\mathbf{A} \tag{13.182}$$

The variable A_{ij} is defined as:

$$A_{ij} = \hat{\mathbf{n}}\cdot\mathbf{A} \tag{13.183}$$

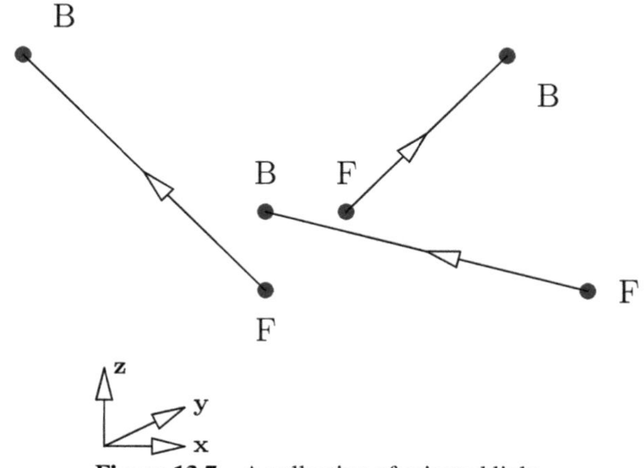

Figure 13.7 A collection of oriented links.

and the transformation leads to a new variable $\mathcal{E}$ which is given by

$$\mathcal{E}_{ij} = \mathbf{E} \cdot \hat{\mathbf{n}} \tag{13.184}$$

The potential term is by construction, with ΔL the length of the link:

$$\hat{\mathbf{n}} \cdot \nabla V = \frac{V_B - V_F}{\Delta L} \tag{13.185}$$

Combining all terms, this gives the following transformation rule:

$$\mathcal{E}_{ij} = \frac{V_F - V_B}{\Delta L} - \mathrm{i}\, \omega\, A_{ij} \tag{13.186}$$

We may compare this expression with Equation (13.181) which we multiply with sgn_{ij}

$$\mathrm{sgn}_{ij} E_{ij} = \mathrm{sgn}_{ij} \frac{V_i - V_j}{h_{ij}} - \mathrm{i}\, \omega\, A_{ij} \tag{13.187}$$

We observe that $\mathcal{E}_{ij} = \mathrm{sgn}_{ij} E_{ij}$ with

$$\begin{aligned} \mathrm{sgn}_{ij} &= 1 \quad \mathrm{i} = \mathrm{front}, \quad \mathrm{j} = \mathrm{back} \\ \mathrm{sgn}_{ij} &= -1 \quad \mathrm{i} = \mathrm{back}, \quad \mathrm{j} = \mathrm{front} \end{aligned} \tag{13.188}$$

Thus we find that

$$\mathcal{E}_{ij} = \mathrm{sgn}_{ij} E_{ij} \tag{13.189}$$

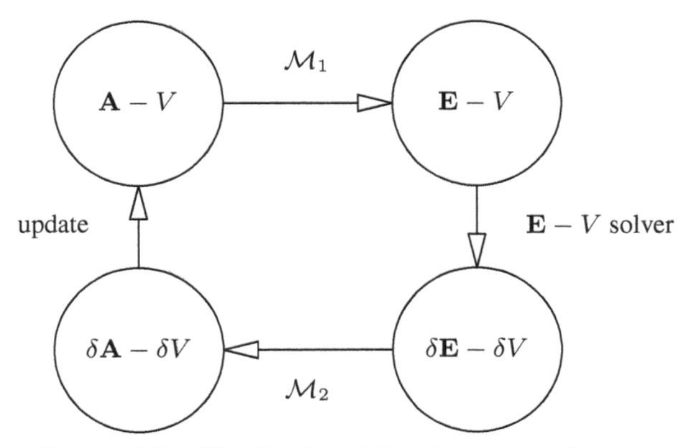

Figure 13.8 Visualization of the solution procedure.

13.11.3 Solution Strategy of the *EV* System

The solution strategy consists of four steps:

- map $\mathbf{A} - V$ onto $\mathbf{E} - V$. This is the map: $\mathcal{M}_1$
- apply the $\mathbf{E} - V$ solver
- map the update $(\delta\mathbf{E}, \delta V)$ onto the update $(\delta\mathbf{A}, \delta V)$ This is the map: $\mathcal{M}_2$
- update the $\mathbf{A} - V$ system

This approach is illustrated in Figure 13.8.

13.12 The Scharfetter-Gummel Discretization

In order to show how the Sharfetter-Gummel trick works, we explain it step by step. Let us start with the following problem along a link:

$$J = \alpha c + \beta \nabla c \tag{13.190}$$

We start with a local coordinate system along the link for which we get:

$$c(0) = c_i \tag{13.191}$$

$$c(h_{ij}) = c_j \tag{13.192}$$

In that case we can see (13.190) as a first order differential equation with a fixed RHS. To solve this we first solve the homogeneous equation:

$$\alpha c_H + \beta \frac{\partial c_H}{\partial x} = 0 \tag{13.193}$$

with the solution:

$$c_H(x) = K \exp\left[\frac{-\alpha}{\beta}x\right] \tag{13.194}$$

The next step is to solve the inhomogeneous equation

$$\alpha c_I + \beta\frac{\partial c_I}{\partial x} = J \tag{13.195}$$

starting with the following solution:

$$c_I(x) = K(x) \exp\left[\frac{-\alpha}{\beta}x\right] \tag{13.196}$$

Introducing (13.196) in (13.195), we get:

$$K(x) = \frac{J}{\alpha}\exp\left[\frac{\alpha x}{\beta}\right] + L \tag{13.197}$$

With solution:

$$\begin{aligned} c_I(x) &= \left[\frac{J}{\alpha}\exp\left[\frac{\alpha x}{\beta}\right] + L\right]\exp\left[\frac{-\alpha}{\beta}x\right] && (13.198)\\ &= \frac{J}{\alpha} + L\exp\left[\frac{-\alpha}{\beta}x\right] && (13.199)\end{aligned}$$

When we introduce the boundary conditions:

$$\begin{aligned} c(0) &= c_i && (13.200)\\ c(h) &= c_j && (13.201)\end{aligned}$$

After some math we come to:

$$c(x) = \frac{c_i\left(\exp\left[\frac{-\alpha x}{\beta}\right] - \exp\left[\frac{-\alpha h}{\beta}\right]\right) + c_j\left(1 - \exp\left[\frac{-\alpha x}{\beta}\right]\right)}{1 - \exp\left[\frac{-\alpha h}{\beta}\right]} \tag{13.202}$$

The next step is to get an expression for the current density J, by substituting the solution for c in (13.195).

$$J = \alpha\frac{-\exp\left[-\frac{\alpha h}{\beta}\right]c_i + c_j}{1 - \exp\left[-\frac{\alpha h}{\beta}\right]} \tag{13.203}$$

In order to make the solution symmetric for a change in the indices i and j, we rewrite (13.203) as:

$$J = \frac{\beta}{h}\left\{\frac{c_i\left[\frac{\alpha h}{\beta}\right]}{1-\exp\left[\frac{\alpha h}{\beta}\right]} - \frac{c_j\left[\frac{-\alpha h}{\beta}\right]}{1-\exp\left[\frac{-\alpha h}{\beta}\right]}\right\} \quad (13.204)$$

$$= \frac{\beta}{h}\left\{-c_i B\left[\frac{\alpha h}{\beta}\right] + c_j B\left[\frac{-\alpha h}{\beta}\right]\right\} \quad (13.205)$$

with the standard definition of the Bernoulli function:

$$B(x) = \frac{x}{e^x - 1} \quad (13.206)$$

The time evolution is described in the Fourier domain. All variables now become dependent on the operation frequency, ω. In the Fourier domain the potential description becomes for the selected gauge

$$\nabla \cdot (\varepsilon \nabla V + j\varepsilon\omega \mathbf{A}) = -\rho \quad (13.207)$$

$$\nabla \times \nabla \times \mathbf{A} - \gamma \nabla \chi = \mu_0 \mathbf{J} - \mathrm{i}\omega\mu_0\varepsilon\nabla V + \mu_0\varepsilon\omega^2 \mathbf{A} \quad (13.208)$$

$$\nabla \cdot \mathbf{A} + \nabla^2 \chi = 0 \quad (13.209)$$

13.12.1 The Static and Dynamic Parts

The electrostatic field, V_0, is obtained by solving the Poisson equation

$$\nabla \cdot (\varepsilon \nabla V_0) = \rho(V_0), \quad (13.210)$$

and the corresponding charge distribution $\rho(V_0)$ must be calculated self-consistently for (a) bounded surface charges on the boundary surfaces of the dielectric regions taking into account the appropriate boundary conditions, (b) free surface charges on the boundaries of a conductor and (c) space charge in the doped semiconductor volume. The current density $\mathbf{J}_0$, gives rise to the vector potential $\mathbf{A}_0$, being the solution of

$$\nabla \times \nabla \times \mathbf{A}_0 - \gamma \nabla \chi_0 = \mu_0 \mathbf{J}_0 (V_0) \quad (13.211)$$

and submitted to the gauge condition

$$\nabla \cdot \mathbf{A}_0 + \nabla^2 \chi_0 = 0 \quad (13.212)$$

Inside conducting media the latter equation is supplemented by

$$\rho = 0 \tag{13.213}$$

$$\mathbf{J}_0 = \sigma \mathbf{E}_0 \tag{13.214}$$

$$\mathbf{E}_0 = -\nabla V_0 \tag{13.215}$$

$$\nabla \cdot \mathbf{J}_0 = 0 \tag{13.216}$$

whereas in the semiconducting regions the following equations apply

$$\rho_0 = q(p_0 - n_0 + N_D - N_A) \tag{13.217}$$

$$\mathbf{J}_{n0} = q\mu_n n_0 \mathbf{E_0} + kT\mu_n \nabla n_0 \tag{13.218}$$

$$\mathbf{J}_{p0} = q\mu_p p_0 \mathbf{E_0} - kT\mu_p \nabla p_0 \tag{13.219}$$

$$\nabla \cdot \mathbf{J}_{n0} - U(n_0, p_0) = 0 \tag{13.220}$$

$$\nabla \cdot \mathbf{J}_{p0} + U(n_0, p_0) = 0 \tag{13.221}$$

The equations that determine the amplitudes and phases of the harmonic perturbations are obtained as linear perturbations of the full system. Returning to Equations (13.207–13.209), one obtains

$$\nabla \cdot (\varepsilon \nabla V_R - \varepsilon \omega \mathbf{A}_I) + \rho_R = 0 \tag{13.222}$$

$$\nabla \cdot (\varepsilon \nabla V_I + \varepsilon \omega \mathbf{A}_R) + \rho_I = 0 \tag{13.223}$$

$$\nabla \times \nabla \times \mathbf{A}_R - \mu_0 \varepsilon \omega^2 \mathbf{A}_R - \mu_0 \mathbf{J}_R - \mu_0 \varepsilon \omega \nabla V_I - \gamma \nabla \chi_R = 0 \tag{13.224}$$

$$\nabla \times \nabla \times \mathbf{A}_I - \mu_0 \varepsilon \omega^2 \mathbf{A}_I - \mu_0 \mathbf{J}_I + \mu_0 \varepsilon \omega \nabla V_R - \gamma \nabla \chi_I = 0 \tag{13.225}$$

$$\nabla^2 \chi_R + \nabla \cdot \mathbf{A}_R = 0 \tag{13.226}$$

$$\nabla^2 \chi_I + \nabla \cdot \mathbf{A}_I = 0 \tag{13.227}$$

where the sources $\mathbf{J}_R$, $\mathbf{J}_I$, ρ_R and ρ_I must be determined by the non-linear constitutive equations.

13.13 Using Unstructured Grids

So far, we have been mainly illustrating the approach using structured grids. However, this was done mainly for 'didactic' reasons. It should be emphasized that the implementation can also be obtained on unstructured grids using a variant of the finite-integration method. In particular, the Stokes theorem applies on general Delaunay grids. It should be emphasized that this is a highly non-trivial observation, since for obtuse volume elements links can have a negative dual area. The evaluation of the circulation around links with negative dual area is compensated by link contributions with large positive dual area such that the net result is in full agreement with the results obtained in grids which do not contain obtuse volume element. One must adapt the discretization scheme to the differential geometrical character of the equation that needs to be solved. The finite-volume method (FVM) naturally matches to discretize the current continuity equation. In an analogous way, a finite-surface method (FSM) is suitable to discretize the Maxwell-Ampere equation.

We demonstrate the validity of our claim by computing the inductance of an integrated spiral inductor as shown in Figure (13.9), using different meshes. The first mesh has a structured grid while the second grid contains generalized elements, but all elements are acute. The third grid also contains obtuse elements which is in general the result from a Delaunay algorithm.

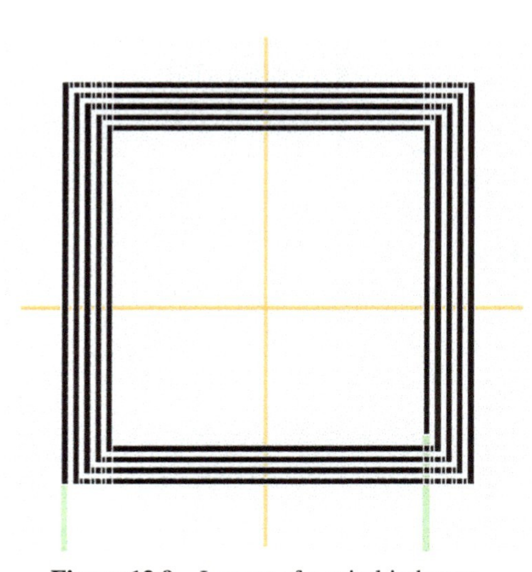

Figure 13.9 Layout of a spiral inductor.

Figure (13.10) shows a detailed view on one of the unstructured grids that is used. In Figure (13.11), the results for the inductance are shown using a structured grid as well as an unstructured grid. Excellent agreement is observed. The computation is achieved by solving the Maxwell-Ampere equation as well as the Poisson equation with voltage boundary conditions. After obtaining balance of the currents, the S-parameters are computed from a two-port system. Each port contains one contact at the inductor and one contact attached to the substrate. Finally, the inductance (and resistance) is extracted from the S-matrix.

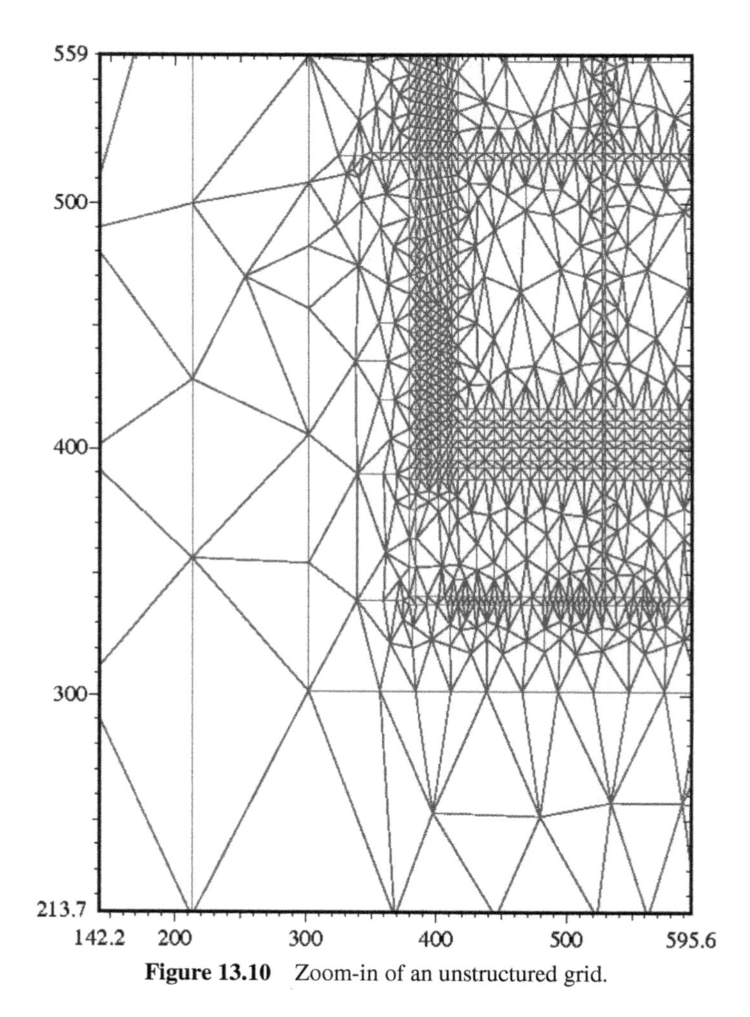

Figure 13.10 Zoom-in of an unstructured grid.

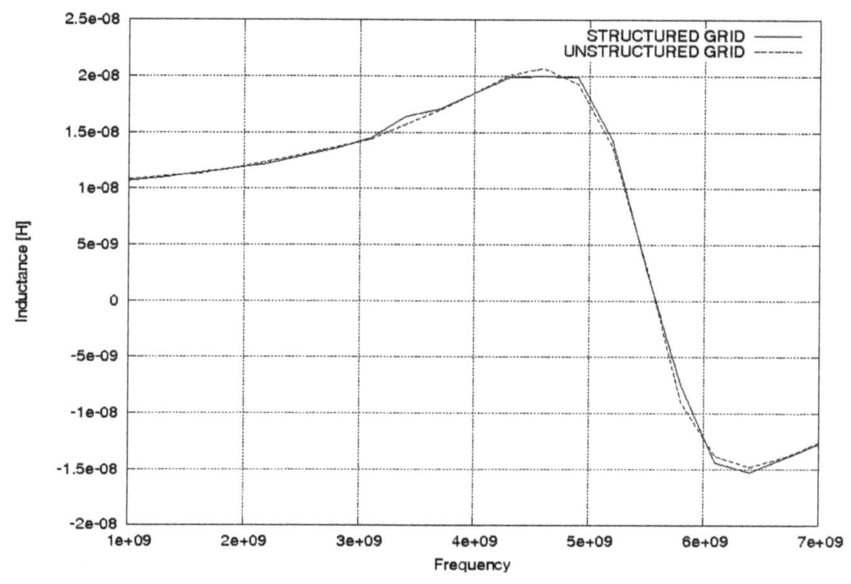

Figure 13.11 Inductances obtained on a structured and unstructured grid.

14

Finite Volume Method and the Transient Regime

This chapter describes the mathematical approach that transient regime. In order to support compact model building, we need ab-initio field solving to verify/falsify approximations that are made, while building the compact models. In the frequency domain, a compact modeling procedure was developed the prior chapter. The main stumbling block in the transient regime is the so called Courant limit[8] which requires unrealistically small time steps for progressing in time.

In this chapter, the transient regime is addressed by implicit methods. This means that a time step is integrated using both the end-point as well as the begin point of the time interval. Our method is inspired by the transient simulation technique which is used in Technology Computer Aided Design (TCAD). In particular, there a backward finite difference is combined with the trapezoidal rule, leading to a rather enhanced time-step size. The following topics will be discussed

- The electrical scalar potential V and the magnetic vector potential $\mathbf{A}$. The computation of $\mathbf{A}$ is complicated by the singular character of the equation for $\mathbf{A}$, leading to the requirement of a gauge condition.
- The discretization scheme demands that variables are placed on grid entities. Here, we present the method that the following grid entities contain fundamental variables: nodes contain the scalar potential V, and the semiconductor variables p, n or ϕ_p, ϕ_n. The links contain the fundamental variable $A = \mathbf{A} \cdot \mathbf{n}$, that means A is the projection of the vector potential along the link direction $\mathbf{n}$.

[8]condition for convergence while solving certain PDEs numerically, arises when using explicit time-stepping methods.

- The ports of the simulation structure, are excited by circuits. Therefore, the set of all equations for the field system is extended with a set of equations that control the port conditions or boundary conditions.
- The material interface conditions must be analyzed in the time domain.

The following questions pop up:

- Can we stay consistent with causality requirements?
- Is the final system of linear equations solvable by iterative methods?

A first technical approach consists of the following scheme. Using the MAGWEL solver, the matrices for the the system-state equations are exported. We consider first field systems that are linear in the time-differential operator. That means that a single system dump suffices for addressing the field problem. After having this problem settled, the more generic approach will be addressed where variables depend in a non-linear way on the state variables. These cases occur when semi-conductors are present.

The goal is to give simulation support to the model developers concerning device behavior in the transient regime. Whereas the frequency regime has already been developed in a rather advanced stage in earlier projects, the transient regime is less exploited. The need for simulations in the transient regime comes from the desire to handle large-signal response.

14.1 The Electromagnetic Drift-Diffusion Solver in the Time Domain

Note: In this section, we will present our equations as $lhs = rhs$, for didactic reason. In a later stage we will collect our results as $lhs = 0$, for computational reasons. Our starting point will be the equations of Maxwell:

$$\text{Gauss' law:} \quad \nabla \cdot \mathbf{D} = \rho \tag{14.1}$$

$$\text{Absence of magnetic monopoles:} \quad \nabla \cdot \mathbf{B} = 0 \tag{14.2}$$

$$\text{Maxwell-Faraday:} \quad \nabla \times \mathbf{E} = -\frac{\partial \mathbf{B}}{\partial t} \tag{14.3}$$

$$\text{Maxwell-Ampère:} \quad \nabla \times \mathbf{H} = \mathbf{J} + \frac{\partial \mathbf{D}}{\partial t} \tag{14.4}$$

where $\mathbf{D}$, $\mathbf{E}$, $\mathbf{B}$, $\mathbf{H}$, $\mathbf{J}$ en ρ are the electric induction, the electric field, magnetic induction and magnetic field, current density and charge density.

The following constitutive laws are used:

$$\mathbf{B} = \mu \mathbf{H}, \qquad \mathbf{D} = \varepsilon \mathbf{E} \tag{14.5}$$

The charge density ρ and current density $\mathbf{J} = 0$ in insulating materials and charge density ρ consists of a fixed background charge. In conductive domains we rely on the current continuity:

$$\nabla \cdot \mathbf{J} + \frac{\partial \rho}{\partial t} = 0 \tag{14.6}$$

If these conductive domains are metallic, then we apply Ohm's law for the connection between electric field intensity and current density. It should be noted that this is not the most general expression and for Hall devices a magnetic field dependence must also be included. However, here we limit ourselves to situations where the magnetic fields are sufficiently weak in order to ignore Hall currents:

$$\mathbf{J} = \sigma \mathbf{E} \tag{14.7}$$

It should also be noted that the charge density and current densities are determined by the physical character of the materials under study. For example leakage currents can flow in insulating layers and the current-field relation is highly non-linear since tunneling mechanisms play an important role. Semiconductors also have more general current-field relations as given above and these will be discussed later.

We introduce the scalar potential V en the magnetic vector potential $\mathbf{A}$ that satisfy

$$\mathbf{B} = \nabla \times \mathbf{A} \tag{14.8}$$

$$\mathbf{E} = -\nabla V - \frac{\partial \mathbf{A}}{\partial t} \tag{14.9}$$

then the Maxwell equation become in these variables;

- *for insulators:*

$$-\nabla \cdot \left[\varepsilon \left(\nabla V + \frac{\partial \mathbf{A}}{\partial t} \right) \right] = 0 \tag{14.10}$$

$$\nabla \times \frac{1}{\mu} (\nabla \times \mathbf{A}) = -\varepsilon \frac{\partial}{\partial t} \left(\nabla V + \frac{\partial \mathbf{A}}{\partial t} \right) \tag{14.11}$$

- *for conductors:*

$$-\nabla \cdot \sigma \left(\nabla V + \frac{\partial \mathbf{A}}{\partial t} \right) = \frac{\partial}{\partial t} \left(\nabla \cdot \varepsilon \left(\nabla V + \frac{\partial \mathbf{A}}{\partial t} \right) \right) \tag{14.12}$$

$$\nabla \times \frac{1}{\mu} (\nabla \times \mathbf{A}) = -\sigma \left(\nabla V + \frac{\partial \mathbf{A}}{\partial t} \right) - \varepsilon \frac{\partial}{\partial t} \left(\nabla V + \frac{\partial \mathbf{A}}{\partial t} \right) \tag{14.13}$$

For the description in the Fourier domain we replace each differentiation w.r.t. time by a factor $\mathrm{i}\omega$, with i the imaginary unit and $\omega = 2\pi f$ the angular velocity and f is the operational frequency.

These differential equations are with the MAGWEL software discretized in 3D-space using the finite-volume method (FVM) and finite-surface method (FSM). Whereas the FVM is based on averaging variables over cells to obtain discrete variables the FSM averages method obtains discrete variables by averaging over surfaces. These averaging procedures apply Gauss' law (FVM) and Stokes' law (FSM). The success of the method is based on respecting the geometrical origin of the various variables that are encountered in the mathematical set up.

For the description in the time domain we can reuse the spatial discretization methods. However, we will end up with a second-order differentiation in time. Part of the work in this chapter will be to give a procedure for a correct treatment of these terms. Physically, the second order time-derivative terms illustrate the wave delay that is found in the Maxwell equations. A popular argument for handling these terms is based on considering the scales of application. In particular, if we are operating in the below 100 GHz range, the wave length is 3 mm=3000 microns. Assuming that typical lengths inside the chip is below this value, it is argued that these 2nd order time derivatives could be ignored. Thus a 'crude' method just ignores the second order time differential on $\mathbf{A}$. As a consequence we arrive at some form of a quasi-static approximation and one is not accounting for the delay induced by wave propagation. Another way of looking at this approximation is to assume that speed of light is infinity. In the PEEC method this approximation is exploited (*PEEC* – Partial Element Equivalent Circuit).

Dropping terms out of equations should be done with care. Although a term may seem irrelevant on an instantaneous view, it must be considered also from the view how it participates in keeping the global behavior physically correct. For instance, spontaneous creation of charge is prohibited and dropping one term might induce the need to drop corresponding terms for physical consistency. In order to explore these pitfalls, we propose another representation.

The second approach introduces a new variable the *pseudo-canonical momentum* $\Pi = \partial \mathbf{A}/\partial t$. We may rewrite the system of equations as (e.g. for a conductor):

$$-\nabla \cdot \sigma \left(\nabla V + \mathbf{\Pi}\right) = \nabla \cdot \varepsilon \left(\nabla \frac{\partial V}{\partial t} + \frac{\partial \mathbf{\Pi}}{\partial t}\right) \tag{14.14}$$

$$\nabla \times \left(\frac{1}{\mu}\nabla \times \mathbf{A}\right) = -\sigma\left(\nabla V + \frac{\partial \mathbf{A}}{\partial t}\right) - \varepsilon\left(\nabla\frac{\partial V}{\partial t} + \frac{\partial \mathbf{\Pi}}{\partial t}\right) \tag{14.15}$$

$$\mathbf{\Pi} = \frac{\partial \mathbf{A}}{\partial t} \tag{14.16}$$

We refer to the variable $\mathbf{\Pi}$ as a pseudo-canonical momentum, because when deriving the field equations from an Lagrange action, the electric field $-\mathbf{E}$ is found as the canonical conjugate to $\mathbf{A}$. The difference between the pseudo-canonical momentum and the canonical momentum is

$$\mathbf{\Pi} = -\mathbf{E} - \nabla V \tag{14.17}$$

14.2 Gauge Conditions

The Equations (14.12–14.13) do not uniquely determine $\mathbf{A}$ and V. A solution can be $\mathbf{A}$, V can be adapted with a arbitrary scalar field χ

$$\begin{aligned} \mathbf{A}' &= \mathbf{A} + \nabla\chi \\ V' &= V - \frac{\partial \chi}{\partial t} \end{aligned}$$

which results into an equally valid solution of (14.12–14.13). After discretization the coefficient-matrix is singular. Additional equations must be added to elevate this singularity. These extra equations are the gauge conditions. For the Coulomb gauge the following constraint is applied:

$$\nabla \cdot \mathbf{A} = \mathbf{0} \tag{14.18}$$

The Lorentz gauge is inspired by dealing with the term $\epsilon\nabla\left(\frac{\partial V}{\partial t}\right)$ in the Ampere-Maxwell equation:

$$\frac{1}{\mu}\nabla\left(\nabla \cdot \mathbf{A}\right) + \epsilon\nabla\left(\frac{\partial V}{\partial t}\right) = 0 \tag{14.19}$$

Usually we encounter the Lorenz gauge condition as

$$\nabla \cdot \mathbf{A} + \mu\epsilon\frac{\partial V}{\partial t} = 0 \tag{14.20}$$

However, we prefer to keep it in the form (14.19) because we need it in this way and moreover the form (14.20) is only equivalent to (14.19) for μ and ϵ constant. The pre-factor $1/\mu$ is a choice which is convenient in our field of application of electronics design automation (EDA). We are dealing mostly with materials that are not ferromagnetic. Then $\mu_r = 1$, which allows an efficient evaluation of the Maxwell-Ampere equation.

Above argument is somewhat 'naive'. Actually, we do skip two terms in the Ampere-Maxwell equation by first adding and subtracting $1/\mu\nabla(\nabla\cdot\mathbf{A})$ and use one to regularize the double curl operator.

For later use, we write the Lorentz gauge condition as

$$\epsilon\nabla\left(\frac{\partial V}{\partial t}\right) = -\frac{1}{\mu}\nabla(\nabla\cdot\mathbf{A}) \tag{14.21}$$

The Maxwell-Ampere equation can be written as

$$\epsilon\frac{\partial\mathbf{\Pi}}{\partial t} = -\nabla\times\left(\frac{1}{\mu}\nabla\times\mathbf{A}\right) + \frac{1}{\mu}\nabla(\nabla\cdot\mathbf{A}) - \sigma\nabla V - \sigma\mathbf{\Pi} \tag{14.22}$$

Furthermore, remember that

$$\frac{\partial\mathbf{A}}{\partial t} = \mathbf{\Pi} \tag{14.23}$$

then we arrive at the following condensed notation for the transient description

$$\varepsilon\frac{\partial}{\partial t}\begin{bmatrix} V \\ \mathbf{A} \\ \mathbf{\Pi} \end{bmatrix} = \mathcal{K} * \begin{bmatrix} V \\ \mathbf{A} \\ \mathbf{\Pi} \end{bmatrix} + \mathcal{B} * \begin{bmatrix} V_{dbc} \\ \mathbf{A}_{dbc} \\ \mathbf{\Pi}_{dbc} \end{bmatrix} \tag{14.24}$$

where $\mathcal{K}$ is a 3x3 matrix that can be generated in the MAGWEL solver, and $\mathcal{B}$ is an operator acting on the boundary-condition prescribed values. The V_{dbc}, $\mathbf{A}_{dbc}$ and $\mathbf{\Pi}_{dbc}$ are given values Dirichlet boundaries.

The $\mathcal{K}$ matrix represents the following operators:

$$\mathcal{K} = \begin{bmatrix} 0 & -\frac{1}{\mu}\nabla\cdot & 0 \\ 0 & 0 & \epsilon \\ -\sigma\nabla & -\nabla\times\left[\frac{1}{\mu}\nabla\times\ \right] + \frac{1}{\mu}\nabla\left[\nabla\cdot\ \right] & -\sigma \end{bmatrix} \tag{14.25}$$

It should be noted that a formal solution exists that can be found by the method of 'variations of constants'. The solution is

$$\begin{bmatrix} V \\ \mathbf{A} \\ \mathbf{\Pi} \end{bmatrix} = \int_0^t \mathrm{d}t' \mathrm{e}^{\frac{\mathcal{K}}{\varepsilon}(t-t')} * \frac{\mathcal{B}}{\varepsilon} * \begin{bmatrix} V_{dbc}(t') \\ \mathbf{A}_{dbc}(t') \\ \mathbf{\Pi}_{dbc}(t') \end{bmatrix} + \begin{bmatrix} V_0 \\ \mathbf{A}_0 \\ \mathbf{\Pi}_0 \end{bmatrix} \tag{14.26}$$

where the last represents the field configuration at $t = 0$. An interesting case corresponds to switch-on response. In that case the initial solution is zero and the boundary terms are:

$$\begin{bmatrix} V_{dbc}(t) \\ \mathbf{A}_{dbc}(t) \\ \mathbf{\Pi}_{dbc}(t) \end{bmatrix} = \begin{bmatrix} V_{dbc} \\ \mathbf{A}_{dbc} \\ \mathbf{\Pi}_{dbc} \end{bmatrix} \Theta(t) \tag{14.27}$$

and $\Theta(t)$ is the step function. Then the solution is:

$$\begin{bmatrix} V \\ \mathbf{A} \\ \mathbf{\Pi} \end{bmatrix} (t) = \mathcal{K}^{-1} * \left(1 - e^{\frac{\mathcal{K}}{\epsilon}t}\right) * \mathcal{B} * \begin{bmatrix} V_{dbc} \\ \mathbf{A}_{dbc} \\ \mathbf{\Pi}_{dbc} \end{bmatrix} \tag{14.28}$$

Note that in the evolution of the system, the gauge condition is needed to describe the time dependence of V. Quite remarkably, the Gauss' equation is not used for that purpose. Gauss' law can be written as

$$\mathcal{G} * \begin{bmatrix} V \\ \mathbf{A} \\ \mathbf{\Pi} \end{bmatrix} = \mathcal{Q} \tag{14.29}$$

The $\mathcal{G}$ matrix is represented by the 1×3 operator:

$$\mathcal{G} = \begin{bmatrix} \nabla[-\epsilon[\nabla]], & 0, & -\nabla[\epsilon \ ] \end{bmatrix} \tag{14.30}$$

In more familiar language:

$$\nabla \cdot [\epsilon \left(-\nabla V - \mathbf{\Pi}\right)] = \rho \tag{14.31}$$

Gauss' law is a constraint between 'canonical' coordinates and momenta. Every update from one time instant to the next must be compliant with this constraint. In order to have a correct physical evolution it is needed that the Gauss' law is also obtained after a forward sweep in time. We suspect that this is guaranteed if the operators $\mathcal{K}$ and $\mathcal{G}$ commute. Thus a discrete implementation requires that

$$[\mathcal{K}, \mathcal{G}] = 0 \tag{14.32}$$

The discretization procedure generates explicit expression for the matrices $\mathcal{K}$ and $\mathcal{G}$ and therefore their commutation can be checked.

It is important to realize that a *transient* simulator solves the temporal evolution in terms of:

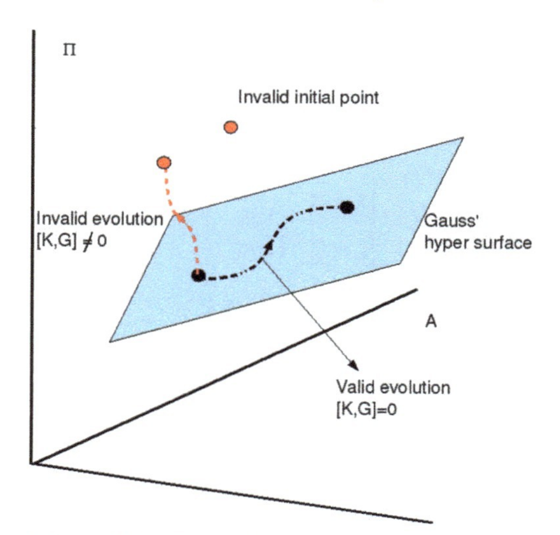

Figure 14.1 Illustration of the use of Gauss' law.

- The *gauge condition* for computing the changes in time of V
- The *definition* of the pseudo-canonical momentum $\mathbf{\Pi}$ for computing the changes in time of $\mathbf{A}$
- The *Maxwell-Ampere equation* for computing the changes in time of $\mathbf{\Pi}$.
- Gauss law does *not* play a role for determining the time-evolution, but informs us that we can not start from an arbitrary $(V, \mathbf{A}, \mathbf{\Pi})$ – configuration, but one that satisfies Gauss' law.

The last item is illustrated in Figure 14.1. In the $(V, \mathbf{A}, \mathbf{\Pi})$ – configuration space there is a hyper surface compliant with Gauss' law. An initial point should be located at this hyper surface. Next the time-evolution operator $\mathcal{K}$ should guarantee that the flow remains on this hyper surface.

14.3 Semiconductor Treatment

For the simulation of semiconductor regions we use the current-continuity equations

$$\nabla \cdot \mathbf{J}_{\mathrm{n}} - q\frac{\partial n}{\partial t} = U(n,p), \tag{14.33}$$

$$\nabla \cdot \mathbf{J}_{\mathrm{p}} + q\frac{\partial p}{\partial t} = -U(n,p). \tag{14.34}$$

Here, n and p are the electron and hole concentrations and $U(n,p)$ represent the generation/recombination mechanisms. The drift-diffusion model provides us with explicit relations between the electron and hole current densities and the electric field intensity as well as the carrier concentration

$$\mathbf{J}_\mathrm{n} = q\mu_\mathrm{n} n\mathbf{E} + qkT\mu_\mathrm{n}\nabla n, \tag{14.35}$$

$$\mathbf{J}_\mathrm{p} = q\mu_\mathrm{p} p\mathbf{E} - qkT\mu_\mathrm{p}\nabla p. \tag{14.36}$$

The charge density ρ is

$$\rho = q(p - n + N_\mathrm{D} - N_\mathrm{A}) \tag{14.37}$$

The carrier concentrations are modeled using the Boltzmann distribution, i.e. the amount of carriers with a given amount of energy is proportional to the exponential of the $-E/kT$. Introducing the Fermi potentials ϕ_n and ϕ_p the following relation holds:

$$n = n_i\, \mathrm{e}^{\frac{q}{kT}(V-\phi_n)}, \qquad p = n_i\, \mathrm{e}^{\frac{q}{kT}(\phi_p - V)},$$

and n_i is the intrinsic carrier concentration.

We can add the following set of equations for the description of semiconductors (14.38)–(14.41)

$$-\nabla\cdot\varepsilon\left(\nabla V + \frac{\partial\mathbf{A}}{\partial t}\right) = qn_i\,\mathrm{e}^{\frac{q}{kT}(\phi_p - V)} - qn_i\,\mathrm{e}^{\frac{q}{kT}(V-\phi_n)} + N_\mathrm{D} - N_\mathrm{A} \tag{14.38}$$

$$\begin{aligned}\nabla\times\frac{1}{\mu}(\nabla\times\mathbf{A}) = {} & q\mu_p n_i\,\mathrm{e}^{\frac{q}{kT}(\phi_p - V)}\left(\nabla V + \frac{\partial\mathbf{A}}{\partial t}\right)\\ & - qkT\mu_p\nabla n_i e^{\frac{q}{kT}(\phi_p - V)}\\ & + q\mu_n n_i\,\mathrm{e}^{\frac{q}{kT}(V-\phi_n)}\left(\nabla V + \frac{\partial\mathbf{A}}{\partial t}\right)\\ & + qkT\mu_n\nabla n_i e^{\frac{q}{kT}(V-\phi_n)}\\ & - \varepsilon\frac{\partial}{\partial t}\left(\nabla V + \frac{\partial\mathbf{A}}{\partial t}\right)\end{aligned} \tag{14.39}$$

$$\nabla \cdot \left(q\mu_p \, n_i e^{\frac{q}{kT}(\phi_p - V)} \left(\nabla V + \frac{\partial \mathbf{A}}{\partial t} \right) - qkT\mu_p \nabla e^{\frac{q}{kT}(\phi_p - V)} \right) \\ = -U(n,p) - qn_i \frac{\partial}{\partial t} \left(e^{\frac{q}{kT}(\phi_p - V)} \right) \quad (14.40)$$

$$\nabla \cdot \left(q\mu_n \, n_i e^{\frac{q}{kT}(V - \phi_n)} \left(\nabla V + \frac{\partial \mathbf{A}}{\partial t} \right) + qkT\mu_n \nabla e^{\frac{q}{kT}(V - \phi_n)} \right) \\ = U(n,p) + qn_i \frac{\partial}{\partial t} \left(e^{\frac{q}{kT}(V - \phi_n)} \right) \quad (14.41)$$

We have presented here all expressions explicitly in terms of the potentials in order to illustrate the non-linear character of the equation system.

The set of unknowns $(V, \mathbf{A}, \mathbf{\Pi})$ needs to be extended with the Fermi levels ϕ_p and ϕ_n or alternatively with p and n. In this case we can add the following set of equations for the description of semiconductors (14.38–14.41)

$$-\nabla \cdot \varepsilon \left(\nabla V + \frac{\partial \mathbf{A}}{\partial t} \right) = q(p - n) + N_{\mathrm{D}} - N_{\mathrm{A}} \quad (14.42)$$

$$\nabla \times \frac{1}{\mu} (\nabla \times \mathbf{A}) = \underbrace{-q\mu_p p \left(\nabla V + \frac{\partial \mathbf{A}}{\partial t} \right) - qkT\mu_p \nabla p}_{:=\mathbf{J}_p} \\ \underbrace{-q\mu_n n \left(\nabla V + \frac{\partial \mathbf{A}}{\partial t} \right) + qkT\mu_n \nabla n}_{:=\mathbf{J}_n} \\ - \varepsilon \frac{\partial}{\partial t} \left(\nabla V + \frac{\partial \mathbf{A}}{\partial t} \right) \quad (14.43)$$

$$\nabla \cdot \left(q\mu_p p \left(\nabla V + \frac{\partial \mathbf{A}}{\partial t} \right) - qkT\mu_p \nabla p \right) = -U(n,p) - q\frac{\partial p}{\partial t} \quad (14.44)$$

$$\nabla \cdot \left(q\mu_n n \left(\nabla V + \frac{\partial \mathbf{A}}{\partial t} \right) + qkT\mu_n \nabla n \right) = U(n,p) + q\frac{\partial n}{\partial t} \quad (14.45)$$

In the frequency domain we have linearized the equation at some operation point

$$X_{op} = (\mathbf{E}_0, p_0, n_0) \quad (14.46)$$

$$\mathbf{J}_{\mathrm{n}} = q\mu_{\mathrm{n}}n_0\mathbf{E} + q\mu_{\mathrm{n}}n\mathbf{E}_0 + qkT\mu_{\mathrm{n}}\nabla n \tag{14.47}$$

$$\mathbf{J}_{\mathrm{p}} = q\mu_{\mathrm{p}}p_0\mathbf{E} + q\mu_{\mathrm{p}}p\mathbf{E}_0 - qkT\mu_{\mathrm{p}}\nabla p, \tag{14.48}$$

where the operation point is determined by a solution of the static bias conditions $(\mathbf{E}_0, p_0, n_0)$. In the time domain, we are interested in a direct integration in time to explore the solution. Linearization methods will have only restricted value.

14.4 Implementation of Numerical Methods for Solving the Equations

The discretization procedure will be separated into two parts. First we will described the handling of the fields on the discrete spatial grid. After the spatial discretization, we have transform our problem definition from having a finite set of fields $\psi_n(\mathbf{x}, t)$ to a lattice of variables $\psi_{n,k}(t)$, where n is an index to the field under consideration e.g. $V, \mathbf{A}, \phi_p, \phi_n$ and k labels a grid object, i.e. a node or a link. So far each variable is still a continuous function of time. In the second part, we will discuss how the transient problem will be numerically addressed.

14.5 Spatial Discretization

In order to solve the continuous partial differential equations a transition must be made to a discretization grid. How this is done in general is well-known in the literature. Here, we will emphasize the main ingredients as well as the subtle details that come with our specific set of equations under consideration.

A convenient mental picture is to address the discretization in three steps: First we replace the continuous 'universe' by a lattice of discrete points. Next we replace the continuous differential operators by local coupling between neighboring lattice points, and thirdly, we cut out of the 'entire universe' a finite portion and give conditions (usually idealized ones) how the rest of universe interferes with the portion under consideration. We refer to this finite portion as the simulation domain, Ω which has an enclosing surface $\partial\Omega$. In a nutshell this is what discretization is about. Already right at the start we encounter a subtle detail. Cutting out a piece out of large geometrical entity and remembering that our starting equations have a local character, it would mean that the impact of the external domain would take place only via the surface of the simulation domain. In other words boundary conditions are expected on the surface $\partial\Omega$. Although this makes sense from a mathematical perspective, it has been shown beneficial to allow for 'internal' boundary conditions also. Since the external part is not part of our computation problem ("by definition") all the lattice variables are corresponding to this part have no representation in the computer memory. There presence is absorbed in a series of restrictions for the lattice variables of the internal, i.e. those variables that have a chunk of memory allocated. Within this view, one may consider boundary conditions as a set of limitations to construct a solvable problem only the word 'boundary' should not be taken too literally.

14.6 Discretization of Gauss' Law

As an illustration of the discretization method, we will here discretize Gauss' law[9]. The starting equation is

$$-\nabla\left[\epsilon\left(\nabla V+\frac{\partial \mathbf{A}}{\partial t}\right)\right]-\rho=0 \tag{14.49}$$

On a discretization grid we consider the Voronoi cells around each node, say i, of the grid and take take the volume integral of Equation (14.49). Applying Gauss' law we find that

$$\int_{\partial(\Delta v)}\mathbf{dS}\cdot(-\epsilon)\left(\nabla V+\frac{\partial \mathbf{A}}{\partial t}\right)-Q(\Delta v)=0 \tag{14.50}$$

Using the geometrical information of the Voronoi cells this amounts to adding all contribution from each link that ends or begins in node i

$$\sum_j \epsilon\left[\frac{S_{ij}}{h_{ij}}(V_i-V_j)-\sigma_{ij}\frac{\mathrm{d}A_{ij}}{\mathrm{d}t}S_{ij}\right]-Q_i=0. \tag{14.51}$$

In this expression, the sum j is over all neighboring nodes. S_{ij} is the perpendicular area for the link connecting node i and node j. The variable h_{ij} is the length of the link ij. Furthermore, A_{ij} is the projection of the vector potential on link (ij) and is also a degree of freedom. The variable $\sigma_{ij}=\pm 1$. It represents the relative orientation of the vector potential with respect to the grid orientation.

[9]It should be emphasized that the purpose of this section is to show typical discretization steps. Remember that the role of Gauss' law is to provide constraints on the initial values and that the time evolution should be compliant with it.

14.7 Boundary Conditions for Gauss' Discretized Law

We are now in the position to consider the boundary conditions for above set of equations. Let $\mathcal{N}_\infty$ be the collection of all lattice nodes and links for the 'universe', let $\mathcal{N}_{sim}$ be the collection of all lattice nodes and link that participate in the simulation problem. How can we get rid off $\mathcal{N}_\infty - \mathcal{N}_{sim}$? A simple idea is that there are parts of the surface of the simulation domain where the interaction is not present. This means that the perpendicular displacement at $\partial\Omega$ is zero, i.e $\mathbf{D}.\mathbf{n} = 0$. As a consequence, we can assemble (14.51) just as if the the rest of the universe does not exists. The assembling is illustrated in the introduction. This approach is known as putting Neumann boundary conditions.

Another elimination procedure is to assume that the impact of the rest of the universe is screened off by prescribed values at (other) segments of $\partial\Omega$. Voltage sources are a physical realization. These are the Dirichlet's boundary conditions.

We can now consider Equation (14.51) in more detail. Actually as it stands it reads as follows:

$$\sum_j \epsilon \left[\frac{S_{ij}}{h_{ij}} (V_i - V_j) - \sigma_{ij} \frac{\mathrm{d}A_{ij}}{\mathrm{d}t} S_{ij} \right] - Q_i = 0 \qquad i \subset \mathcal{N}_\infty \tag{14.52}$$

This is now further specified as follows: The set $\mathcal{N}_{sim}$ consists of two groups. The first set contains the true degrees of freedom. We refer to this set as $\mathcal{N}_{sim}^{dof}$. The second set contains the Dirichlet boundary condition set, denoted as $\mathcal{N}_{sim}^{dbc}$. Let us also introduce a notation convenience: time derivatives will be denoted with a dot.

$$\sum_j \epsilon \left[\frac{S_{ij}}{h_{ij}} (V_i - V_j) - \sigma_{ij} \dot{A}_{ij} S_{ij} \right] - Q_i = 0 \qquad i \subset \mathcal{N}_{sim}^{dof} \tag{14.53}$$

Next we can consider the sum at the left-hand side. For each degree-of-freedom, we can separate the sum into two sets also. The first set contains the coupling to other degrees of freedom ('internal' nodes) and the second set contains the coupling to the Dirichlet boundary nodes. Thus we obtain

$$\begin{aligned} &\sum_{j \subset \mathcal{N}_{sim}^{dof}} \epsilon \left[\frac{S_{ij}}{h_{ij}} (V_i - V_j) - \sigma_{ij} \dot{A}_{ij} S_{ij} \right] \\ &+ \sum_{j \subset \mathcal{N}_{sim}^{dbc}} \epsilon \left[\frac{S_{ij}}{h_{ij}} (V_i - V_j) - \sigma_{ij} \dot{A}_{ij} S_{ij} \right] - Q_i = 0 \quad i \subset \mathcal{N}_{sim}^{dof} \end{aligned} \tag{14.54}$$

Finally, the second sum can be split as shown below:

$$\sum_{j \subset \mathcal{N}_{sim}^{dof}} \epsilon \left[\frac{S_{ij}}{h_{ij}} (V_i - V_j) - \sigma_{ij} \dot{A}_{ij} S_{ij} \right] + V_i \sum_{j \subset \mathcal{N}_{sim}^{dbc}} \epsilon \frac{S_{ij}}{h_{ij}}$$

$$- \sum_{j \subset \mathcal{N}_{sim}^{dbc}} \sigma_{ij} \dot{A}_{ij} S_{ij} - Q_i = \sum_{j \subset \mathcal{N}_{sim}^{dbc}} \epsilon \frac{S_{ij}}{h_{ij}} V_j, \qquad i \subset \mathcal{N}_{sim}^{dof}$$

Recognizing all V_i as degrees of freedom and all V_j as prescribed values we see that the discretized system finally take the form:

$$\mathbf{A} * \mathbf{x} = \mathbf{b} \tag{14.55}$$

Here, we assumed that ρ and therefore Q_i is independent of the voltages. For semiconductors being present this is not true, and we end up with a system that is non-linear in the voltages. Then Newton-Raphson schemes are needed to solve the full system.

We have not mentioned yet the detailed discretization for the projected vector potentials however, as for as the discussion for the Gauss' law is concerned, all A_{ij} are degrees of freedom. This is because the links on $\partial\Omega$ have a simple Dirichlet-type boundary condition, $A_{ij} = 0(ij) \subset \partial\Omega$. In fact, this matrix consists of two parts in this example. To be more precise, it takes the form[10]

$$\left(\mathbf{A}_0 + \mathbf{A}_1 \frac{\mathrm{d}}{\mathrm{d}t} \right) * \mathbf{x} = \mathbf{b} \tag{14.56}$$

where $\mathbf{A}_1$ acts on the link degrees of freedom.

In a first phase of the project, we will make available the matrix $\mathbf{A}$ explicitly for linear problems. To be more precise; starting from the existing MAGWEL solver in the frequency domain, we can assemble the equations as is done for the harmonic analysis and extract explicitly the matrices $\mathbf{A}_0$ and $\mathbf{A}_1$. That means that no semiconductors are present in this phase. It also means that the matrices $\mathbf{A}$ can be used at all time steps.

[10]Beware that this formulation is misleading! Writing down Gauss' law by making d/dt explicit, suggests that it represents a time evolution equation after all. However, it does not as is discussed below.

The right-hand side will be made explicitly available as

$$\mathbf{b} = \mathbf{B} * \mathbf{u}, \tag{14.57}$$

where $\mathbf{u}$ is a vector of size of the number of contacts to which voltages can be assigned and $\mathbf{B}$ is a matrix describing the right-hand side of (14.55). The number of rows is equal to the number of degrees of freedom. Furthermore, just as for $\mathbf{A}$, the matrix $\mathbf{B}$ can be extracted from the MAGWEL solver in for harmonic analysis.

The time derivative in (14.56) appears because we kept the variables in the basic representation. When transforming to the first-order formulation, the link variables $\dot{A}_{ij}$ will become equal to the canonical momenta assigned to the links, i.e.

$$\Pi_{ij} = \dot{A}_{ij} \tag{14.58}$$

Thus Equation (14.56) can also be written as

$$\mathbf{A} * \mathbf{x} = \mathbf{b} \qquad \mathbf{x} = \begin{bmatrix} V_1 \\ \cdot \\ \cdot \\ V_i \\ \cdot \\ \cdot \\ \Pi_{12} \\ \cdot \\ \cdot \\ \Pi_{ij} \\ \cdot \\ \cdot \end{bmatrix} \tag{14.59}$$

This is in agreement with Equation (14.31) in which neither an explicit time differentiation is found. We will now proceed with the spatial discretization of the magnetic sector. For future considerations, we will introduce here a new notation. Instead of referring to a link variable A_{ij}, we will now work with a unique ID (identity) for each link which is an integer. A link degree-of-freedom will be simply labeled with this ID. Besides the assumption that there are n nodes that are generating voltage degrees of freedom, we assume that there are m links that generate degrees of freedom. Furthermore, referring to Equation (14.30), we may rewrite (14.59) as

$$\mathbf{A}' * \mathbf{x} = \mathbf{b} \qquad \mathbf{x} = \begin{bmatrix} V_1 \\ \cdot \\ \cdot \\ V_n \\ A_1 \\ \cdot \\ \cdot \\ A_m \\ \Pi_1 \\ \cdot \\ \cdot \\ \Pi_m \end{bmatrix} \tag{14.60}$$

where $\mathbf{A}'$ represent the operator $\mathcal{G}$ of (14.30).

14.8 Discretization of the Maxwell-Ampere System

Just as for nodal variables, the entities associated to other geometrical objects, such as links or surfaces, we must first limit ourselves to a finite subset of links. Whereas Equation (14.22) is written down for the complete 'universe', a domain restriction is required. Suppose, we have a finite domain Ω selected. Furthermore, a grid is built using the nodes that were identified in the foregoing section. Next, we focus on all the links that connect these nodes. Again, some links will be found on the surface of Ω, to be precise, $(ij) \in \partial\Omega$. The construction of the equations of motion (and/or constrain equations) requires special care, because the finite-integration methods around such links we bring us outside Ω and that falls outside the region for which we compute information. We will first consider the situation when (ij) is an internal link, i.e. the link is not at the surface of the simulation domain.

Let us start with Equation (14.22) and consider for each link its dual surface. We will take the integral of this equation over the dual surface. Furthermore, we multiply the results with the length L of the link and obtain

$$\begin{aligned} &\epsilon L \frac{\partial}{\partial t} \int_{\Delta S} d\mathbf{S} \cdot \mathbf{\Pi} + L \int_{\Delta S} d\mathbf{S} \cdot \nabla \times \left(\frac{1}{\mu} \nabla \times \mathbf{A} \right) \\ &- L \int_{\Delta S} d\mathbf{S} \cdot \frac{1}{\mu} \nabla \left(\nabla \cdot \mathbf{A} \right) \\ &+ L \int_{\Delta S} d\mathbf{S} \cdot \sigma \nabla V + L \int_{\Delta S} d\mathbf{S} \cdot \sigma \mathbf{\Pi} \;=\; 0 \end{aligned} \tag{14.61}$$

Here, we put the equation in the appearance $lhs = 0$. The discretization of each term will now be discussed. Starting at the left-hand side, we define a link variable Π_{ij} for the link going from node i to node j. The surface integral is approximated by taking $\mathbf{\Pi}$ constant over the dual area. Thus

$$\epsilon L \frac{\partial}{\partial t} \int_{\Delta S} \mathrm{d}\mathbf{S} \cdot \mathbf{\Pi} \simeq \epsilon L \, \Delta S_{ij} \frac{d\Pi_{ij}}{dt} \tag{14.62}$$

We can assign to each link a volume being $\Delta v_{ij} = L \, \Delta S_{ij}$. The second term on the right-hand side is dealt with using Stokes theorem twice in order to evaluate the circulations.

$$L \int_{\Delta S} \mathrm{d}\mathbf{S} \cdot \nabla \times \left(\frac{1}{\mu} \nabla \times \mathbf{A} \right) = L \oint_{\partial(\Delta S)} \mathrm{d}\mathbf{l} \cdot \left(\frac{1}{\mu} \nabla \times \mathbf{A} \right) \tag{14.63}$$

The circumference $\partial(\Delta S)$ consists of N segments. Each segment corresponds to a dual link that pierces through a *primary* surface. Therefore, we may approximate the right-hand side of (14.63) as

$$L \oint_{\partial(\Delta S)} \mathrm{d}\mathbf{l} \cdot \left(\frac{1}{\mu} \nabla \times \mathbf{A} \right) = L \sum_{k=1}^{N} \Delta l_k \frac{1}{\mu_k} \left(\nabla \times \mathbf{A} \right)_k \tag{14.64}$$

where the sum goes over all primary surfaces that were identified above as belonging to the circulation around the starting link. Note that we also attached an index on μ. This will guarantee that the correct value is taken depending in which material the segment Δl_k is located.

Next we must obtain an appropriate expression for $(\nabla \times \mathbf{A})_k$. For that purpose, we consider the primary surfaces. In particular, an approximation for this expression is found by using

$$(\nabla \times \mathbf{A})_k \simeq \frac{1}{\Delta S_k} \int_{\Delta S_k} \mathrm{d}\mathbf{S} \cdot \nabla \times \mathbf{A} = \frac{1}{\Delta S_k} \oint_{\partial(\Delta S_k)} \mathrm{d}\mathbf{l} \cdot \mathbf{A} \tag{14.65}$$

The last contour integral is evidently replaced by the collection of primary links variables around the primary surface. As a consequence, the second term at the right-hand side of (14.65) becomes

$$L \sum_{k=1}^{N} \Delta l_k \frac{1}{\mu_k} \frac{1}{\Delta S_k} \left(\sum_{l=1}^{N'} \Delta l_{\langle kl \rangle} A_{\langle kl \rangle} \right) \tag{14.66}$$

where we distinguished the link labeling from node labeling (ij) to surface labeling $\langle kl \rangle$.

There is a subtlety related to the use of Equations (14.64) and (14.65). Note that we apply the value of magnetic induction that is evaluated at the center of the dual surface for all positions at the dual link. Our discretization picture is that once that we moved to discrete variables we may imagine these variables to be located at the nodes and at the links. For the currents this does mean that the discretized link currents are located at the links themselves. The computation of the B-field contour integral around the dual surface can be corrected for the distance away from this link current. In (14.65) we use the maximum value of $|\mathbf{B}|$ along the link. This value can be corrected for the distance away from the primary link. This leads to a correction factor of

$$\alpha(\theta) = \left(\int_0^\theta \frac{d\theta'}{\cos\theta'}\right) / \left(\int_0^\theta \frac{d\theta'}{\cos^2\theta'}\right) \tag{14.67}$$

The function is shown in Figure 14.2. The figure shows the inverse of $\alpha(\theta)$ how the correction factor depends on the acuteness of θ.

For $\theta = \pi/4$ this gives $\alpha = 0.88$. Therefore, we will discretize the Maxwell-Ampere system by including this correction factor. It is possible

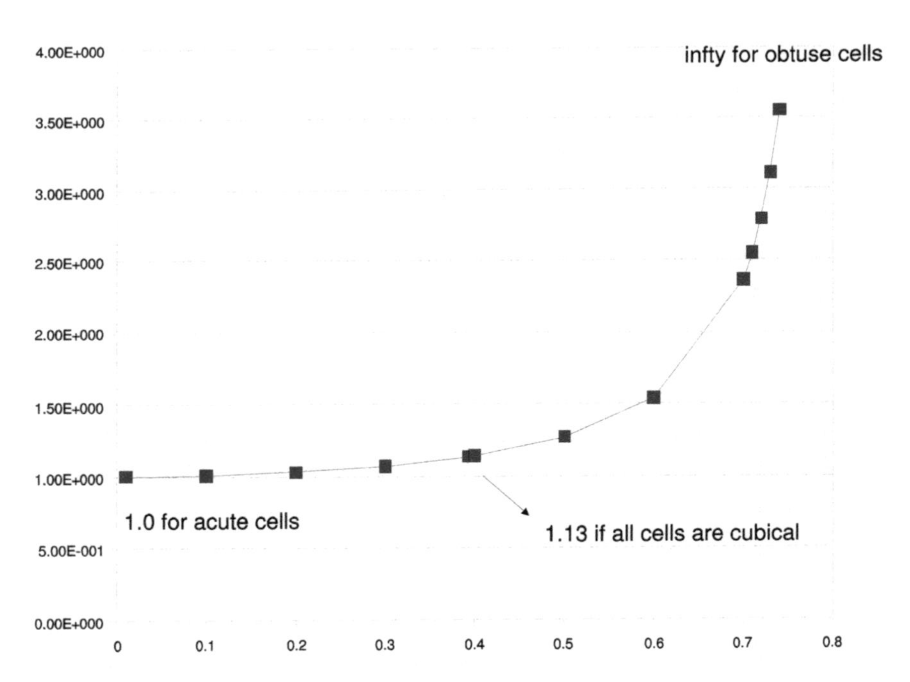

Figure 14.2 Correction factor as a function of the acuteness of the discretization cells.

to compute θ for each dual surface segment. However, this fine tuning is not further considered. So finally, we obtain

$$L \int_{\Delta S} d\mathbf{S} \cdot \nabla \times \left(\frac{1}{\mu} \nabla \times \mathbf{A} \right) = L \sum_{k=1}^{N} \Delta l_k \frac{\alpha}{\mu_k} \frac{1}{\Delta S_k} \left(\sum_{l=1}^{N'} \Delta l_{\langle kl \rangle} A_{\langle kl \rangle} \right) \tag{14.68}$$

Next we consider the third term of (14.65). Now we use the fact that each link has a specific orientation from 'front' to 'back'.

$$\begin{aligned} -L \int_{\Delta S} d\mathbf{S} \cdot \frac{1}{\mu} \nabla (\nabla \cdot \mathbf{A}) \simeq & - \int_{\Delta S} d\mathbf{S} \cdot \frac{1}{\mu} (\nabla \cdot \mathbf{A})_{back} \\ & + \int_{\Delta S} d\mathbf{S} \cdot \frac{1}{\mu} (\nabla \cdot \mathbf{A})_{front} \end{aligned} \tag{14.69}$$

The link orientation coding is illustrated in Figure 14.3. The two term in (14.69) are now discretized as

$$\begin{aligned} \int_{\Delta S} d\mathbf{S} \cdot \frac{1}{\mu} (\nabla \cdot \mathbf{A}) &= \frac{\Delta S}{\mu \Delta v} \int_{\Delta v} dv \nabla \cdot \mathbf{A} \\ &= \frac{\Delta S}{\mu \Delta v} \oint_{\partial(\Delta v)} d\mathbf{S} \cdot \mathbf{A} \\ &= \frac{\Delta S}{\mu \Delta v} \sum_{j}^{n} \Delta S_{ij} A_{ij} \end{aligned} \tag{14.70}$$

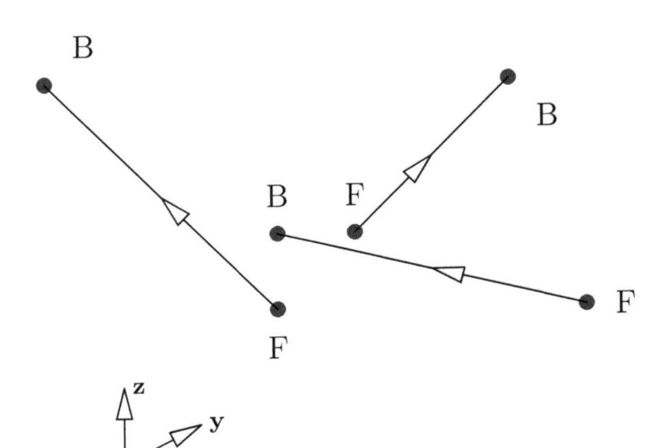

Figure 14.3 A collection of oriented links.

where the sum is now from the front or back node to their corresponding neighbor nodes. The boundary conditions enter this analysis in a specific way. Suppose the front or back node is on the surface of the simulation domain. Then the closed surface integral around such a node will require a dual area contribution from a dual area outside the simulation domain. These surfaces are by definition not considered.

However, we can go back to the gauge condition and use

$$\int_{\Delta S} \mathrm{d}\mathbf{S} \cdot \frac{1}{\mu} (\nabla \cdot \mathbf{A}) = -\Delta S \, \epsilon \frac{\partial V}{\partial t}. \tag{14.71}$$

At first sight this looks weird: First we insert the gauge condition to get rid of the singular character of the curl-curl operation and now we 'undo' this for nodes at the surface. This is however fine because for Dirichlet boundary conditions for $\mathbf{A}$ there are no closed circulations around primary surfaces and there there is no uniqueness problem and therefore the operator is well-defined.

The last two terms are rather straightforward: For the fourth term we consider ∇V constant over the dual surface. Thus we obtain

$$L \int_{\Delta S} \mathrm{d}\mathbf{S} \cdot \sigma \nabla V = (V_{back} - V_{front}) \left(\sum \Delta S_i \sigma_i \right). \tag{14.72}$$

The variation of σ is taken into account by looking at each volume contribution separately.

The fourth term can be dealt with in a similar manner.

$$L \int_{\Delta S} \mathrm{d}\mathbf{S} \cdot \sigma \mathbf{\Pi} = L \, \Pi_{ij} \left(\sum \Delta S_i \sigma_i \right). \tag{14.73}$$

Collecting all terms, we come to the following structure for the discretized Maxwell-Ampere equation:

- Assuming Dirichlet's boundary conditions for the vector potential on the simulation domain boundary, we are dealing only with link degrees of freedom (DOF) corresponding to link that are inside the simulation domain.
- Assuming Neumann's boundary conditions, the links at the surface of the simulation domain also generate degrees of freedom.
- Each DOF-generating link induces two variables: A and Π, where $A = \mathbf{A} \cdot \mathbf{n}$ and $\Pi = \mathbf{\Pi} \cdot \mathbf{n}$ where $\mathbf{n}$ is the intrinsic link orientation,
- The Maxwell-Ampere equation is a time-evolution equation for $\mathbf{\Pi}$

- The time-evolution equation depends on V, A and Π as is summarized below.

$$\hat{\epsilon}\frac{\mathrm{d}}{\mathrm{d}t}\mathbf{\Pi} + M * \mathbf{V} + N * \mathbf{A} + \hat{\sigma} * \mathbf{\Pi} = 0$$

$$\mathbf{V}* = \begin{bmatrix} V_1 \\ \cdot \\ \cdot \\ V_n \end{bmatrix} \quad \mathbf{A} = \begin{bmatrix} A_1 \\ \cdot \\ \cdot \\ A_m \end{bmatrix} \quad \mathbf{\Pi} = \begin{bmatrix} \Pi_1 \\ \cdot \\ \cdot \\ \Pi_m \end{bmatrix} \tag{14.74}$$

Here, $\hat{\epsilon}$ and $\hat{\sigma}$ are diagonal matrices that take care of the material and geometrical weighting of the permittivity and conductivity.

$$\hat{\epsilon} = \begin{bmatrix} \epsilon_1 & \cdot & \cdot & \cdot & \cdot \\ \cdot & \epsilon_2 \cdot & \cdot & \cdot & \\ \cdot & \cdot & \cdot & \cdot & \cdot \\ \cdot & \cdot & \cdot & \cdot & \cdot \\ \cdot & \cdot & \cdot & \cdot & \epsilon_m \end{bmatrix} \quad \hat{\sigma} = \begin{bmatrix} \sigma_1 & \cdot & \cdot & \cdot & \cdot \\ \cdot & \sigma_2 \cdot & \cdot & \cdot & \\ \cdot & \cdot & \cdot & \cdot & \cdot \\ \cdot & \cdot & \cdot & \cdot & \cdot \\ \cdot & \cdot & \cdot & \cdot & \sigma_m \end{bmatrix} \tag{14.75}$$

The matrix N represents the discretization of the operator $\nabla \times [1/\mu\nabla\times\ ] - 1/\mu\nabla[\nabla\cdot\ ]$ as described above. The matrix N is of size $m \times m$. The matrix M describes the coupling to the voltage degrees of freedom and is of size $m \times n$.

Just as for Gauss' law, we can extract the matrices M, N, $\hat{\epsilon}$ and $\hat{\sigma}$ using the MAGWEL harmonic analysis solver.

14.9 Boundary Conditions for the Maxwell-Ampere Equation

First of all, we emphasize that there are two classes of boundary conditions. We already mentioned that the vector potential consists of *three* components (fields) and each component requires its own boundary condition. Just for convenience, suppose we have a domain boundary parallel to the (x,y)-plane. The Dirichlet boundary conditions are that the x and y component of the vector potential are vanishing at the boundary, i.e. $A_x = 0$ and $A_y = 0$. We still have two more fields to consider: the potential V and the third component A_z. At the surface we should also respect the gauge condition. In particular, in the Lorentz gauge, we obtain at the surface that

$$\frac{1}{\mu}\frac{\partial A_z}{\partial z} + \epsilon\frac{\partial V}{\partial t} = 0 \tag{14.76}$$

Since the the tangential components of A vanish at the surface, and therefore also their partial derivatives with respect to x and y. The Dirichlet boundary condition for A_x and A_y physically corresponds to a magnetic field arriving tangential at the surface. This field is described by the curl of the z-component of $\mathbf{A}$, i.e.

$$B_x = \frac{\partial A_z}{\partial x} \qquad B_y = -\frac{A_z}{\partial y} \qquad B_z = \frac{\partial A_y}{\partial x} - \frac{\partial A_x}{\partial y} = 0 \tag{14.77}$$

This is just the magnetic field corresponding the a current impinging perpendicular at the surface. Such a field is needed since we can not carry a current to a contact without also impinging a magnetic field on that surface. So a good usage of the Dirichlet boundary condition is to use them for describing domain boundaries where contacts are found. Examples are pairs of contacts (ports) that are impinged by a TEM wave. Contrary to Dirichlet boundary conditions, we also can imagine Neumann type of boundary conditions. We look at a different class of boundary conditions for A_x and A_y. We will now invent something that resembles Neumann type boundary conditions for these fields. What could that be? Let us think of the surface parallel to the (x,y)-plane again. Now we want to say something about

$$\frac{\partial A_x}{\partial z} \qquad \frac{\partial A_y}{\partial z} \tag{14.78}$$

What can we say for these partial derivatives without using a primal link outside the simulation domain? Should we use links that by definition are not in the memory of the computer? No, we can say something about these partial derivatives. In Figure 14.4 the red line outside the simulation domain is shown. When $\delta \to 0$ then one obtains above derivatives. So one obtains a component of the tangential magnetic field at the surface

$$\frac{\partial A_x}{\partial z} = -B_y = -\mathbf{B} \cdot \tau_y \qquad \frac{\partial A_y}{\partial z} = B_x = \mathbf{B} \cdot \tau_x \tag{14.79}$$

where τ is a tangential vector at the surface. For later use we define ν as a normal vector to the surface. So far, we did not achieve much. We substituted one piece of desired knowledge by another piece, since what to take for this $\mathbf{B}$? Here we may however apply some physics, Suppose that the boundary is located in some insulating region (air). Then we know that the solution satisfies the Maxwell equation in free space. Such solutions describe transverse polarized electromagnetic waves. *Assuming* an outgoing wave perpendicular to the surface (in the z direction or in Figure 14.4, in the direction of δ), we can assert the value of these partial derivatives in terms of the link variables A_x and A_y themselves.

$$\mathbf{A}(\mathbf{x}, t) = (A_x \mathbf{e}_x + A_y \mathbf{e}_y)\ \mathrm{e}^{\mathrm{i}\omega t - \mathrm{i}k_z z} \tag{14.80}$$

The plane TEM wave is characterized by a Dirichlet boundary condition on V, in particular $V = 0$. Together with the gauge condition we find that $\mathbf{k} \cdot \mathbf{A} = 0$, which explain (14.80). The wave satisfies

$$\left(\frac{1}{c^2}\frac{\partial^2}{\partial t^2} - \frac{\partial^2}{\partial z^2}\right)\mathbf{A} = 0 \tag{14.81}$$

In one dimension this is equivalent to

$$\left(\frac{1}{c}\frac{\partial}{\partial t} - \frac{\partial}{\partial z}\right)\left(\frac{1}{c}\frac{\partial}{\partial t} + \frac{\partial}{\partial z}\right)\mathbf{A} = 0 \tag{14.82}$$

For outgoing waves this boils down to

$$\left(\frac{1}{c}\frac{\partial}{\partial t} - \frac{\partial}{\partial z}\right)A_x = 0 \qquad \left(\frac{1}{c}\frac{\partial}{\partial t} - \frac{\partial}{\partial z}\right)A_y = 0 \tag{14.83}$$

This can be summarized as

$$\tau \cdot \mathbf{\Pi} = c\,(\nu \cdot \nabla)\tau \cdot \mathbf{A} \tag{14.84}$$

The Neumann boundary conditions for the surface links can be obtained from

$$\left(\frac{\partial}{\partial t} - c\,(\nu \cdot \nabla)\right)(\tau \cdot \mathbf{A}) = 0 \qquad \left(\frac{1}{c}\frac{\partial}{\partial t} - c\,(\nu \cdot \nabla)\right)(\tau \cdot \mathbf{\Pi}) = 0 \tag{14.85}$$

So far, we have specified the boundary condition for three fields: V, $\mathbf{A} \cdot \tau_x$ and $\mathbf{A} \cdot \tau_y$. Finally, we must discuss the normal component: $\mathbf{A} \cdot \nu$. Remember that for domain boundaries, where contacts are located we have give preference to the boundary conditions

- V Dirichlet type at contacts
- V Neumann type not at contact
- $\mathbf{A}_{//}$ Dirichlet type: $\mathbf{A} \cdot \tau_x = \mathbf{A} \cdot \tau_y = 0$
- $\mathbf{A}_{\perp}$ match of the gauge condition

The complementary set of boundary conditions will be applied only to domain boundaries where no contact areas are found. These boundaries will

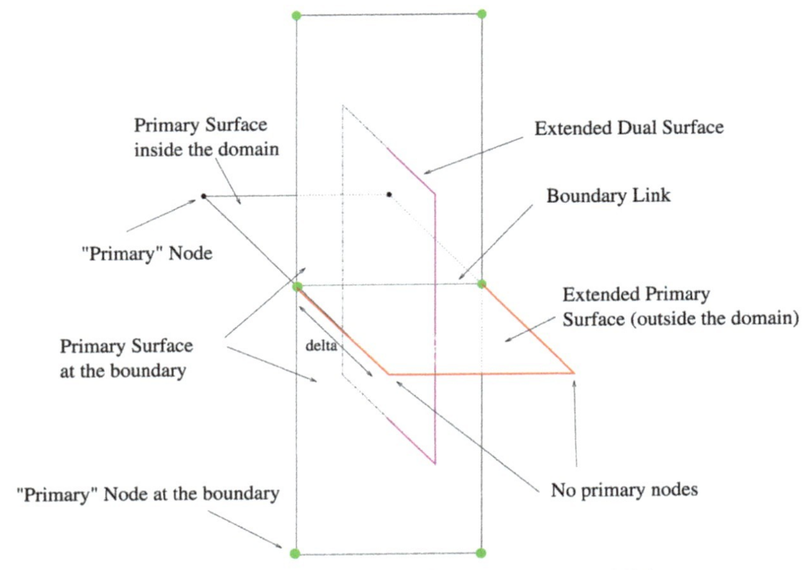

Figure 14.4 Sketch of the boundary nodes and links.

allow for incoming or outgoing EM radiation. From the the Maxwell-Ampere equation with $\mathbf{J}_{\text{cond}} = 0$ we obtain when *using* the gauge condition:

$$\left(\epsilon \frac{\partial^2}{\partial t^2} - \frac{1}{\mu} \nabla^2 \right) (\mathbf{A} \cdot \nu) = 0 \tag{14.86}$$

The conditions read:

- V Dirichlet type at the domain boundary ($V = 0$)
- $\mathbf{A}_{//}$ Neumann type
- $\mathbf{A}_{\perp}$ match of the gauge condition, e.g. $\mathbf{A}_{\perp} = 0$

These conditions will likely to be fine for a very large domain such that a 'far-field' limit is valid. If the simulation domain is such that we have to enforce a truncation of the normal electric field then the following conditions apply:

- V Neumann type at the domain boundary: $(\nabla V) \cdot \nu = 0$
- $\mathbf{A}_{//}$ Neumann type
- $\mathbf{A}_{\perp}$ Dirichlet type, e.g. $\mathbf{A}_{\perp} = 0$

Note: Above description of the Neumann-type of boundary conditions can be seen as a first realization of the Mur boundary conditions that are applied for FDTD algorithms. In the latter case the underlying variables are the electric field $\mathbf{E}$ and $\mathbf{B}$. Here we applied the underlying ideas to the vector potential $\mathbf{A}$. We have restricted ourselves to perpendicular changes with respect to the surface of the simulation domain. We also implemented the approach for waves arriving under a non-perpendicular angle to the surface, using Mur's idea of expanding the square-root operator in terms of a power series. However, numerical experiments showed a serious convergence flaw.

14.10 Generalized Boundary Conditions for the Maxwell-Ampere Equation

When discussing Gauss' law we have discussed the two options for closing the equation at the surface of the simulation domain. At the surface either we applied Neumann boundary conditions if no contacts are present or Dirichlet boundary conditions. These two cases are not exhaustive and when discussing devices in interaction we must allow for various generalizations. First of all, although from a mathematical point of view it is convenient to think of boundary condition as belonging to the boundary of the simulation domain, from an engineering point of view the boundary conditions can be imposed at surface elements located *inside* the simulation domain. Whereas such a perception of boundary conditions for so-called internal contacts is can be imagined for Dirichlet boundary conditions one should realize that the internal contacts can also be used for current boundary conditions. In the latter case, the terminal currents are used to complete the system of equation leading to a well-defined problem. This is nicely illustrated in Figure 14.5.

In this set-up, the contacts are grouped in pairs. In order to extract one column of the S-matrix, one pair of contacts, i.e. "port", is excited with a frequency voltage signal, whereas all other ports (pair of contacts) are closed with a reference impedance, Z_0.

To evaluate the scattering matrix, say of an N-port system, we iterate over all ports and put a voltage difference over one port and put an impedance load over all other ports. Thus the potential variables of the contacts belonging to all but one port, become degrees of freedom that need to be evaluated. The following variable are required to understand the scattering matrices, where Z_0 is a real impedance that is usual taken to be 50 Ohms.

$$a_i = \frac{V_i + Z_0 I_i}{2\sqrt{Z_0}} \tag{14.87}$$

$$b_i = \frac{V_i - Z_0 I_i}{2\sqrt{Z_0}} \tag{14.88}$$

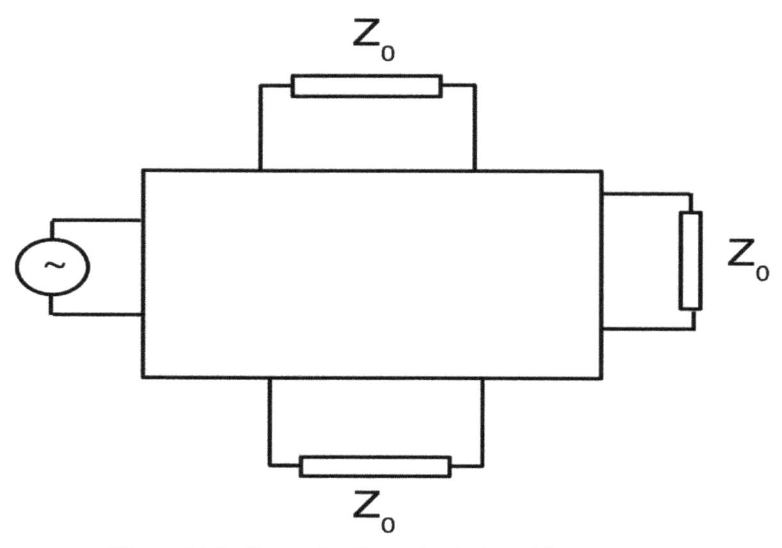

Figure 14.5 Set up for direct simulation of S-parameters.

The variables a_i represent the voltage waves incident on the ports labeled with index i and the variables b_i represent the reflected voltages at ports i. The scattering parameters s_{ij} describe the relationship between the incident and reflected waves.

$$b_i = \sum_{j=1}^{N} s_{ij} a_j \tag{14.89}$$

The scattering matrix element s_{ij} can be found by putting a voltage signal on port i and place an impedance of Z_0 over all other ports. Then a_j is zero by construction, since for those ports we have that $V_j = -Z_0 I_j$ and therefore,

$$s_{ij} = \frac{b_i}{a_j} \tag{14.90}$$

Note that the simulation set up extracts at each port the complex currents and voltages. Moreover, these values are assigned to contacts. Finally these contacts can very well be internal. Thus this modeling approach ignores the presence of currents in the simulation domain once that they have been captured by the ports. How can that be made physically understood? The paradox is eliminated by considering in more detail how the ports are attached to the measurement equipment. The reference impedance represents the impedance of the coax wire of the measurement equipment. The simulation domain is

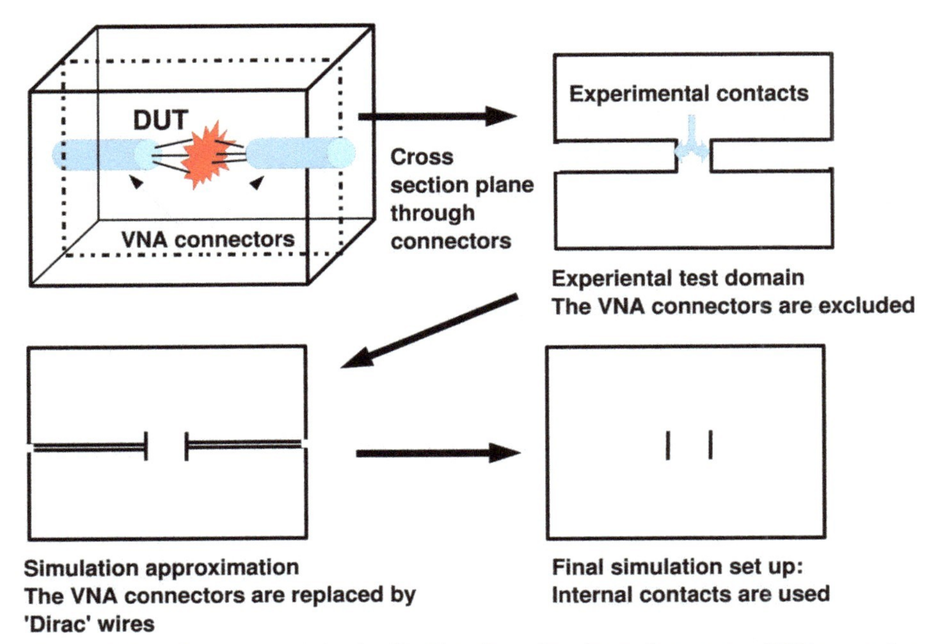

Figure 14.6 Illustration of the justification for using internal contacts. (VNA – vector network analyser, e.g. frequency domain measurement equipment.).

restricted to the device environment volume *minus* the volume occupied by the coax wires. Finally, we may consider idealized (infinitely thin) wires such that the full volume may be considered. The ports are connected to the measurement equipment with a string as a variation of a "Dirac string" being an infinitely-thin coax wire. This idealization leads to a simulation domain of which the probe wires occupy an infinitesimal amount of space, which is neglected in setting up the domain. The results are internal contacts. This is illustrated in Figure 14.6.

14.11 Discretization of the Gauge Condition

The Maxwell-Ampere equation is discretized with use of the gauge condition. Therefore, we better make sure that the gauge condition is really satisfied. In steady-state or in AC small-signal analysis, this was achieved implicitly by solving the current-continuity equations and Gauss' law, which were solved explicitly. In the set up of the transient simulation flow, we arrived at a different picture. To summarize:

- Solve the gauge condition explicitly: $\frac{1}{\mu}\nabla \cdot \mathbf{A} + \epsilon \frac{\partial V}{\partial t} = 0$
- Apply the validity of the gauge condition when discretizing the Maxwell-Ampere (MA) equations (see 14.22): $\epsilon \frac{\partial \mathbf{\Pi}}{\partial t} + \nabla \times \left(\frac{1}{\mu}\nabla \times \mathbf{A}\right) - \frac{1}{\mu}\nabla\left(\nabla \cdot \mathbf{A}\right) + \sigma \nabla V + \sigma \mathbf{\Pi} = 0$
- Solve the current continuity equations: $\nabla \cdot \mathbf{J} + \frac{\partial \rho}{\partial t} = 0$
- Then Gauss' law is *implicitly satisfied* in the following way: $\frac{\partial \rho}{\partial t} = \frac{\partial}{\partial t}\left(\nabla \cdot \mathbf{D}\right)$
  Therefore, if Gauss' law is satisfied at $t = 0$, it will also be at $t > 0$.

This summary suggests that an alternative approach may be possible. We will start from Gauss' law and solve it explicitly at every time instance. This gives us the value of the charge density. The current-continuity equations are not regarded as a by product of the MA equation, but are considered as an independent set of equation to be solved, together with the MA equations. Then we arrive at the following scheme:

- Solve Gauss' law at each time step: $\rho + \nabla \cdot \left(\epsilon \left[\nabla V + \Pi\right]\right) = 0$ ([11])
- Solve the current-continuity equation: $\frac{\partial \rho}{\partial t} - \nabla \cdot \left(\sigma \left[\nabla V + \Pi\right]\right) = 0$
- Solve the Maxwell-Ampere equation (as before).
- Then the gauge condition is *implicitly satisfied* as can be easily demonstrated by contracting the MA equation with the divergence operator and using the other two equations.

Both schemes are equivalent, but the second scheme is more difficult to implement if we want to express every variable in terms of ρ, $\mathbf{A}$ and $\mathbf{\Pi}$, i.e. eliminating V in favor of ρ, which can be done by the variable transformation:

$$\rho = \nabla \cdot \mathbf{D} \;\Rightarrow V = -\int \mathrm{d}\mathbf{r} \;\nabla^{-2}\left(\frac{\rho}{\epsilon} + \nabla \cdot \mathbf{\Pi}\right) \tag{14.91}$$

We have implemented the second scheme, not by using the non-local Equation (14.91) but solving Gauss' law in differential form for obtaining V.

14.12 Temporal Discretization

There exists several views to address above system of equations in the temporal regime. We can distinguish between the linear case (no semiconductors present) and the non-linear case (semiconductors included). In the latter case

[11]This is the reason that the discretization method using Gauss' law was discussed in detail.

it is not useful to isolate the time differentiations from the equations. Therefore, we write the system of equations as a differential-algebraic equation (DAE):

$$M * \frac{\mathrm{d}}{\mathrm{d}t}\mathbf{X}(t) + H(\mathbf{X}(t), t) + F * \mathbf{X}_{bc}(t) = 0 \qquad (14.92a)$$

$$G(\mathbf{X}) = 0 \qquad (14.92b)$$

Here, $\mathbf{G}$ is determined by the Gauss' equation. $\mathbf{X}$ is the vector of unknowns $(V, \mathbf{A}, \mathbf{\Pi}, n, p)$ and $\mathbf{X}_{bc}(t)$ is the vector of boundary-condition values, that are coupled into the system via the operator F. We assume that the boundary condition can be linked into the system using a linear operator. This is definitely the case for linear materials (insulators and metals) but this assumption must be revised for semi-conductors.

The discretized Maxwell equations for conductor / insulator systems lead to a linear system

$$M * \frac{\mathrm{d}}{\mathrm{d}t}\mathbf{X}(t) + H * \mathbf{X}(t) + F * \mathbf{X}_{bc}(t) = 0 \qquad (14.93a)$$

$$G(\mathbf{X}) = 0 \qquad (14.93b)$$

14.13 BDF for DAEs

First, we explain the standard way of using the k-step backward-differentiation formula (BDF) method when integrating differential-algebraic equations (DAEs) in standard form

$$f(x', x, t) = 0, \quad t \in [t_0, t_F].$$

Suppose that the approximations $x_{n-j} \approx x(t_{n-j}),\ j = 1, 2, \ldots, k$ have already been calculated. We denote $\tau_m = t_m - t_{m-1},\ m = 1, 2, \ldots$ as the m-th time step. An approximation x_n to $x(t_n)$ will be obtained by solving the nonlinear equation

$$f\left(\frac{1}{\tau_n}\sum_{j=0}^{k} \alpha_{j,n} x_{n-j}, x_n, t_n\right) = 0.$$

The BDF coefficients $\alpha_{0,n}, \alpha_{1,n}, \ldots, \alpha_{k,n}$ depend on the step sizes $\tau_n, \tau_{n-1}, \ldots, \tau_{n-k+1}$ (if the time step size is not constant). They are determined such that the derivative $x'(t_n)$ is approximated by

$$x'(t_n) \approx \frac{1}{\tau_n} \sum_{j=0}^{k} \alpha_{j,n} x_{n-j}.$$

with order k. In case of index-1 DAEs, the order of convergence of the BDF method equals k. In case of index-2 DAEs, one may lose one order of τ. In case of index-3 DAEs, the BDF method may fail completely. Therefore, we should check whether our coupled system has at most index 2 as the circuit equations have.

In order to obtain an approximation X_n to the value of $X(t_n)$, where $X(t)$ is the exact solution of (14.92b), we proceed in a similar way. In this case, the derivative $d'(X(t_n), t_n)$ is replaced by a sum approximating it, X_n is then the solution of the nonlinear equation

$$A\left(\frac{1}{\tau_n} \sum_{j=0}^{k} \alpha_{j,n} d\left(X_{n-j}, t_{n-j}\right)\right) + b(X_n, t_n) = 0.$$

14.14 State-Space Matrices and Linking Harmonic to Transient Analysis

We have already indicated that the MAGWEL harmonic field solver is capable of generating matrix information that can be used for transient analysis. The connection between the harmonic and transient solver approach is provided by the state-space description of the fields. As a reminder, we first summarize here the set-up of the MAGWEL solver for frequency analysis. First of all it should be noted that the solver in the frequency domain solves

- Solving Gauss' law.
- Solving Current-continuity equations.
- Solving Maxwell-Ampere equations,
- Respecting the gauge condition is a side product of solving above set.

These equations are solved with the following sign conventions

$$\nabla \cdot \mathbf{D} - \rho = 0 \tag{14.94}$$

This choice leads to discretization matrices with positive diagonal entries The current-continuity equation for conductors is solved for the same reason in the following form:

$$\nabla \cdot \mathbf{J} + \frac{\partial \rho}{\partial t} = 0 \tag{14.95}$$

In the frequency domain this becomes

$$\nabla \cdot \mathbf{J} + \mathrm{i}\omega\rho = 0 \tag{14.96}$$

since any small-signal (AC) solution is assumed to have the following appearance: $X(t) = X_0 + \mathrm{e}^{\mathrm{i}\omega t}\ X_1$ The small-signal parts of the hole- and electron current continuities are solved in the following form

$$\nabla \cdot \mathbf{J}_{\mathrm{p}} + \mathrm{i}q\omega p + q(r - G) = 0 \tag{14.97}$$

$$-\nabla \cdot \mathbf{J}_{\mathrm{n}} + \mathrm{i}q\omega p + q(r - G) = 0 \tag{14.98}$$

The choice of the sign convention is motivated by having positive diagonal matrix entries. In here, q denotes the positive unit of charge and p, n, R, G the carrier densities for holes and electrons, the recombination and the generation. Finally, the small-signal parts of the Maxwell-Ampere equation is addressed in the following form:

$$\nabla \times \left(\frac{1}{\mu}\nabla \times \mathbf{A}\right) - \frac{1}{\mu}\nabla\left(\nabla \cdot \mathbf{A}\right) + \sigma\nabla V + \mathrm{i}\omega\sigma\ \mathbf{A} - \epsilon\omega^2\mathbf{A} = 0 \tag{14.99}$$

where the minus sign of the last term is the result of i^2. In the last equation we have not included the diffusive part of the carrier currents but these ones can be easily taken into account. However, their inclusion has limited validity since for non-linear materials the method that is described in this section is not usable. Thus, this section applies to insulators and inductors only. At this stage we should compare Equation (14.99) with Equation (14.22). Transforming the latter to the frequency domain gives

$$\epsilon(\mathrm{i}\omega)(\mathrm{i}\omega)\mathbf{A} + \nabla \times \left(\frac{1}{\mu}\nabla \times \mathbf{A}\right) - \frac{1}{\mu}\nabla\left(\nabla \cdot \mathbf{A}\right) + \sigma\nabla V + \mathrm{i}\omega\sigma\ \mathbf{A} = 0 \tag{14.100}$$

We now proceed with the discretization of above equations. The result will be that a linear problem is given for a set of variables

$$A(\omega)\mathbf{X} + \mathbf{b} = 0 \tag{14.101}$$

Here $\mathbf{X} = \{V_1, V_2, ...V_n, A_1, A_2, ...A_m\}$ in which V and A have the same meaning as before on the discretized grid. The matrix A is frequency dependent and the vector $\mathbf{b}$ describes the coupling to the contact potentials.

We now come to a key observation of the harmonic solver. *There is a one-to-one mapping of the system of equations in the time domain with the equations in the frequency domain, provided that the materials are linear.*

The time-dependent system

$$\left[A_0 + A_1\frac{\mathrm{d}}{\mathrm{d}t} + A_2\frac{\mathrm{d}^2}{\mathrm{d}t^2}\right]\mathbf{X}(t) + \mathbf{b}(t) = 0 \tag{14.102}$$

maps with the insertion $X(t) = X_0 + \mathrm{e}^{\mathrm{i}\omega t}\, X_1$ onto

$$\left[A_0 + \mathrm{i}\omega A_1 + (\mathrm{i}\omega)^2 A_2\right]\mathbf{X} + \mathbf{b}_0 + (\mathrm{i}\omega)\mathbf{b}_1 + (\mathrm{i}\omega)^2\mathbf{b}_2 = 0 \tag{14.103}$$

We will now elaborate on the vector $\mathbf{b}$. For that purpose we remind us of the coupling of the field degrees of freedom to the boundary conditions. The voltage boundary conditions at the contacts, are collected in a set $\mathbf{U} = \{u_1, u_2, , , , u_{n^c}\}$. The coupling is a $n \times n^c$ matrix B such that

$$\mathbf{b}(t) = B\,\mathbf{U}(t) \tag{14.104}$$

In order to obtain an explicit expression for the matrix B, we consider the situation that the contacts is attached to conductors and moreover, that the current-continuity equations must be solved. We use similar methods as described in detail in Section 14.7. The assembling rule for any conductor or conductor/insulator interface node is:

$$\sum_j \sigma\left[\frac{S_{ij}}{h_{ij}}(V_i - V_j) - \sigma_{ij}\frac{\mathrm{d}A_{ij}}{\mathrm{d}t}S_{ij}\right] + \frac{\mathrm{d}}{\mathrm{d}t}\sum_j \epsilon\left[\frac{S_{ij}}{h_{ij}}(V_i - V_j) - \sigma_{ij}\frac{\mathrm{d}A_{ij}}{\mathrm{d}t}S_{ij}\right] = 0 \quad i \subset \mathcal{N}_\infty \tag{14.105}$$

Here we apologize for the multiple use of the symbol σ. The one without labels stands for the conductance, whereas the one with labels keeps track of the direction of $\mathbf{A}$ with respect to the link orientation $\mathbf{n}$. Furthermore, we can divide the sums over internal nodes, i.e. degrees of freedom and contact nodes. There are no link degrees of freedom that are 'external'. For Dirichlet's boundary conditions, the links on the surface of the simulation domain have zero vector potential. As a consequence, the B matrix has an expansion in $\mathrm{d}/\mathrm{d}t$ up to first order. Therefore, Equation (14.102) becomes

$$\left[A_0 + A_1\frac{\mathrm{d}}{\mathrm{d}t} + A_2\frac{\mathrm{d}^2}{\mathrm{d}t^2}\right]\mathbf{X}(t) + \left[B_0 + B_1\frac{\mathrm{d}}{\mathrm{d}t}\right]\mathbf{U}(t) = 0 \tag{14.106}$$

and as a consequence, Equation (14.102) can be written as

$$\left[A_0 + \mathrm{i}\omega A_1 + (\mathrm{i}\omega)^2 A_2\right]\mathbf{X} + \left[B_0 + \mathrm{i}\omega\, B_1\right]\mathbf{U} = 0 \tag{14.107}$$

The matrices A_0, A_1, A_2, B_0, B_1 are explicitly constructed using the harmonic analysis field solver. In particular, there are also matrix elements of B that couple into the Maxwell-Ampere equation, due to the presence of the $\sigma \nabla V$ -term and the treatment of the surface nodes. When combining the field solver with some other program, e.g. a circuit solvers of which the unknowns are collected in a vector $\mathbf{Y}$ then besides the state-space variables $\mathbf{X}$, the observables $\mathbf{Y}$ are part of the complete model. We take the currents at the contacts as the observables of the state space. Furthermore, the currents are defined as positive (> 0), when the current is outgoing of the simulation domain. In particular, $\mathbf{Y} = \{I_1, I_2, ...I_{n^c}\}$. Thus the full picture takes the following form:

$$\begin{bmatrix} A & B \\ C & D \end{bmatrix} * \begin{bmatrix} \mathbf{X} \\ \mathbf{U} \end{bmatrix} + \begin{bmatrix} \mathbf{0} \\ \mathbf{Y} \end{bmatrix} = 0 \tag{14.108}$$

We will now present the detailed derivation of the matrices C and D. Our starting point will be the current-continuity equations for the contact nodes. The current continuity can be expressed as:

$$\nabla \cdot \mathbf{J} = 0 \quad \text{where} \quad \mathbf{J} = \mathbf{J}_{\text{cond}} + \mathbf{J}_{\text{disp}} \tag{14.109}$$

For a contact node the discrete assembling gives

$$\sum_j \Delta S_{ij} J_{ij} + I_i^{\text{out}} = 0 \tag{14.110}$$

In Figure (14.7) this is illustrated.

The total current for the contact is obtained by summing over all the contact nodes:

$$I^{\text{out}} = \sum_i I_i^{\text{out}} \tag{14.111}$$

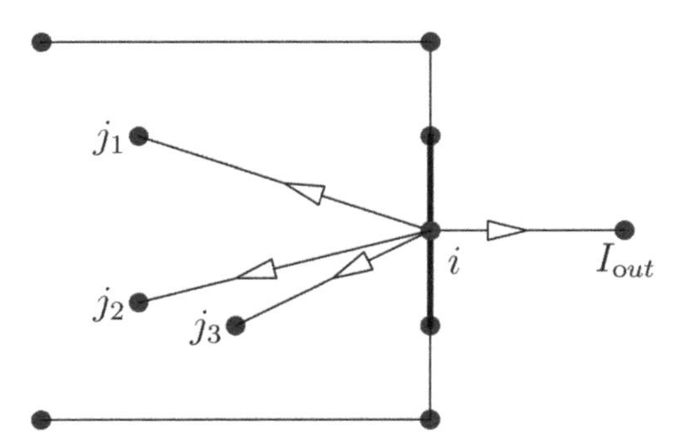

Figure 14.7 Illustration of the currents at a contact node, i. The nodes j are field nodes.

The assembling of the current-continuity equations for internal nodes, j, runs over all links that are connected to j, including the contact nodes i. Whereas in the stand-alone field solving approach these contributions are not considered for the matrix building, we can evaluate these contributions for the construction of the matrices C and D. Using the fact that $J_{ij} = -J_{ji}$, the assembling leads to the following result:

$$\begin{aligned}
-\sum_j \frac{\Delta S_{ij}}{h_{ij}}\sigma V_j + \sum_j \frac{\Delta S_{ij}}{h_{ij}}\sigma V_i + \sum_j \Delta S_{ij}\sigma\sigma_{ij}\frac{\mathrm{d}}{\mathrm{d}t}A_{ji} \\
-\sum_j \frac{\Delta S_{ij}}{h_{ij}}\frac{\mathrm{d}}{\mathrm{d}t}\epsilon V_j + \sum_j \frac{\Delta S_{ij}}{h_{ij}}\frac{\mathrm{d}}{\mathrm{d}t}\epsilon V_i \\
+\sum_j \Delta S_{ij}\epsilon\sigma_{ij}\frac{\mathrm{d}^2}{\mathrm{d}t^2}A_{ji} + I_i^{\mathrm{out}} = 0 \quad (14.112)
\end{aligned}$$

For each term we can read off their contribution to the matrices C and D. Remembering that V_i is an entry in the stimulus vector $\mathbf{U}$ and V_j is one of the entries of $\mathbf{X}$, we find that $-\sum_j \Delta S_{ij}\sigma$ contributes to the zeroth-order term C_0, i.e.

$$C = C_0 + \mathrm{i}\omega C_1 + (\mathrm{i}\omega)^2 C_2 \quad (14.113)$$

Furthermore $\sum_j \Delta S_{ij}\sigma$ contributes to D_0

$$D = D_0 + \mathrm{i}\omega D_1 + (\mathrm{i}\omega)^2 D_2 \quad (14.114)$$

The fifth term in Equation (14.112), being $-\sum_j \Delta S_{ij}\epsilon$, induces a contribution in D_1. We have observed that the use of second order time differentiation can be completely avoided by using the variable $\mathbf{\Pi}$ in Equation (14.112). Then this equation becomes

$$\begin{aligned}
-\sum_j \frac{\Delta S_{ij}}{h_{ij}}\sigma V_j + \sum_j \frac{\Delta S_{ij}}{h_{ij}}\sigma V_i + \sum_j \Delta S_{ij}\sigma\sigma_{ij}\frac{\mathrm{d}}{\mathrm{d}t}A_{ji} \\
-\sum_j \frac{\Delta S_{ij}}{h_{ij}}\frac{\mathrm{d}}{\mathrm{d}t}\epsilon V_j + \sum_j \frac{\Delta S_{ij}}{h_{ij}}\frac{\mathrm{d}}{\mathrm{d}t}\epsilon V_i \\
+\sum_j \Delta S_{ij}\epsilon\sigma_{ij}\frac{\mathrm{d}}{\mathrm{d}t}\Pi_{ji} + I_i^{\mathrm{out}} = 0 \quad (14.115)
\end{aligned}$$

and the second order term in the expansion of C is zero.

The second order term in D is *always* zero since there are no second-order time derivatives acting on voltages (neither internal nor applied ones.) As a consequence, in the linear interface we produce the terms proportional to $(\mathrm{i}\omega)$ and $(\mathrm{i}\omega)^2$. The latter are connected to the Π state-space variables.

14.15 A Technical Detail: Link Orientations

The orientation of the link direction is not related to some preferred direction of the coordinate system. In Figure (14.8) a collection of links is shown, including their orientation. As is seen the orientation of a link is just an convention of which node is used as the front node and which node is seen as the back node.
We may now take the unit vector $\hat{\mathbf{n}}$ which points from the front node to the back node and project the electric field on this vector. This gives:

$$\mathbf{E} \cdot \hat{\mathbf{n}} = -\hat{\mathbf{n}} \cdot \nabla V - \mathrm{i}\omega\, \hat{\mathbf{n}} \cdot \mathbf{A} \tag{14.116}$$

The variable A_{ij} is defined as:

$$A_{ij} = A_{ji} = \hat{\mathbf{n}} \cdot \mathbf{A} \tag{14.117}$$

The indices, ij just remind us between which nodes the link is located.

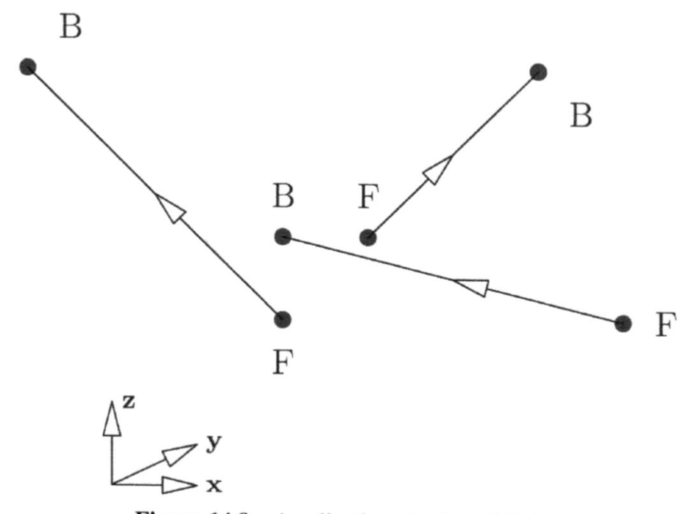

Figure 14.8 A collection of oriented links.

The potential term is by construction, with ΔL the length of the link:

$$\hat{\mathbf{n}} \cdot \nabla V = \frac{V_B - V_F}{\Delta L} \tag{14.118}$$

In order to incorporate this convention into the construction of the discretized versions of the continuity equations and Gauss' law, we start from some node i and consider the link to node j. Then there are two possibilities:

- i is the front node F and j is the back node B. Then $\sigma_{ij} = +1$
- i is the back node B and j is the front node F. Then $\sigma_{ij} = -1$

and $E_{ij} = -\frac{V_j - V_i}{h_{ij}} - \sigma_{ij}(\mathrm{i}\omega)A_{ij}$ where $h_{ij} = \Delta L$

14.16 Scaling

We will have a closer look at scaling and unscaling of the field equations. For that purpose we will start from Equations (14.10)–(14.13). The Maxwell equations are:

- *for insulators:*

$$-\nabla \cdot \left[\varepsilon\left(\nabla V + \frac{\partial \mathbf{A}}{\partial t}\right)\right] = 0 \tag{14.119}$$

$$\nabla \times \frac{1}{\mu}(\nabla \times \mathbf{A}) = -\varepsilon\frac{\partial}{\partial t}\left(\nabla V + \frac{\partial \mathbf{A}}{\partial t}\right) \tag{14.120}$$

- *for conductors:*

$$-\nabla \cdot \sigma\left(\nabla V + \tfrac{\partial \mathbf{A}}{\partial t}\right) = \tfrac{\partial}{\partial t}\left(\nabla \cdot \varepsilon\left(\nabla V + \tfrac{\partial \mathbf{A}}{\partial t}\right)\right) \tag{14.121}$$

$$\nabla \times \tfrac{1}{\mu}(\nabla \times \mathbf{A}) = -\sigma\left(\nabla V + \tfrac{\partial \mathbf{A}}{\partial t}\right) - \varepsilon\tfrac{\partial}{\partial t}\left(\nabla V + \tfrac{\partial \mathbf{A}}{\partial t}\right) \tag{14.122}$$

14.16.1 Scaling the Poisson Equation

Starting from Gauss' law, the actual expression that is programmed is:

$$\int_{\partial(\Delta v)} \mathbf{dS} \cdot (-\epsilon)\left(\nabla V + \frac{\partial \mathbf{A}}{\partial t}\right) - Q(\Delta v) = 0 \tag{14.123}$$

Let λ be the scaling of the length. The scaling constant has dimension [Length] = meter. Since the surface integral introduces a scaling factor λ^2, and the gradient a factor λ^{-1}. The voltage is scaled using $\sigma_V = 0.02585\ [V]$.

Moreover, since the vacuum permittivity is absorbed in the scaling process, the first term is:

$$\lambda\epsilon_0\sigma_V \int_{\partial(\Delta\tilde{v})} \mathrm{d}\tilde{S} \cdot (-\epsilon_r)\tilde{\nabla}\tilde{V}$$

in which all tildes indicate scaled variables. The corresponding term that is assembled in the program is:

$$\int_{\partial(\Delta\tilde{v})} \mathrm{d}\tilde{S} \cdot (-\epsilon_r)\tilde{\nabla}\tilde{V}$$

The matrix element of the scaled equations are computed in the MAGWEL software. These matrix elements can be viewed as the proportionality factor of the contributions originating from voltage degrees of freedom. Therefore, the scaling factor of these matrix elements do not contain σ_V, and the proper way to unscale these element is by multiplying them with $\lambda\epsilon_0$. The permittivity of vacuum is in units $\epsilon_0 = 8.85418 \times 10^{-12}$ $[C^2/Nm^2]$ and the length-scaling parameter is $\lambda = 1.19527e - 05[m]$.

For the matrix elements describing the coupling to the vector potential, we consider the term

$$\int_{\partial(\Delta v)} \mathbf{dS} \cdot (-\epsilon) \left(\frac{\partial \mathbf{A}}{\partial t}\right)$$

Following the same reasoning as above, we get scale this term as:

$$\epsilon_0\lambda^2\sigma_A/\sigma_t \int_{\partial(\Delta\tilde{v})} \mathrm{d}\tilde{S} \cdot (-\epsilon_r) \left(\frac{\partial \tilde{\mathbf{A}}}{\partial \tilde{t}}\right)$$

Since $\sigma_t = \sigma_\omega^{-1} = 1.42867 \times 10^{-10}$ [sec] and since in the expansions of the matrices, the frequency (or time differentiations) is made explicit, we do not have to incorporate this scaling factor in the transition in going from scaled to unscaled matrices, provided that we also use unscaled times. The scaling factor σ_A is also excluded, provided that unscaled vector potentials are used. Therefore, the scaling factor for the A-coupling is $\epsilon_0\lambda^2 = 1.26498\times 10^{-21}[\mathrm{C}^2/\mathrm{N}]$.

14.16.2 Scaling the Current-Continuity Equations

For the current-continuity equation, the expression that is programmed is:

$$\int_{\partial(\Delta v)} \mathbf{dS} \cdot (-\sigma) \left(\nabla V + \frac{\partial \mathbf{A}}{\partial t} \right) - \frac{\partial Q(\Delta v)}{\partial t} = 0 \tag{14.124}$$

The same reasoning as above applies, keep in mind that we now use the conductance scaling $\sigma_c = 0.06197$ [S/m]. The V term leads to a scaling factor $\sigma_c \lambda = 7.40767 \times 10^{-07}$ [S].

14.16.3 Scaling the Maxwell-Ampere Equation

The Maxwell-Ampere equation is programmed starting from:

$$\begin{aligned}
&\frac{L}{\mu_0} \int_{\Delta S} \mathrm{d}\mathbf{S} \cdot \nabla \times \left(\frac{1}{\mu_r} \nabla \times \mathbf{A} \right) \\
&+ \epsilon L \frac{\partial}{\partial t} \int_{\Delta S} \mathrm{d}\mathbf{S} \cdot \mathbf{\Pi} - L \int_{\Delta S} \mathrm{d}\mathbf{S} \cdot \frac{1}{\mu} \nabla (\nabla \cdot \mathbf{A}) \\
&+ L \int_{\Delta S} \mathrm{d}\mathbf{S} \cdot \sigma \nabla V + L \int_{\Delta S} \mathrm{d}\mathbf{S} \cdot \sigma \mathbf{\Pi} \ = \ 0
\end{aligned} \tag{14.125}$$

which is next multiplied with μ_0. As a consequence the starting point for programming is:

$$\begin{aligned}
&L \int_{\Delta S} \mathrm{d}\mathbf{S} \cdot \nabla \times \left(\frac{1}{\mu_r} \nabla \times \mathbf{A} \right) \\
&+ \mu_0 \, \epsilon L \frac{\partial}{\partial t} \int_{\Delta S} \mathrm{d}\mathbf{S} \cdot \mathbf{\Pi} - L \int_{\Delta S} \mathrm{d}\mathbf{S} \cdot \frac{1}{\mu_r} \nabla (\nabla \cdot \mathbf{A}) \\
&+ L \, \mu_0 \int_{\Delta S} \mathrm{d}\mathbf{S} \cdot \sigma \nabla V + L \, \mu_0 \int_{\Delta S} \mathrm{d}\mathbf{S} \cdot \sigma \mathbf{\Pi} \ = \ 0
\end{aligned} \tag{14.126}$$

In order to decide on the appearance of the unscaled equation, we will work our way to the equivalent of the Maxwell-Ampere equation in the following form:

$$\oint \mathbf{H} \cdot \mathrm{dl} = \int \mathrm{d}\mathbf{S} \cdot \left(\mathbf{J}_c + \frac{\partial \mathbf{D}}{\partial t} \right) \tag{14.127}$$

However, this is almost correct. The precise form is:

$$\mu_0 \, L \oint \mathbf{H} \cdot \mathrm{dl} = \mu_0 \, L \int \mathrm{d}\mathbf{S} \cdot \left(\mathbf{J}_c + \frac{\partial \mathbf{D}}{\partial t} \right) \tag{14.128}$$

The right-hand side can be further elaborated:

$$\mu_0 \, L \oint \mathbf{H} \cdot \mathrm{dl} = \mu_0 \int \mathrm{d}\mathbf{S} \cdot \nabla \times \mathbf{H} = L \int \mathrm{d}\mathbf{S} \cdot \nabla \times \frac{1}{\mu_r}\mathbf{B}$$
$$= L \int \mathrm{d}\mathbf{S} \cdot \nabla \times \left(\frac{1}{\mu_r} \nabla \times \mathbf{A} \right)$$

From the last term we observe that in order to get from the unscaled equation to the scaled equation we have a scaling of the matrix entries according to $\lambda * \lambda^2 * \lambda^{-1} * \lambda^{-1}$. As a consequence, the scaling factor for the second term in (14.126) is: $\lambda = 1.19527 \;\; 10^{-5}$ [m].

Just as for the Poisson and the current-continuity equations, the unscaled matrix elements should be such that when multiplied with the unscaled ω or ω^2 and unscaled variable V or A, the result should be an unscaled contribution to the unscaled Maxwell-Ampere Equation (14.128).

Let us now proceed with the second term in (14.126).

$$\mu_0 \, \epsilon_0 L \, \frac{\partial}{\partial t} \int_{\Delta S} \mathrm{d}\mathbf{S} \cdot \epsilon_r \, \mathbf{\Pi} = \frac{\mu_0 \, \epsilon_0}{k} \, L \int_{\Delta S} \mathrm{d}\mathbf{S} \cdot \epsilon_r \, k \, \frac{\partial^2}{\partial t^2}\mathbf{A} \qquad (14.129)$$

We have inserted a factor k into the integral since this factor is found in the (scaled) discretized equations. The desired scaling parameter is obtained by ignoring the term $\frac{\partial^2}{\partial t^2}\mathbf{A}$. Therefore, the scaling is given by

$$\frac{\mu_0 \, \epsilon_0}{k} * \lambda^3 = \lambda \, \sigma_t^2 \qquad (14.130)$$

The numerical value is $1.70765e - 15$ [msec^2].

For the third term of Equation (14.126), we can use the same de-scaling parameter as for the first term. For the fifth term, we consider

$$L \, \mu_0 \int_{\Delta S} \mathrm{d}\mathbf{S} \cdot \sigma \mathbf{\Pi} \qquad (14.131)$$

Since the implementation has explicitly the factor k in the assembling we rewrite this as:

$$\frac{L \, \mu_0}{k} \int_{\Delta S} \mathrm{d}\mathbf{S} \cdot \, k \, \sigma \mathbf{\Pi} \qquad (14.132)$$

We can now “read off” the scaling parameter by ignoring the scaling of Π, and assign the appropriate scaling factor to the remaining terms. It becomes:

$$\frac{\mu_0 \lambda^3 \, \sigma_c}{k} = \lambda \, \sigma_t \tag{14.133}$$

Finally, the coupling of the voltage V into the unscaled MA equation is extracted from

$$L \, \mu_0 \int_{\Delta S} \mathrm{d}\mathbf{S} \cdot \sigma \nabla V = \frac{L \, \mu_0}{k} \int_{\Delta S} \mathrm{d}\mathbf{S} \cdot \, k \, \sigma \nabla V \tag{14.134}$$

This allows us to "read off" the scaling parameter

$$\frac{\mu_0 \lambda^2 \, \sigma_c}{k} = \sigma_t \tag{14.135}$$

The numerical value is: $1.42867 \;\; 10^{-10}$ [sec]. This completes our scaling considerations.

Summary

This chapter described the mathematical details and subtleties in order to successfully perform transient field solver approaches. The key concern is to end up after discretization with a non-singular problem such that the linear solvers are confronted with a well-defined problem. Moreover, the discretization should respect basic physical requirement such as for example charge conservation. Our discretization scheme respects these requirements. Having a well-defined problem from the mathematical perspective provides an answer in a finite run time, but not necessarily the *correct* answer. For that purpose it is needed that a sufficient amount of structural detail is included. In practice it requires that the the computational mesh is fine enough to capture the relevant details.

PART III

Applications

In Part III we will apply the developed discretization methods to several design problems. Since the author's background is in microelectronics the examples are borrowed from this field.

15

Simple Test Cases

15.1 Examples

We present a few examples demonstrating that the proposed potential formulation in terms of the Poisson scalar field V, the vector potential field $\mathbf{A}$ and the ghost field χ, is a viable method to solve the Maxwell field problem. All subtleties related to that formulation, i.e. the positioning of the vector potential on links, and the introduction of the ghost field χ, are already encountered in constructing the solutions of the static Equations (Schoenmaker and Meuris [2002a]).

15.1.1 Crossing Wires

The first example concerns two crossing wires and thereby addresses the three-dimensional features of the solver. The structure is depicted in Figure 15.1 and has four ports. In the simulation we put one port at 0.1 Volts and kept the other ports grounded. The current is 4 Amps. The simulation domain is $10 \times 10 \times 14\ \mu\mathrm{m}^3$. The metal lines have a perpendicular cross section of $2 \times 2\mu\mathrm{m}^2$. The resistivity is $10^{-8}\ \Omega m$. In Table 15.1, some typical results are presented.

The energies have been calculated in two different ways and good agreement is observed. This confirms that the methods underlying the field solver are trustworthy. The χ-field is zero within the numerical accuracy, i.e. $\chi \sim O(10^{-14})$.

15.1.2 Square Coaxial Cable

To show that also inductance calculations are adequately addressed, we calculate the inductance per unit length (L) of a square coaxial cable as depicted in Figure 15.2. The inductance of such a system with inner dimension a and outer dimension b, was calculated from

$$l \times \frac{1}{2} L I^2 = \frac{1}{2\mu_0} \int_\Omega B^2 \mathrm{d}\tau = \frac{1}{2} \int_\Omega \mathrm{d}\tau \, \mathbf{J} \cdot \mathbf{A} \tag{15.1}$$

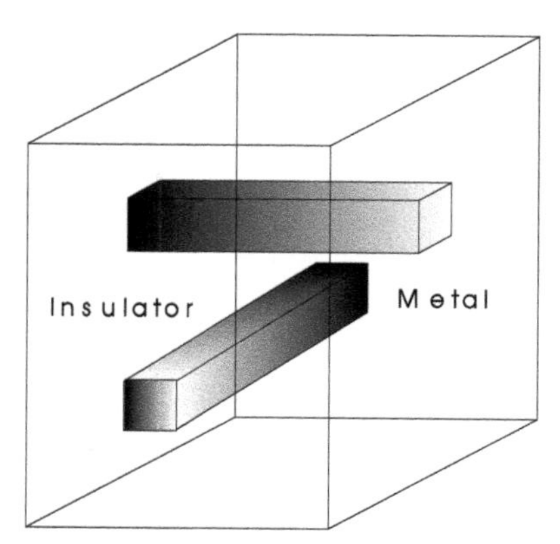

Figure 15.1 Layout of two crossing wires in insulating environment.

Table 15.1 Some characteristic results for two crossing wires

Electric Energy (J)		Magnetic Energy (J)	
$\frac{1}{2}\epsilon_0 \int_\Omega \mathrm{d}\tau \, E^2$	1.03984×10^{-18}	$\frac{1}{2\mu_0} \int_\Omega \mathrm{d}\tau \, B^2$	2.89503×10^{-11}
$\frac{1}{2} \int_\Omega \mathrm{d}\tau \, \rho\phi$	1.08573×10^{-18}	$\frac{1}{2} \int_\Omega \mathrm{d}\tau \, \mathbf{J} \cdot \mathbf{A}$	2.92924×10^{-11}

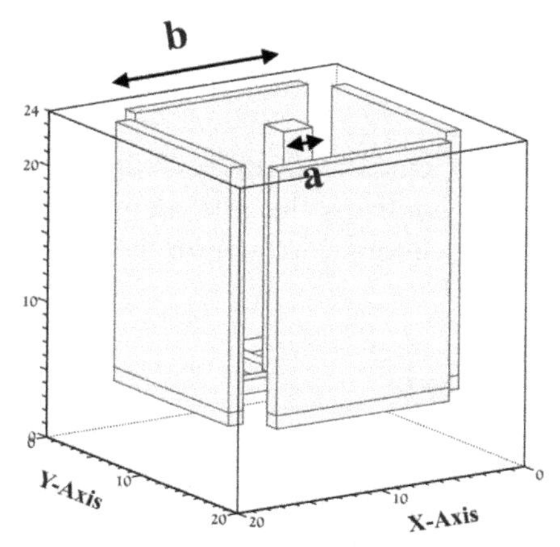

Figure 15.2 Layout of the square coax structure.

Table 15.2 Some characteristic results for a square coaxial cable

a μm	b μm	b/a	L (cylindrical) (nH)	L (square) (nH)
2	6	3	220	255
1	5	5	322	329
1	7	7	389	390
1	10	10	461	458

with l denoting the length of the cable. As expected, for large values of the ratio $r = b/a$, the numerical result for the square cable approaches the analytical result for a cylindrical cable, $L = (\mu_0/2\pi)\ln(b/a)$.

15.1.3 Spiral Inductor

A spiral inductor, as shown in Figure 15.3 was simulated. This structure also addresses the three-dimensional features of the solver. The cross-section of the different lines is 1 μm $\times$ 1 μm. The overall size of the structure is 8 μm $\times$ 8 μm and the simulation domain is 23 $\times$ 20 $\times$ 9 μm^3. The resistance is evaluated as $R = V/I$ and equals 0.54 Ω. In Figure 15.4, the intensity

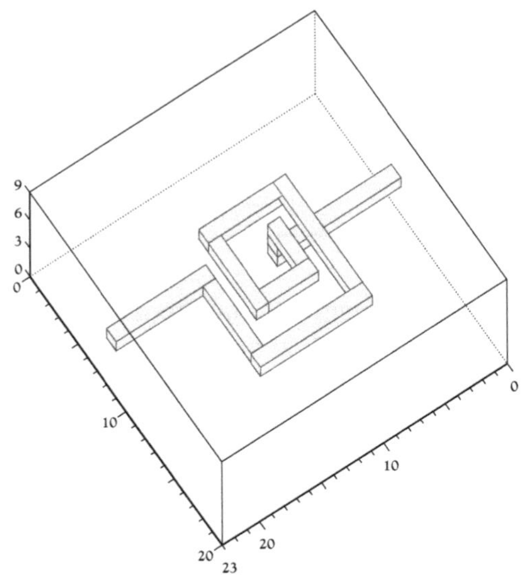

Figure 15.3 Layout of the spiral inductor structure.

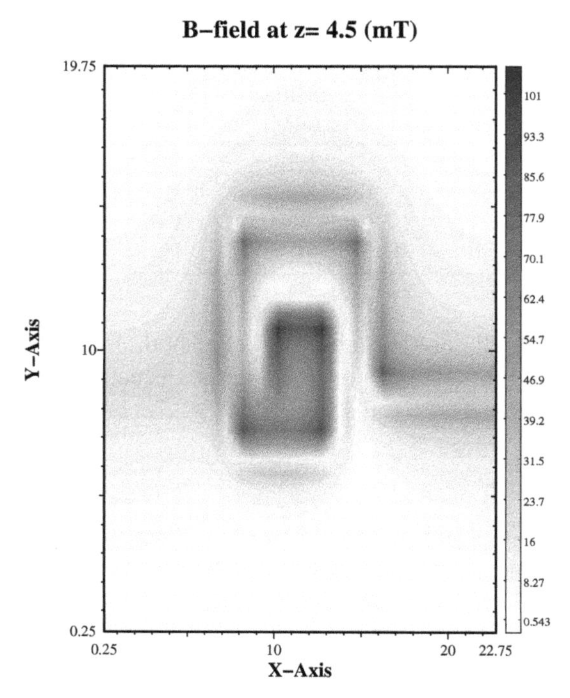

Figure 15.4 Magnetic field strength in the plane of the spiral inductor.

Table 15.3 Some characteristic results for the spiral inductor

Electric Energy (J)		Magnetic Energy (J)	
$\frac{1}{2}\epsilon_0 \int_\Omega \mathrm{d}\tau\, E^2$	2.2202×10^{-18}	$\frac{1}{2\mu_0} \int_\Omega \mathrm{d}\tau\, B^2$	3.8077×10^{-13}
$\frac{1}{2} \int_\Omega \mathrm{d}\tau\, \rho\phi$	2.3538×10^{-18}	$\frac{1}{2} \int_\Omega \mathrm{d}\tau\, \mathbf{J}\cdot\mathbf{A}$	3.9072×10^{-13}

of the magnetic field is shown at height 4.5 μm. From the results in Table 15.3 we obtain that the inductance of the spiral inductor is 4.23×10^{-11} Henry.

By computing a collection of structures taken from various resources using different simulation tools, we obtain a good insight into the positioning of the source codes that are available in the project in comparison to tools from other providers. Moreover, we learn about the strengths and weaknesses of the tools and gain understanding on which aspects need to be improved irrespective of the task related to the building of the transient solver. The cases are selected on the following criteria: (1) availability, (2) "intriguing" (the benchmark challenges insight) and (3) compact model availability. Using these criteria we have selected the following set a benchmarks:

- Simple conductive rod
- Two parallel transmission lines
- Strip line above a conductive plate
- Coax configuration
- Inductor with grounded guard ring
- Inductor with narrow winding above a patterned semiconductor layer

It is important to realize that some of these benchmarks have been acquired under a non-disclosure agreement. As a consequence we can can not release the data that deals with the processing of the structure. However, it suffices to say that the structures are loaded with equal setting of the technology files into the different simulators. The simulations are done in the frequency regime.

15.2 S-Parameters, Y-Parameters, Z-Parameters

In order to determine the S-matrix, a rather straightforward procedure is followed. For that purpose a collection of ports is needed and each port consists of two contacts. A contact is defined as a collection of nodes that are electrically identified. A rather evident appearance of a contact is a surface segment on the boundary of the simulation domain. A slightly less trivial contact consists of two or more of these surfaces on the boundary of the simulation domain. The nodes that are found on these surfaces are all at equal potential. Therefore, although there may be many nodes assigned to a single contact, all these nodes together generate only one potential variable to the system of unknowns. Of course, when evaluating the current entering or leaving the contact, each node in the contact contributes to the total contact current. Assigning prescribed values for the contact potential can be seen as applying Dirichlet's boundary conditions to these contacts. This is a familiar technique in technology CAD. Outside the contact regions, Neumann boundary conditions are applied. Unfortunately, since we are now dealing with the full system of Maxwell equations, providing boundary conditions for the scalar potential will not suffice. We also need to provide boundary conditions for the vector potential. Last but not least, since the set of variables V and $\mathbf{A}$ are not independent, setting a boundary condition for one variable has an impact on the others. Moreover, the choice of the gauge condition also participates in the appearance of the variables and their relations. A convenient set of boundary conditions is given by the following set of rules:

- contact surface: $V = V|_c^i$. To each contact area is assigned a prescribed potential value.
- outside the contact area on the simulation domain: $\mathbf{D}_n = 0$. There is no electric field component in the direction perpendicular to the surface of the simulation domain.
- For the complete surface of the simulation domain, we set $\mathbf{B}_n = 0$. There is no magnetic induction perpendicular to the surface of the simulation domain.

We must next translate these boundary conditions to restrictions on $\mathbf{A}$. Let us start with the last one. Since there is no normal $\mathbf{B}$, we may assume that the vector potential has only a normal component on the surface of the simulation domain. That means that the links at the surface of the simulation domain do not generate a degree of freedom. It should be noted that more general options exist. Nevertheless, above set of boundary conditions provides the minimal extension of the TCAD boundary conditions if vector potentials are present.

In order to evaluate the scattering matrix, say of an N-port system, we iterate over all ports and put a voltage difference over one port and put an impedance load over all other ports. Thus the potential variables of the contacts belonging to all but one port, become degrees of freedom that need to be evaluated. The following variables are required to understand the scattering matrices, where Z_0 is a real impedance that is usual taken to be 50 Ohms.

$$a_i = \frac{V_i + Z_0 I_i}{2\sqrt{Z_0}} \tag{15.2}$$

$$b_i = \frac{V_i - Z_0 I_i}{2\sqrt{Z_0}} \tag{15.3}$$

The variables a_i represent the voltage waves incident on the ports labeled with index i and the variables b_i represent the reflected voltages at ports i. The scattering parameters s_{ij} describe the relationship between the incident and reflected waves.

$$b_i = \sum_{j=1}^{N} s_{ij} a_j \tag{15.4}$$

The scattering matrix element s_{ij} can be found by putting a voltage signal on port i and place an impedance of Z_0 over all other ports. Then a_j is zero by construction, since for those ports we have that $V_j = -Z_0 I_j$. Note that I_j is defined positive if the current is ingoing. In this configuration $s_{ij} = b_i/a_j$. In

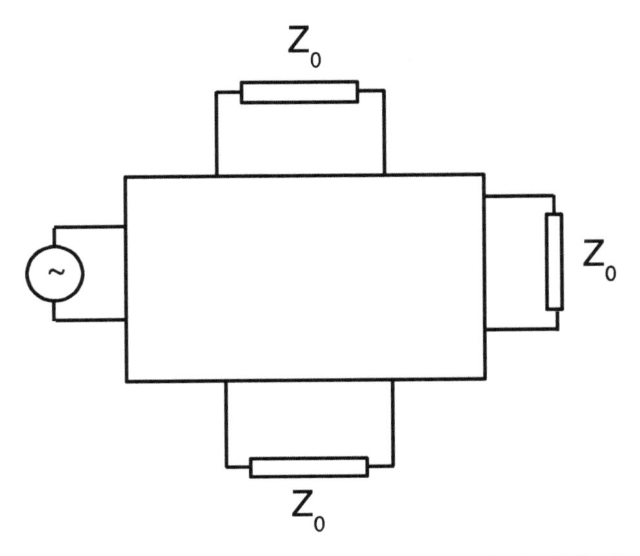

Figure 15.5 Set up of the S-parameter evaluation: 1 port is excited and all others are floating.

a simulation setup, we may put the input signal directly over the contacts that correspond to the input port. This would imply that the input load is equal to zero. The S-parameter evaluation set up is illustrated in Figure 15.5. Once the S parameter matrix is known, the admittance matrix (Y) and impedance matrix (Z) can be computed from it (Orfanidis [2002]).

15.3 A Simple Conductive Rod

The simple conductive rod is a circular bar with constant conductance. In the static regime, a magnetic field is created, that behaves as $1/r$ outside the rod. As a consequence, the total magnetic energy diverges, i.e.

$$E_{\text{magn}} = \int \mathbf{dr}\ \mathbf{B} \cdot \mathbf{H} \propto \int_0^{\infty} drr^2(1/r)^2 \qquad (15.5)$$

Therefore, we would like to learn if the simulation tool can deal with such configurations. Furthermore, the MAGWEL solver has a unique discretization algorithm based on a two-fold application of Stokes' law in order to deal with the term $\nabla\ (1/\mu\ \times \nabla \times \mathbf{A})$. Recently this term was programmed also for unstructured grids. This benchmark will test if these implementations can accurately produce analytic known results.

Simulation set up The simulation set up is done with the following parameters:

- Wire cylindrical
- Radius 2 micron
- Length 100 micron
- Sigma=10^8 S/m
- $\mu_0 = 4\pi * 10^{-7}$ H/m

An illustration of the wire is shown in Figure 15.6.

Results Check if the B field matches the analytic expectation.
Outside the wire we have:

$$B(r) = \frac{4\pi \times 10^{-7} * I}{2\pi r} \tag{15.6}$$

We put a voltage of V=10 volt over the wire. This gives the following values for the current and resistance.

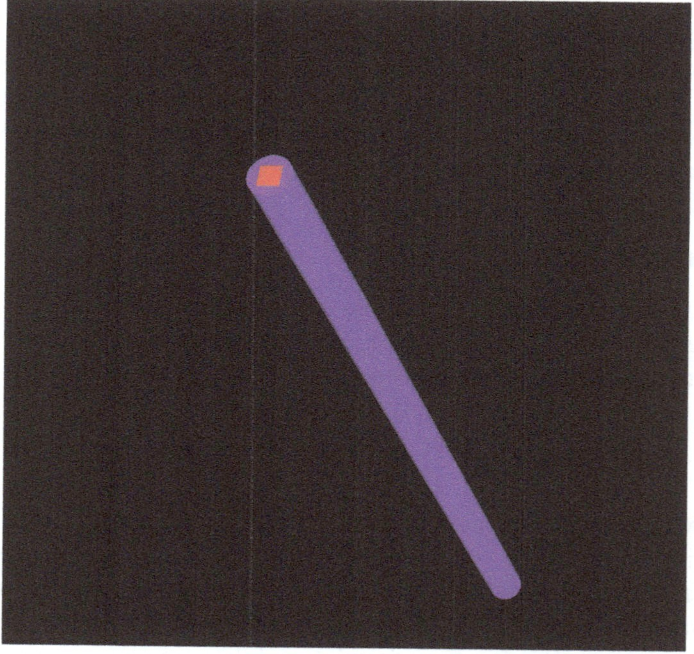

Figure 15.6 Illustration of the wire in the MAGWEL editor; the red square is a contact area.

Parameter	Value
Analytic current	1.257861 10^2 A
Analytic resistance	7.95 10^{-2} Ohm

The meshing is done using a two-dimensional Delaunay scheme with 3D extrusion. The mesh in shown in Figure 15.7 and a zoom-in of the mesh is shown in Figure 15.8. The result of the simulation is:

Parameter	Value
Numerical current	1.23323376 10^2 A
Numerical resistance	8.10876 10^{-2} Ohm

We plot the magnitude of the B-field along a line through the center of the wire using plotmtv. The result is shown in Figure 15.9. The results (red) are the outcome of a full numerical simulation on this mesh.

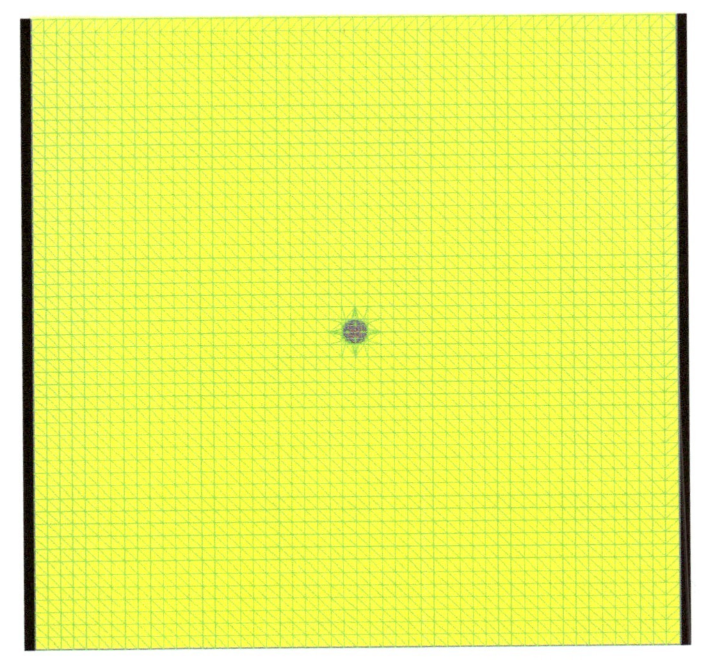

Figure 15.7 Illustration of the mesh in the MAGWEL editor.

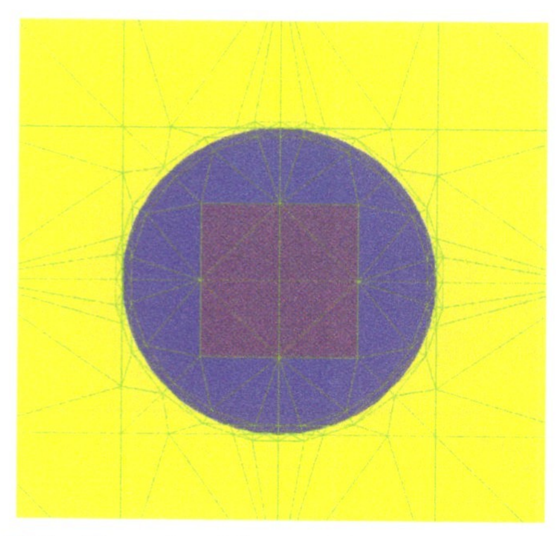

Figure 15.8 Zoom-in of the mesh in the MAGWEL editor.

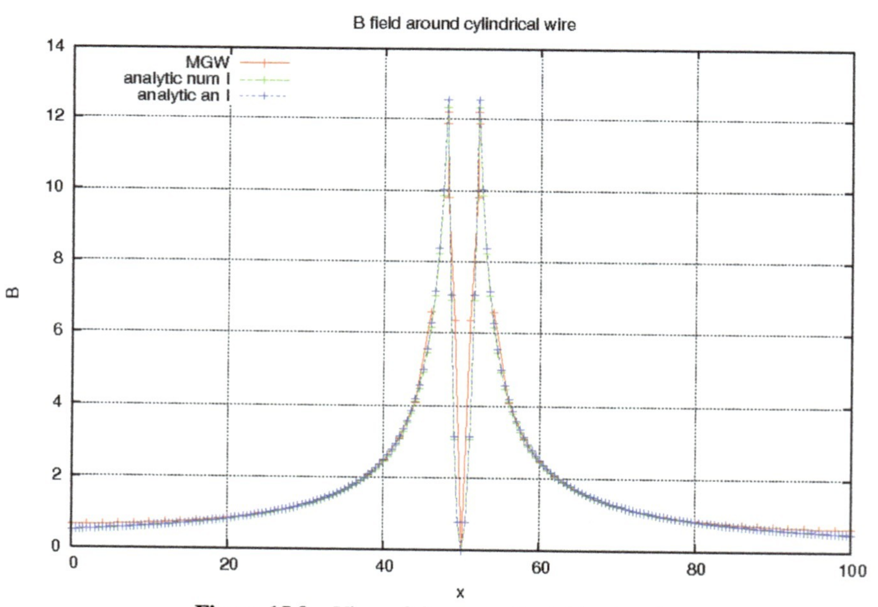

Figure 15.9 View of the magnetic field strength.

Then there are two analytic curves. The green curve (analytic num I) refers to an analytic computation using the current as computed numerically on this mesh, whereas the blue curves also refers to the current computed analytically.

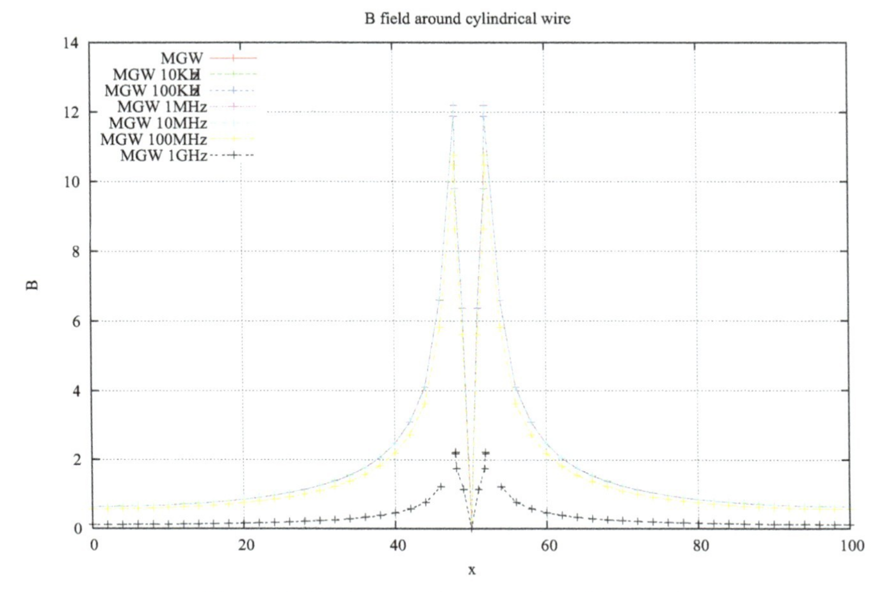

Figure 15.10 View of the magnetic field strength.

The $1/r$ drop is very well reproduced. This is only possible if the curl-curl operator is correctly implemented (ignoring finite-size effects), no loss of B field propagation of the inverse operator is observed. The numerical B field (red) is in agreement with the analytic current (blue). The numerical current value induces an analytic B field (green) in agreement with the numerical B field (red). We expect that at 1 GHz the conclusion is still more or less valid, the skin depth = 1.59 micron < R= 2 micron. The plot in Figure 15.10 is a scan over frequencies from 1KHz to 1GHz. It is very similar up to 100 MHz and then there is a collapse.

From this benchmark we see that the numerical treatment of the curl-curl operator is leading to correct analytic results in the static regime. No specific techniques for correcting or calibrating the outcome is needed.

15.4 Strip Line above a Conductive Plate

When computing capacitances using the MAGWEL field solver there is a tendency for overestimation. This problem is not seen for capacitors which extend in two dimensions all the way over the simulation domain, i.e. for infinitely large parallel plates, but when finite segments are taken, then the measured capacitance is below the computed capacitance. Therefore, this benchmark is of interest to get a grip on this systematical computational error which is related to fringe capacitance. The benchmark is of interest for support and selection of adaptive meshing strategies. The strip-line theory is explained in full detail in an online book: http://www.ece.rutgers.edu/ orfanidi/ewa/ch10.pdf:pp. 401–403 (Orfanidis [2002]). The modeling is due to Hammerstad and Jensen (Hammerstad and Jensen [1980] Itoh [1987]). The model has the following ingredients:

- Effective relative permeability: ε_{eff}
- Characteristic impedance: Z
- Relations between Z, L and C

The capacitance, C, per unit length is $C = \varepsilon\frac{\eta}{Z}$, where $\varepsilon = \varepsilon_0 * \varepsilon_r$ and $\varepsilon_0 = 8.85418e^{-12}$ F/m. A cross sectional view of the strip line above a plate is shown in Figure 15.11. The parameter w is the width and h is the height of the dielectric layer. Furthermore the parameters η and η_0 are defined as:

$$\eta = \sqrt{\frac{\mu}{\varepsilon}}, \quad \eta_0 = \sqrt{\frac{\mu_0}{\varepsilon_0}}, \tag{15.7}$$

$$\tag{15.8}$$

where $\mu = \mu_0 = 4\pi * 10^{-7}$ H/m. The contacting is shown in Figure 15.12. The modeling is based on the geometrical parameter $u = w/h$ and the effective material parameter ε_{eff}:

$$\varepsilon_{eff} = \frac{\epsilon_r + 1}{2} + \frac{\varepsilon_r - 1}{2} \times (1 + \frac{10}{u})^{-ab} \tag{15.9}$$

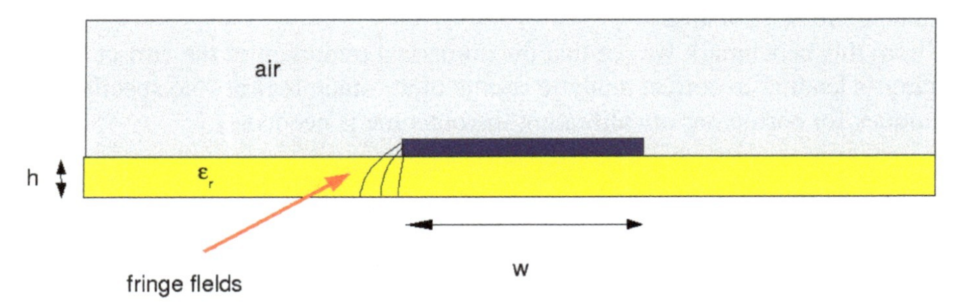

Figure 15.11 Layout of the parallel strip above a conductive plate.

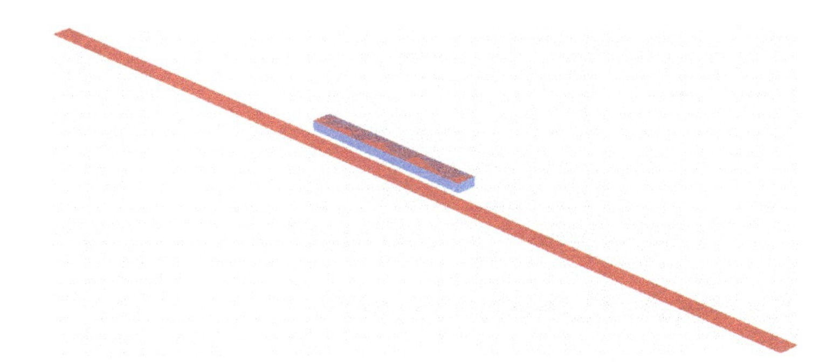

Figure 15.12 Illustration of the contacts.

$$Z = \frac{\eta_0}{2\pi\sqrt{\varepsilon_{eff}}} ln\left[\frac{f(u)}{u} + \sqrt{\frac{1+4}{u^2}}\right] = \frac{Z_{01}(u)}{\sqrt{\varepsilon_{eff}}} \tag{15.10}$$

The variables a and b are given below:

$$a = 1 + \frac{1}{49} ln[\frac{u^4 + (\frac{u}{52})^2}{u^4 + 0.432}] + \frac{1}{18.7} \ln\left[1 + (\frac{u}{18.1})^3\right] \tag{15.11}$$

$$b = 0.564\left(\frac{\varepsilon_r - 0.9}{\varepsilon_r + 3}\right)^{0.053} \tag{15.12}$$

These expression were programmed and the text book examples are confirmed (page 403 of the on-line book S.J. Orfanidis (Orfanidis [2002])) For example:

Input		Output	
Width	10	w/h	10.0000000000000
Height	1	a	1.00833298546091
ε_r	3.9	b	0.539644218604637
		Z	15.6369203475650
		ε_{eff}	3.44440939659828
		C	3.958997E-010

The value of C is in F/m.

Simulation set up We take a strip line with length of 1 micron and width of 10 micron.

Variable	Value
h	1 micron
w	10 micron
ε_r	3.9
T_{metal}	0.5
T_{air}	1 micron

The input variables for the numerical experiment

Results The results of this set up are shown below:

Variable	Value
MAGWEL	3.951870e-16 F
Hammerstad and Jensen	3.958997e-016 F
Parallel plate model $C = \varepsilon A/h$	3.4531302e-16 F

Static capacitance results

Note that the result from the simple parallel-plate model has a substantial error. A few plots of the electric field are shown in Figure 15.13 and Figure 15.14. At first sight, the result for the capacitance looks very acceptable. However, we have not accounted for the finite thickness of the strip

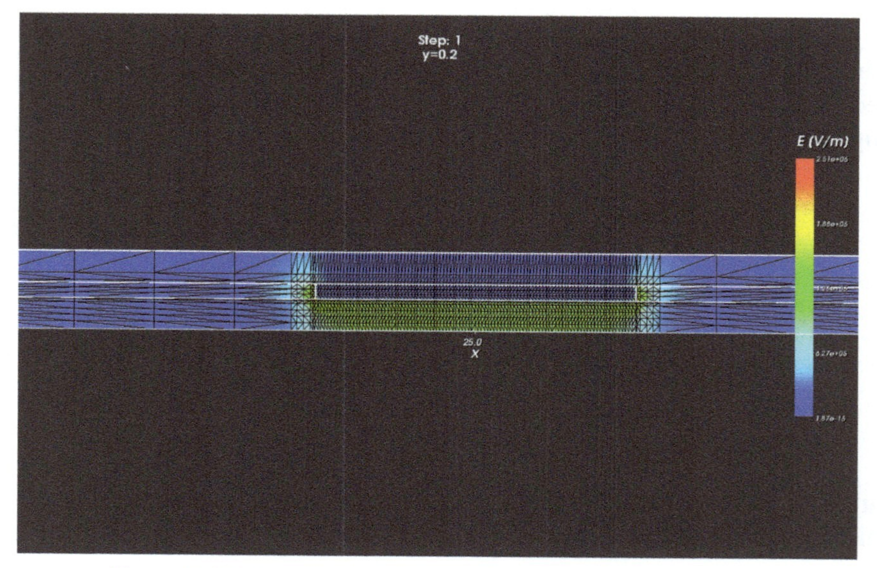

Figure 15.13 Layout of the parallel strip above a conductive plate.

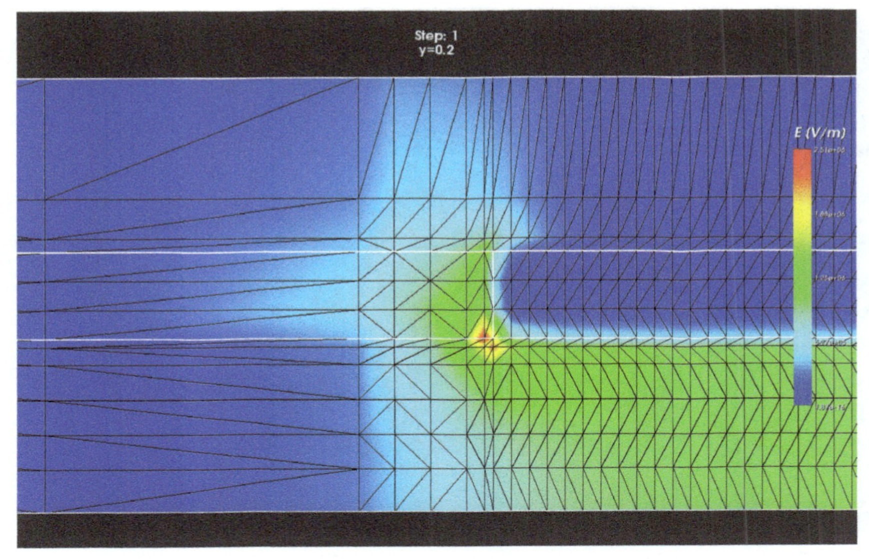

Figure 15.14 Layout of the parallel strip above a conductive plate.

(here we used 0.5 micron) nor did we take into account a substantial layer of air (1 micron). So far, the height of the strip is not accounted for by Hammerstad and Jensen. By taking very thin strips, we can deduce from the equations of Hammerstad and Jensen (HJ) that $\varepsilon_{eff}^{HJ} < \varepsilon_r$. When $w/h \to \infty$ then $\varepsilon_{eff}^{HJ} \to \varepsilon_r$. We have studied this limit and found serious deviations between the analytic and numerical results. The reason is that there also needs to be included a sufficient amount of air in the field solver set up. We re-computed the capacitance with t_{metal} =0.01 micron, t_{air} = 10 micron. In Figure 15.15, the spread out of the electric field is illustrated.

The mesh is shown in Figure 15.16. Also an impression of the mesh is given in Figure 15.17.

Effects of finite strip thickness A follow-up paper of Hammerstad and Jensen by Itoh (Itoh [1987]) deals width the impact of the finite strip thickness. The modified theory is given below. The input parameters are:

- w: strip width,
- h: thickness of the dielectric layer,
- t_M: thickness of the metal strip.

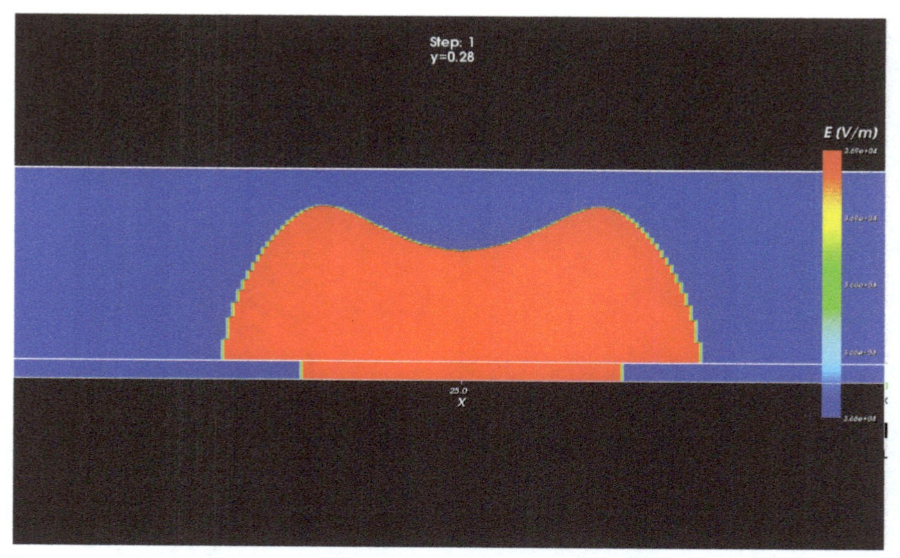

Figure 15.15 Separation in two domains of the electric field. red: $E > 0.01E_{max}$, blue: $E < 0.01E_{max}$.

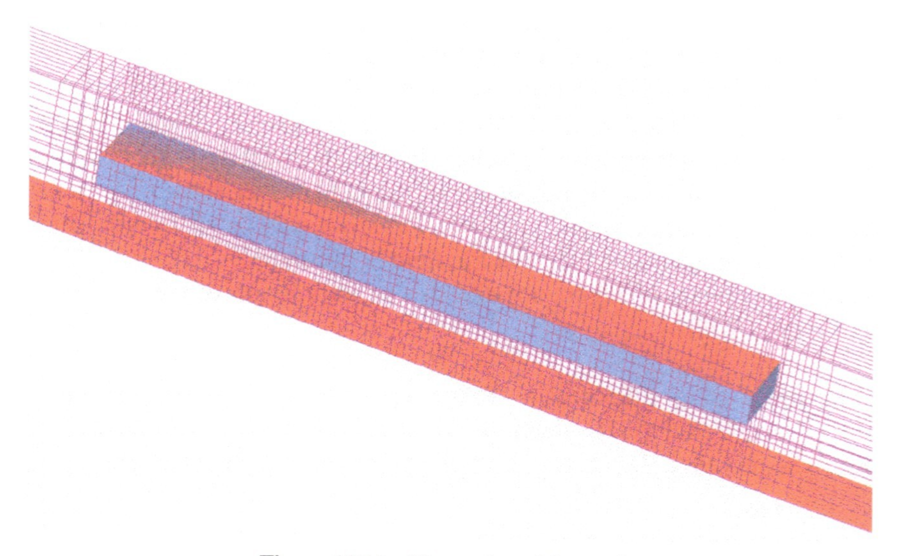

Figure 15.16 Illustration of the mesh.

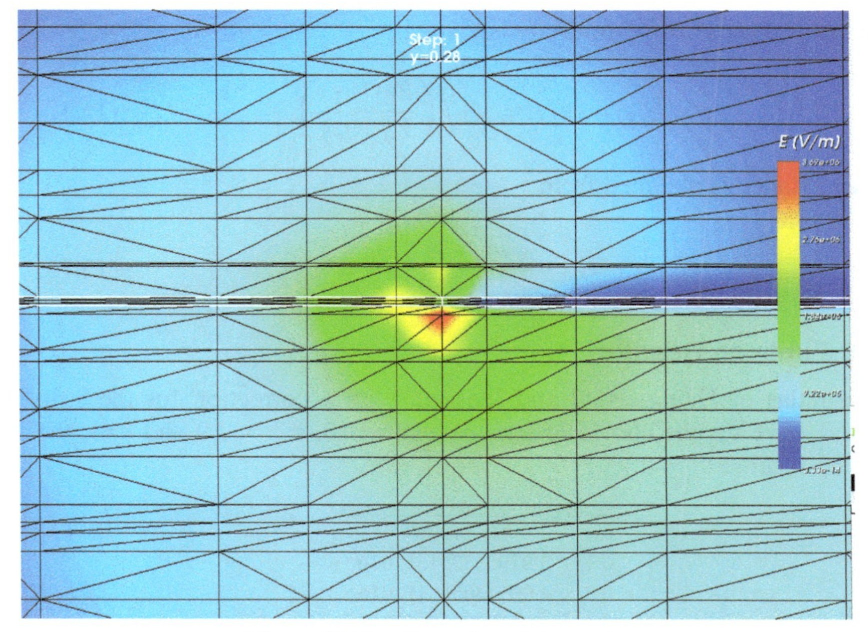

Figure 15.17 Illustration of the mesh.

The normalized thickness is $t = t_M/h$. Furthermore we define:

$$u_1 = u + \delta u_1$$
$$u_r = u + \delta u_r$$

in which:

$$\delta u_1 = \frac{t}{\pi} ln \left(1 + \frac{4e^{(1)}}{t \cdot \coth^2(\sqrt{6.517u}}\right)$$
$$\delta u_r = \frac{1}{2}\left(1 + \frac{1}{\cosh\sqrt{\varepsilon_r - 1}}\right)\delta u_1$$

The parameter $Z_0(u, t, {}_r)$ is defined as

$$Z_0(u, t, {}_r) = \frac{Z_{01}(u_r)}{{}_{eff}(u_r, {}_r)} \tag{15.13}$$

where the function Z_{01} was defined as (see Equation (15.10)

$$Z_{01}(u) = \frac{\eta_0}{2\pi} \ln\left[\frac{f(u)}{u} + \sqrt{1 + \frac{4}{u^2}}\right] \tag{15.14}$$

A thickness-dependent permittivity is given by

$$\varepsilon_{eff}^{t_M} = \varepsilon_{eff}(u_r, \varepsilon_r)\left(\frac{Z_{01}(u_1)}{Z_{01}(u_r)}\right)^2 \tag{15.15}$$

The capacitance is obtained from

$$C = \frac{\sqrt{\mu_0\varepsilon_0} \times \varepsilon_{eff}^{t_M}}{Z} \tag{15.16}$$

15.4.1 Finite t_M Results

The formulas of Hammerstad and Jenssen, with a correction for the strip thickness, give for the choice of parameters: w=10, $\varepsilon_r = 3.9$ and h=1, the following results:

t_M	C in F/m
0.1	3.9790333515E-016
0.3	4.0054405631E-016
0.4	4.0162279229E-016
0.5	4.0260254404E-016

t_M capacitances

We refined the capacitance computation. Indeed, the results that where obtained in our first experiment are *accidentally* in agreement. The Hammerstad and Jenssen model is an *underestimation* because the thickness of the metal was not included and the MAGWEL simulation was an *underestimation* because the air layer was only 1 micron (instead of 'infinity'). By taking a layer of air of 10 micron we obtain the following numerical results:

t_M	C_{MGW} in F/m
0.01	3.964446958408e-16 F/m
0.5	4.015557092879e-16 F/m

Numerical results

This last number is in close agreement with the *corrected* HJ result:

$$t_M = 0.5, \qquad C_{corrected}^{HJ} = 4.02602544041718e - 016 \text{ F/m}$$

15.5 Running the Adapter

Start from the most-coarse mesh and running adaptive meshing over several sweeps, e.g. by using more refined meshes we find that the result moves towards the analytic result.

Capacitance	Refinement Level
1.038975073324e-15	no refinement
6.966594046402e-16	1 sweep
5.296263512904e-16	2 sweeps
4.525636287160e-16	3 sweeps
4.211632499841e-16	4 sweeps

Capacitances

Adaptation was based on HF testing. The C plots were for static capacitances, but they are very similar to 1 GHz plots for V and E.

15.6 Simulations with Opera – VectorFields

The commercial software of Vector Fields (Opera [1984]) provides design solutions for a wide variety of electromagnetic applications. Opera is a state-of-the-art software package for the modeling of static and time varying electromagnetic fields, using 2D and 3D finite elements. For benchmarking purposes for the strip line model, the 2D electrostatic adaptive solver was used, where capacitance calculations can be performed in two ways:

- Energy calculation: $C_1 = 2E_{\text{elec}}/V^2$
- Gauss' law: $C_2 = \oint \mathbf{D} \cdot d\mathbf{A}$

For the adaptive meshing stopping criterion, we use the Internal Error $< 0.5\%$ criterion, which means that the relative difference between the electric field solution in each mesh element adjacent to a node and the averaged field in a node is smaller than 0.5%. The meshing is base on a triangular Delauney algorithm. Furthermore, we put the electrostatic potential V=0 on the bottom of the substrate (ground plane) and V=1 on the strip line, while the rest of the boundary has a Neumann condition. Figure 15.18 shows the model (using different scales for the X- and Y-axis). The next figures, Figures 15.19–15.20, where X- and Y-axis now have the same scale, shows the potential distribution and electric fields (arrows). In the table we plot the different capacitance values C_1 and C_2 where we vary the Error and air height.

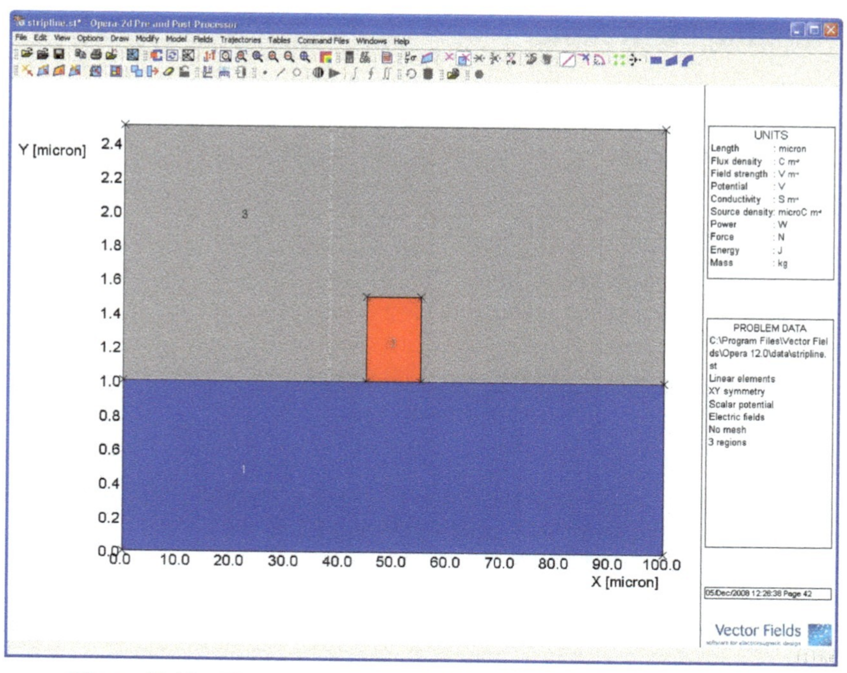

Figure 15.18 Illustration of the model with X and Y at different scales.

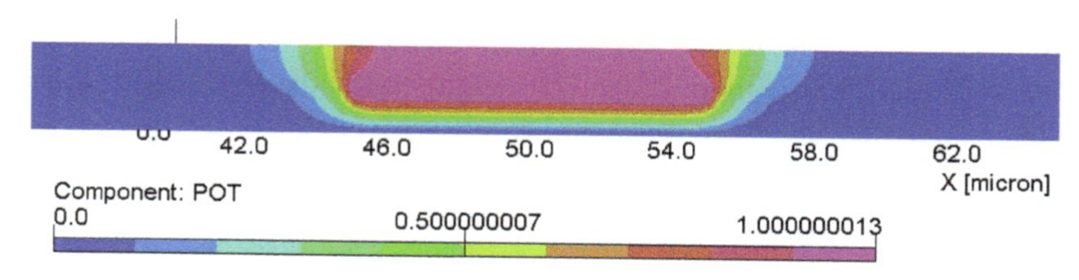

Figure 15.19 Illustration of the potential with X and Y at the same scale.

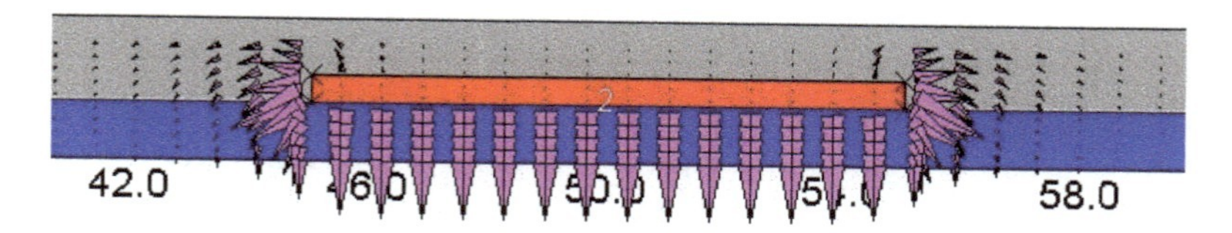

Figure 15.20 Illustration of the electric field with X and Y at the same scale.

The Hammerstad and Jensen value is 4.0260e-016 F/m for the strip line case with thickness T= 0.5 μm. On average a number of 15 iterations in the mesh adaptation process were needed to get an internal accuracy of less than 0.5%, with a final average (of C_1 and C_2) difference of 0.2% compared to the Hammerstad & Jensen value, which of course may also contain some error itself.

T= 0.5 μm T= air height	Internal Error	C_1 in e-16 F/m	C_2 in e-16 F/m	# elem. # elem.	# nodes # nodes
2.5 μm	2%	3.9089	3.9152	11151	5726
2.5 μm	1%	3.9093	3.9121	28745	14558
2.5 μm	0.5%	3.9096	3.9108	96719	48611
2.5 μm	0.2%	3.9106	3.9111	285335	143043
6.5 μm	0.5%	3.9821	3.9831	116517	58468
11.5 μm	0.5%	4.0054	4.0061	189073	94736
16.5 μm	0.5%	4.0121	4.0127	149111	74716
50 μm	0.5%	4.0181	4.0189	165182	82744

Numerical results of the OPERA simulations

In the next table we plot C_1 and C_2 and also the Hammerstad & Jensen values for various values of the strip line thickness, with average (of C_1 and C_2) differences of 0.02–0.2% with the H-J values.

Air Height = 50 μm Error = 0.5%	H&J	C_1	C_2	Aver. Diff. %
T= 0.4 μm	4.01623	4.01016	4.0115	0.13
T= 0.3 μm	4.00544	4.0010	4.0024	0.09
T= 0.2 μm	3.99330	3.99046	3.9916	0.06
T= 0.1 μm	3.97033	3.97802	3.9787	0.20
T= 0.01 μm	–	3.96148	3.9638	–
T= 0 μm	3.95900	3.95926	3.9600	0.02

Numerical results of the OPERA simulations for various strip line thicknesses

To get an idea of the adaptation process and the meshes, see the Figures 15.21–15.35:

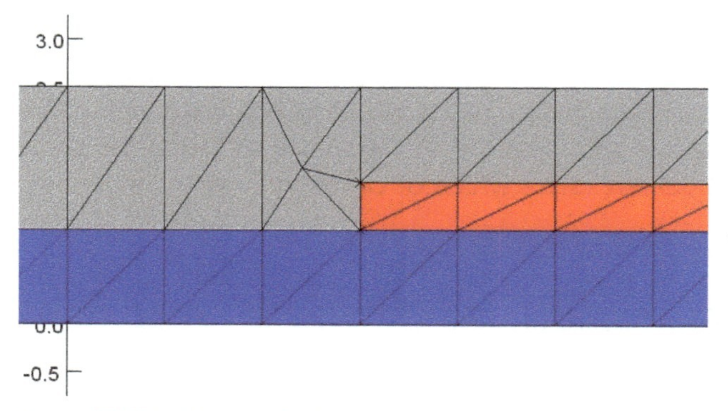

Figure 15.21 Mesh1: 436 elements, evaluated error 18.21371%.

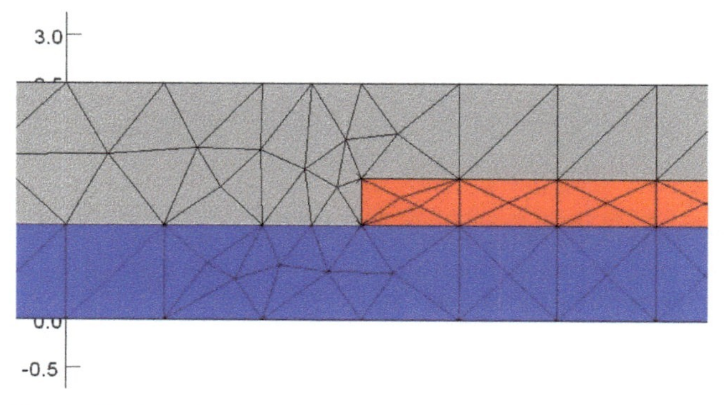

Figure 15.22 Mesh2: 532 elements, evaluated error 15.04066%.

Figure 15.23 Mesh3: 671 elements, evaluated error 17.95522%.

Figure 15.24 Mesh4: 881 elements, evaluated error 10.40742%.

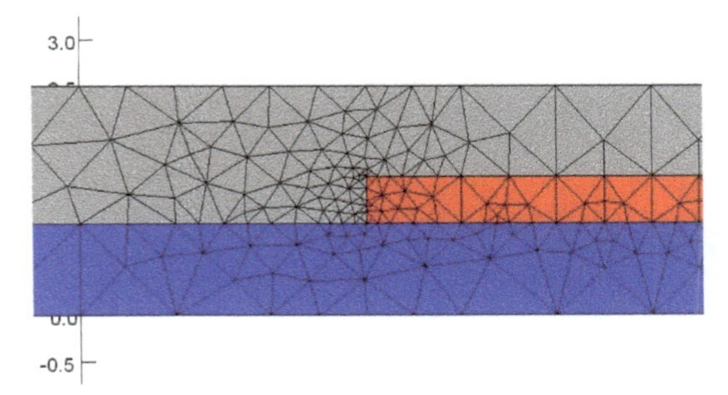

Figure 15.25 Mesh5: 1079 elements, evaluated error 8.484277%.

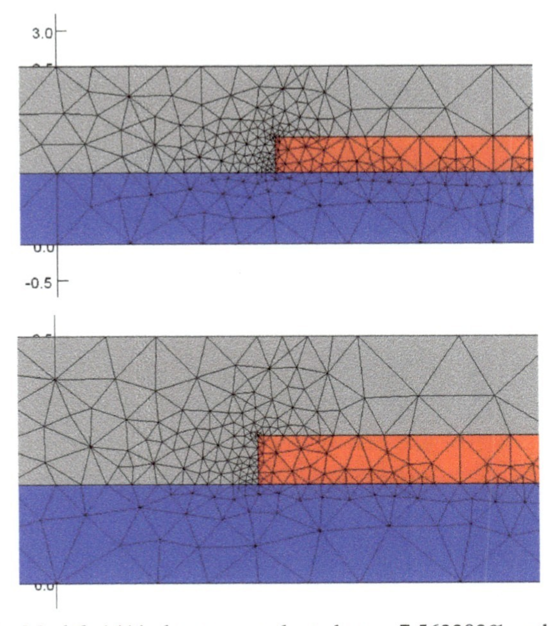

Figure 15.26 Mesh6: 1411 elements, evaluated error 7.563303% and with zoom-in.

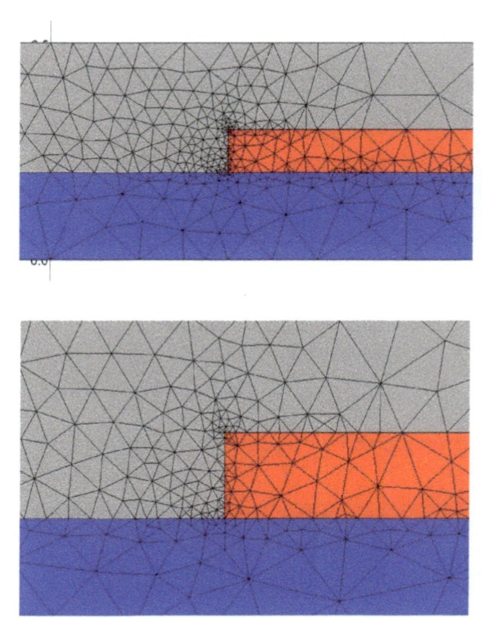

Figure 15.27 Mesh7: 1955 elements, evaluated error 6.604616% and with zoom-in.

Figure 15.28 Mesh8: 2706 elements, evaluated error 4.726004% and with zoom-in.

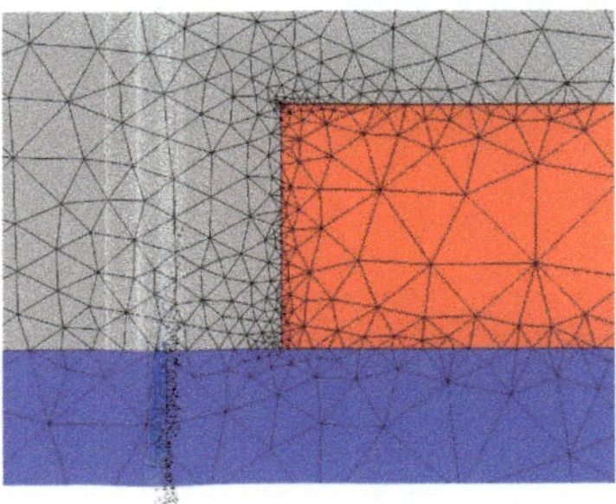

Figure 15.29 Mesh9: 3647 elements, evaluated error 3.773939%.

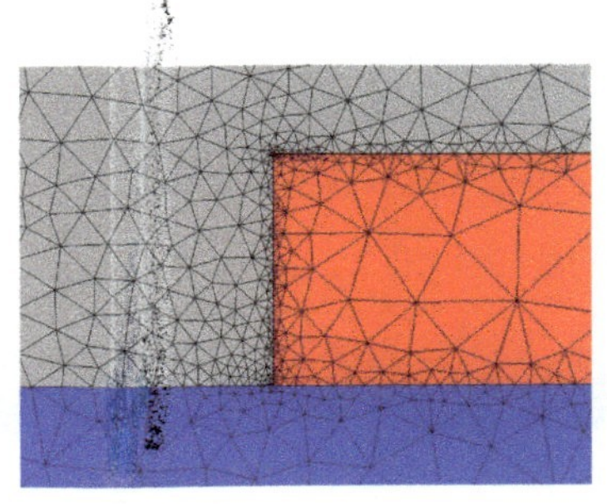

Figure 15.30 Mesh10: 4967 elements, evaluated error 3.268908%.

Figure 15.31 Mesh11: 7100 elements, evaluated error 2.078323% and with zoom-in.

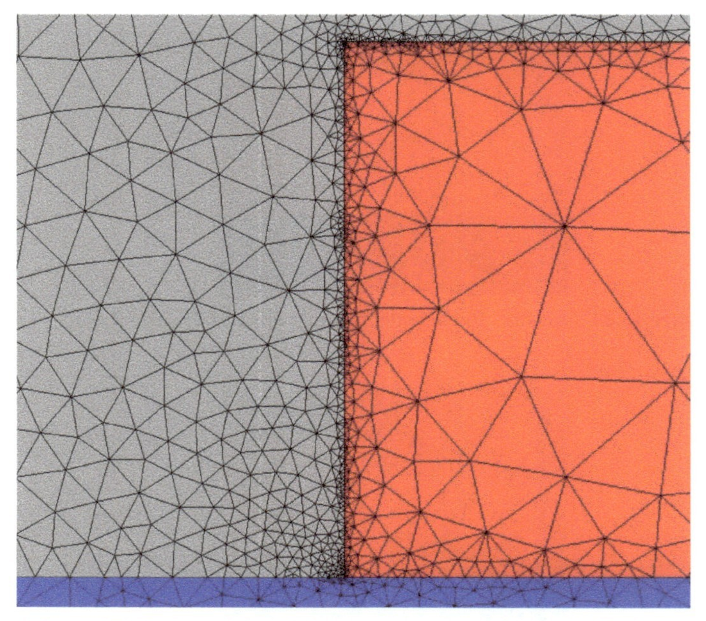

Figure 15.32 Mesh12: 9894 elements, evaluated error 2.127898%.

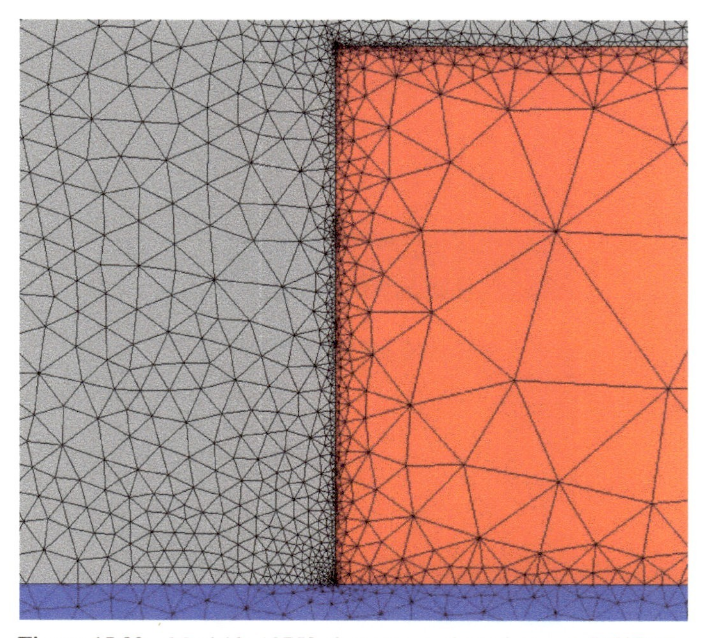

Figure 15.33 Mesh13: 13752 elements, evaluated error 1.517278%.

Figure 15.34 Mesh14: 18945 elements, evaluated error 1.209618%.

Figure 15.35 Mesh15: 24736 elements, evaluated error 0.908549%.

Many effects influence the final result of the fringe capacitance, including air layer, mesh size, strip thickness, domain size, etc. Refined analytic and refined numerical results converge towards each other. The difference is: $(C_{\mathrm{MGW}} - C_{\mathrm{HJ}})/C_{\mathrm{HJ}} = 0.011/4.02 = 0.003 = 0.3\%$. We demonstrated that the MAGWEL solver gives results similar to OPERA (and therefore with any other finite-element based field solver) as far as accuracy vs. node consumption is concerned.

15.7 Coax Configuration

The simple coax example is used for testing a circular geometry. In the MAGWEL solver (devEM [2003]) these circles are converted to polygons, and either a staircase Manhattan approximation is used or a general unstructured Delaunay mesh. This example enables us to compare with analytical expressions. The simulation set up is done with the following parameters:

- Cylindrical coax
- Inner radius 1 micron (R_0)
- Outer radius of outer conductor 5.5 micron (R_2)
- Length 10 micron
- $\mu 0$ =4 $\pi \times 10^{-7}$ H/m
- μ_r=1

An illustration of the wire is shown in Figure 15.36

Results We compare the static simulation results with the analytical magnetic field (Figure 15.37).

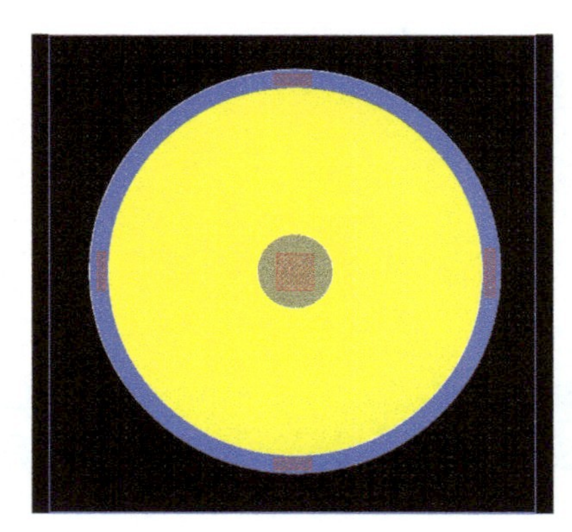

Figure 15.36 Illustration of the coax wire in the MAGWEL editor.

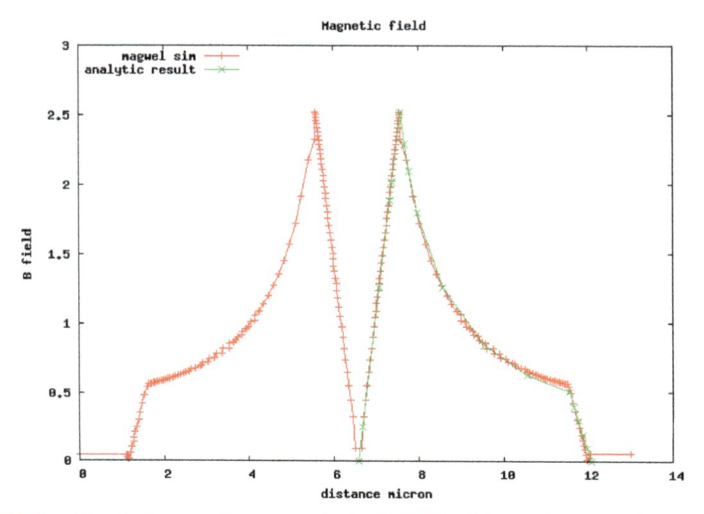

Figure 15.37 Simulation results and analytical MAGWEL results for a Manhattan mesh.

$$0 \leq r \leq R_0 \;:\; B(r) = \frac{B_{max} r}{R_0} \tag{15.17}$$

$$R_0 \leq r \leq R_1 \;:\; B(r) = \frac{B_{max} R_0}{r} \tag{15.18}$$

$$R_1 \leq r \leq R_2 \;:\; B(r) = \frac{B_{max} R_0}{r} \left(\frac{R_2^2 - r^2}{R_2^2 - R_1^2} \right) \tag{15.19}$$

Also the resulting inductance per unit length is analytically known as:

$$L = \frac{\mu_o}{2\pi} \ln \frac{R_1}{R_0} + \frac{\mu_o}{2\pi} \left(\frac{R_2^2}{R_2^2 - R_1^2} \right)^2 \ln \frac{R_2}{R_1} - \frac{\mu_o}{4\pi} \left(\frac{R_2^2}{R_2^2 - R_1^2} \right) \tag{15.20}$$

and for the given configuration this results in a value of $L_{analytical}$=3.7837 10^{-7} H/m. This result is reproduced by MAGWEL simulations using different meshes (Manhattan and Delaunay), to give the following results:

Nodes	L_{sim}	Error	Mesh Type
8448	3.7709 10^{-7} H/m	0.38%	Delaunay
31000	3.7768 10^{-7} H/m	0.18%	Delaunay
13920	3.7372 10^{-7} H/m	1.40%	Manhattan
209760	3.7950 10^{-7} H/m	0.15%	Manhattan

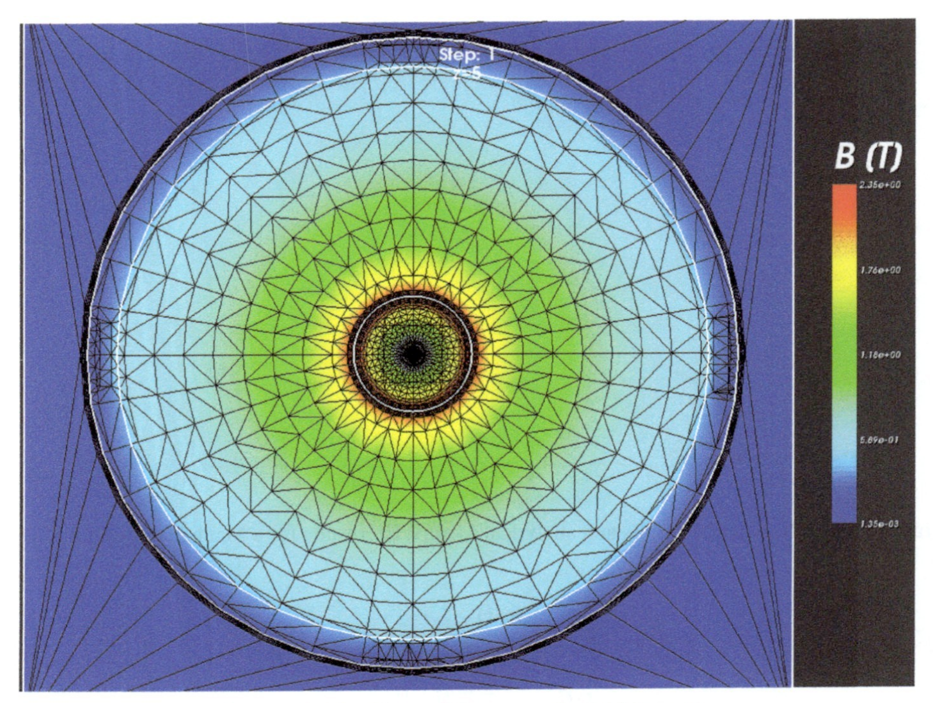

Figure 15.38 View of the magnetic field strength.

The RF simulations for the same structure get exactly the same resulting inductance, when we stay below a frequency of 1 GHz. For higher frequencies, extra phenomena play a role, like skin effect, current re-distributions, etc. From this benchmark we learned that the numerical treatment of the magnetic operator is leading to correct analytic results in the static regime. Typical circular geometries are adequately simulated with the MAGWEL software.

15.8 Inductor with Grounded Guard Ring

This design considers an inductor that is shielded by a closed-loop grounded guard ring. The layout is shown in Figure 15.39. When an alternating current is injected into the inductor, an induced current in opposite phase is induced in the closed loop. Although the loop is grounded at both ends, the vector

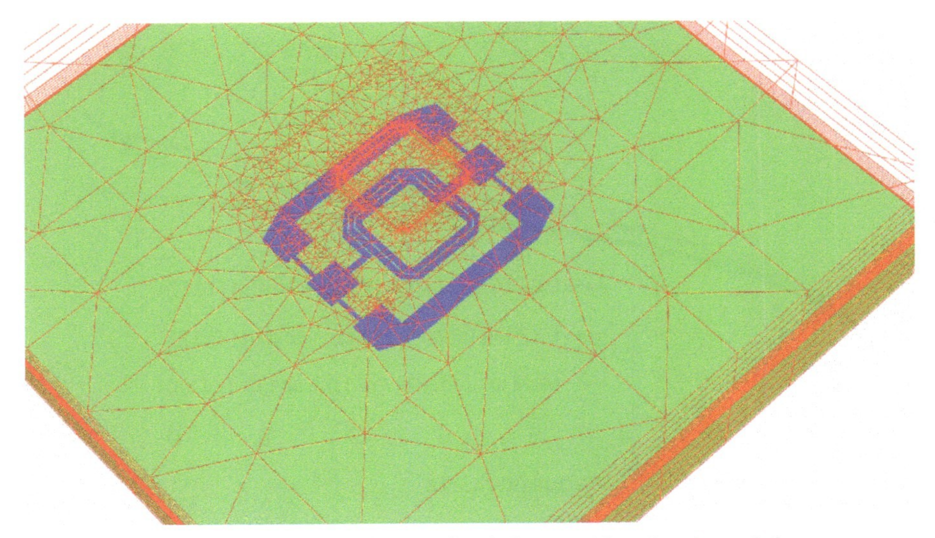

Figure 15.39 Inductor layout of an inductor with a closed guard ring.

potential is still present and therefore the induced current exists despite the fact that the loop is grounded. This configuration is also instructive to obtain an in-depth understanding of the measurement set-up for S-parameters.

Simulation set up The structure is simulated using the Y-parameter extraction method. This design has a high-resistive substrate. Therefore, the decay of the field strength is rather slow and a full stack of 625 micron substrate is included in the simulation. The experimental data of the substrate is: $\sigma = 0.1$ S/m. The computed curves are obtained by using 1) the MAGWEL solver with calibrated substrate resistance, 2) the results obtained using Agilent Momentum (Fach et al. [1984]) are shown and also the experimental results are shown. The curves are identified with the labels MGW, MOM and EXP respectively. The technique based on the integrating factor in Chapter 27 is not applied.

Results In Figure 15.40 the Y-parameters are shown. The curves of MAGWEL and the measurements agree for Y11 and Momentum disagrees. The curves for Im(Y11) and Re(Y12) from MAGWEL and Momentum agree but both deviate from the measured data. Finally for Y12 all curves agree. In Figures 15.41–15.43 the S-parameters are shown.

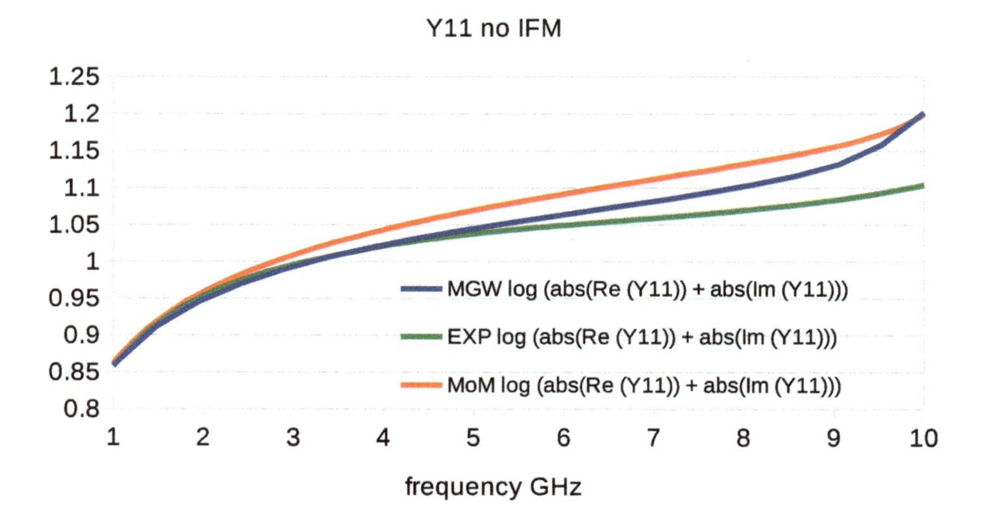

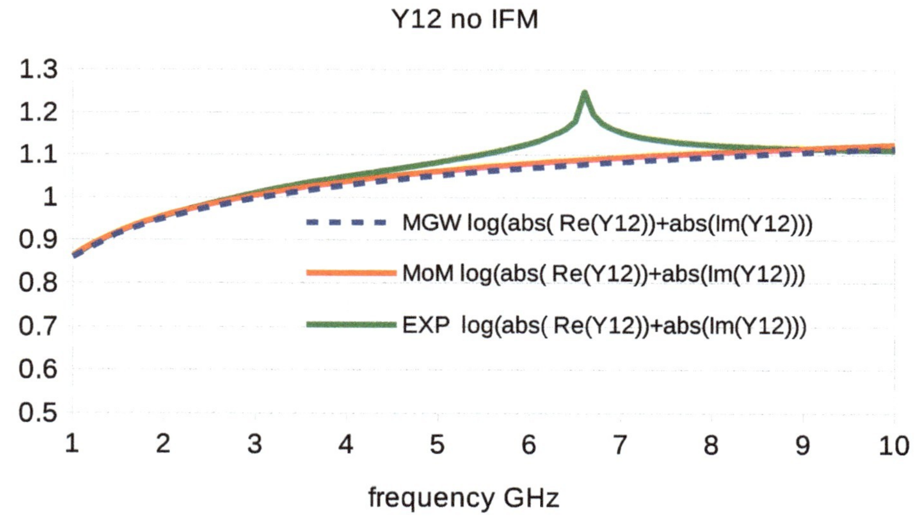

Figure 15.40 log(abs(Re(Y11)) + abs(Im((Y11))) and log(abs(Re(Y12)) + abs(Im(Y12))).

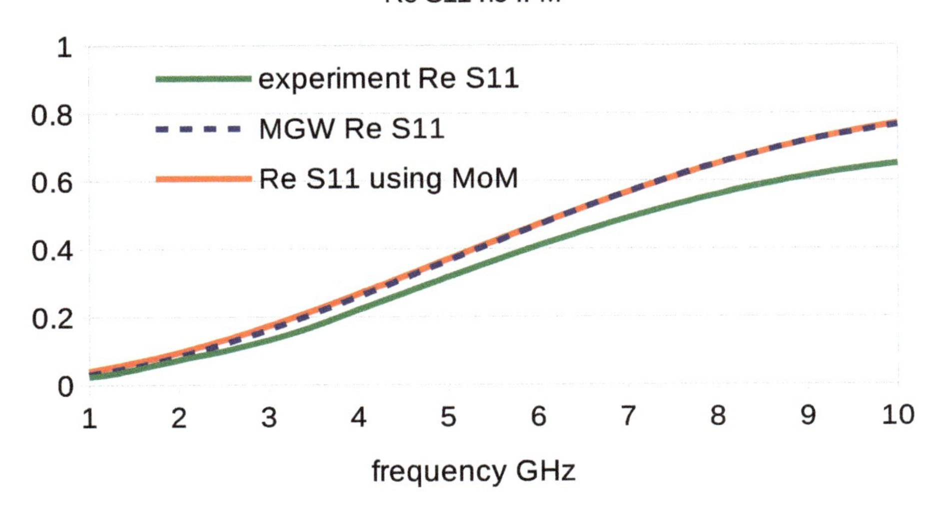

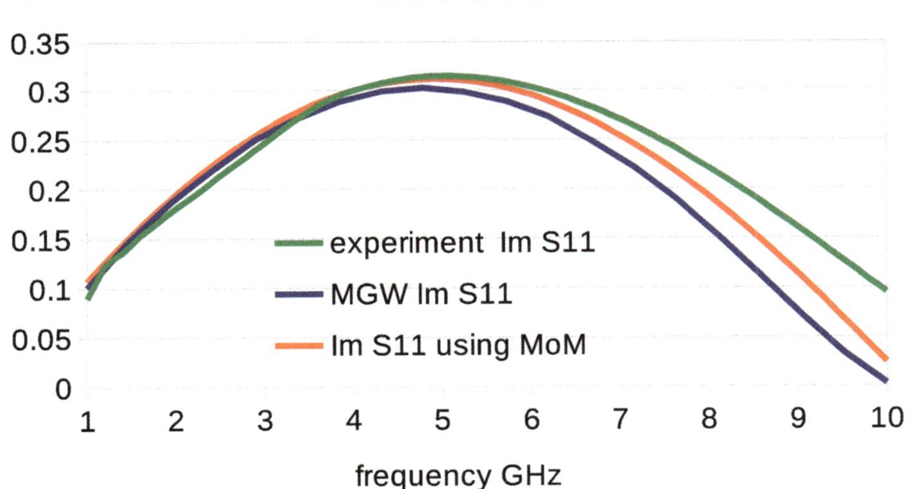

Figure 15.41 Re (S11) and Im (S11).

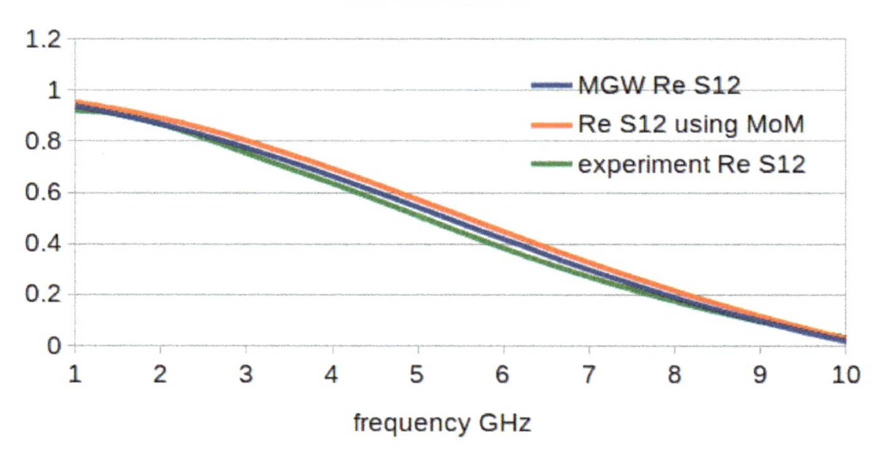

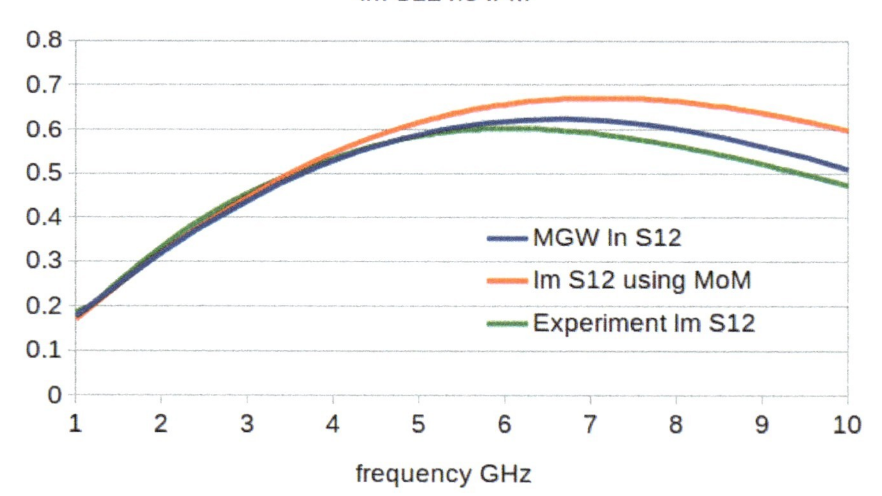

Figure 15.42 Re (S12) and Im (S12).

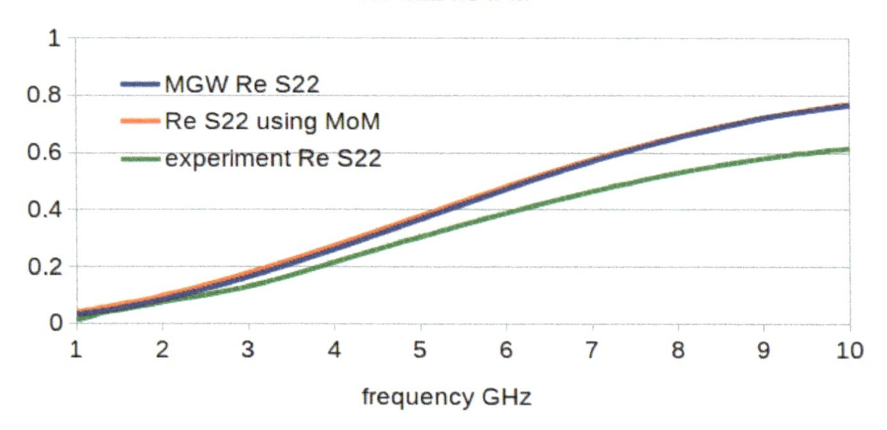

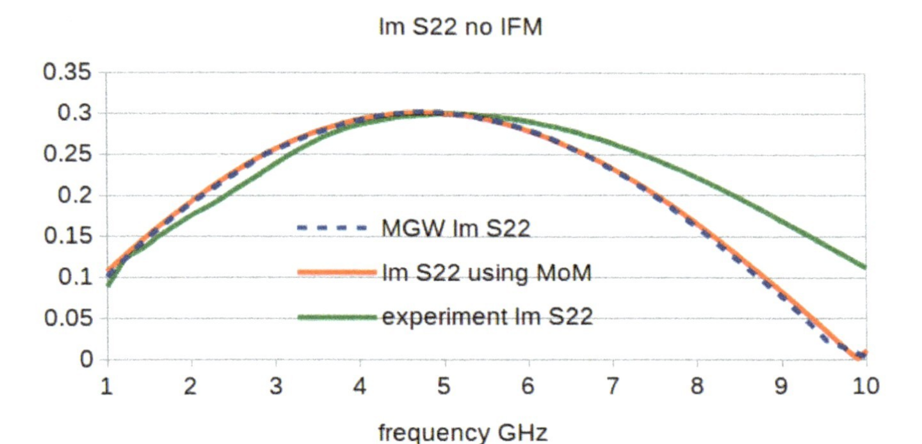

Figure 15.43 Re (22) and Im (S22).

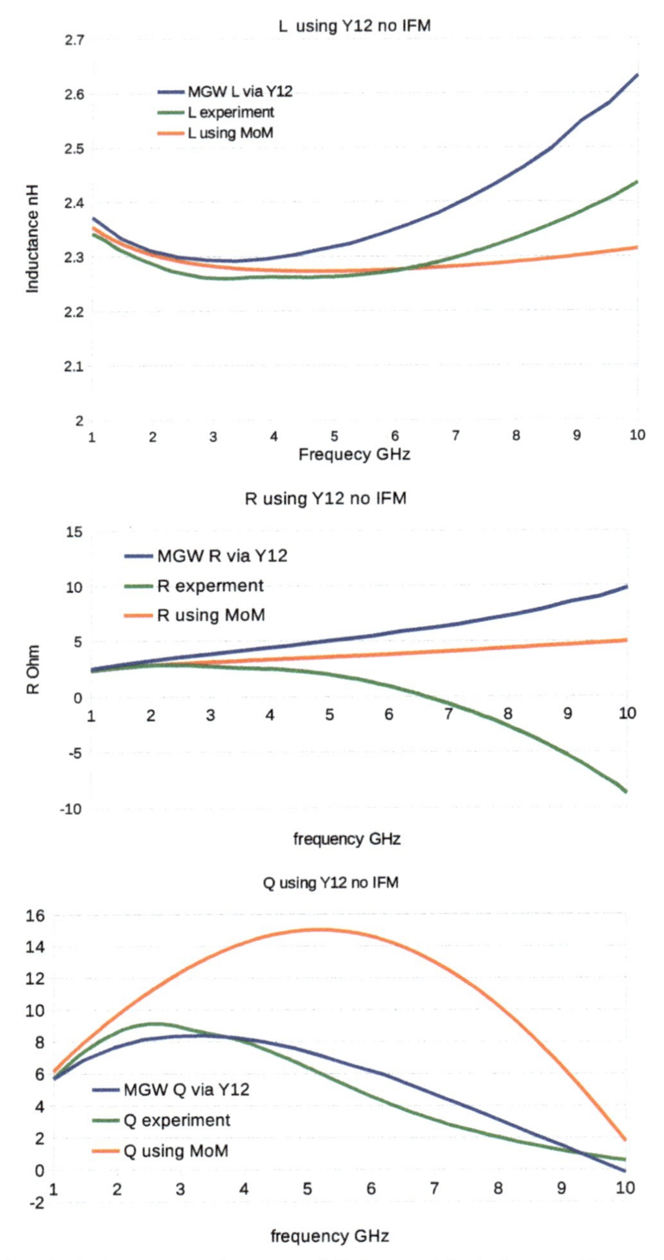

Figure 15.44 The inductance, resistance and Q factor of the inductor with grounded closed-guard ring.

The Y-parameter information can be transformed into compact-model parameter information. For details, we refer to the next section. Here we plot the 2-port resistance, the 2-port inductance and the Q-factor. The results are shown in Figure 15.44.

From this simulation work we observe that the implementation of the Hodge operators is very important for getting results that are within 1% accuracy for the inductance. A naive implementation gives errors of 10%. These errors can be counteracted by using an enhancement factor for the permeability μ_r. This factor is tuned once at zero frequency. Furthermore, in order to get results that are very accurate for inductors with wide windings, the integrating factor method improves the results at higher frequency. This is in agreement with the expectations since the integrating factor becomes important for high frequency and large mesh elements.

15.9 Inductor with Narrow Winding above a Patterned Semiconductor Layer

This inductor is interesting because the narrow windings require a rather advanced meshing strategy. There are numerous calibration issues involved for the simulation of these type of inductors. The layout of the inductor is shown in Figure 15.45.

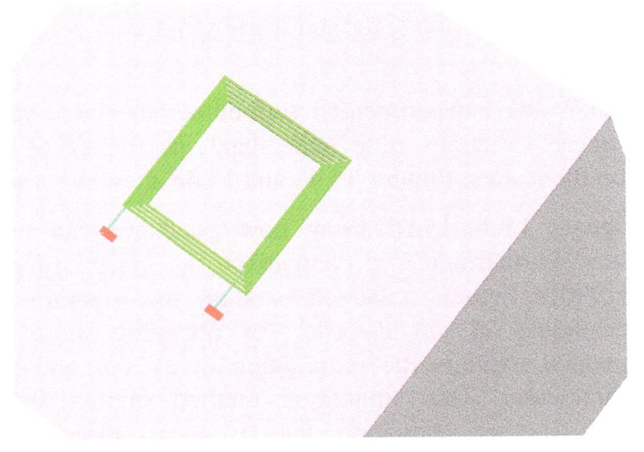

Figure 15.45 View of the inductor (no details).

Figure 15.46 2D View of the Nwell pattern.

However, the details of this design are found in the N-well pattern that is supposed to limit the eddy currents in the substrate. The N-well pattern is shown in Figure 15.46. The underlying idea is that conductive patterns will induce an equipotential in the top layer and thereby induce a collapse of the eddy currents.

Simulation set up We have experimented with numerous layout variations. The purpose of these variations is to understand the impact of different contact designs on the results. Figures 15.47 and 15.48 show two variations.

Results The Figures 15.49–15.66 show how good we can reach the experimental data. Furthermore, a comparison is given using HFSS results (HFSS [1990]). Figures 15.49–15.54 show the S-parameters. We have several techniques at out disposal for obtaining these parameters. A direct method exploits the termination of all but one port and extract the S-parameters. This simulation method mimics the experimental situation. Another technique that is more robust for simulation, grounds all but one port and extracts the Y-parameters. The latter method is applied here. Both methods give equal results.

Figure 15.47 View of the design using vertical contacts.

Figure 15.48 View of the small contacts.

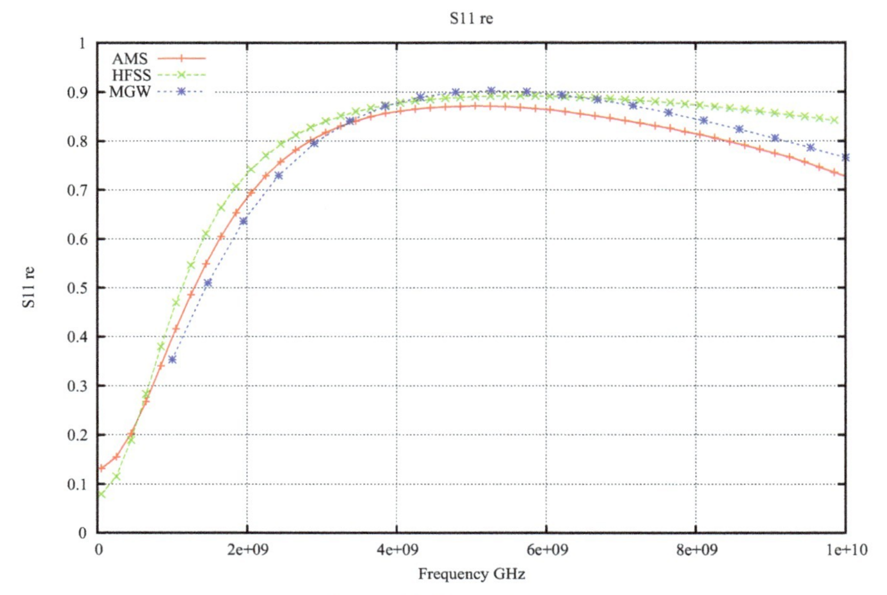

Figure 15.49 Re(S11).

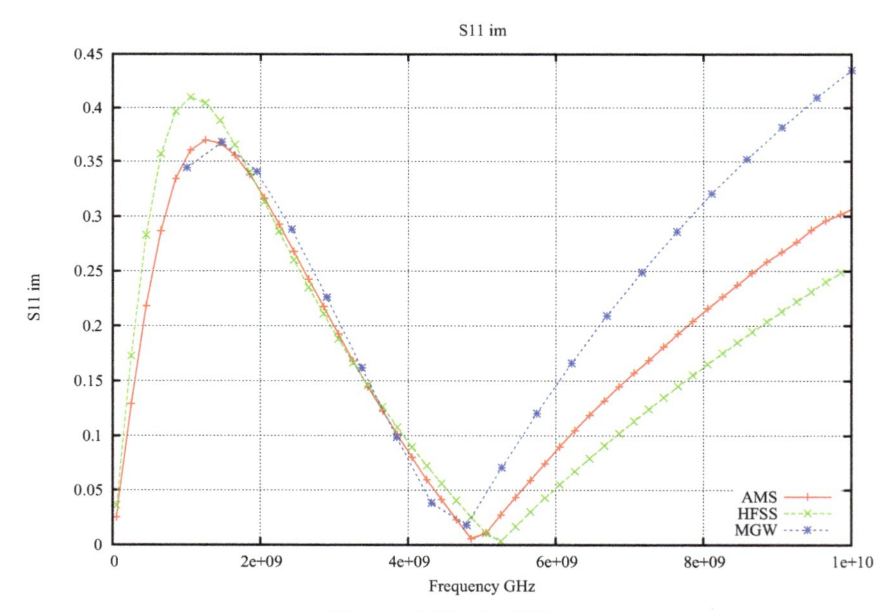

Figure 15.50 Im(S11).

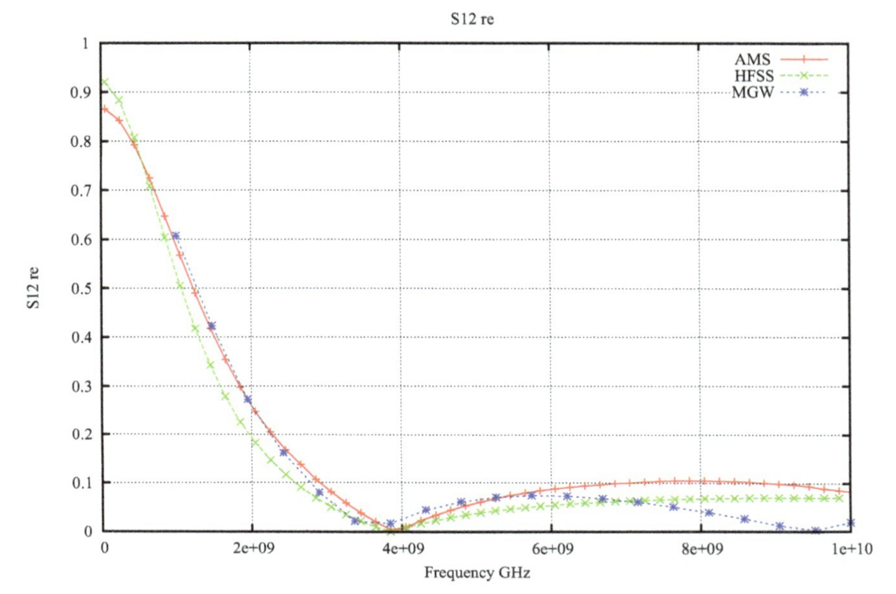

Figure 15.51 Re(S12).

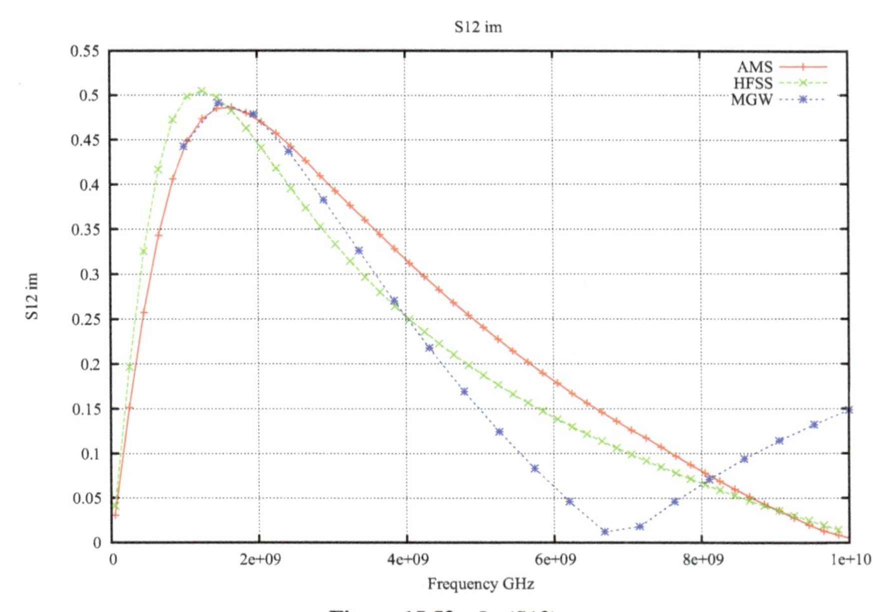

Figure 15.52 Im(S12).

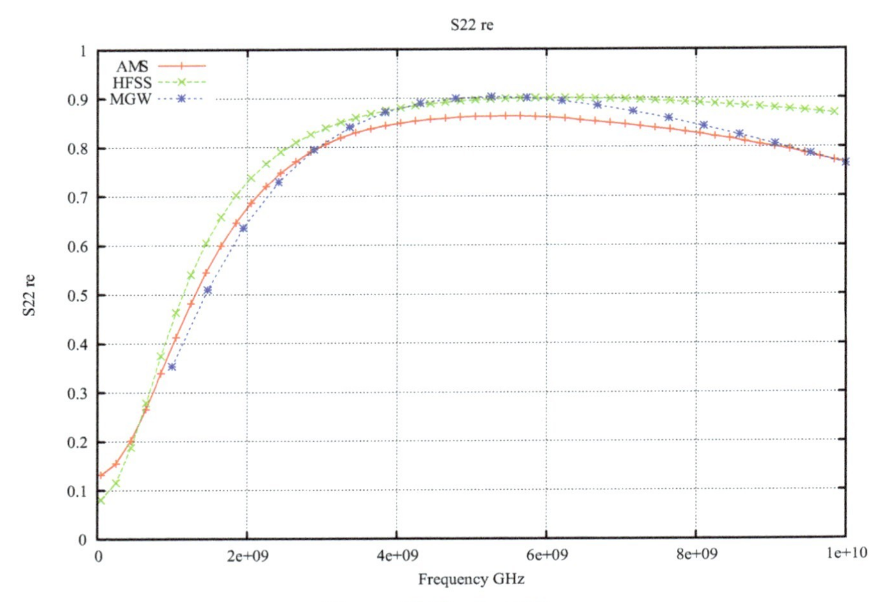

Figure 15.53 Re(S22).

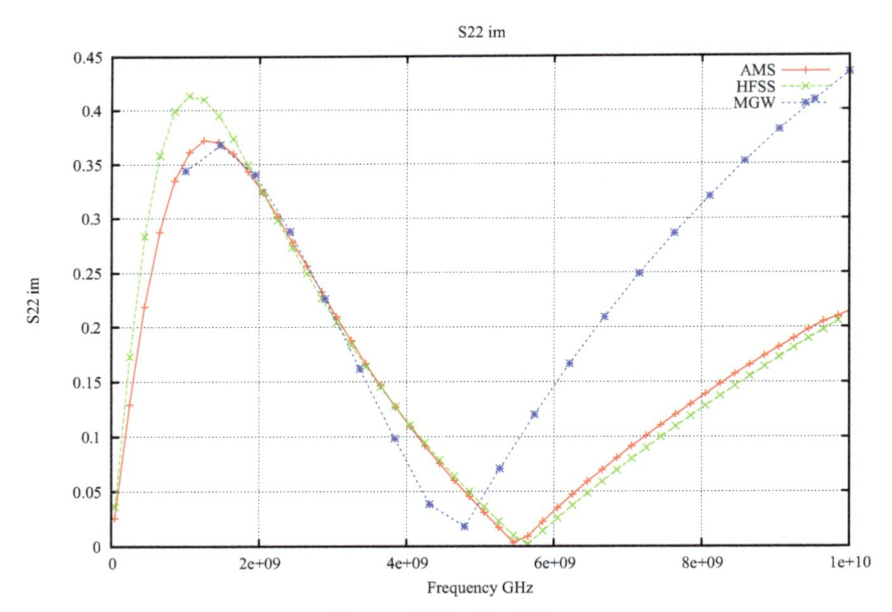

Figure 15.54 Im(S22).

Of more interest are the compact-model parameters that can be extracted from these S-parameters. For that purpose we convert the S-matrix to the Y- and Z-matrix. Depending on the way of using the inductor in a circuit, there are two definitions of the inductance. In the CODESTAR project (CODESTAR [2003]) the inductance was defined using a 1-port approach:

$$R = \mathrm{Re}\left(\frac{1}{Y_{11}}\right) \tag{15.21}$$

$$L = \frac{1}{\omega}\,\mathrm{Im}\left(\frac{1}{Y_{11}}\right) \tag{15.22}$$

For the 2-port approach we use the following definitions as was also done in the CHAMELEON-RF project (Chameleon-RF [2006])

$$R = -\mathrm{Re}\left(\frac{1}{Y_{12}}\right) \tag{15.23}$$

$$L = -\frac{1}{\omega}\,\mathrm{Im}\left(\frac{1}{Y_{12}}\right) \tag{15.24}$$

These formulas can be easily applied using python scripts.

```
python script input lines
# this script uses Y-parameters
      and converts to S- and Z-parameters
#!/usr/bin/env python2.4
import os
import math
import scipy
from scipy.linalg import inv, det, eig
import math

Yparam = scipy.zeros((2,2), complex)
Zparam = scipy.zeros((2,2), complex)
Sparam = scipy.zeros((2,2), complex)

# help matrices
M1 = scipy.zeros((2,2), dtype = complex)
M2 = scipy.zeros((2,2), dtype = complex)
refImpedance =50;
```

```
F = scipy.zeros((2,2), dtype = complex)
F[0,0]=refImpedance
F[1,1]=refImpedance
ONES = scipy.zeros((2,2), dtype = complex)
ONES[0,0]=1.0
ONES[1,1]=1.0
times=50
Y0 = scipy.zeros((2,2), dtype = complex)
Z0 = scipy.zeros((2,2), dtype = complex)
Y0ref = 1/refImpedance
Zparam = scipy.linalg.inv(Yparam)
M1 = -Z0+Zparam
M2 = scipy.linalg.inv(Z0+Zparam)
Sparam = scipy.dot(M1, M2)
R = - (1.0/Yparam[0,1]).real
L = - (1.0/Yparam[0,1]).imag/(2.0*3.1415*freq)
R1 = (1.0/Yparam[0,0]).real
L1 = (1.0/Yparam[0,0]).imag/(2.0*3.1415*freq)
Q= (1.0/Yparam[0,0]).imag /(1.0/Yparam[0,0]).real
```

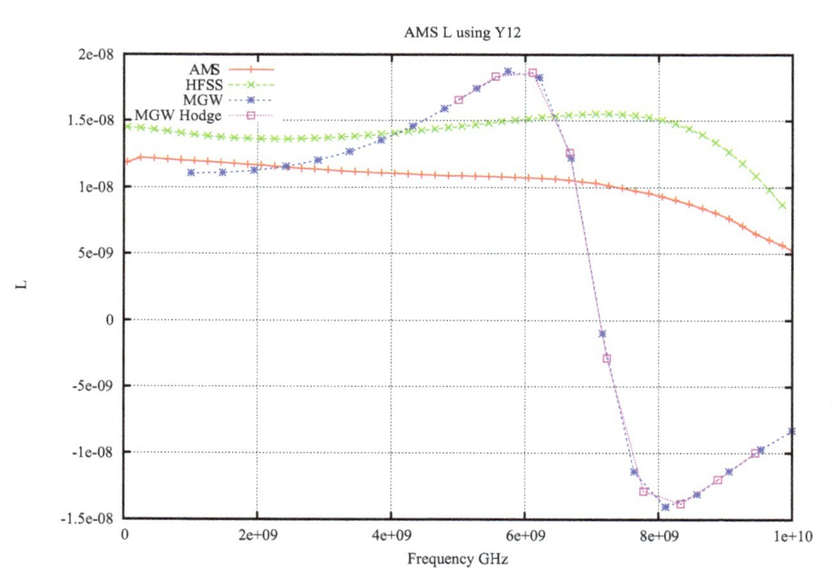

Figure 15.55 L using Y12.

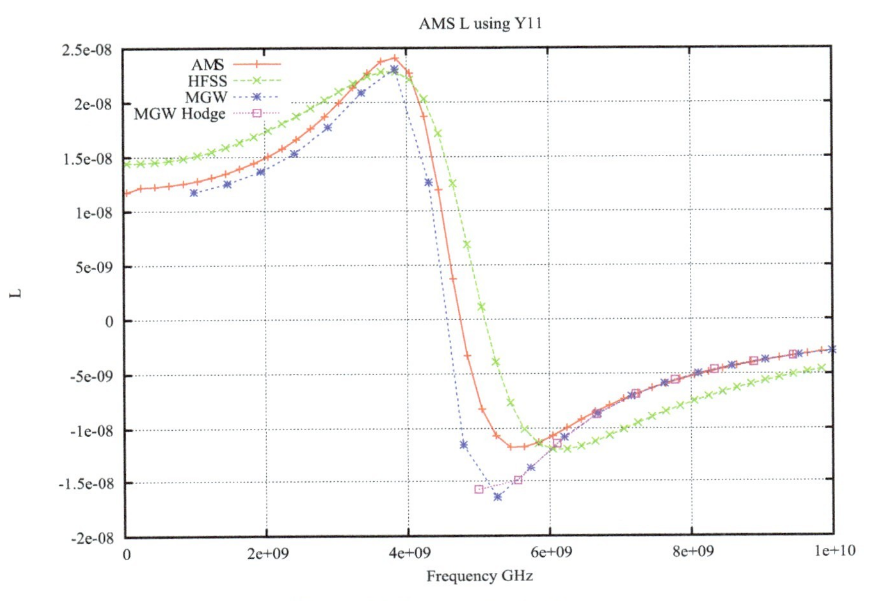

Figure 15.56 L using Y11.

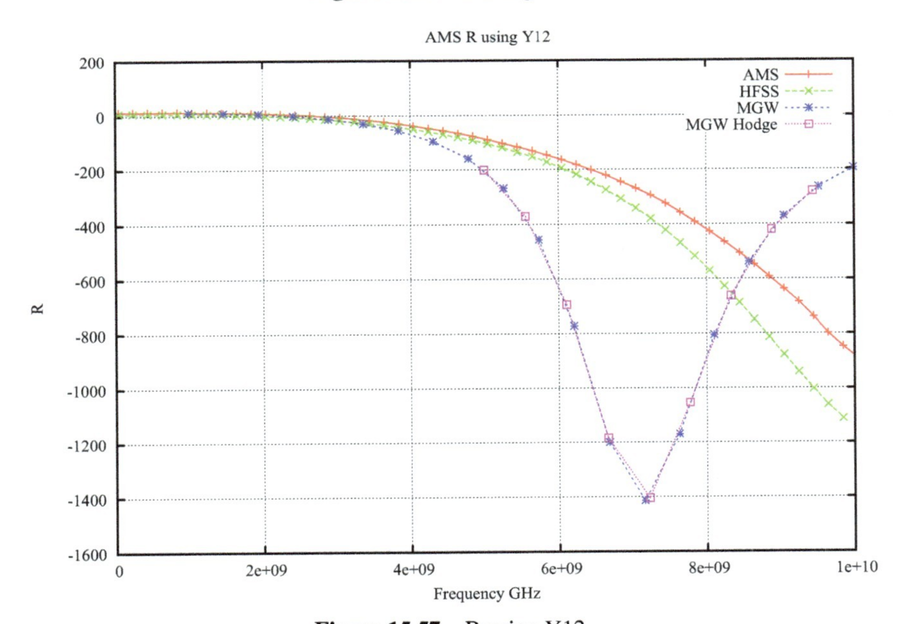

Figure 15.57 R using Y12.

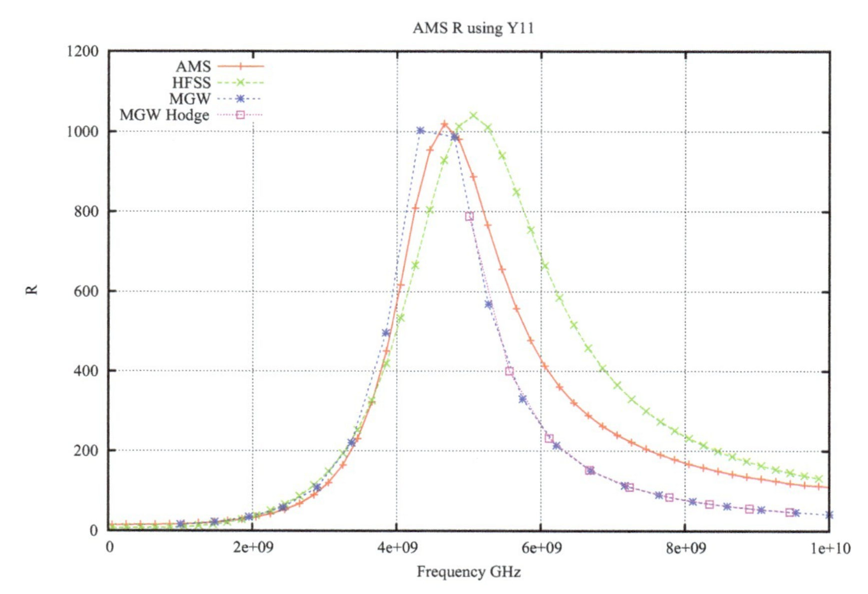

Figure 15.58 R using Y11.

The Q-factor is shown in Figure 15.59.

Finally, we also plot the Y-parameters.

This simulation work has learned us numerous facts. First of all, in order to position the simulation tools with respect to each other it is important to eliminate all disputable uncertainties. In this example, there were two important aspects. The first deals with guard ring design and the second one deals with the contact design. Both uncertainties led to some guess work and in particular, this guess work manifests it self in the definition of the ports. As a consequence, both the MAGWEL solver as well as HFSS produce outcomes that do not match the measurements over the full frequency range. Nevertheless, by pursuing this simulation case, we have obtained a much better understanding of which aspects play an important role in obtaining successful Q-factor results. Moreover, we were able to define a calibration strategy for good inductor simulation. The strategy is based on first getting the static inductance and resistance in place. Next the determination of the Q-factor is set by an accurate modeling of the substrate. In practice it means that the substrate resistance and permittivity must be accurately determined. Finally, the high-frequency behavior must be settled by obtaining a good value for the inter-winding capacitive coupling. Based on these observations we conclude that accurate simulation at high frequency requires

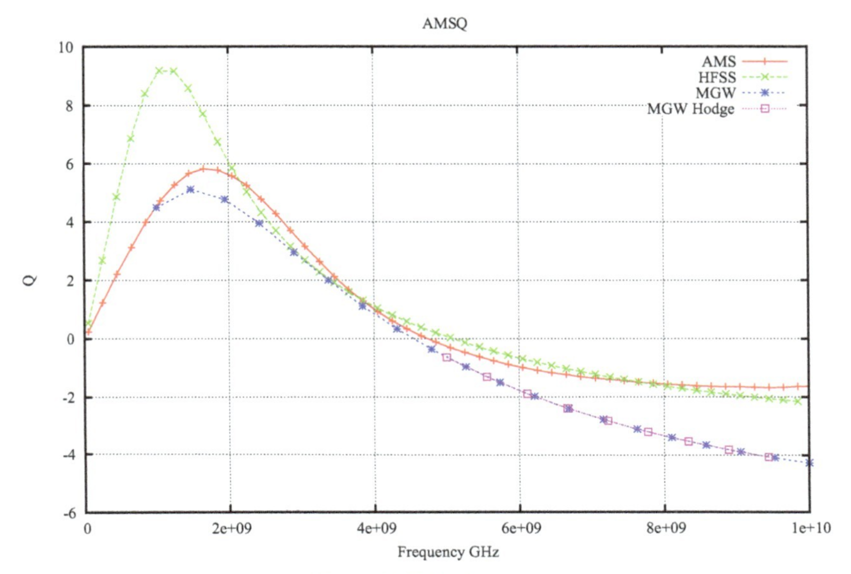

Figure 15.59 Q factor.

adaptive meshing. Without adaptive meshing the capacitive couplings are over-estimated. We suspect that this is the reason why the 2-port inductance of Figure 15.60 shows a resonance at a too low resonance frequency. Since HFSS is equipped with adaptive measurement facilities, this resonance is shifted to a higher frequency since the over-estimation of the inter-winding capacitance is avoided. In order to verify/falsify this interpretation of the results we repeated the simulation and replaced the permittivity of the oxide with 1.0, The Y parameter results are shown in Figures 15.61–15.63. This leads to a smaller inter-winding capacitance and that should give a higher self-resonance frequency. In Figures 15.64–15.66, the compact model parameters are shown. These plots demonstrate that the above explanation is correct. Of course, other deviations are seen because the oxide also is important for the coupling to the substrate.

It should be noted however, that it remains a challenge to get all the curves right over the full frequency range. Besides, the specific knowledge that was generated for simulating inductors above substrates, a few more interesting facts were gathered: At low frequencies, the computation of inductance requires that Hodge operator effects are taken into account. At high frequencies a modified discretization is needed. S-parameter and Y-parameter based simulations provide equivalent compact model results.

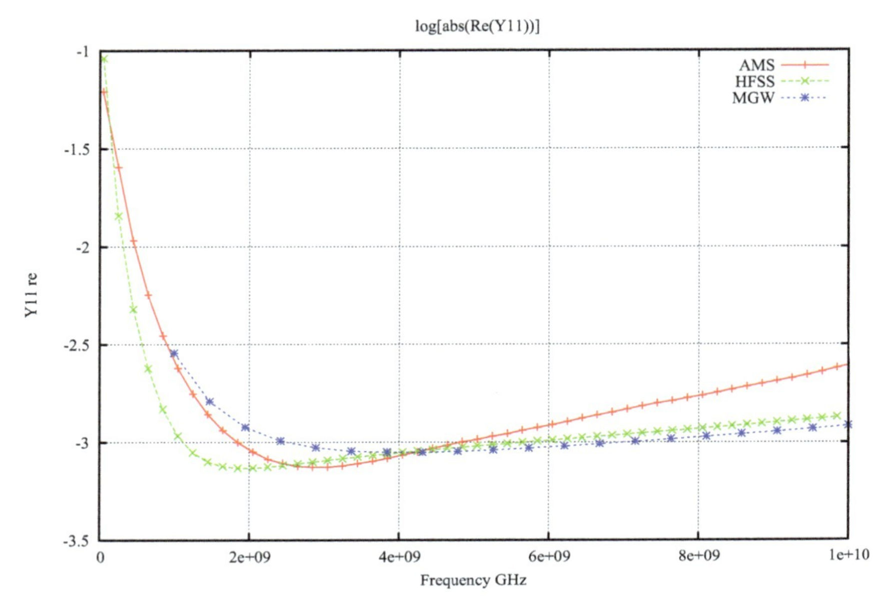

Figure 15.60 Re(Y11).

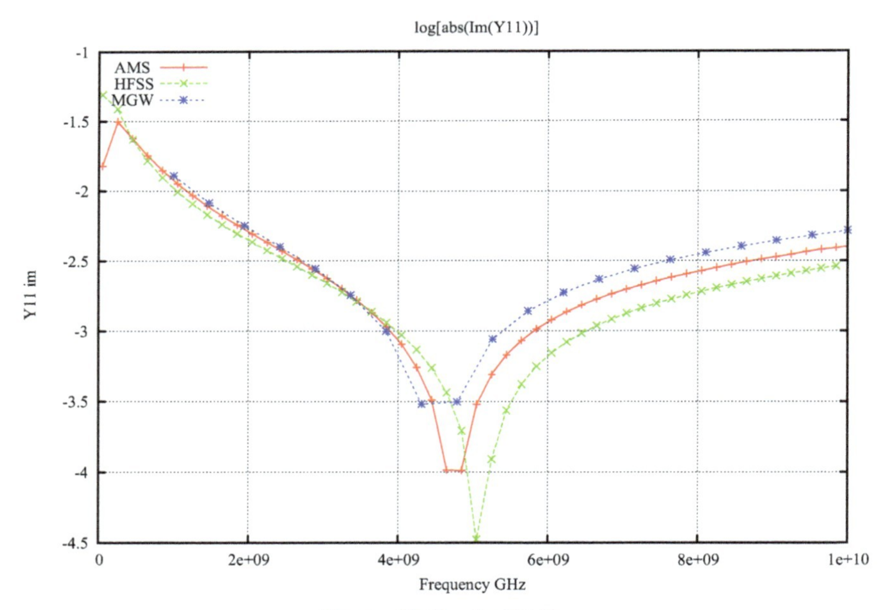

Figure 15.61 Im(Y11).

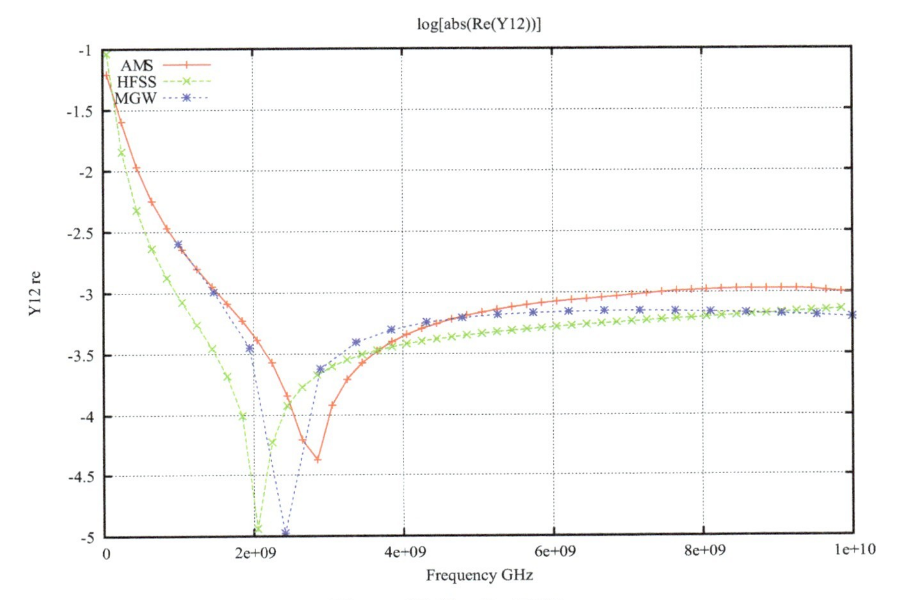

Figure 15.62 Re(Y12).

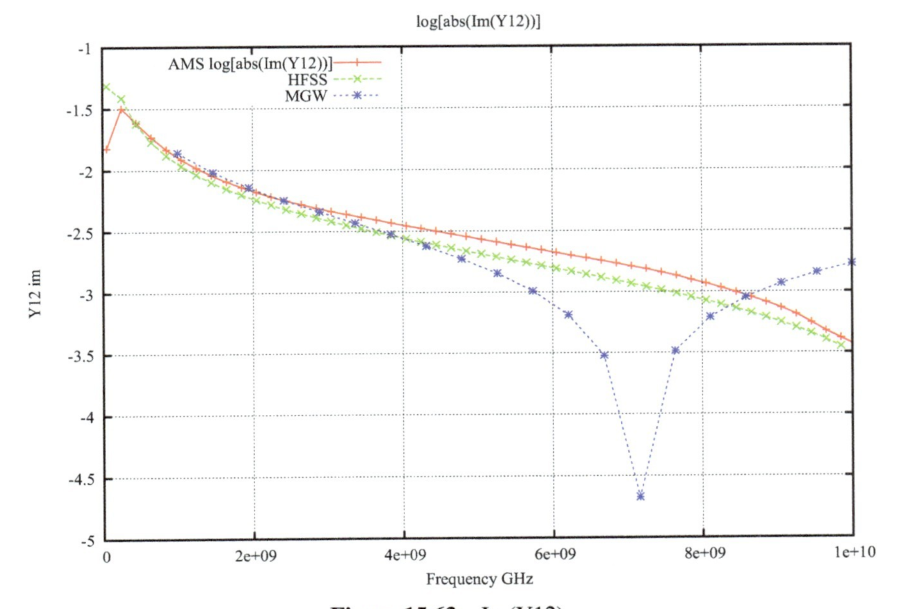

Figure 15.63 Im(Y12).

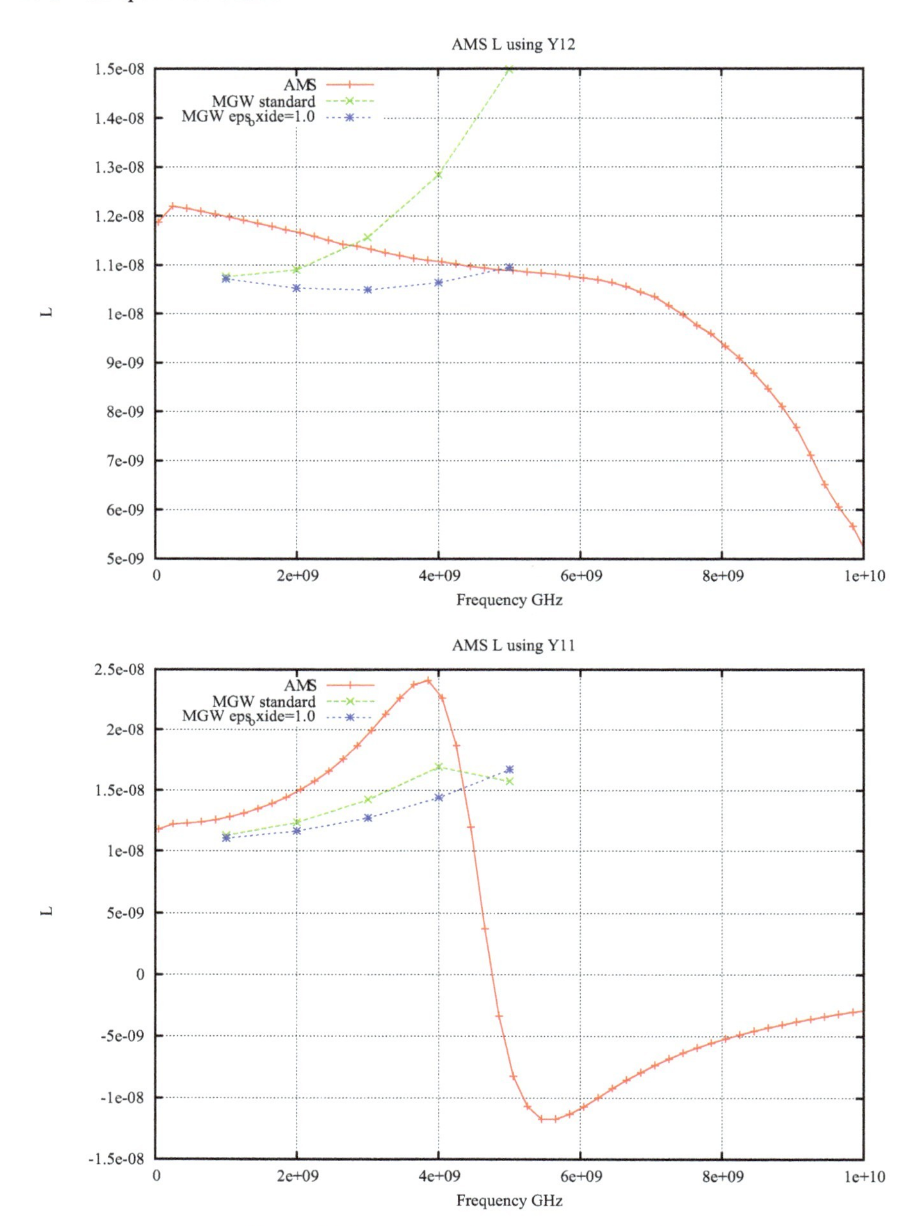

Figure 15.64 L using Y12 and L using Y11.

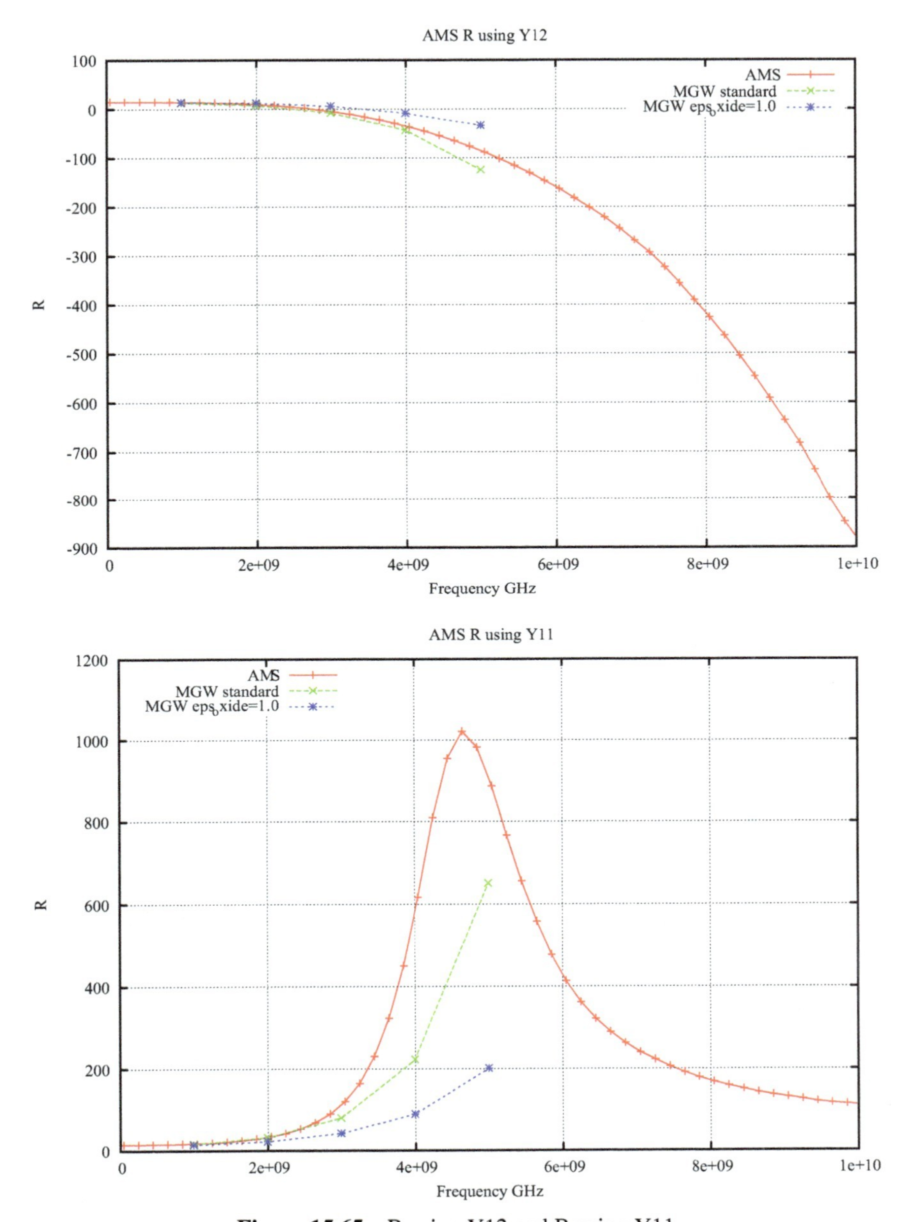

Figure 15.65 R using Y12 and R using Y11.

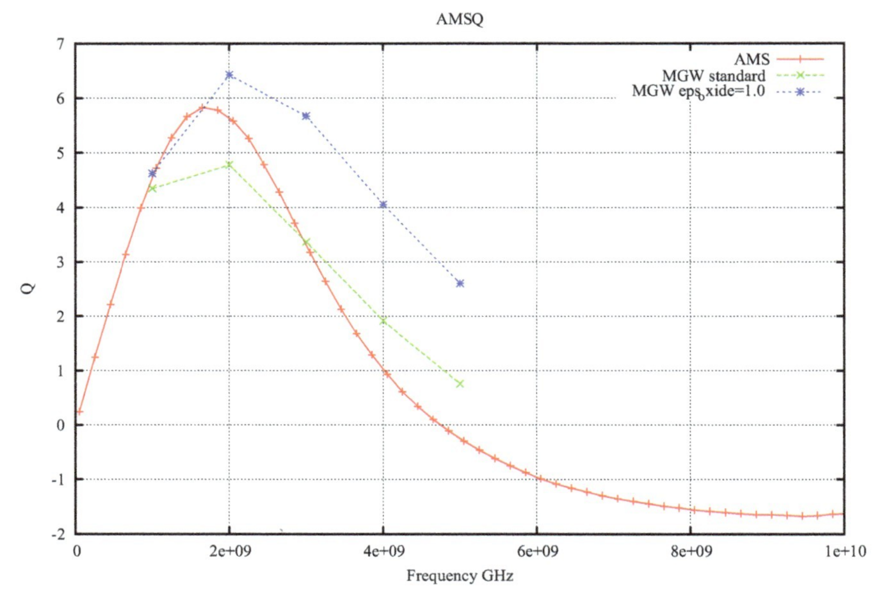

Figure 15.66 Q-factor.

Summary

We observed that the test cases designed for testing elementary qualities give excellent results. However, for industrial examples, the situation is more involved. We have compared results from MAGWEL with results from Momentum (Fach et al. [1984]) and HFSS (HFSS [1990]). The details of the results differ, but the overall conclusion is that a solver based on potentials is able to generate outcomes of comparable quality as Momentum and HFSS. In particular, the simulation of high-resistive substrate inductors has been a serious challenge for many tools. The Q-factor computed by potential method are in general somewhat too small at the peak value, giving a 'pessimistic' result, whereas other field solvers generate a too large (optimistic) result. Nevertheless, an important area of attention is definitely adaptive meshing which is essential for obtaining accurate results in the RF regime.

16

Evaluation of Coupled Inductors

With the use of increasing frequency ranges, electromagnetic coupling becomes a more pronounced design concern because induced electric fields are proportional to the rate of change of the magnetic induction[12]. However, not only the pace of time variations are determining for including electromagnetic coupling but also the problem scale and the intensity of the currents that are responsible for the induced fields must be considered. An overall picture of the scaling arguments is presented in Section 16.1 which helps to identify the needed steps and inclusion of non-negligible effects. Once we note from scale considerations that electromagnetic coupling terms represent a non-negligible contribution to the full system of equations, we move to the the solution of these equations. In Section 16.2, we review and update the approach that was proposed some years ago by the author and co-workers (Meuris et al. [2001b] Schoenmaker et al. [2002a] Schoenmaker and Meuris [2002a]). We will refer to this approach as 'computational electrodynamics'.

At several occasions we were inquired if this method is equivalent to the method based on Nedelec's edge elements (Nedelec [1980] Lee et al. [1991]). The main difference is that we do not refer to test functions at all. Our method is more related to finite-integration techniques (FIT) (Weiland [1977] Schuhmann and Weiland [1998]).

Scale considerations are not the only an issue for deciding if some terms in the full system of Maxwell equations and constitutive laws can be neglected. When discussing the coupling of devices, it is also important to realize that different devices can have intrinsic or geometrical scales that differ orders of magnitude. In such scenarios the coupled problem is most easily split in computational domains. Computational electrodynamics gives,

[12]This chapter is a reprint of Schoenmaker et al. [2010b]

rather straightforwardly, a series of prescriptions for matching the interface conditions of the various domains.

Electromagnetic coupling of microelectronic devices is an RF issue and is most conveniently measured using s-parameters. In Section 16.4, we present our method to compute these matrix elements. In fact, s-parameter extraction is straightforwardly achieved as a post-processing of the results of a computational electrodynamics problem with the appropriate setting of the boundary conditions.

In Section 16.5 we will present an example of a coupled problem.

16.1 Scaling Rules for the Maxwell Equations

The use of scaling arguments is definitely not new to the field of computing in electromagnetic modeling. Well-known approximations are the so-called EQS (electro-quasi-static) and MQS (magneto-quasi-static) approximations. Approximations can be put in a different perspective by considering the scaling step that is necessary when converting the full set of equations to dimensionless equations before the actual computing can start. For our present argument it suffices to consider insulators and metals only. Diffusive currents in semiconductors can easily be added to the equations. Therefore, we start from the Maxwell equations in which $\mathbf{J}_c$ is the conductive current:

$$\mathbf{J}_{\mathrm{c}} = \sigma\mathbf{E}, \quad \mathbf{D} = \varepsilon_0\varepsilon_r\mathbf{E}, \quad \mathbf{H} = \frac{1}{\mu_0\mu_r}\mathbf{B}, \tag{16.1}$$

$$\mathbf{E} = -\nabla V - \frac{\partial \mathbf{A}}{\partial t}, \quad \mathbf{B} = \nabla \times \mathbf{A}. \tag{16.2}$$

We consider the Maxwell equations in the potential formulation. The Poisson equation is used to solve the scalar field in insulators and semiconducting regions and the current-continuity equation is used in metals to find the scalar potential. The electric system is:

$$\nabla . \left[\varepsilon\left(\nabla V + \mathrm{i}\omega\mathbf{A}\right)\right] + \rho = 0, \quad \nabla . \left[\left(\sigma + \mathrm{i}\omega\varepsilon\right)\left(\nabla V + \mathrm{i}\omega\mathbf{A}\right)\right] = 0. \tag{16.3}$$

The Maxwell-Ampere equation is:

$$\nabla \times \left(\frac{1}{\mu}\nabla \times \mathbf{A}\right) - \left(\sigma + \mathrm{i}\omega\varepsilon\right)\left(-\nabla V - \mathrm{i}\omega\mathbf{A}\right) = 0. \tag{16.4}$$

This system must be completed with a gauge condition

$$\nabla \cdot \mathbf{A} + \mathrm{i}\omega\xi\varepsilon\mu V = 0, \tag{16.5}$$

where ξ is a parameter that allows us to slide over different gauge conditions. Now let L be the 'natural' length scale of the problem that is considered. For example $L = 1\mu$m. Furthermore, let T be the natural time scale, for example $T = 10^{-9}$ sec. It is possible to reformulate the Equations (16.3) and (16.4) in *dimensionless* variables V and $\mathbf{A}$ and the set of equations is controlled by two dimensionless variables, K and ν

$$\nabla . \left[\varepsilon_r \left(\nabla V + \mathrm{i}\omega \mathbf{A}\right)\right] + \rho = 0, \qquad \nabla . \left[\left(\sigma + \mathrm{i}\omega\varepsilon_r\right)\left(\nabla V + \mathrm{i}\omega \mathbf{A}\right)\right] = 0, \tag{16.6}$$

and

$$\nabla \times \left(\frac{1}{\mu_r} \nabla \times \mathbf{A}\right) - K\omega^2 \left(\varepsilon_r - \mathrm{i}\,\nu\right)\mathbf{A} - \mathrm{i}\,\omega K \left(\varepsilon_r - \mathrm{i}\nu\right)\nabla V = 0. \tag{16.7}$$

The constants $K = \varepsilon_0\mu_0 L^2/T^2$ and $\nu = \sigma T/\varepsilon_0$. Note that for $\sigma = 10^4$ S/m we obtain $K\nu = 10^{-5}$. This value corresponds to the conductance of an inversion layer in the on-state of a transistor. This number enters into the Maxwell-Ampere equation and suggests that in this scenario the magnetic sector is negligible. For a single transistor finger this is a valid conclusion, but one should be aware that in actual designs many fingers may operate in a parallel mode therefore the value of K could increase since L must be adapted to this situation. Taking into account the presence of the back-end processing, one encounters metallic conductance of 10^7 S/m, such that magnetic effects are important.

16.2 Discretization

In our earlier work, we presented a discretization method that decided for each variable where on the grid it belongs. It was concluded that the geometrical and physical meaning of variables plays a key role. For instance, a scalar variable, e.g. the Poisson potential, V, is a number assigned to each space location and for a computational purpose, its discretized value should be assigned to the nodes of the grid. On the other hand the vector potential $\mathbf{A}$ is a variable of the same character as ∇V and should therefore be assigned to the links of the computational grid. Geometrical considerations have been an important guide for correctly discretizing Maxwell's equations, as was also elaborated by Bossovit (Bossavit [2005, 2003]).

The conversion of continuous variables to discrete variables on the computation grid also has consequences for the particular discretization route that is followed when implementing discrete versions of the Maxwell equations. Gauss' law is discretized by considering elementary volumes around the nodes of the grid and one next perform an integration of Gauss' law over these volume cells. The flux assigned to each segment of the enclosing surface is assumed to be constant which allows for expressing this (constant) flux in terms of the node variables and link variables. This scheme has been the key to the success of the simulation of the semiconductor devices. The Scharfetter-Gummel formulation of the discretized currents can be set up following the above approach (Scharfetter and Gummel [1969]). Since links variables are fundamentally different from node variables, we expect that the discretization of the Maxwell-Ampere equation has to be done taking this geometrical aspect into account. Whereas it was quite 'natural' to regard node variables as a representative of some volume element, in the same way we consider a link variable representing some area element. Thus to each link is associated an area element and in order to discretize the Maxwell-Ampere equation on a grid we now apply Stokes' law to arrive at the discretized equations.

After having obtained a scheme to discretize the Maxwell equations, we proceed with expanding them into a small signal analysis. This means that each variable is written as a time-independent part and an harmonic part

$$X = X_0 + X_1 e^{i\omega t}. \tag{16.8}$$

If we apply boundary conditions of a similar form and collect terms independent of ω and terms proportional to $e^{i\omega t}$ and omit terms proportional to X_1^2 then we obtain a system of equations for the phasors X_1. Of particular interest is the treatment of the spurious modes in the fields. These modes can be eliminated by selecting a 'gauge tree' in the mesh, adding a ghost field to the equation system or apply a projection method while iterating towards the solution. We can also apply a gauge condition and construct discrete operators that resemble the continuous operators as close as possible including having a semi-definite spectrum. Using a two-fold application of Stokes' law, the term $\nabla \times \left(\frac{1}{\mu_r}\nabla \times \mathbf{A}\right)$ appears in the discretized formulation as a collection of closed-loop circulations. By subtracting a discretized version of $\nabla(\nabla \cdot \mathbf{A})$ we arrive at an operator that resembles $-\nabla^2\mathbf{A}$. However, since $\mathbf{A}$ is a vector field, the latter can only have meaning in terms of the foregoing expressions. The discretization of the first term in (16.4) can be illustrated as shown in Figure 16.1. The primary link PQ has a dual area assigned to it. This area is denoted with the links a, b, c and d. The curl-curl operator is realized as

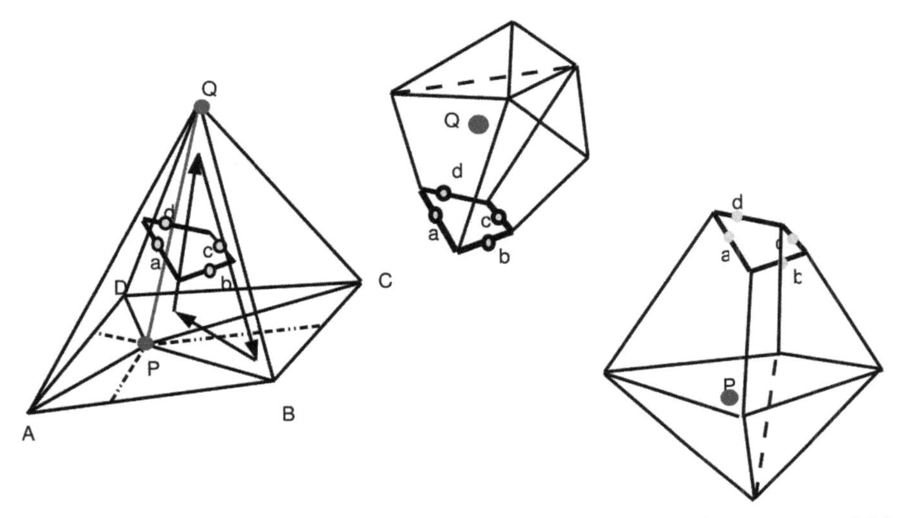

Figure 16.1 Discretized version of the regularized curl-curl operator acting on a vector field.

a sum of circulations around all primary surfaces that contain this link. The most-left picture of Figure 16.1 illustrates this aspect. The subtraction of the grad-div operator is done in two steps: The grad means that both at P and at Q a divergence is evaluated. The center- and right drawing show these divergences. Next, these terms are added with opposite sign.

16.3 The E*V* Solver

Besides scaling and geometrical considerations, another important ingredient for a successful discretization is to avoid unnecessary matrix fill when selecting dynamical variables. In this section, we present a method to reduce the cross coupling between the V and $\mathbf{A}$ system. Let us consider the Ampere-Maxwell equation. For notational convenience we will introduce the notation: $\phi = \sigma + \mathrm{i}\omega\varepsilon_{\mathrm{r}}$. Then we can write (16.7) as

$$\nabla \times \left(\frac{1}{\mu_{\mathrm{r}}} \nabla \times \mathbf{A} \right) + K\,\phi\,(\nabla V + \mathrm{i}\omega \mathbf{A}) - K\,\mathbf{J}_{\mathrm{diff}} = 0, \tag{16.9}$$

where $\mathbf{J}_{\mathrm{diff}}$ is the diffusive part of the current. Furthermore, we will need the gauge condition

$$\nabla \cdot \mathbf{A} + \mathrm{i}\,\omega\xi K\,\varepsilon_{\mathrm{r}} V = 0, \tag{16.10}$$

where ξ is the slider between 0 (Coulomb gauge) and 1 (Lorentz gauge). The crucial observation now is that for any scalar field, the equation $\nabla \times \nabla V = 0$ is valid. This leads to

$$\frac{1}{\mathrm{i}\omega} \nabla \times \left(\frac{1}{\mu_\mathrm{r}} \nabla \times \left[\mathrm{i}\omega \, \mathbf{A} + \nabla V \right] \right) + K \, \phi \, \left(\nabla V + \mathrm{i}\omega \mathbf{A} \right) - K \, \mathbf{J}_\mathrm{diff} = 0. \tag{16.11}$$

We recognize $\mathrm{i}\omega \mathbf{A} + \nabla V = -\mathbf{E}$ and therefore we find that

$$\nabla \times \left(\frac{1}{\mu_\mathrm{r}} \nabla \times \mathbf{E} \right) + K \, \mathrm{i} \, \omega \, \phi \, \mathbf{E} + K \, \mathrm{i} \, \omega \, \mathbf{J}_\mathrm{diff} = 0. \tag{16.12}$$

Of course, this equation could have been straightforwardly obtained from the Maxwell equations by noting that $\mathbf{B} = -1/(\mathrm{i}\omega) \nabla \times \mathbf{E}$. However, here we consider $\mathbf{E}$ as a variable transformation of $\mathbf{A}$. Just as for the $\mathbf{A}$ system, we must regularize the operator $\nabla \times \nabla \times \mathbf{E}$. This is achieved by subtracting the gauge condition. Using

$$\mathbf{A} = \frac{\mathrm{i}}{\omega} \left(\mathbf{E} + \nabla V \right), \tag{16.13}$$

we obtain

$$\nabla \cdot \left\{ \frac{\mathrm{i}}{\omega} \left[\mathbf{E} + \nabla V \right] \right\} + \mathrm{i} \, \omega \, K \xi \, \varepsilon_\mathrm{r} \, V = 0. \tag{16.14}$$

This is equivalent to the following expression:

$$\nabla \cdot \mathbf{E} + \nabla^2 \, V + \omega^2 \, K \xi \, \varepsilon_\mathrm{r} \, V = 0. \tag{16.15}$$

The regularization is now achieved by subtraction of the gradient of Equation (16.15) from Equation (16.12).

$$\nabla \times \left(\frac{1}{\mu_\mathrm{r}} \nabla \times \mathbf{E} \right) - \nabla \left(\nabla \cdot \mathbf{E} \right) + K \, \mathrm{i} \, \omega \, \phi \, \mathbf{E}$$
$$- \nabla \left(\nabla^2 \, V \right) - \omega^2 \, K \xi \, \nabla \left(\varepsilon_\mathrm{r} \, V \right) + K \, \mathrm{i} \, \omega \, \mathbf{J}_\mathrm{diff} = 0. \tag{16.16}$$

As is seen from this equation the coupling to the variables V has a strength of order one and is not growing with σ. Furthermore it should be noticed that the Poisson equation is not part of the set of equations that must be solved. It is an implicit consequence of the Ampere-Maxwell system. Therefore, the equation to be used for determining V, is the gauge condition:

$$\nabla^2 V + \nabla \cdot \mathbf{E} + K \, \xi \, \omega^2 \, \varepsilon_\mathrm{r} \, V = 0. \tag{16.17}$$

With Equation (16.16) for the solution of $\mathbf{E}$ and (16.17) for the solution of V, we can compute the full $\mathbf{E}V$ system. The cross couplings will not explode for large σ in the bulk of the material. Thus we expect that this set-up of equations would have lead to linear systems that will solve faster at high high-frequencies in comparison with the system of equations based on the $\mathbf{A}V$ formulation. However, it should be noted that a third-order derivative term is present. As a consequence the matrix fill increases substantially. We were able to solve (16.16) and (16.17) self-consistently for a series of applications at the cost of using *direct* solvers. Finally we note that a full-wave solution needs again *four* fields, i.e. E_x, E_y, E_z and V, to be solved.

16.3.1 Boundary Conditions

Although no strong coupling exists in the bulk of the material, the boundary conditions introduce again this coupling in some circumstances.
The boundary conditions for the vector Equation (16.16) can be deduced from the boundary conditions for the vector potential $\mathbf{A}$. Since for each link in the surface of the simulation domain we have put the boundary condition $\mathbf{A} \cdot \hat{\mathbf{t}} = 0$, and $\hat{\mathbf{t}}$ is a tangential unit vector, we obtain

$$\mathbf{E} \cdot \hat{\mathbf{t}} = -\hat{\mathbf{t}} \cdot \nabla V. \tag{16.18}$$

The boundary conditions for the scalar Equation (16.17) can be deduced from the condition that for surface regions outside the contacts, the outward pointing electric field component is taken equal to zero, i.e. $\mathbf{E} \cdot \hat{\mathbf{n}} = 0$ where $\hat{\mathbf{n}}$ is a normal unit vector. However, this will not be sufficient to determine the boundary condition for V, since an additional unknown, $\partial V/\partial n$ needs to be given outside the contact regions. Fortunately, there is still room for further restriction. The boundary condition for $\mathbf{A}$ was only provided for the tangential components of $\mathbf{A}$. We will now include also a boundary condition for the normal component of $\mathbf{A}$ that consists of stating that the normal component of $\mathbf{A}$ will have be continuous when crossing the simulation surface

$$\hat{\mathbf{n}} \cdot \mathbf{A}_{\text{inside}} = \hat{\mathbf{n}} \cdot \mathbf{A}_{\text{outside}}. \tag{16.19}$$

This can also be written as $\partial A_{\perp}/\partial\, n = 0$, or in other words: a Neumann boundary condition is used for the perpendicular component of $\mathbf{A}$. However, the surface nodes of the simulation domain can also be determined by applying the Poisson equation and/or current continuity equation for these nodes.

$$\nabla \cdot (\phi\, \mathbf{E}\,) = 0. \tag{16.20}$$

For internal nodes, this equation is a consequence of the Maxwell-Ampere system. However, at the surface it must explicitly be enforced by the boundary condition. Thus for the boundary nodes, we apply the Poisson and current-continuity equations, using the inwards pointing link variables E_{ij}. This enables one to get boundary conditions for the V variables on the simulation boundary.

16.4 Scattering Parameters

In order to determine the S matrix, a rather straightforward procedure is followed. For that purpose a collection of ports is needed and each port consists of two contacts. A contact is defined as a collection of nodes that are electrically identified. A rather evident appearance of a contact is a surface segment on the boundary of the simulation domain. A slightly less trivial contact consists of two or more of these surfaces on the boundary of the simulation domain. The nodes that are found on these surfaces are all at equal potential. Therefore, although there may be many nodes assigned to a single contact, all these nodes together generate only one potential variable to the system of unknowns. Of course, when evaluating the current entering or leaving the contact, each node in the contact contributes to the total contact current. Assigning prescribed values for the contact potential can be seen as applying Dirichlet's boundary conditions to these contacts. This is a familiar technique in technology CAD. Outside the contact regions, Neumann boundary conditions are applied. Unfortunately, since we are now dealing with the full system of Maxwell equations, providing boundary conditions for the scalar potential will not suffice. We also need to provide boundary conditions for the vector potential. Last but not least, since the set of variable V and $\mathbf{A}$ are not independent, setting a boundary condition for one variable has an impact on the other. Moreover, the choice of the gauge condition also participates in the appearance of the variables and their relations. A convenient set of boundary conditions is given by the following set of rules:

- Contact surface $V = V|_c^i$. To each contact area a prescribed potential value is assigned.
- Outside the contact area on the simulation domain $\mathbf{D}_n = 0$. There is no electric field component in the direction perpendicular to the surface of the simulation domain.

- For the complete surface of the simulation domain, we set $\mathbf{B}_n = 0$. There is no magnetic induction perpendicular to the surface of the simulation domain.

We must next translate these boundary condition to restrictions on $\mathbf{A}$. We start with the last one. Since there is no normal component $\mathbf{B}$, we may assume that the vector potential is perpendicular to the surface of the simulation domain. That means that the links at the surface of the simulation domain do not generate a degree of freedom. It should be noted that more general options exist. Nevertheless, the above set of boundary conditions provide the minimal extension of the TCAD boundary conditions if vector potentials are present.

In order to evaluate the scattering matrix, say of an N-port system, we iterate over all ports and put a voltage difference over one port and put an impedance load over all other ports. Thus the potential variables of the contacts belonging to all but one port, become degrees of freedom that need to be evaluated. The following variables are required to understand the scattering matrices, where Z_0 is a real impedance that is usual taken to be 50 Ohms

$$a_i = \frac{V_i + Z_0 I_i}{2\sqrt{Z_0}} \quad (16.21)$$

$$b_i = \frac{V_i - Z_0 I_i}{2\sqrt{Z_0}}. \quad (16.22)$$

The variables a_i represent the voltage waves incident on the ports labeled with index i. The variables b_i represent the reflected voltages at ports i. The scattering parameters s_{ij} describe the relationship between the incident and reflected waves

$$b_i = \sum_{j=1}^{N} s_{ij} a_j. \quad (16.23)$$

The scattering matrix element s_{ij} can be found by putting a voltage signal at port i and place an impedance of Z_0 over all other ports. Then a_j is zero by construction, since for those ports we have that $V_j = -Z_0 I_j$. Note that I_j is defined positive if the current is ingoing. In this configuration $s_{ij} = b_i/a_j$. In a simulation setup, we may put the input signal directly over the contacts that correspond to the input port. This would imply that the input load is equal to zero. The s-parameter evaluation set up is illustrated in Figure 16.2.

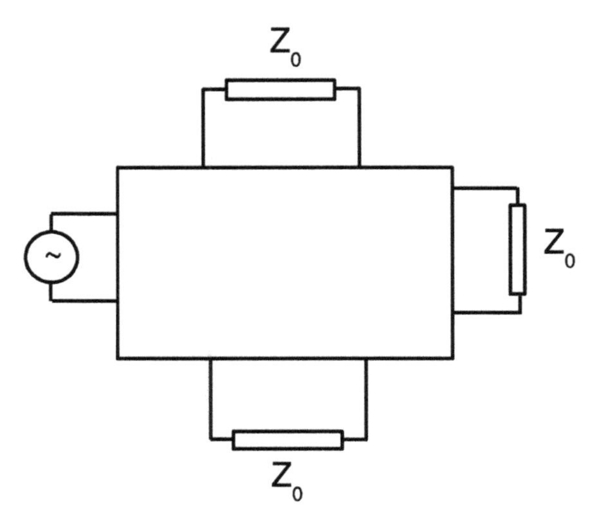

Figure 16.2 Set up of the s-parameter evaluation: 1 port is excited and all others are floating.

16.5 Application to Compute the Coupling of Inductors

Using the solver based on computational electrodynamics, we are able to compute the s-parameters by setting up a field simulation of the full structure. This allows us to study in detail the physical coupling mechanisms. As an illustration, we consider two inductors which are positioned on a substrate layer separated by a distance of 14 micron. This structure was processed and characterized and the s-parameters were obtained. It is quite convenient when studying a compact model parameters to obtain a quick picture of the behavior of the structure. For this device a convenient variable is the 'gain', which corresponds to the ratio of the injected power and the delivered power over an output impedance (Niknejad and Meyer [1998])

$$G = \frac{P_{in}}{P_{out}}. \tag{16.24}$$

The structure is shown in Figure 16.3. and a three-dimensional view using editEM (MAGWEL) is shown in Figure 16.4.

When computing the s-parameters, we put the signal source on one spiral (port 1) and place 50 Ohm impedance over the contacts of the second spiral (port 2).The s_{11}-parameter is shown in Figure 16.5 and the s_{12}-parameter is shown in Figure 16.6. Finally, the gain plot is shown in Figure 16.7. This results shown here have been obtained without any calibration of the material

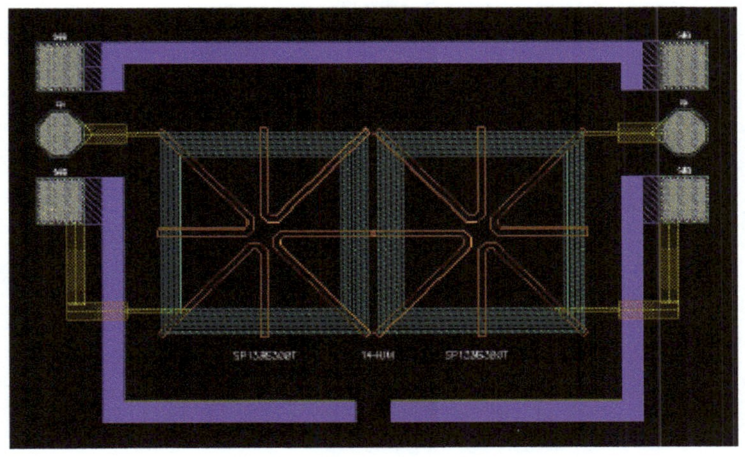

Figure 16.3 View on the coupled spiral inductor using the Virtuosa design environment.

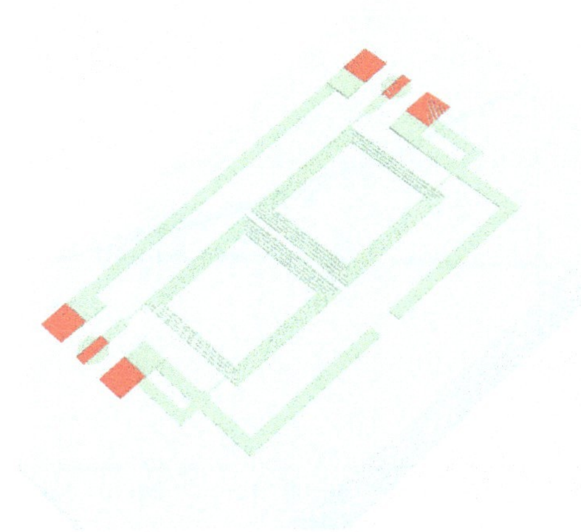

Figure 16.4 View on the coupled spiral inductor using the MAGWEL editor.

parameters. The silicon is treated 'as-is'. This means that the substrate and the eddy current suppressing n-wells are dealt with as doped silicon.

We presented a version of computational electrodynamics which is based on the scalar and vector potential formulation. Whereas the finite-integration technique directly deals with the field intensity quantities $\mathbf{E}$ and $\mathbf{B}$, our formulation deals with the more fundamental gauge fields. It should be

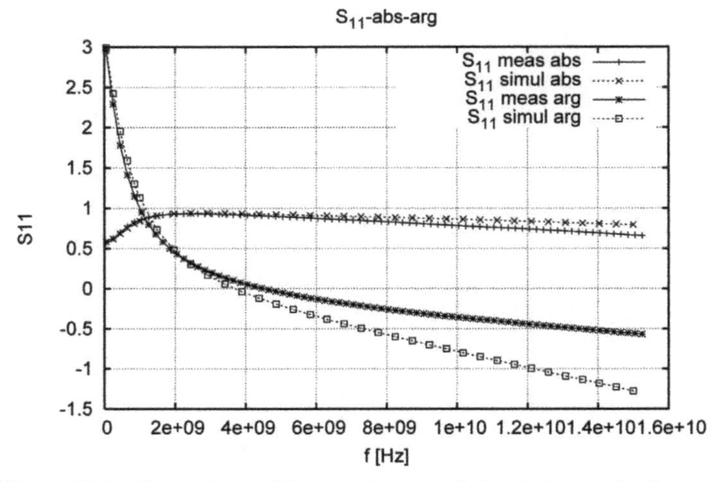

Figure 16.5 Comparison of the experiment and simulation results for s_{11}.

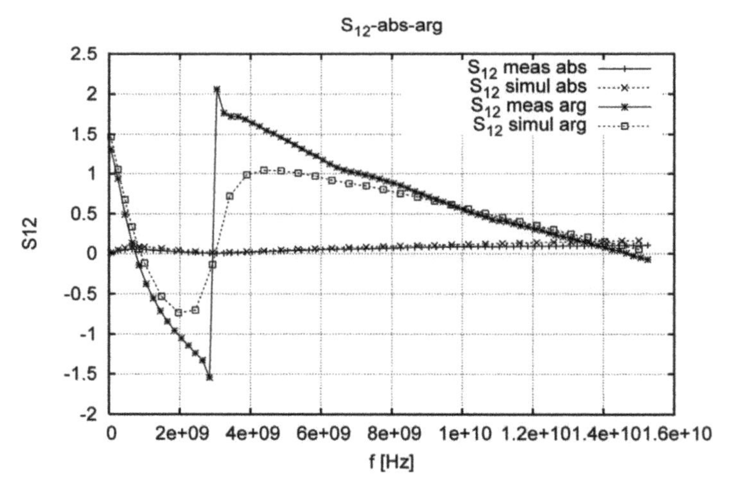

Figure 16.6 Comparison of the experiment and simulation results for s_{12}.

emphasized that the field quantities are derived variables and once that the potentials have been computed, whereas all other variables are obtained by 'post-processing'. Our approach is a discrete implementation of the geometrical interpretation of electrodynamics (Frankel [1997]). According to this interpretation, the field intensities correspond to the curvature and the potentials are connections in the geometrical sense. The practical capabilities

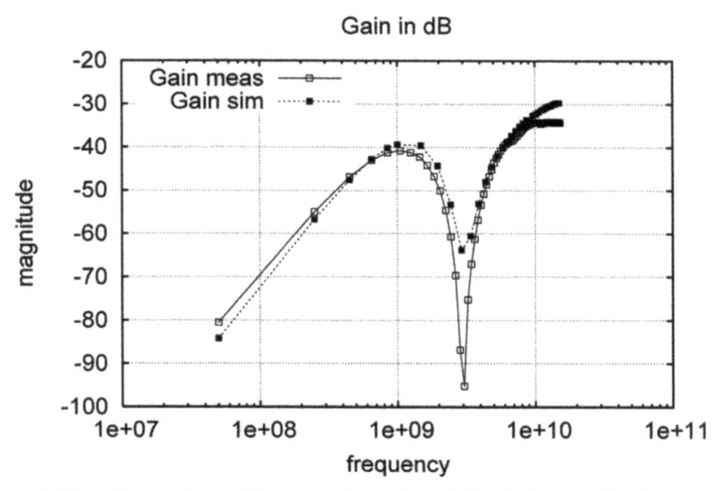

Figure 16.7 Comparison of the experimental and simulation results for the gain.

of our method are comparable to other field solvers that focus directly on the fields $\mathbf{E}$ and $\mathbf{B}$, with one exception: if the potentials are needed in the evaluation of the constitutive relations then our method has a clear advantage. This happens if semiconductor modeling is needed and one can not mimic the semiconductor with moderately conductive material. Another area of application is the unified solving of quantum problems and magnetic induction problems where the potential approach is definitely the most natural choice. We have shown with a realistic application that the method is capable of producing fairly good results. The deviations at higher frequency are an indication that adaptive meshing methods are mandatory.

17

Coupled Electromagnetic-TCAD Simulation for High Frequencies

In this chapter, we present an effective formulation tailored for electromagnetic-TCAD (technology computer-aided design) coupled simulations for extremely high frequency ranges and beyond (>50GHz)[13]. A transformation of variables is exploited from the starting $\mathbf{A}-V$ formulation to the $\mathbf{E}-V$ formulation, combined with adopting the gauge condition as the equation of V. The transformation significantly reduces the cross-coupling between electric and magnetic systems at high frequencies, providing therefore much better convergence for iterative solution. The validation of such transformations is ensured through a careful analysis of redundancy in the coupled system and material properties. Employment of advanced matrix permutation technique further alleviates the extra computational cost introduced by the variable transformation. Numerical experiments confirm the accuracy and efficiency of the proposed E-V formulation.

Advanced high-speed integrated circuits (ICs), such as RF/photonic ICs, represent a complicated electromagnetic (EM) system generally consisting of metal or polysilicon interconnects, various semiconductor material systems, and surrounding media, as shown in Figure 17.1. Conventional approaches for simulating such complex systems rely on conducting separate characterizations of active devices by technology computer-aided design (TCAD) device models without consideration of electrodynamic effects and finite conductivity of metals, and the passive interconnects by pure EM models simplifying semiconductors to equivalent conducting or dielectric materials. Whereas this "decoupled" characterization suffices at low to medium frequencies, it becomes increasingly questionable beyond extremely high frequency (EHF), e.g., >50GHz, where the interplay between semiconductor carrier dynamics and EM wave dynamics (Willis et al. [2009]) is prominent.

[13]This chapter is a reprint of Chen et al. [2011b].

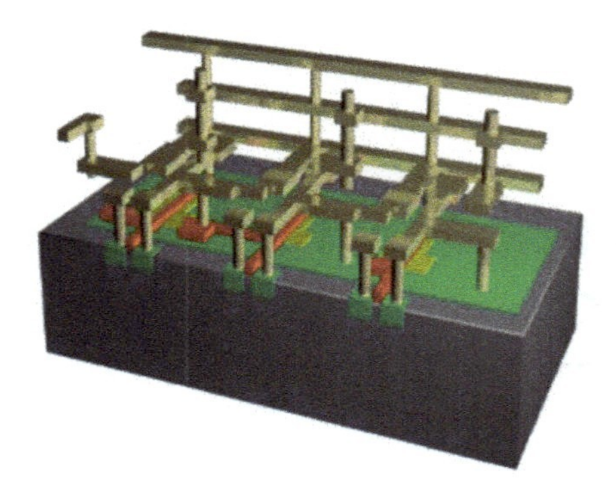

Figure 17.1 A typical on-chip structure consisting of metallic interconnects, semiconductor devices and substrate (Kapora et al. [2009]).

Examples with such strong interplay include metal-insulator-semiconductor (MIS) interconnects (Wang et al. [2002a]), substrate noise isolation structures (Yeh et al. [2008]) and through-silicon-via (TSV) in 3D integrations (Xu et al. [2009]). It is therefore suggested that the on-chip actives and passives should no longer be analyzed separately; instead, the critical mixed-signal/RF block must be treated as an entity, and simulated with the full-wave EM physics coupled with the semiconductor carrier dynamics in the design phase to avoid costly mismatch leading to simulation failures or silicon re-spins. This motivates the development of EM-TCAD coupled simulation approach. It should be noted that a coupled full-wave EM-TCAD approach is not demanded in all circumstances. The necessity for the inclusion of the magnetic effects depends on the scale of the structure under consideration. For nano-scale device designs the self-induced magnetic fields may be safely ignored. However, while designing larger parts of the IC, the induced magnetic fields can have a noticeable effect. The eddy current effects in substrates, which take place at still larger scales, are essentially a magnetic field phenomenon. More extensive discussions can be found in (Schoenmaker et al. [2007a] Nastos and Papananos [2006]).

The objective of coupled simulation is to model and emulate various components (devices, interconnects, substrates and dielectrics) of on-chip structures within a uniform framework without differentiation in the level of abstraction and/or modeling methodology. Some works have been done on the combination of time-domain full-wave EM analysis and different semiconductor models, mostly via the finite-difference time-domain method

(FDTD) (Grondin et al. [1999] Willis et al. [2009]). Yet the choice of basic variables in FDTD (**E**, **H**) is *not* fully compatible with that in TCAD modeling (potentials V) forcing different solution strategies being adopted in different models, which corresponds to a "loosely coupled" scheme. In the frequency domain, the finite-element method (FEM) has been applied to couple the full-wave Maxwell equations with the semiconductor transport equations (Wang et al. [2002a] Bertazzi et al. [2006a]). Nevertheless, standard FEM aiming at building solution with minimal Galerkin residue may not be able to guarantee exact charge conservation and thus may cause spurious oscillations in the numerical solution (Schoenmaker et al. [2002b]).

A more sophisticated frequency-domain technique for multi-domain (metal, semiconductor and insulator) coupled simulation was proposed in (Meuris et al. [2001b] Schoenmaker and Meuris [2002b]), based on the finite-volume method (FVM) that has built-in guarantee of charge conservation. Instead of the conventional use of electric and magnetic fields (**E** and **H**), the technique uses the scalar potential V and vector potential $\mathbf{A}$ (with $\mathbf{B} = \nabla \times \mathbf{A}$) as fundamental variables (denoted as A-V formulation or A-V solver hereafter), and as a consequence provides a convenient, physically consistent and "tightly coupled" interfacing between the full-wave EM model and the TCAD device model. The A-V formulation has been validated for a number of cases in which the simulator results were compared with measured data on test structures developed in industry (CODESTAR [2003] Chameleon-RF [2006]). The solver has been transferred into a series of tools of MAGWEL (devEM [2003]).

Despite the attractive performance from DC to tens of GHz, the A-V solver suffers from a slow iterative solution in the high microwave and even terahertz (THz) regimes wherein EM-TCAD coupled simulation capacity is pressingly demanded. The high-frequency difficulty of A-V solver is attributed to the strong cross-couplings between electric and magnetic systems at high frequencies and when metallic materials are involved, which together lead to linear systems with significant off-diagonal dominance that largely affects the convergence of iterative solvers.

In this chapter, we present a framework of coupled simulation characterized by using V and $\mathbf{E}$ as basic variables, called the E-V formulation or solver henceforth, in the problem formulation, and by using a modified gauge condition as the equation of V in metals and insulators. The transformation removes much of the undesirable dependency of the cross-couplings on frequency and metal conductivity which is in general a large variable. In this way the diagonal dominance of resultant Jacobian matrices is improved and the performance of iterative solution is greatly enhanced for EHF problems. Validity of the proposed E-V formulation is proved though examining the redundancy problem specific to coupled simulation and the influence of material properties. Additional cost in computations introduced by the reformulation is well alleviated by the column approximate minimum degree (COLAMD) permutation, rendering the E-V solver an effective tool for generic integrated simulation tasks at sub-THz and THz frequency ranges.

It should be emphasized that the E-V formulation obtained by performing a variable transformation has interesting consequences for microelectronic applications. First, the minimal procedure to identify voltages and currents at contacts and ports is preserved. In other words, the connection to the Kirchhoff variables is straightforward, as was discussed in (Schoenmaker and Meuris [2002b]). Furthermore, the planar technology implies that current elements in the vertical direction are usually over the lengths of the via heights. Consequently, it is a valuable assumption that the vertical component of the vector potential is negligible. This approximation leads to a much smaller size of the set of degrees of freedom and this property is also preserved after the transformation. We also note that the resulting system is not identical to the E-formulation that is often exploited in finite element solvers. The latter often works in the temporal gauge for which $V = 0$ everywhere.

17.1 Review of A-V Formulation

17.1.1 A-V Formulation of the Coupled System

In the A-V framework for coupled simulation, the Gauss law is used to solve for the scalar potential V in insulating and semiconducting regions, and the current-continuity equation $\nabla \cdot J + \mathrm{i}\omega\rho = 0$ is to employed to find the V in metals.

$$\begin{cases} \nabla \cdot [\varepsilon_r (\nabla V + \mathrm{i}\omega \mathbf{A})] + \rho = 0 & \text{insul. \& semi.} \\ \nabla \cdot [(\sigma + \mathrm{i}\omega\varepsilon_r)(\nabla V + \mathrm{i}\omega \mathbf{A})] = 0 & \text{metal} \end{cases} \tag{17.1}$$

where σ, ε_r and ω denote respectively the conductivity, relative permittivity, and frequency. The free charge density is denoted by ρ and in semiconductors $\rho = n + p + N_d$ where N_d is the net doping concentration. This set of equations is often regarded as the electric system.

The current-continuity equation is exploited to solve the electron and hole charge carrier densities, n and p, in the semiconductor region.

$$\nabla \cdot \mathbf{J}_\chi - \mathrm{i}\omega q\chi \mp R(n,p) = 0, \quad \chi \in \{n,p\}, \tag{17.2}$$

in which $R(n,p)$ refers to the generation/recombination of carriers and q the elementary charge. The sign $\mp$ is for electrons and holes, respectively. Provided the drift-diffusion model is employed, the semiconductor current is determined by $\mathbf{J}_\chi = q\mu_\chi\chi\,(-\nabla V - \mathrm{i}\omega\mathbf{A}) \pm kT\mu_\chi\nabla\chi,\ \chi \in \{n,p\}$, where μ, k and T denote the carrier mobility, Boltzmann constant and temperature, respectively. The Scharfetter-Gummel scheme is applied to discretize (17.2) (Scharfetter and Gummel [1969]).

To solve the magnetic vector potential $\mathbf{A}$, we consider the Maxwell-Ampére equation, which is regarded as the magnetic system

$$\begin{aligned}\nabla \times \left(\frac{1}{\mu_r}\nabla \times \mathbf{A}\right) + K(\sigma + \mathrm{i}\omega\varepsilon_r)(\nabla V + \mathrm{i}\omega\mathbf{A})& \\ -K\,\mathbf{J}_{semi} = 0,&\end{aligned} \tag{17.3}$$

where $\mathbf{J}_{semi} = \mathbf{J}_n + \mathbf{J}_p$ denotes the semiconductor current, and K is the dimensionless constant in the scaling scheme (Schoenmaker and Meuris [2002b]). For generic materials of on-chip structures it is safe to set $\mu_r = 1$. Equation (17.3) itself is not well-defined since the operator $\nabla \times (\nabla \times)$ is intrinsically singular when discretized by FVM. A special treatment in the A-V solver is to subtract (17.3) by the divergence of the gauge condition

$$\nabla \cdot \mathbf{A} + \xi K \mathrm{i}\omega\varepsilon_r V = 0, \tag{17.4}$$

which yields

$$\begin{aligned}\nabla \times (\nabla \times \mathbf{A}) - \nabla(\nabla \cdot \mathbf{A}) + K(\sigma + \mathrm{i}\omega\varepsilon_r)\mathrm{i}\omega\mathbf{A}& \\ +K(\sigma + \mathrm{i}\omega\varepsilon_r)\nabla V - \xi K\mathrm{i}\omega\varepsilon_r\nabla V - K\mathbf{J}_{semi} = 0,&\end{aligned} \tag{17.5}$$

where ξ is the gauge slider ranging from 0 (Coulomb gauge) to 1 (Lorenz gauge). This regularization procedure recovers a Laplacian-like operator and thus eliminates the singularity.

The task of coupled simulation is to find the simultaneous solution of (17.1), (17.2) and (17.5), which are represented in a condensed notation as

$$\begin{cases}\mathbf{F}(V,\{n,p\},\mathbf{A}) = 0 \\ \mathbf{H}(V,\{n,p\},\mathbf{A}) = 0. \\ \mathbf{G}(V,\{n,p\},\mathbf{A}) = 0\end{cases} \tag{17.6}$$

17.2 Origin of the High-Frequency Breakdown of the A-V Solver

The coupled system of Equation (17.6) is intrinsically non-linear when semiconducting regions are present and preferably solved by the Newton's method. The non-linearity arises from the discretization of the current flux along the links of the computational grid. As has been shown in (Meuris et al. [2001b]), the discretized carrier current associated to a link of the grid is

$$J_{\chi ij} = \chi_i B(\mp X_{ij}) - \chi_j B(\pm X_{ij}), \quad \chi \in \{n, p\} \tag{17.7}$$

where $\chi_i$and χ_j are the carrier concentrations of the begin and end nodes of the link, and

$$B(z) = \frac{z}{\exp(z) - 1} \tag{17.8}$$

is the Bernoulli function. The argument $X_{ij} = V_i - V_j + \mathrm{sgn}(ij)\ \mathrm{i}\omega h_{ij} A_{ij}$ and V_i, V_j are the nodal voltages and A_{ij} is the projection of the vector potential $\mathbf{A}$ on the link $<$ ij $>$. Finally $\mathrm{sgn}(ij)$ is $\pm$ depending on the orientation of the link with respect to its begin and end nodes. Evidently, the presence of the Bernoulli function turns the problem into a highly non-linear one.

Starting from some initial guess, for example the DC solution, the update vector in each Newton's iteration is obtained by solving the sparse linear system

$$MX = b, \tag{17.9}$$

which in details reads

$$\begin{bmatrix} \frac{\partial \mathbf{F}}{\partial V} & \frac{\partial \mathbf{F}}{\partial \{n,p\}} & \frac{\partial \mathbf{F}}{\partial \mathbf{A}} \\ \frac{\partial \mathbf{H}}{\partial V} & \frac{\partial \mathbf{H}}{\partial \{n,p\}} & \frac{\partial \mathbf{H}}{\partial \mathbf{A}} \\ \frac{\partial \mathbf{G}}{\partial V} & \frac{\partial \mathbf{G}}{\partial \{n,p\}} & \frac{\partial \mathbf{G}}{\partial \mathbf{A}} \end{bmatrix} \begin{bmatrix} \Delta V \\ \Delta \{n,p\} \\ \Delta \mathbf{A} \end{bmatrix} = - \begin{bmatrix} \mathbf{F}(V, \{n,p\}, \mathbf{A}) \\ \mathbf{H}(V, \{n,p\}, \mathbf{A}) \\ \mathbf{G}(V, \{n,p\}, \mathbf{A}) \end{bmatrix}. \tag{17.10}$$

The numerical difficulty of the A-V solver can be revealed by analyzing the magnitudes of the matrix entries in (17.10), which are mainly dependent on the electric properties of materials and the frequency under consideration. The differential operators ($\nabla, \nabla\cdot, \nabla\times$) are usually of order one with spatial scaling. In the electric sector (17.1) the magnitude of the cross-coupling of $\mathbf{A}$ to V is

$$\frac{\partial \mathbf{F}}{\partial \mathbf{A}} = \nabla \cdot [\mathrm{i}\omega (\sigma + \mathrm{i}\omega\varepsilon_r)] \sim O(\mathrm{i}\omega(\sigma + \mathrm{i}\omega\varepsilon_r)), \tag{17.11}$$

where $O(\cdot)$ denotes the order of the magnitude, and in the magnetic sector (17.5) the magnitude of the cross-coupling of V to $\mathbf{A}$ is

$$\frac{\partial \mathbf{G}}{\partial V} = [K(\sigma + \mathrm{i}\omega\varepsilon_r)\nabla - \xi K \mathrm{i}\omega\varepsilon_r \nabla] \sim O\left(K(\sigma + \mathrm{i}\omega\varepsilon_r)\right). \quad (17.12)$$

Although the cross-couplings between $\mathbf{A}$ and V tend to vanish at zero frequency, they become dominant for frequencies in the higher GHz range, especially when the structure includes metallic conductors that have large conductivity σ and skin effects are desired to be computed such that surface-impedance models can be obtained. When solved by the widely-used Krylov subspace methods such as GMRES, these significant off-diagonal blocks will impose negative effects to the convergence rate through inducing undesirable spectral distribution in the preconditioned system matrix. For instance, the popular ILU preconditioner and its variants compute the incomplete L and U factors of M such that

$$M = LU + E, \quad (17.13)$$

where E is the error matrix. Then an iterative solver effectively deals with the preconditioned matrix

$$(LU)^{-1} M = I + U^{-1}L^{-1}E, \quad (17.14)$$

For diagonally dominant matrices, L and U are well conditioned and the size (2-norm) of $U^{-1}L^{-1}E$ remains reasonably bounded, which confines the eigenvalues of the preconditioned matrix within a small neighborhood of 1 and allows a fast convergence of Krylov subspace methods. When the matrix M lacks diagonal dominance, L^{-1} or U^{-1} may have large norms, rendering the "preconditioned" error matrix $U^{-1}L^{-1}E$ of large size and thus adding large perturbations to the identity matrix (Saad [1996]). This large perturbation causes the eigenvalues dispersed and spread far away from each other, resulting in slow down or even failure of the convergence of iterative solution. As a consequence, the A-V solver becomes increasingly inefficient, if not impossible, in EHF scenarios wherein coupled simulation is demanded to capture the complicated interplay between EM wave and semiconductor carrier transport.

17.3 E-V Formulation

From a modeling perspective, the above intensive cross-coupling arise from explicitly separating the electric field (more precisely the electric field in the metals) into its static component from V and dynamic component from $\mathbf{A}$, and associating them with equal weightings that have magnitudes depending on the frequency and metal conductivity. To reduce the weight of cross-coupling terms, we reformulate the coupled system using the scalar potential V and the electric field $\mathbf{E}$ via the variable transformation

$$\mathbf{E} = -\nabla V - \mathrm{i}\omega\mathbf{A}. \tag{17.15}$$

The coupled system of (17.1), (17.2) and (17.5) under this transformation changes into

$$\mathbf{F}' : \begin{cases} \nabla \cdot (\varepsilon_r \mathbf{E}) + \rho = 0, & \text{semi. \& insul.} \\ \nabla \cdot [(\sigma + \mathrm{i}\omega\varepsilon_r)\mathbf{E}] = 0, & \text{metal} \end{cases} \tag{17.16a}$$

$$\mathbf{H}' : \nabla \cdot \mathbf{J}_\chi - \mathrm{i}\omega q\chi \mp R(n,p) = 0, \quad \chi \in \{n,p\}, \tag{17.16b}$$

$$\mathbf{G}' : \nabla \times (\nabla \times \mathbf{E}) - \nabla(\nabla \cdot \mathbf{E}) + K\mathrm{i}\omega(\sigma + \mathrm{i}\omega\varepsilon_r)\mathbf{E} \tag{17.16c}$$

$$-\nabla\left(\nabla^2 V\right) - \xi K\omega^2 \nabla(\varepsilon_r V) + K\mathrm{i}\omega\mathbf{J}_{semi} = 0,$$

where $\mathbf{J}_\chi = q\mu_\chi\chi\mathbf{E} \pm kT\mu_\chi\nabla\chi,\ \chi \in \{n,p\}$ accordingly. Note that here we only consider $\mathbf{E}$ as a variable transformation of $\mathbf{A}$ instead of an independent physical quantity.

The transformation from A-V to E-V immediately removes the conductance-dependent cross-coupling of V from (17.16c). Yet little improvement has been made to (17.16a) wherein the cross-coupling coefficients of $\mathbf{E}$ still have undesirable dependence on σ and ω. A detailed analysis of the coupled system in the next subsection however indicates that we are allowed to exploit, instead of the Gauss law (17.16a) (as well as the current-continuity), the transformed gauge condition to determine the scalar potential in the metal and insulator regions (but not in the semiconductor regions), which reads

$$\nabla^2 V + \xi K\omega^2 \varepsilon_r V + \nabla \cdot \mathbf{E} = 0. \tag{17.17}$$

This way, the cross-couplings of $\mathbf{E}$ have magnitudes of order one, and are not growing any more with the large value of σ and frequency in the bulk of metallic materials. Whereas the conversion from A-V solver to E-V solver looks promising, there are certain subtleties that require special attention to guarantee a correct implementation of the E-V solver.

17.3.1 Redundancy in Coupled System

It is a unique feature for coupled EM-TCAD simulation to look for a simultaneous solution of the following system, which consists of the Gauss law, the current-continuity law, and the Maxwell-Ampére law

$$\nabla \cdot \mathbf{D} - \rho = 0, \tag{17.18a}$$

$$\nabla \cdot \mathbf{J} = 0, \tag{17.18b}$$

$$\nabla \times \mathbf{H} - \mathbf{J} = 0, \tag{17.18c}$$

where $\mathbf{J}$ represents the total current including the conduction and the displacement parts. In the A-V formulation all unknowns are expressed in terms of potentials V and $\mathbf{A}$, and a gauge condition is required to eliminate the well-known gauge freedom to ensure a unique solution.

$$\nabla \cdot \mathbf{A} + f = 0 \tag{17.19}$$

where f can be an arbitrary function of V and $\mathbf{A}$.

It is straightforward to see the system (17.18), more exactly (17.18b) and (17.18c), is redundant by taking the divergence on both sides of (17.18c) (Enders [2009]). Conventional TCAD device simulation or full-wave EM simulation alone, is free of this redundancy, in that the former uses only (17.18a) and (17.18b) to solve for V and ρ (n and p) without $\mathbf{A}$, while the latter uses only (17.18c) and (17.19) to look for V and $\mathbf{A}$ in the absence of ρ (which is recovered later via (17.18a)). When the A-V solver has to deal with the combined system of (17.18b) and (17.18c) that is fundamental to describe the field-carrier interaction, it employs a specific technique to address the redundancy problem, which, as mentioned above, is to subtract (17.18c) by the divergence of the gauge condition, yielding

$$\nabla \times \mathbf{H} - \mathbf{J} - \nabla \cdot (\nabla \cdot \mathbf{A} + f) = 0. \tag{17.20}$$

There are several points in (17.20) that deserve attention: 1) The system of (17.18b) and (17.20) is not redundant as long as the gauge condition is not explicitly involved in the system of equations; 2) Though the gauge condition does not participate in the solution procedure, it should serve as an implicit constraint and be recovered from the solution thereby obtained. Taking divergence on (17.20) and together with (17.18b), we have

$$\nabla^2 (\nabla \cdot \mathbf{A} + f) = 0. \tag{17.21}$$

which is essentially a Laplace's equation. In numerical theory, it is known that for a Laplace's equation $\nabla^2\phi = 0$ in some domain Ω, the solution will be zero everywhere in Ω provided the boundary condition $\phi|_{\partial\Omega} = 0$ is applied. Hence the requirement of the second point is that the gauge condition must be set equal to zero at the boundary of the simulation domain. This is done in the discretization of (17.20), wherein the evaluations of $\nabla \cdot \mathbf{A} + f$ are forced to be equal to zero by the discretization scheme for the nodes bouncing on the boundary of the simulation domain. In other words, the gauge condition will be automatically recovered over the usual Maxwell-Ampére equation for the whole domain provided that the current-continuity equation is solved.

Above discussion shows the way how the A-V solver deals with the redundancy arising when different systems of equations are coupled together, wherein the current-continuity equation is solved explicitly for V with the gauge condition being an implicit constraint. Alternatively, one could choose the gauge condition as the equation of V constrained by the current conservation. This change requires the current-continuity equation being removed from the system of equations, otherwise the system will become redundant again since the gauge condition is enforced explicitly and implicitly at the same time. The removal of current-continuity equation as a consequence requires the removal of charge density ρ from the unknown list for an equal counting of equations and unknowns. Such removal is applicable in the metallic and insulating regions in which ρ is able to be recovered by the Gauss law, while not in the semiconductors in which the carrier concentrations n and p are of fundamental interest and cannot be recovered from merely the field variables.

As a result, in the E-V solver the equation (17.17) can be exploited to solve for V in the metals and insulators (though not entirely, see the next section), while (17.16a) remains the one we should use in the semiconducting regions. Note that the gauge condition will still be recovered in the semiconductors given the fact that the gauge condition is set zero at all nodes surrounding the semiconductors. In addition, using (17.16) in the semiconductors will not introduce large cross-coupling terms as the relative permittivities of semiconducting materials are generally of order one.

17.3.2 Issues of Material Properties

Depending on the electric properties of the materials under investigation (metal or insulator or material interfaces), there are still subtle distinctions in the appropriate formulation of gauge condition that should be employed as the equation of V i.e., (17.17) in the E-V solver. For nodes in the bulk of metals as well as material interfaces except semiconductor/insulator interfaces, the governing equation is essentially the current-continuity equation; the Gauss equation will come into use only when the (surface) charge density is demanded in post-process steps. Therefore, the equation in the E-V solver for these nodes is exactly (17.17), which together with (17.16c) are equivalent to the current-continuity Equation (17.1) in the A-V solver.

The situation is slightly different for the insulating regions. The current-continuity is now trivial ($0 = 0$) and the (small-signal) free charge density is zero by definition. Therefore, there is no need to recover the current-continuity equation and only the Gauss's equation must be recovered. Although applying the gauge condition to find out V remains possible, such choice will induce certain numerical difficulty, especially for the insulators with homogeneous dielectric constants. This is because, due to the homogeneity of dielectric constants, solution of (17.17) in the bulk of insulators has to obey a stronger requirement of demanding $\nabla^2 V + Ki\omega\varepsilon_r V = 0$ and $\nabla \cdot \mathbf{E} = 0$ simultaneously. Direct application of (17.17) will cause non-uniqueness in the solution and render the system matrix highly ill-conditioning. As a result, the ordinary Gauss equation remains an appropriate choice to determine the scalar potential in the insulators, no matter homogeneous or inhomogeneous:

$$\nabla \cdot (\varepsilon_r \mathbf{E}) = 0. \tag{17.22}$$

The nodes at semiconductor/insulator interfaces are classified as semiconductor nodes for which it is natural to apply the original Gauss Equation (17.16a).

17.3.3 Boundary Conditions

Boundary conditions for the E-V solver are derived from its A-V counterpart through the transformation relation (17.15). The underlying principle is to minimize the coupling between the simulation domain and the rest of the world.

As done in the A-V solver (Meuris et al. [2001b] Schoenmaker and Meuris [2002b]), we divide the boundary of the simulation domain into two parts: contact regions and non-contact regions. The contacts allow currents, and thus energy, to enter and leave the simulation domain, wherein the constant voltage condition is applied. The remainder, the non-contact boundary, is characterized by demanding that the outward-pointing normal component of $\mathbf{E}$ vanishes, i.e., $\mathbf{E}_n = 0$. This leads to the boundary condition of the electric system for non-contact boundary nodes:

$$\nabla \cdot [(\sigma + i\omega\varepsilon_r)\,\mathbf{E}] = 0. \tag{17.23}$$

The boundary condition (17.23) introduces again a σ-dependent coupling in some circumstances, i.e., for boundary nodes attached by metallic cubes. Their contribution to undesirable cross-couplings, however, is much lower than that in the A-V formulation, wherein all nodes attached by metallic cubes have to be taken into account.

For the magnetic sector, the similar requirement of $\mathbf{B}_n = 0$ is applied to keep all magnetic fields remain inside the simulation domain. This forces a zero tangential component of $\mathbf{A}$ since $\mathbf{B} = \nabla \times \mathbf{A}$, viz. $\mathbf{A}_t = 0$, which holds for both the contact and non-contact boundaries. In light of (17.15), the boundary condition of magnetic system should be

$$\mathbf{E}_t = -\nabla V_t. \tag{17.24}$$

The above condition implies the unknowns associated to the boundary links, which are part of degrees of freedom in the E-V solver but not in the A-V solver, can be substituted by the corresponding nodal unknowns of V in the solution phase and recovered later by post-processing. This way, the total number of unknowns are identical for both the A-V and E-V solvers.

It should be emphasized that above selection of the boundary conditions represents a particular choice which was motivated by upgrading standard TCAD simulations into the electromagnetic regime. However, there is nothing "fundamental" about this choice. One may equally well choose radiative boundary conditions or Neumann-type boundary conditions for the vector potential. The preferred choice depends on the problem under consideration. Here, it is important to note that when discussing the A-V solver and the E-V solver, the *same* physical boundary conditions are used.

17.3.4 Implementation Details

The full system of equations of the E-V solver is laid out in (17.25).

$$\mathbf{F}' : \begin{cases} \nabla \cdot [(\sigma + i\omega\varepsilon_r)\,\mathbf{E}] - \rho = 0, & \text{boundary.} \\ \nabla \cdot (\varepsilon_r \mathbf{E}) - \rho = 0, & \text{Semi. and insul.} \\ \nabla^2 V + \xi K \omega^2 \varepsilon_r V + \nabla \cdot \mathbf{E} = 0, & \text{remaining regions.} \end{cases} \tag{17.25a}$$

$$\mathbf{H}' : \nabla \cdot \mathbf{J}_\chi - \mathrm{i}\omega q \chi \mp R(n,p) = 0, \quad \chi \in \{n, p\}, \tag{17.25b}$$

$$\begin{aligned} \mathbf{G}' : \nabla \times (\nabla \times \mathbf{E}) - \nabla (\nabla \cdot \mathbf{E}) + K\mathrm{i}\omega(\sigma + \mathrm{i}\omega\varepsilon_r)\mathbf{E} \\ -\nabla \left(\nabla^2 V\right) - \xi K \omega^2 \nabla \left(\varepsilon_r V\right) + K\mathrm{i}\omega \mathbf{J}_{semi} = 0, \end{aligned} \tag{17.25c}$$

Similar to (17.10), a linear system $MX = b$ is solved at each Newton's step, in which M results from the FVM discretization of the Jacobian of (17.25).

It is convenient to upgrade the A-V solver to include an E-V solver by exploiting the following 4-step solution strategy:

1. Map $\mathbf{A} - V$ variables onto $\mathbf{E} - V$ variables via (17.15).
2. Apply the E-V solver to compute the update vector $[\Delta V, \Delta \mathbf{E}]^T$ in the Newton's iteration.
3. Map $[\Delta V, \Delta \mathbf{E}]^T$ onto $[\Delta V, \Delta \mathbf{A}]^T$
4. Update the A-V system.

Using this approach, the data structure in the original A-V solver is unaltered and the switching between the A-V and E-V solvers is easy to realize. This is beneficial in that, as shown in Section 17.4, the A-V and E-V solvers are suitable to work complementarily with the former at low frequencies and the latter at high frequencies, and thus switching between solvers may be needed in a wide-band simulation.

17.3.5 Matrix Permutation

Despite the attractive reduction in the magnitudes of cross-couplings, the E-V transformation introduces more non-zero fill-ins into the off-diagonal block of $\frac{\partial \mathbf{G}}{\partial V}$ through the term $\nabla \left(\nabla^2 V\right)$ in (17.25c). The increased amount of fill-ins from A-V to E-V formulations is approximately $10\times$ the number of boundary links. Inferior sparsity burdens the construction of preconditioner as well as subsequent iterations, limiting the applications of E-V solver to relatively small-sized problems.

To enhance the performance of the E-V solver, the column approximate minimum degree (COLAMD) permutation (Davis et al. [2004b]) is applied to the system matrix M, which computes a permutation vector p such that the (incomplete) LU factorization of $M(:, p)$ tends to be much sparser than that of M. This way, building popular ILUT preconditioners with small threshold becomes possible for M in the order of tens of thousands. The subsequent iteration also speeds up owing to the improved sparsity of the computed LU factors.

17.4 Numerical Results

The proposed E-V solver as well as its A-V counterpart are implemented in the MAGWEL software as well as in MATLAB. The "de Mari" scaling scheme (Schoenmaker and Meuris [2002b]) is adopted. For simplicity, the Coulomb gauge ($\xi = 0$) is adopted throughout the numerical experiments. Three structures are tested to demonstrate the efficiency of the E-V solver: 1) a cross wire structure consisting only passives as shown in Figure 17.2; 2) a metal plug structure consisting of both passives and actives as shown in Figure 17.3; 3) a practical substrate noise isolation (SNI) structure as shown in Figure 17.4. The iterative solutions are computed by GMRES with ILUT preconditioners. All programming and simulations were done on a 3.2GHz 16Gb-RAM Linux-based server.

17.4.1 Accuracy of E-V Solver

Figure 17.5 verifies the accuracy of E-V solver in comparison with the A-V solver at frequencies ranging from 10^6Hz to 10^{15}Hz for the three test benches. This frequency range covers a wide spectrum from medium radio frequency to visible light. Direct solver (Gaussian elimination) is used to solve the linear equations at each Newton's iteration. The accuracy is measured by the relative error between the whole internal state space solution of the A-V and E-V solvers err $= \|X_{EV} - X_{AV}\|_2 \,/\, \|X_{AV}\|_2$, $X = [V; n; p; \mathbf{A}]$. It is seen that the E-V solver is in an excellent agreement with the A-V solver throughout the testings,

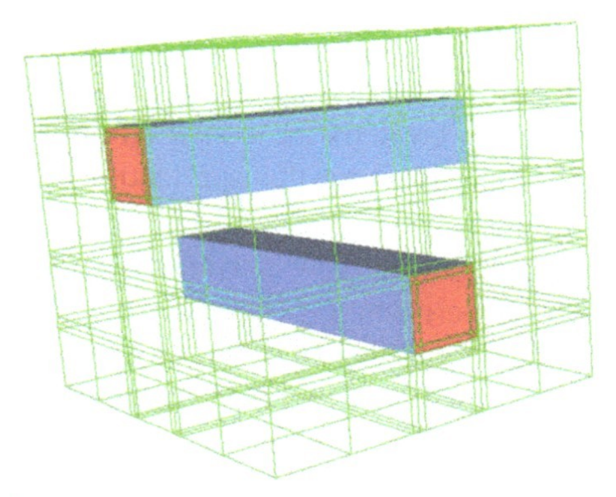

Figure 17.2 Cross wire structure. Simulation domain is $10 \times 10 \times 10\mu m^3$ and the cross sections of metal wires are $2 \times 2\mu m^2$. $\sigma = 5.96 \times 10^7 S/m$. FVM discretization generates 1400 nodes and 3820 links.

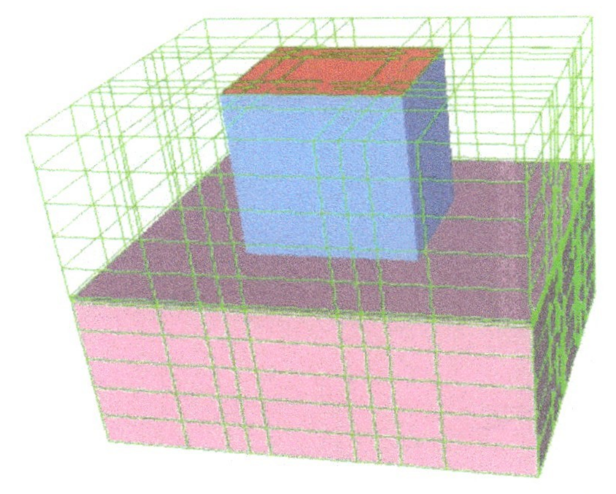

Figure 17.3 Metal plug structure. Simulation domain is $10 \times 10 \times 10 \mu m^3$ the cross section of metal plug is $4 \times 4 \mu m^2$. $\sigma = 3.37 \times 10^7 S/m$. A uniform doping of $N_D = 1 \times 10^{24}$ is used. FVM discretization generates 1300 nodes and 3540 links.

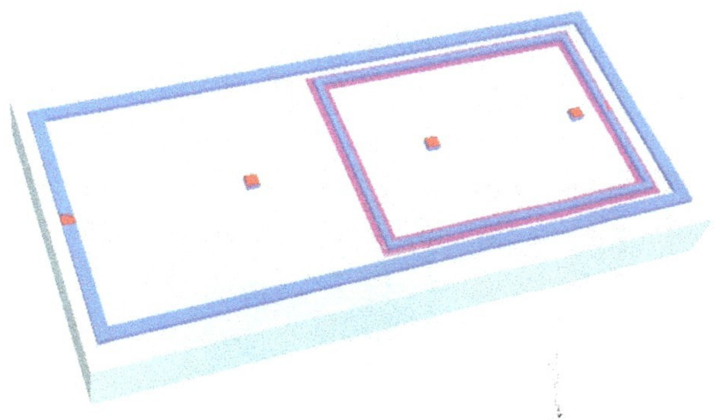

Figure 17.4 Substrate noise isolation structure. A deep n-well (DNW) (pink region) is implanted in the p-type substrate to isolate analog circuits from digital noise sources. Simulation domain is $100 \times 50 \times 11 \mu m^3$. $\sigma = 3.37 \times 10^7$ S/m. A user-defined doping profile is adopted. FVM discretization generates 6300 nodes and 13540 links.

which is expected from a mathematical perspective since no approximation is introduced in the variable and equation transformations. The slight fluctuations in the curves are due to the final precisions of Newton's method and vary among solvers even when the same convergence criterion is applied. Figure 17.6 visualizes the current density inside the substrate of the SNI structure, demonstrating a clear isolation effect for the part of analog circuits from external digital noises.

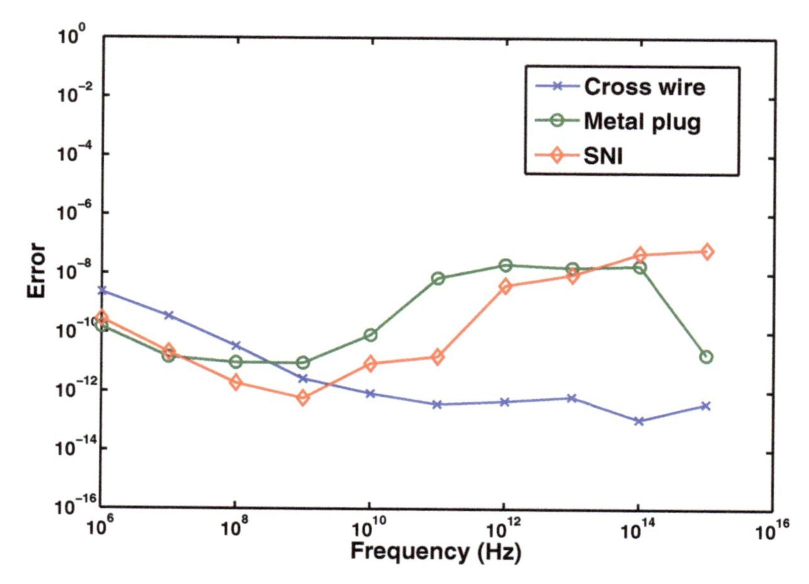

Figure 17.5 Differences between the A-V and E-V solvers for the testing structures (with direct solver).

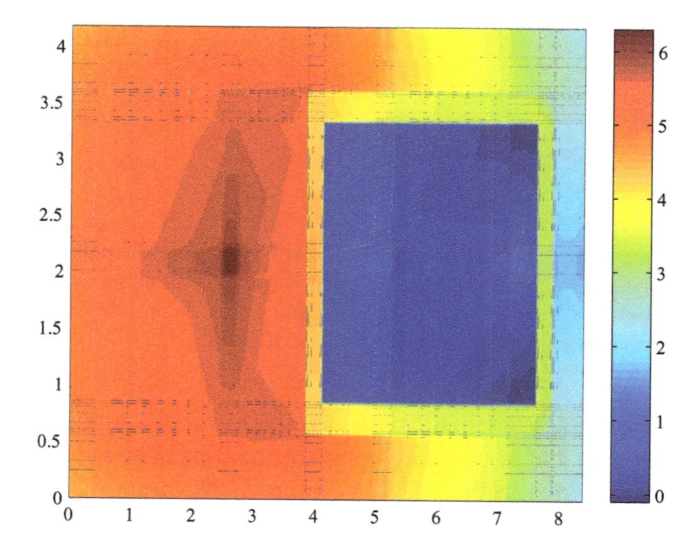

Figure 17.6 Current density at the middle layer of the substrate of SNI structure (shown in log10 scale).

17.4.2 Spectral Analyses

To investigate the influence of increasing frequency on the A-V and E-V solvers, we plot in Figure 17.7 and Figure 17.8 the eigenvalues of the Jacobian matrices preconditioned by ILUT(10^{-6}) for the metal plug structure at four different frequencies. The norms of L^{-1} and U^{-1} of each solver are also computed in Table 17.1.

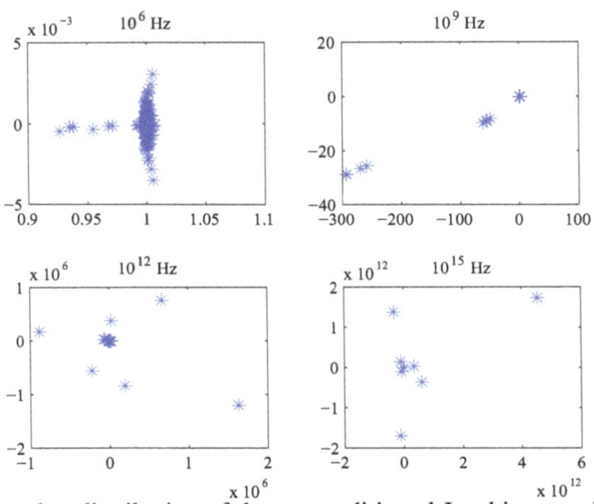

Figure 17.7 Eigenvalue distribution of the preconditioned Jacobian matrices of A-V solver at different frequencies.

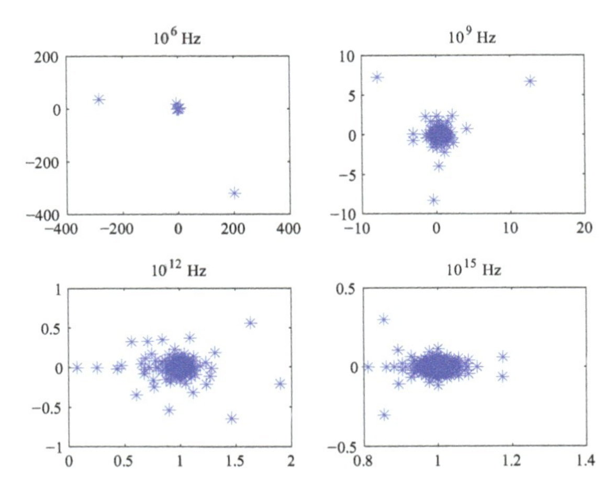

Figure 17.8 Eigenvalue distribution of the preconditioned Jacobian matrices of E-V solver at different frequencies.

Table 17.1 Norms of L^{-1} and U^{-1} computed for the A-V and E-V solvers (ILUT(10^{-6}))

	A-V		E-V	
Frequency	L^{-1}	U^{-1}	L^{-1}	U^{-1}
10^6Hz	1.55×10^1	4.82	9.15×10^1	1.67×10^7
10^9Hz	1.32×10^1	4.87×10^2	4.37×10^1	3.82×10^4
10^{12}Hz	2.50×10^1	1.75×10^7	2.89×10^1	1.16×10^2
10^{15}Hz	2.42×10^2	1.15×10^{11}	1.52×10^1	5.15

At relatively low frequency (10^6Hz), the off-diagonal blocks in (17.11) and (17.12) are of small sizes and so are the norms of the inverses of incomplete factors L and U in (17.14). The eigenvalues of the preconditioned matrix from the A-V solver are then tightly clustered around the point 1 in favor of a fast convergence of iterative solution. As frequency increases, the spectra of the preconditioned matrices have eigenvalue clusters with continuously enlarging radii and separations among each other, reflecting an increasing perturbation to the identity matrix as the result of increasing off-diagonal dominance. This is also confirmed by looking into the growing sizes of U^{-1} in Table 17.1. It therefore suggests a poor iterative performance for the A-V solver at high frequencies when Krylov subspace methods are applied, whose convergence behaviors are closely related to the relative radii of eigenvalue clusters and their separations of the preconditioned matrix (Campbell et al. [1996]). The enlargements of cluster radii and separations are proportional to the increases of frequency.

Compared to that of the A-V solver, the spectral distribution of the preconditioned system of E-V solver has a roughly opposite trend along with the rising frequency. The eigenvalues are clustered more loosely at low frequencies, while becoming increasingly concentrated around 1 at higher frequencies, due to the growing contribution from the term of $Ki\omega(\sigma + i\omega\varepsilon_r)\mathbf{E}$ in (17.25c) improving diagonal dominance. Such concentration of eigenvalues greatly facilitates the convergence of iterative methods and thus suggests enhanced performances for the E-V solver at high-frequency scenarios.

Similar experiments are conducted in Figure 17.9 and Figure 17.10 to examine the role of metal conductivity. The metal plug structure is used again, but with the metallic part replaced by insulating materials, rendering the structure consisting of only insulators and semiconductors. Horizontal comparisons for the A-V solver show that, whereas at sufficiently high frequencies the eigenvalues still disperse, the degree of dispersion is reduced

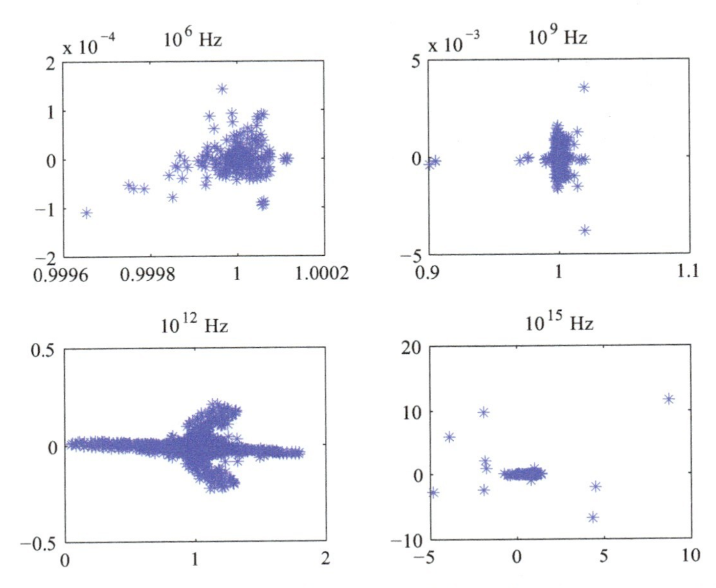

Figure 17.9 Eigenvalue distribution of the preconditioned Jacobian matrices of A-V solver at different frequencies (no metal).

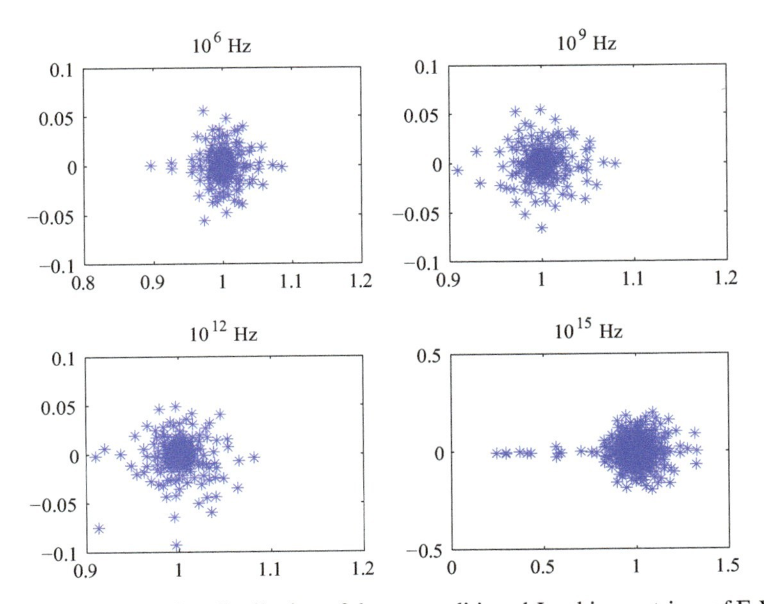

Figure 17.10 Eigenvalue distribution of the preconditioned Jacobian matrices of E-V solver at different frequencies (no metal).

by a great extent. This confirms that the presence of metallic conductors with large conductivity is one origin of the numerical difficulty of A-V solver at high frequencies. The eigenvalues of E-V solver are in a similar distribution over the four testing frequencies, suggesting roughly constant performance in iterative solution.

17.4.3 Performance Comparisons

Detailed comparisons between the A-V and E-V solvers for their performances in iterative solution are tabulated in Table 17.2–17.4. A relative tolerance of 10^{-10} and a maximum number of iterations of 1000 are used in GMRES. The high accuracy used in iterative solution is intended to minimize the number of necessary Newton iterations to as closed as with direct solver, so as not to complicate the analysis by the issues related to the convergence rate of Newton's method. Successful solutions with convergence achieved are shown with

Table 17.2 Iterative performance of A-V and E-V solvers for the cross wire structure. (ILUT(10^{-6}), time unit: second)

	A-V(ORIG)			A-V(AMD)			A-V	E-V(ORIG)			E-V(AMD)			E-V
Freq.	t_{pre}	N_{it}	$t_{per\ it}$	t_{pre}	N_{it}	$t_{per\ it}$	DS	t_{pre}	N_{it}	$t_{per\ it}$	t_{pre}	N_{it}	$t_{per\ it}$	DS
10^6	48.34	5	0.37	11.22	5	0.38			NC			NC		
10^7	62.15	6	0.34	12.68	7	0.37		179.16	546	1.51	49.88	515	1.01	
10^8	62.96	11	0.41	12.91	12	0.39		179.49	130	1.11	45.73	91	0.71	
10^9	77.59	27	0.42	14.80	35	0.41		163.00	17	1.09	43.15	14	0.67	
10^{10}	105.06	43	0.73	20.77	167	0.50	12.76	163.73	9	1.09	38.01	8	0.65	15.08
10^{11}		NC			NC			155.41	8	1.03	38.09	7	0.63	
10^{12}		NC			NC			154.45	7	1.01	38.50	7	0.61	
10^{13}		IP			NC			154.52	7	1.08	38.77	6	0.57	
10^{14}		IP			IP			150.29	6	0.98	34.55	6	0.57	
10^{15}		IP			IP			127.37	5	1.01	28.64	6	0.51	

Table 17.3 Iterative performance of A-V and E-V solvers for the metal plug structure. (ILUT(10^{-6}), time unit: second)

	A-V(ORIG)			A-V(AMD)			A-V	E-V(ORIG)			E-V(AMD)			E-V
Freq.	t_{pre}	N_{it}	$t_{per\ it}$	t_{pre}	N_{it}	$t_{per\ it}$	DS	t_{pre}	N_{it}	$t_{per\ it}$	t_{pre}	N_{it}	$t_{per\ it}$	DS
10^6	100.02	5	0.57	34.83	5	0.62			NC			NC		
10^7	146.18	6	0.54	53.71	7	0.65			NC			NC		
10^8	156.44	10	0.60	62.01	11	0.68		616.12	187	2.11	136.12	141	1.11	
10^9	200.28	22	0.69	62.50	27	0.66		455.02	81	1.81	129.19	51	0.91	
10^{10}	207.16	102	0.73	69.39	165	0.72	37.39	442.65	21	1.73	128.94	18	0.92	50.12
10^{11}		NC			NC			440.56	13	1.72	100.18	12	0.80	
10^{12}		NC			NC			369.19	11	1.39	85.50	11	0.78	
10^{13}		IP			NC			363.20	9	1.32	83.16	10	0.78	
10^{14}		IP			IP			362.21	7	1.23	83.18	7	0.81	
10^{15}		IP			IP			309.64	5	1.10	83.20	6	0.74	

Table 17.4 Iterative performance of A-V and E-V solvers for the SNI structure. (ILUT(10^{-4}), time unit: second)

	A-V(ORIG)			A-V(AMD)			A-V	E-V(ORIG)			E-V(AMD)			E-V
Freq.	t_{pre}	N_{it}	$t_{per\ it}$	t_{pre}	N_{it}	$t_{per\ it}$	DS	t_{pre}	N_{it}	$t_{per\ it}$	t_{pre}	N_{it}	$t_{per\ it}$	DS
10^6	1298.45	8	2.57	252.83	9	2.12			NC			NC		
10^7	1506.15	32	3.54	349.11	41	3.45			NC			NC		
10^8	2112.06	100	4.60	539.78	140	4.08			NC			NC		
10^9		NC			NC			17347	141	11.12	770.81	101	3.62	
10^{10}		NC			NC		1243	13556	71	9.34	598.76	70	3.49	1791
10^{11}		NC			NC			9342	47	9.02	474.05	44	3.28	
10^{12}		IP			NC			8006	40	8.62	427.25	36	3.20	
10^{13}		IP			IP			7681	36	8.00	394.00	32	3.01	
10^{14}		IP			IP			7262	19	7.68	318.64	18	2.85	
10^{15}		IP			IP			7245	16	6.11	301.66	15	2.79	

the time for building preconditioner, the number of iterations and the time per iteration. Unsuccessful solutions, according to the reasons of failure, are marked as "NC" (no convergence within the maximum number of iterations) or "IP" (ill-conditioned preconditioner). The iterative solvers with and without COLAMD pre-processing are labeled as "AMD" and "ORIG", respectively. The run times of direct solver (at 10GHz) are also shown with the label of "DS" for reference.

For all the three structures, the A-V solver exhibits deteriorated performances as frequency increases, and tends to break down at frequencies beyond (tens of) GHz for requiring either a large number of iterations or high-quality preconditioners that are costly to generate ("UV catastrophe"). The E-V solver fails to converge for low frequencies ("IR singularity") but in contrast to the A-V solver, performs increasingly well with growing frequency. These results are consistent with the above spectral analysis and confirm the merit of the E-V solver as a capable tool in EHF applications. Meanwhile, it suggests that the A-V and E-V solvers have complementary preferable frequency ranges and thus are suitable to work together to provide a truly wide-band coupled simulation framework.

As frequency increases, calculating ILUT preconditioners are generally more time-consuming for the A-V solver, while less for the E-V solver, which may be attributed to their opposite behaviors in terms of diagonal dominance. Without matrix permutation, the constructions of preconditioners for the E-V solver is several times slower than that of the A-V solver due to a higher number of matrix fill-ins introduced by the $\nabla(\nabla^2 V)$ term. This prevents the E-V solver from being applied to simulate large-scale problems. As a remedy, application of the COLAMD permutation in the E-V solver provides a remarkable reduction ($\sim$ 20X for the SNI example) in the computational cost of preconditioner and a moderate reduction ($>$ 2X for the SNI example) in the cost of following iterations, rendering the overall cost of E-V solver comparable even with that of the COLAMD-permuted A-V solver, for which the improvement is less significant because of its inherently better sparsity. Besides, an increased number of iterations is observed for the COLAMD-permuted A-V solver compared to the original version, offsetting its gain in preconditioner computations. It turns out that the E-V solver combined with COLAMD is the most robust and efficient tool for coupled simulations at EHF and beyond.

Summary

An effective E-V formulation is proposed, with dedication to the simultaneous simulation of full-wave EM and semiconductor dynamics for EHF regime onwards. The underlying idea is to reformulate the conventional A-V formulation through variable and equation transformations, which removes the undesirable dependencies of cross-couplings on frequency and metal conductivity, and as a consequence brings substantial improvement into the efficiency of iterative solution at high frequencies. From a spectral perspective, the improved diagonal dominance ameliorates the concentric appearance of eigenvalues of the preconditioned Jacobian matrix, by which a fast convergence of iterative solution is achieved. The equation transformation from the Gauss equation to the gauge condition is rigorously validated by a careful investigation of the redundancy in coupled system and the influence of material properties. The COLAMD matrix permutation technique is applied to offset the additional cost introduced by the E-V reformulation, rendering the E-V solver comparably efficient with its A-V counterpart. Numerical experiments have confirmed the superior performance of the proposed method with frequency up to optical range.

18

EM-TCAD Solving from 0–100 THz

We have presented a transformation of the electromagnetic field drift-diffusion system such that the resulting equations become much more attractive to solve at extreme high frequencies[14]. As was demonstrated the incorporation of magnetic effects into the semi-conductor equations demands that these effects are represented by the vector potential (Meuris et al. [2001b] Schoenmaker and Meuris [2002a]). The key argument is that the Poisson potential is required to obtain the carrier densities. We have named the resulting set of equations the AV system or AV formulation. It consists of the drift-diffusion equations completed with the Maxwell equations in the potential formulation. After having discretized this system of equations and applying them to industrial design problems (Schoenmaker et al. [2010b]), we experienced a dramatic drop in convergence behavior if the frequencies go up. Depending on the structure under consideration, iterative solving schemes fail for frequencies above ~50–200 GHz. In (Chen et al. [2011b]) we have identified the causes for this convergence failure and proposed a remedy for it. The purpose of this contribution is two-fold: (1) We present the proposed remedy in a wider perspective. (2) We report on the learning cycles for setting up a successful series of linear solving settings to arrive at good convergence of the newly proposed formulation.

18.1 From AV to EV

As was shown in the central observation for explaining the deterioration of solving the AV system at high frequencies is that the Poisson system (V) couples to the magnetic system represented by $\mathbf{A}$ with a term proportional to the frequency, ω and the conductance σ in metallic domains or $q\mu_{\mathrm{n}}n$ or $q\mu_{\mathrm{p}}p$ in n-doped or p-doped regions respectively. This coupling destroys the

[14]This chapter is a reprint of Chen et al. [2011a]

diagonal dominance of the Newton-Raphson matrices in such a way that at sufficient high frequencies no suitable preconditioner can be found despite using highly sophisticated permutation algorithms (Duff and Koster [2001]). The cure of the convergence failure problem is to apply a transformation of variables in such a way that the coupling is removed. Starting from the AV formulation summarized in Equations (18.1–18.8) with ξ a slider between the Coulomb gauge ($\xi = 0$) and the Lorenz gauge ($\xi = 1$)

$$\nabla \cdot (\epsilon \mathbf{E}) - \rho = 0 \tag{18.1}$$

$$\nabla \cdot \mathbf{J}_{\text{metal}} + \mathrm{i}\omega\rho = 0, \; \mathbf{J}_{\text{metal}} = \sigma \mathbf{E} \tag{18.2}$$

$$\nabla \cdot \mathbf{J}_{\text{p}} + \mathrm{i}\omega q p + q(R - G) = 0, \; \mathbf{J}_{\text{p}} = q\mu_{\text{p}} p \mathbf{E} - qD_{\text{p}} \nabla p \tag{18.3}$$

$$\nabla \cdot \mathbf{J}_{\text{n}} - \mathrm{i}\omega q n - q(R - G) = 0, \; \mathbf{J}_{\text{n}} = q\mu_{\text{n}} n \mathbf{E} + qD_{\text{p}} \nabla n \tag{18.4}$$

$$\nabla \times \left(\frac{1}{\mu} \nabla \times \mathbf{A} \right) - \mathbf{J}_{\text{c}} - \mathrm{i}\omega\epsilon \mathbf{E} = 0 \tag{18.5}$$

$$\nabla \cdot \mathbf{A} + \mathrm{i}\omega\mu\epsilon\xi V = 0 \tag{18.6}$$

$$\mathbf{E} = -\nabla V - \mathrm{i}\omega \mathbf{A} \tag{18.7}$$

$$p = n_{\text{i}} \exp(\phi_{\text{p}} - V), \; n = n_{\text{i}} \exp(V - \phi_{\text{n}}) \tag{18.8}$$

we rewrite this system of equations using (18.7) by performing a transformation of variables $\mathbf{A} = \frac{\mathrm{i}}{\omega}(\mathbf{E} + \nabla V)$. Of course $\mathbf{E}$ is the electric field but the point is that starting from six field degrees of freedom $\{V, \phi_{\text{p}}, \phi_{\text{n}}, A_x, A_y, A_z\}$ we transform to six different field degrees of freedom $\{V, \phi_{\text{p}}, \phi_{\text{n}}, E_x, E_y, E_z\}$. The system of Equations (18.1–18.8) gets adapted for (18.5) and (18.6) leading to

$$\nabla \times \left(\frac{1}{\mu} \nabla \times \mathbf{E} \right) - \mathbf{J}_{\text{c}} - \mathrm{i}\omega\epsilon \mathbf{E} = 0 \tag{18.9}$$

$$\nabla \cdot \mathbf{E} + \nabla^2 V + \mathrm{i}\omega\xi V = 0 \tag{18.10}$$

In (Chen et al. [2011b]) we have named the system of Equations (18.1–18.4) together with (18.9–18.10), the EV system or EV formulation. Before entering the details of solving this system we note that the EV formulation resembles the usual Maxwell system in the frequency domain using $\mathbf{B} = \frac{\mathrm{i}}{\omega} \nabla \times \mathbf{E}$. Thus the question arises why there is a voltage variable V left over at all. Apart from the evident answer that the transformation of variables does not "transform away" degrees of freedom, there is the observation that (18.9) contains the singular operation $\nabla \times \nabla$ that needs regularization by an additional constraint, e.g., Equation (18.10). Finally, it is indeed possible to avoid usage of the voltage variable V completely. In order to achieve this one must refrain from the Coulomb or Lorenz gauge (18.6, 18.10) and select the *temporal gauge* $V = 0$. However, as was argued in the introduction, in this gauge it becomes very awkward (if not impossible) to compute carrier densities in semiconductors.

18.2 Discretization

Although the term $\nabla \times (\frac{1}{\mu}\nabla \times \mathbf{E})$ in (18.9) is singular (non-invertible), when combined with $\mathbf{J}_{\text{total}} = (\sigma + \mathrm{i}\epsilon\omega)\mathbf{E}$, it is regular for $\omega \neq 0$ and $\sigma \neq 0$ and/or $\epsilon \neq 0$. Yet, we can improve its iterative convergence behavior substantially by making it more Laplacian-alike by subtracting $\nabla(\frac{1}{\mu}\nabla \cdot \mathbf{E})$. For constant permeability we may subtract the divergence of (18.10) from (18.9) without altering the solution. The resulting equation is

$$\nabla \times \left(\frac{1}{\mu}\nabla \times \mathbf{E}\right) - \nabla\left[\nabla \cdot \mathbf{E} + \nabla^2 V + \mathrm{i}\omega\xi V\right] - \mathbf{J}_{\text{total}} = 0 \tag{18.11}$$

Equation (18.11) is the starting point for the discretization of the EV system. The discretization is done fully analogous to the discretization of the AV system in (Meuris et al. [2001b]). This approach requires that a grid variable E_k is assigned to every link in the computational grid representing the projection of the electric field in the direction of that link. In regions with constant permittivity and zero charge we obtain from (18.1) that $\nabla \cdot \mathbf{E} = 0$ and therefore in those regions we also could use

$$\nabla \times \left(\frac{1}{\mu}\nabla \times \mathbf{E}\right) + \nabla\left(\nabla \cdot \mathbf{E}\right) - \mathbf{J}_{\text{total}} = 0 \tag{18.12}$$

When addressing structures that contain semiconducting regions, Equations (18.3) and (18.4) are part of the system of equations that need to be solved. With metals included Equation (18.2) can be added to make the solution satisfy current continuity everywhere. Acting with the divergence on (18.11) and (18.12) implies that

$$\nabla^2\left[\nabla \cdot \mathbf{E} + \nabla^2 V + \mathrm{i}\omega\xi V\right] = 0 \tag{18.13}$$

$$\nabla^2\left[\nabla \cdot \mathbf{E}\right] = 0 \tag{18.14}$$

The following theorem plays an important role to complete the discretization:

Theorem: *If $\nabla^2 f = 0$ for some domain Ω and $f = 0$ at the boundary of the domain $\partial\Omega$, then $f = 0$ everywhere inside Ω.*

Thus the discretization requires that (18.11) or (18.12) is imposed for the surface of the simulation domain and as a consequence the gauge condition and/or Gauss' law is obtained everywhere when solving (18.3, 18.4) and (18.11) or (18.12). In this way the redundancy in the formulation of the full system of equations is avoided (Chen et al. [2011b], Endes [2009]). It is interesting to note that the argumentation can be inverted: solving the gauge condition or Gauss' law everywhere as well as (18.11, 18.12), guarantees current continuity. However, although such analytic observations can be easily deduced, their numerical benefits still need to be shown. In our numerical experiments we have found that the first route is the preferred one, i.e. imposing current continuity explicitly is a more robust scheme than imposing the gauge condition or Gauss' law explicitly.

18.3 Simplified EV Schemes

In the numerical experiments discussed below, we have marked the results that are based on the approach using Gauss' law, e.g. Equation (18.12) by the label 'EV-Gauss'. The results that are obtained using the gauge conditions, e.g. Equation (18.11) is based on older work at MAGWEL and is labeled as 'EV-Magwel'. An interest line of reasoning, inspired by the universal validity of Gauss' law is to apply it in the gauge condition (18.10) in regions of zero charge and constant permittivity, i.e. inside metals and insulators :

$$\nabla^2 V + \mathrm{i}\omega\xi V = 0 \tag{18.15}$$

This route was originally explored by Chen and the author. Its appealing feature is that large blocks of entries in the Newton matrix originating from EV mixing empty, thereby speeding up the (iterative) solving considerably. In the numerical experiments discussed below we have marked the corresponding results by the label 'EV-Chen'. Thus the resulting system is based on Equations (18.12) and (18.15) in charge-free regions and keep Equation (18.12) elsewhere, i.e. at interfaces. It should be emphasized that this approach is approximate since it modifies the system of equations in such a way that their physical content changes. This can be understood from the fact that acting with the divergence on the Maxwell-Ampere Equation (18.12) leads to $\nabla^2(\nabla \cdot \mathbf{E}) - \mathrm{i}\mu\omega(\sigma + \mathrm{i}\epsilon)(\nabla \cdot \mathbf{E}) = 0$ or $\nabla^2 f + kf = 0$ for $f = \nabla \cdot \mathbf{E}$ and $k \neq 0$ constant and therefore the theorem is not applicable. Indeed the numerical solutions that were obtained using this approach show deviations from the physical correct ones, meaning that non-zero values for f have been mixed in. The deviations are in some cases rather small and therefore the method can still generate valuable results. However, so far we have no general guidelines when the method is sufficiently accurate.

18.4 Combination of AV and EV Solvers

The AV and EV solvers have a complementary working range. The AV solver works well at low and medium frequencies (0–50GHz) whereas the EV solver behaves competent at high frequencies ($>$ 50GHz). Combining the merits of the two solvers will enable a true wide-band EM-TCAD co-simulator. The following pseudo-code algorithm provides a convenient upgrading of the AV solver to include an EV solver applying a 4-step solution strategy:

1. Map $\mathbf{A} - V$ variables onto $\mathbf{E} - V$ variables via $\mathbf{A} = \frac{\mathrm{i}}{\omega\epsilon}\left(\mathbf{E} + \nabla V\right)$.
2. Apply the EV solver to compute the update vector $[\Delta V, \Delta\mathbf{E}]^T$ in Newton iteration.
3. Map $[\Delta V, \Delta\mathbf{E}]^T$ onto $[\Delta V, \Delta\mathbf{A}]^T$
4. Update the AV system.

Using this approach, the data structure in the original AV solver is unaltered and the switching between the AV and EV solvers is easy to realize.

18.5 Numerical Experiments

In this section we present a numerical experiment to demonstrate the characteristics of the AV and EV schemes. Many more examples exist but the generic trend is similar to the one shown here. We focus on the eigenvalues and condition numbers of the Newton-Raphson matrices. A simple square inductor with 3.5 windings is used as illustrated in Figure 18.1. We applied a simple structured mesh since the focus is here on testing the qualities of the linear system. In Figure 18.2 we plot the solution for the potential field V in the inductor plane computed using the AV scheme, whereas in Figure 18.3 we show the potential field computed according to the EV scheme. Figure 18.4 shows the magnitude of the electric field computed using the AV scheme and Figure 18.5 shows the same variable using the EV scheme. The plots are pair wise identical demonstrating that the transformation from the AV to EV system has no effect on the solution as it should.

In Figure 18.6 the condition number vs. the frequency is shown for the various solution strategies. As is observed, the three EV versions have very large condition numbers at low frequency which drop and the AV solver has

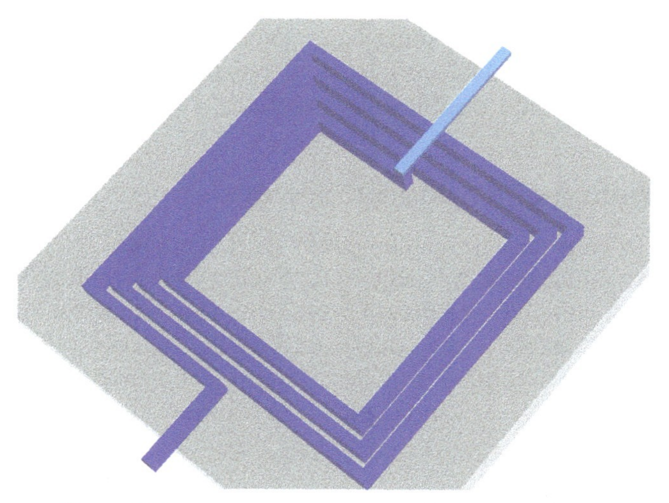

Figure 18.1 Lay out of an integrated inductor problem.

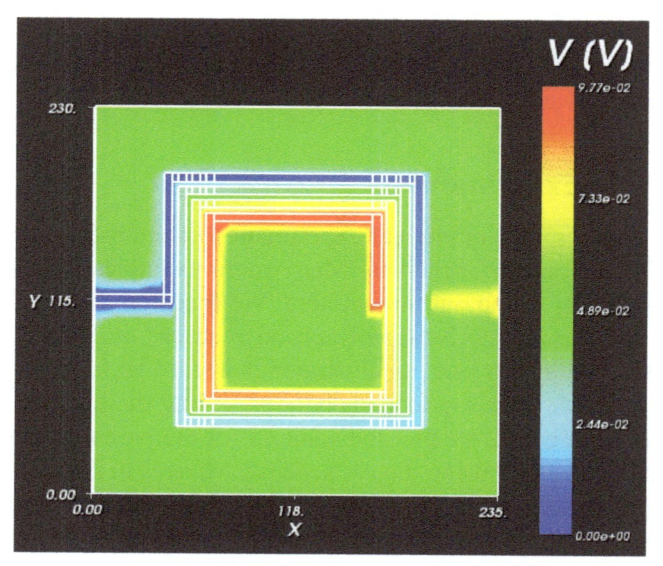

Figure 18.2 Voltage in the inductor plane at 10 GHz using the AV solver.

the smallest condition number at low frequency which gradually goes up. Around 50 GHz the curves do cross meaning that from then onwards it is more favorable to use EV solving.

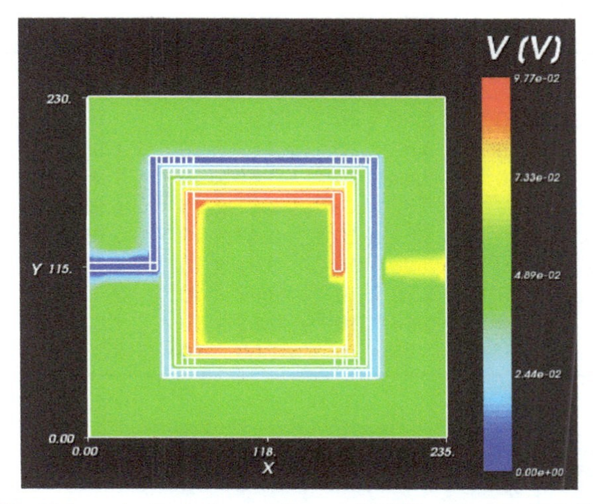

Figure 18.3 Voltage in the inductor plane at 10 GHz using the EV Gauss solver.

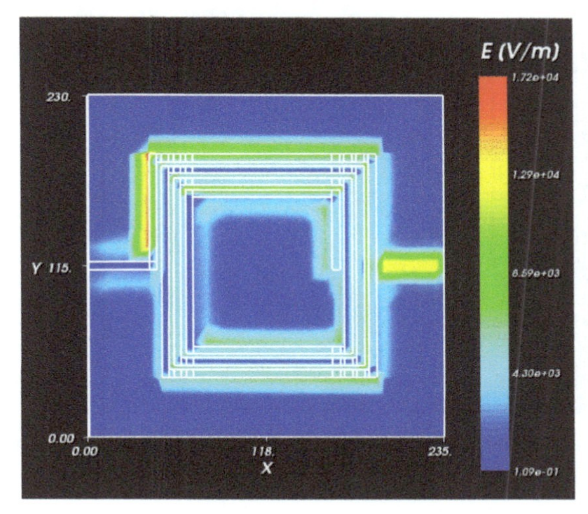

Figure 18.4 Magnitude of the electric field in the inductor plane at 10 GHz using the AV solver.

In order to obtain a better understanding of the patterns that are seen in Figure 18.6 (we observed similar patterns in many different structure setups) we have zoom in at the eigenvalues of the Newton-Raphson matrices. For the AV solver we find that the real part of the smallest (in magnitude) eigenvalue as a function of frequency is $\lambda_0(\omega) \sim \text{const}$ and the constant of order 10^{-2}.

Table 18.1 Condition numbers of the **A**-V and **E**-V- solvers

f (Hz)	**E**-V-Magwel	**E**-V-Chen	**E**-V-Gauss	**A**-V
10^4	1.00×10^{30}	3.82×10^{19}	3.67×10^{15}	1.98×10^{12}
10^5	2.87×10^{28}	3.82×10^{18}	3.67×10^{14}	1.98×10^{12}
10^6	1.26×10^{29}	3.82×10^{17}	3.67×10^{13}	1.98×10^{12}
10^7	2.9×10^{25}	3.82×10^{16}	3.67×10^{12}	1.99×10^{12}
10^8	6.72×10^{20}	3.85×10^{15}	3.43×10^{12}	2.05×10^{12}
10^9	1.47×10^{18}	5.28×10^{14}	2.91×10^{12}	2.59×10^{12}
10^{10}	1.47×10^{16}	2.12×10^{14}	2.02×10^{12}	2.42×10^{13}
10^{11}	1.47×10^{14}	6.56×10^{13}	1.88×10^{12}	1.63×10^{15}
10^{12}	1.48×10^{12}	5.75×10^{11}	1.60×10^{11}	8.38×10^{16}

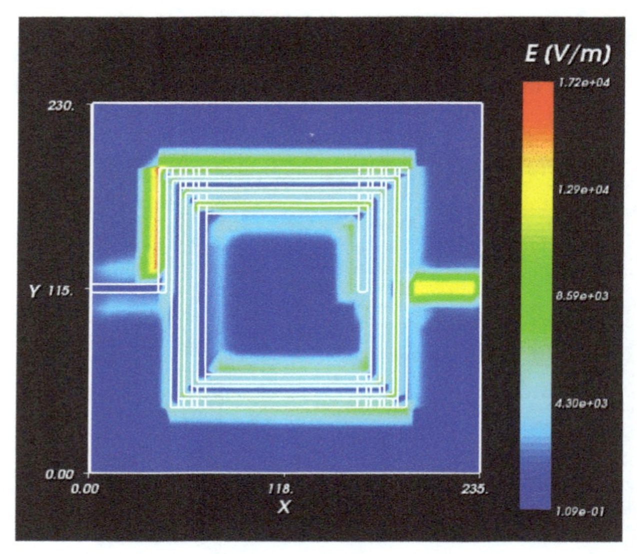

Figure 18.5 Magnitude of the electric field in the inductor plane at 10 GHz using the EV Gauss solver.

On the other hand the smallest eigenvalue for the EV solver behaves as $\lambda_0(\omega) \sim \text{const}'/\omega$ where const′ is also of order 10^{-2}. This explains why EV solving becomes difficult at low frequencies.

On the other hand the modulus largest eigenvalues are constant for both cases. However, for the AV solver the value is $O(10^{10})$, whereas for the EV solver the value is $O(10^6)$. Since the condition number can be seen as a measure for the ratio of the smallest and the largest eigenvalue we identified

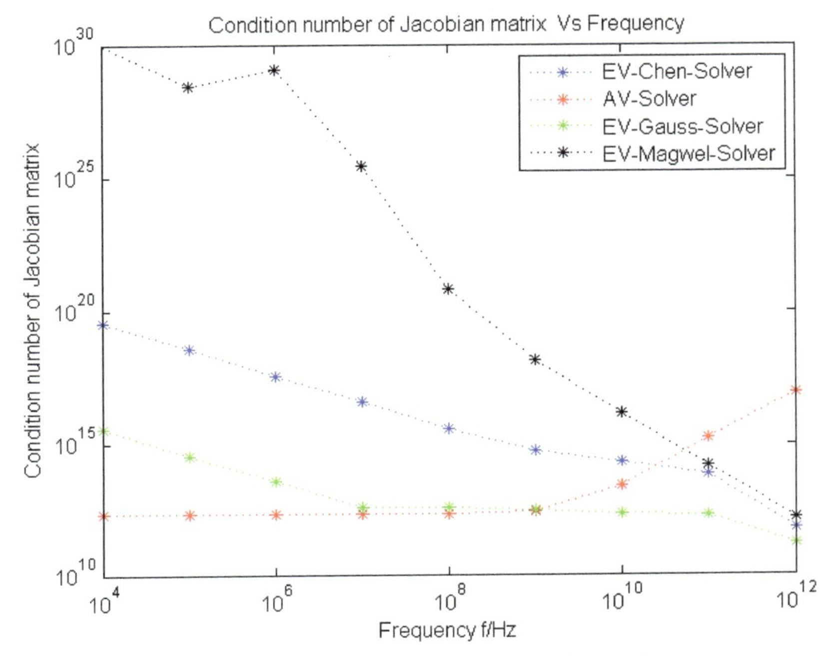

Figure 18.6 Condition number Vs frequency for the integrated inductor problem.

the cause of the pattern observed in Figure 18.6. Moreover, the repetitive character of the pattern allows to identify a switching point in the solver between the AV and EV approach. This point is located around ~50 GHz.

18.6 Best Practices for Iterative Solving

We end the chapter with reporting on our experience of solving the AV and EV systems using iterative solvers. As far as the AV system is concerned the most robust approach that we identified applies matrix row/column permutation using the method of Duff and Koster (Duff and Koster [2001]). Its purpose is to make the diagonal as dominant as possible. Next a standard procedure of incomplete LU decomposition is applied followed by a linear solver such as BiCGStab (van der Vorst [1989]). For the EV system, we have a different starting point: note that in the EV Gauss formulation it is the Maxwell-Ampere equation that implicitly determines V and Gauss' law (18.14) is generating entries in the Newton-Raphson matrix which are fully off-diagonal. It turns out that the method of (Duff and Koster [2001]) does not lead to a good pre-ordering of the linear system. However, with the approximate minimum degree (AMD) method and in particular, the variant COLAMD (Davis et al. [2004a]), the permuted system is suitable for iterative solving. The more robust choice of linear solver is CGS (Sonneveld [1989]).

19

Large Signal Simulation of Integrated Inductors on Semi-Conducting Substrates

We present a formulation of transient field solving that allows for the inclusion of semiconducting materials whose dynamic responses are prescribed by drift-diffusion modeling[15]. The robustness and the feasibility is demonstrated by applying the scheme to compute accurately the large-signal response of an integrated inductor. The most common way to address electromagnetic (full-wave) field problems is by solving the Maxwell equations, i.e. setting up and solving discretized versions of these equations for the electric field $\mathbf{E}$ and the magnetic field $\mathbf{B}$. Moreover, another common ingredient is to solve the equations in the frequency regime. Faraday's law then provides a reduction of unknowns in the discretization using edge elements. The finite-integration technique does not attempt to reduce grid variables and therefore, these variables are a very faithful representation of the continuous degrees of freedom. In the transient regime, the latter has therefore been very successful in comparison to transient field solving based on finite-element methods and generalizations thereof. However, in semiconductor physics, the electric and magnetic fields are coupled to the carrier concentrations in a highly non-linear fashion. This is because the carrier densities depend on the *energy* density via the Boltzmann distribution functions, whereas the fields refer primarily to *forces*. The path dependency of the energy, being a force integrated along some path, will highly complicate the full-wave field solving if semiconductors are involved. This explains that so far, most semiconductor device simulators ignore the magnetic field because in this approximation, the force integral becomes path independent again. Fortunately, there is an appealing solution to avoid above complications. When the EM field problem is solvable in terms of the scalar potential and the vector potential, then it is possible to insert these solutions into the semiconductor equations and

[15]This chapter is a reprint of Schoenmaker et al. [2012].

one has obtained a very straightforward upgrading of the semiconductor device simulation tools into the EM wave regime. In the last decade, we have demonstrated that, at least in the frequency domain, this solution is not only feasible but also leads to accurate results using standard non-linear and linear solver techniques (Schoenmaker et al. [2010b]). This is not at all evident since the potential field formulation leads at first instance to a singular operator that needs to be regularized by gauge fixing. At the same time, the choice of the gauge fixing should not affect the outcome of physical variables such as resistance, inductance, capacitance, etc. The gauge independence of the physical variables is the outcome of a subtle interplay between the differential-geometry based discretization, the formulation of the boundary conditions and the selection of the gauge condition (Schoenmaker et al. [2007b]). The potential field formulation requires only one grid (the 'primary grid') to be built. The scalar and vector potentials are located at the primary grid. The dual grid is a 'conceptional' tool to make the proper differential-geometry identifications.

19.1 Need for Mimetic Formulation

Just as in the frequency regime, the formulation of the transient equations for the potential fields are also subject to gauge fixing and differential-geometry based discretization considerations. Moreover, in the transient regime it becomes also evident that Gauss' law is not a dynamical evolution equation but a constraint, meaning that after each time step the state space vector should be compliant with this constraint. This is illustrated in Figure 19.1. On the other hand in general the gauge condition can be time dependent, for

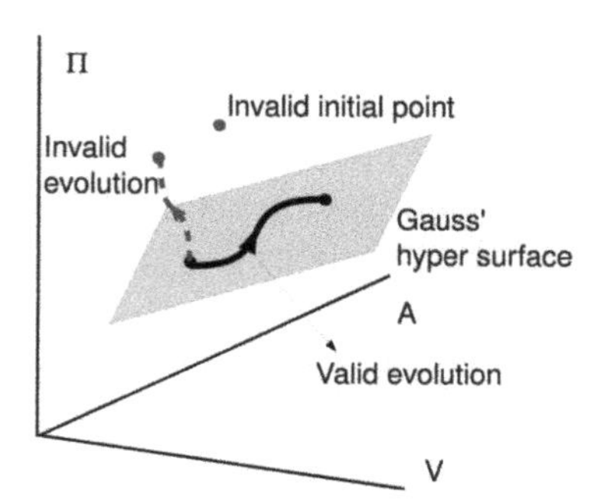

Figure 19.1 'Artist impression' of the Gauss' law-induced constraint for the time evolution of the full wave variables.

example the Lorenz gauge. It can therefore play a role as part of the evolution equations. In this paper we will provide the complete and correct formulation of the transient equations for the potential fields which is fully compliant with mimetic principles (Lipkinov et al. [2004]). Moreover, we show that standard linear (CGS) and non-linear solvers (Newton-Raphson) can be used to unfold the time evolution. The method is applicable to metallic materials that are covered by Ohm's law as well as semiconducting materials for which drift and diffusion contributions determine their voltage responses.

19.2 Field Equations

We start from the Maxwell equations in the potential formulation in the time domain. Then the Maxwell equations in these variables become:
For insulators:

$$-\nabla \cdot \left[\varepsilon \left(\nabla V + \frac{\partial \mathbf{A}}{\partial t} \right) \right] = 0 \tag{19.1}$$

$$\nabla \times \frac{1}{\mu} (\nabla \times \mathbf{A}) = -\varepsilon \frac{\partial}{\partial t} \left(\nabla V + \frac{\partial \mathbf{A}}{\partial t} \right) \tag{19.2}$$

for conductors:

$$-\nabla \cdot \sigma \left(\nabla V + \frac{\partial \mathbf{A}}{\partial t} \right) = \frac{\partial}{\partial t} \left(\nabla \cdot \varepsilon \left(\nabla V + \frac{\partial \mathbf{A}}{\partial t} \right) \right) \tag{19.3}$$

$$\nabla \times \frac{1}{\mu} (\nabla \times \mathbf{A}) = -\sigma \left(\nabla V + \frac{\partial \mathbf{A}}{\partial t} \right) - \varepsilon \frac{\partial}{\partial t} \left(\nabla V + \frac{\partial \mathbf{A}}{\partial t} \right) \tag{19.4}$$

Besides the usual full-wave equations in the potential formulation for metals and insulators we find
for semiconductors:

$$-\nabla \cdot \frac{\varepsilon}{q} \left(\nabla V + \frac{\partial \mathbf{A}}{\partial t} \right) = p - n + N_{\text{dop}} \tag{19.5}$$

$$p = n_i \, \mathrm{e}^{\frac{q}{kT}(\phi_p - V)}, \quad n = n_i \, \mathrm{e}^{\frac{q}{kT}(V - \phi_n)} \tag{19.6}$$

$$\mathbf{J}_p = q \mu_p n_i \left[\mathrm{e}^{\frac{q}{kT}(\phi_p - V)} \left(\nabla V + \frac{\partial \mathbf{A}}{\partial t} \right) - kT \, \nabla \mathrm{e}^{\frac{q}{kT}(\phi_p - V)} \right] \tag{19.7}$$

$$\mathbf{J}_n = q\mu_n n_i \left[\mathrm{e}^{\frac{q}{kT}(V-\phi_n)} \left(\nabla V + \frac{\partial \mathbf{A}}{\partial t} \right) + kT \, \nabla \mathrm{e}^{\frac{q}{kT}(\phi_p - V)} \right] \tag{19.8}$$

The hole current-continuity equation is with U the recombination/generation:

$$\nabla \cdot \mathbf{J}_p = -U(n,p) - \frac{\partial p}{\partial t}, \quad \nabla \cdot \mathbf{J}_n = U(n,p) - \frac{\partial n}{\partial t} \tag{19.9}$$

The Maxwell-Ampere equation becomes

$$\nabla \times \frac{1}{\mu} (\nabla \times \mathbf{A}) = \mathbf{J}_p + \mathbf{J}_n - \varepsilon \frac{\partial}{\partial t} \left(\nabla V + \frac{\partial \mathbf{A}}{\partial t} \right) \tag{19.10}$$

Finally, one needs to provide a *gauge condition:*

$$\frac{1}{\mu} \nabla (\nabla \cdot \mathbf{A}) + \epsilon \nabla \left(\frac{\partial V}{\partial t} \right) = 0$$

Due to the full-wave nature of the problem, there are *second*-order time differentiations. We circumvent these 2nd-order differentiations in time by (1) applying a Legendre transformation to obtain the canonical momentum $\mathbf{\Pi} = \partial \mathbf{A}/\partial t$ conjugate to the vector potential field variable and (2), by putting *two* variables on each link of the primary grid, i.e. besides the usual projections of the vector potential on the links of the computational grid also the canonical momentum is projected on this grid and for each link it becomes an additional unknown (degree of freedom). Finally, since we have transformed the 2nd-order time differentiation into a first-order one we use backward Euler time stepping. The remaining variables V, ϕ_p and ϕ_n are discretized in the conventional way: these discrete variables are placed on grid nodes. In Figure 19.2, the variables are shown for one mesh cell. The presence of semiconducting materials requires that the currents in these regions need to be discretized using the Scharfetter-Gummel discretization method. The presence of the vector potential implies that the transient current becomes:

$$J_{ij} = \mathbf{J} \cdot \mathbf{n} = s^c \mu \, \frac{\Delta A}{h_{ij}} \left(c_i B[s^c \, X_{ij}] - c_j B[-s^c \, X_{ij}] \right) \tag{19.11}$$

where μ is the carrier mobility, $\mathbf{n}$ is the unit vector along the link $\langle ij \rangle$ between nodes i and j and ΔA is the dual area corresponding to the link $\langle ij \rangle$ whereas h_{ij} is the length of the link $\langle ij \rangle$. The sign factor s^c is $+1$ for holes and -1 for electrons. The carrier density $c_{i,j} = p_{i,j}$ for holes and

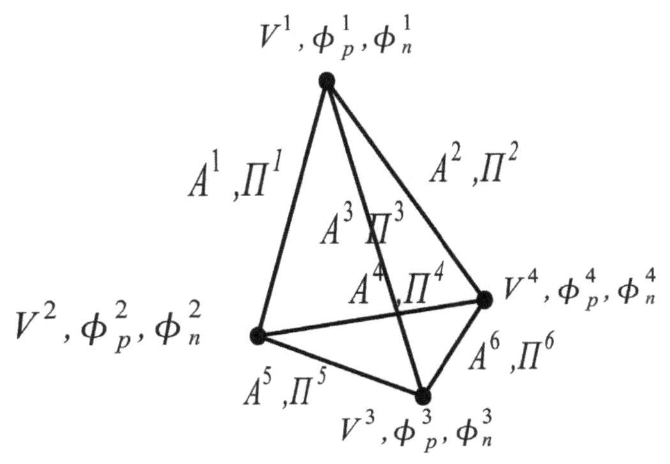

Figure 19.2 Illustration of the discrete variables in one mesh cell.

$n_{i,j}$ for electrons respectively in the nodes i and j. The argument of the Bernoulli function $B(x) = \frac{x}{e^x - 1}$ is $X_{ij} = \frac{q}{kT}(V_j - V_i + s_{ij}\ \Pi_{ij} h_{ij})$. Finally $s_{ij} = \pm 1$, depending on the orientation of the link $\langle ij \rangle$ with respect to its intrinsic orientation. It should be noted that the full system of equations still is redundant despite the fact that the gauge condition is added to regularize the singular (non-invertible) property of the curl-curl operator. The divergence of the Maxwell-Ampere equation leads to the current conservation law, and therefore, the latter is an implicit result after having found a solution of the first one. Alternatively, one may still insist on having the current-continuity equation(s) as part of the full set of equations that needs to be solved. In that case, one must omit the gauge condition as an independent equation to be solved. The gauge condition will be respected as an additional result from solving Gauss' law, the current-continuity equation and the Maxwell-Ampere equation. In fact, the elimination of the redundancy can also be lifted by using either the gauge conditions as a temporal evolution equation or Gauss' law as a constraint on the state variables at the latest time instance. All this is achieved provided that the mimetic principle is respected in the discretization method. The differential-geometry discretization scheme guarantees that either choice implies the other. The reader may wonder why introducing the gauge condition since it may be avoided after all in solving the transient problem. The reason for the inclusion of the gauge condition is found in the fact that the transient problem may have a static solution. In that case one runs into the singular character of the curl-curl operator. Therefore, the gauge condition still is needed for regularization of this operator. Details of the corresponding discretization procedure can be found in (Chen et al. [2011b]).

19.3 Application to An Octa-Shaped Inductor

We demonstrate the feasibility of the method by computing the on-set transient response of the current flow that is induced in the substrate (semiconductor) by switching on the voltage from 0 to 1 Volt in 100 ps. The structure is shown in Figure 19.3 and is isolated from the substrate by 2.48 micron of dielectric consisting of several layers. Integrated inductors are a key-component for RF circuits such as low-noise amplifiers (LNA), voltage controlled oscillators (VCOs), filters and impedance matching networks. In Monolithic Microwave Integrated Circuits (MMICs), inductors still occupy a significant portion of the total area. Furthermore, integrated inductors can induce parasitic couplings. In order to prevent and limit such disturbing couplings, special attention should be paid to placement and radiation-optimization of integrated inductors. Planar spiral inductors, while offering scalable layout canonical architectures (rectangular, octagonal) with ease of manufacturing suffer from low Q-factors. During the last years, considerable efforts have been directed towards finding ways to design inductors with increased quality factors and higher resonant frequencies. The shown octa-shaped configuration (Einziger [2004]), compared to classical rectangular, octagonal, circular topologies, has the advantage of limiting EMC (Electro-Magnetic Compatibility) related issues when symmetrical structures are considered. The EMC reduction is expected to result from the twisted nature of the octa-shaped topology, where the two constitutive loops will lead to equal magnetic field distributions with opposite polarities (Tesson [2008] Fahs et al. [2010]). Here a design improvement is highlighted to show the difference between an octagonal coil and an 8-shaped coil. Figure 19.3 shows the octagonal and 8-shaped coil used for a VCO. Figure 19.4 gives the measurement results of the 2fm spur (in dBc) at the RF output as a function of the output power (Pout). The main graph (VCO-octagonal coil) shows that the spur is above the –40dBc, which is the upper level that is acceptable. Use of the VCO octa-shaped coil improves the spurious level by 10dB. Using 8-shaped coils, the quality factor and the self-inductance (L) of the coil will decrease only slightly, depending on the situation. Refer to Figure 19.5. The inductor is designed in M6-M7-M8. The substrate thickness is 100 micron and is equipped with a ground contact at the back (not shown in Figure 19.6)

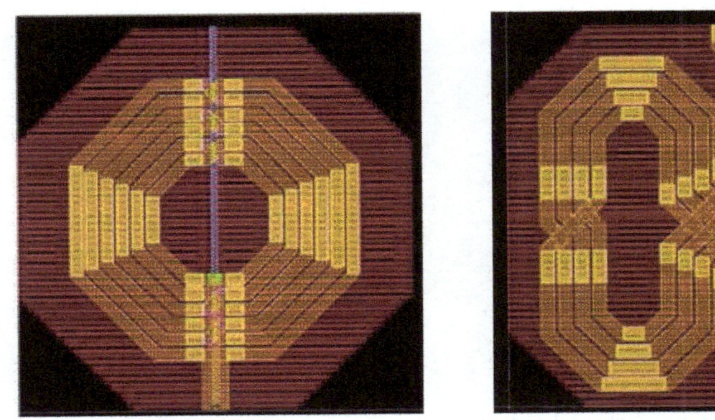

Figure 19.3 Octagonal (left) and 8-shaped (right) VCO coil.

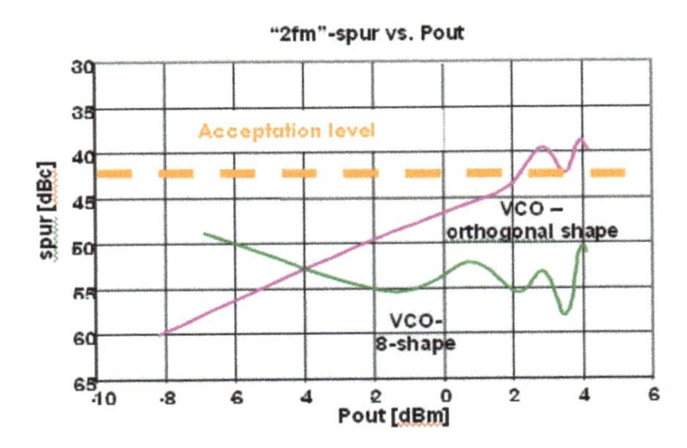

Figure 19.4 Measurement of spur level of octagonal and octa-shaped coil.

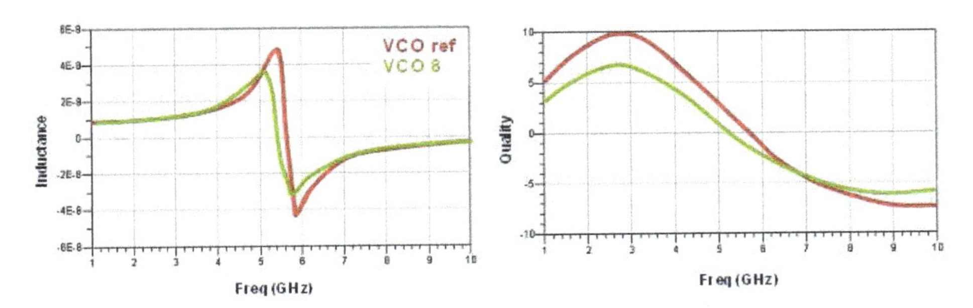

Figure 19.5 Inductance and quality factor of octagonal (red) and 8-shaped (green) coil.

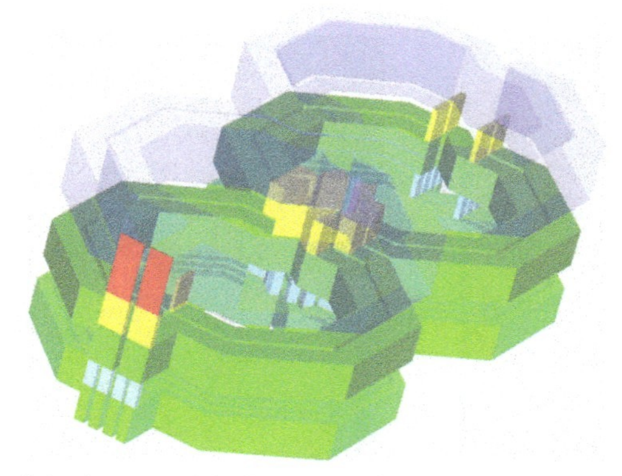

Figure 19.6 View of the integrated 8-shaped inductor from above. The vertical direction is stretched.

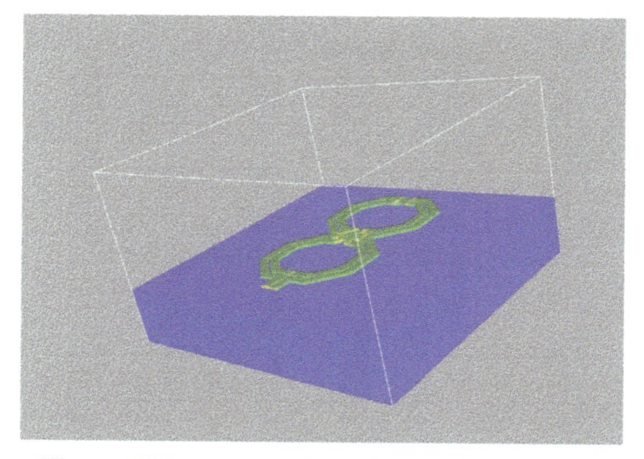

Figure 19.7 View of the full simulation domain.

and is p-type doped with a value of $10^{15}\mathrm{cm}^{-3}$. The full simulation domain is shown in Figure 19.7. Note that a layer of air is included to allow for the electromagnetic field spread around the inductor. The structure is discretized using a grid of 52185 nodes leading to 330983 variables in the linear systems of the Newton-Raphson scheme needed to solve the implicit next time step problem. We performed a time stepping of 1 ns in 10 intervals with a step function potential change at one of the inductor contacts. A capacitive coupling to the substrate is detected and its strength as well as the inductance of the on-chip inductor can be extracted from the results. In Figure 19.8, the currents into and out of the inductor contacts are shown. In Figure 19.9, the transient current in the ground plane contact is shown.

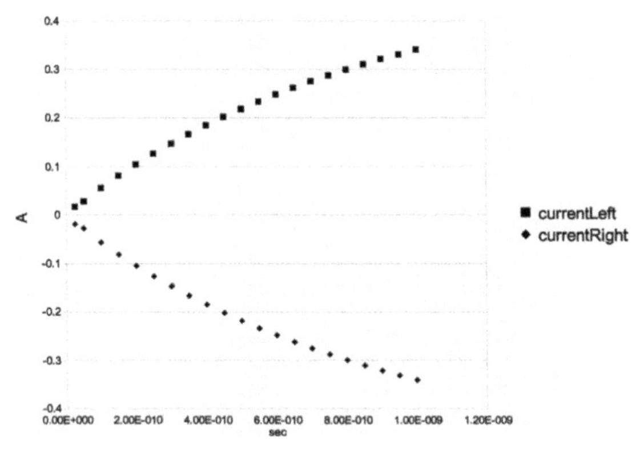

Figure 19.8 Value of currents at the left and right contact of the inductor.

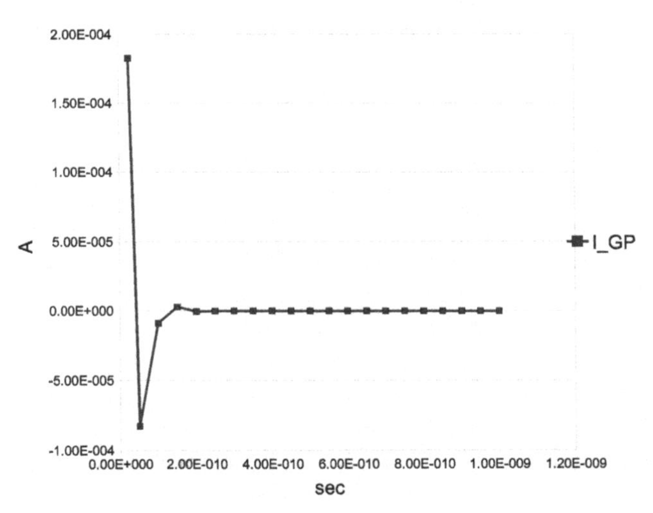

Figure 19.9 Value of the current in the ground plane contact. A transient overshoot is observed.

An overshoot effect is observed. In Figure 19.10, the current of the ground plane contact is shown in a logarithmic plot. Clearly, two time constants are observed. The corresponding 'signal-decay' constants are (1) for 0–4 nsec: $1.5 \times 10^{10}\text{sec}^{-1}$ and (2) for 4–10 nsec: $0.48 \times 10^{9}\text{sec}^{-1}$. In general, the advantage of having a transient design flow in place is to circumvent the problems arising from using frequency domain EM models (S-parameters) in

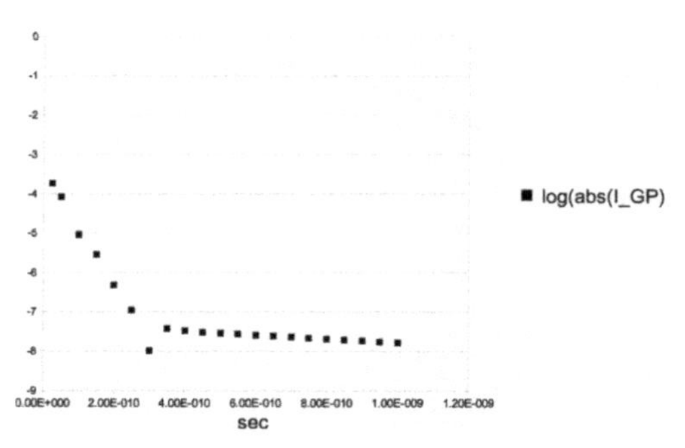

Figure 19.10 Logarithm of the absolute values of the current in the ground-plane contact. Two time scales are observed.

transient circuits. Often these S-parameter models turn out to have a lack of passivity and stability, resulting in transient solutions not converging, arising from failing to capture dominant poles in the right frequency plane. For this approach, as an example, the 8-shaped inductor can be used by comparing a simple circuit with a simple lumped element inductor model in transient with the same circuit coupled with the EM model. In order to assess the results of the current build-up shown in Figure 19.8, we represent the inductor as a simple lumped compact model (see Figure 19.11), consisting of L, R and C to ground. As a first approximation we can take the values from the RF-simulations which give a value of 3 Ω at 1 GHz, but in order to get a good fit, we took R=2 Ω. Using a step magnitude of 1.0 Volt, the result in Figure 19.12 is obtained. The need for this fitting already indicates that the RF values that were obtained with the assumption of small-signal perturbations can not be assumed to be the correct values if large signals are applied. The current build-up is shown to be in good agreement with the currents of Figure 19.8.

Beside the constant time step solutions presented before, we realized and tested a variable time step and variable order implementation of the backward differential formulas (BDF), also known as the implicit Gear formulas, see (Gear [1971]). It is particularly suited for stiff ordinary differential equation systems and widely used for transient circuit simulation, e.g. in all SPICE based circuit simulation packages. After space discretization of the electro-magnetic field equations (19.1)-(19.10), we obtain an ordinary differential equation system of the form

$$g(\frac{\mathrm{d}^2\mathrm{u}(\mathrm{t})}{\mathrm{dt}^2}, \frac{\mathrm{du}(\mathrm{t})}{\mathrm{dt}}, u(t), t) = 0$$

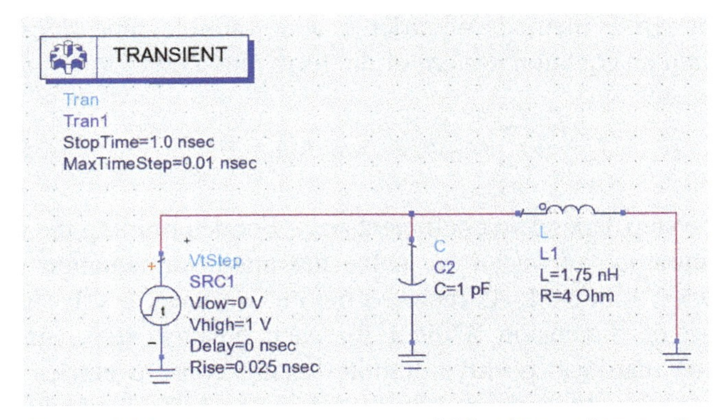

Figure 19.11 Set up of a compact model for the transient results.

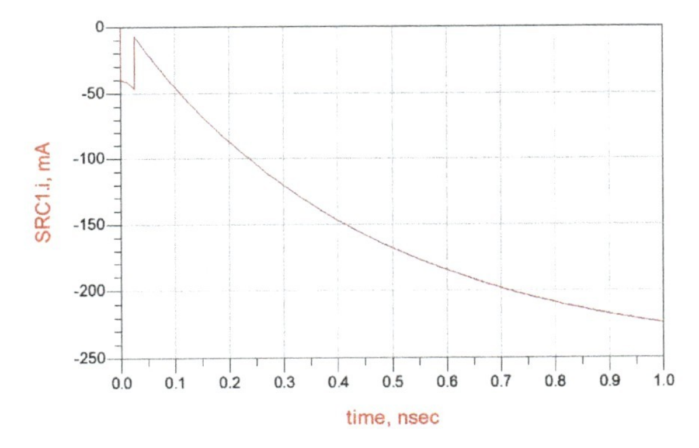

Figure 19.12 Results of a compact model for the transient simulation using a step magnitude of 1Volt.

with $u(t)$ involving the vector potential field variables $\mathbf{A}(t)$ for each link of the primary grid, the nodal potentials $V(t)$ at each node of the primary grid as well as the electron density $n(t)$ and the hole density $p(t)$ at each node of the primary grid belonging to semiconducting material, all evaluated at the time point t. Using the canonical momentum $\mathbf{\Pi} = \partial\mathbf{A}/\partial t$ we arrive at a first order system of the form

$$f(\frac{\mathrm{dw(t)}}{\mathrm{dt}}, w(t), t) = 0 \tag{19.12}$$

with $w(t)$ including $u(t)$ as well as $\mathbf{\Pi}(t)$ at each link of the primary grid.

Applying the BDF methods of order k with variable time steps τ_n to (19.12), we obtain an equation system of the form

$$f(\frac{1}{\tau_n}\sum_{i=0}^{k}\alpha_{ni}w_{n-i}, w_n, t_n) = 0 \qquad (19.13)$$

with certain time step dependent coefficients α_{ni}. For determining the numerical approximation w_n of $w(t_n)$ we solve the nonlinear equation system (19.13) by Newton's method. As reported before, we have to solve a linear equation system of dimension 330983 for each Newton step. Since the resulting Newton matrix is positive definite having nonzero entries on the diagonal, we used the algebraic multigrid package SAMG (Trottenberg and T. Clees [2009]) for a memory saving and time efficient solution of these systems.

Figures 19.13 and 19.14 show the transient results for the 8-shaped inductor when a sinusoidal voltage with 1 GHz frequency is applied between the first and the second contact. The variable stepsize was automatically selected by a predictor - corrector based error estimation guaranteeing an error of the magnitude of 10^{-4}. Figure 19.15 shows two nice results of the quality of our simulation. First, the numerical discretization preserves charges (the sum of all currents through the 8-shaped inductor is almost zero). Secondly, the global error is smaller than 3×10^{-6}, i.e. the prescribed error tolerance has been reached.

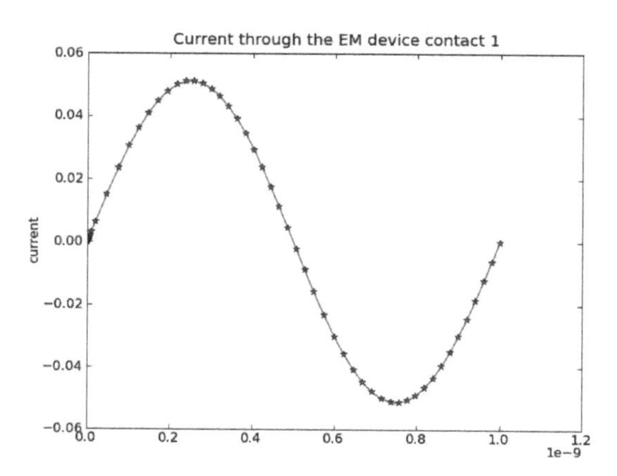

Figure 19.13 Current through the first contact of the 8-shaped inductor after transient simulation with variable order and variable time step size.

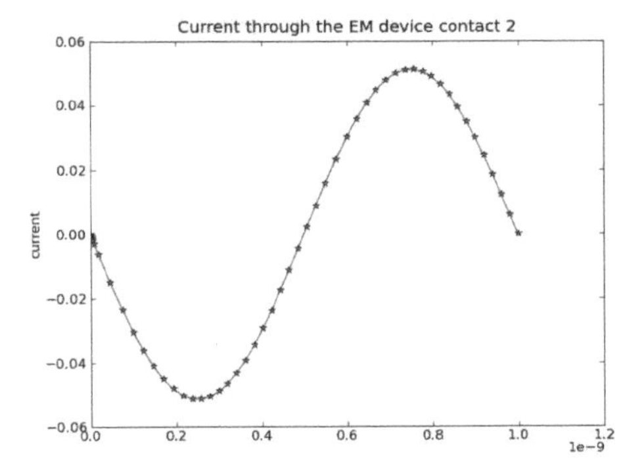

Figure 19.14 Current through the second contact of the 8-shaped inductor after transient simulation with variable order and variable time step size.

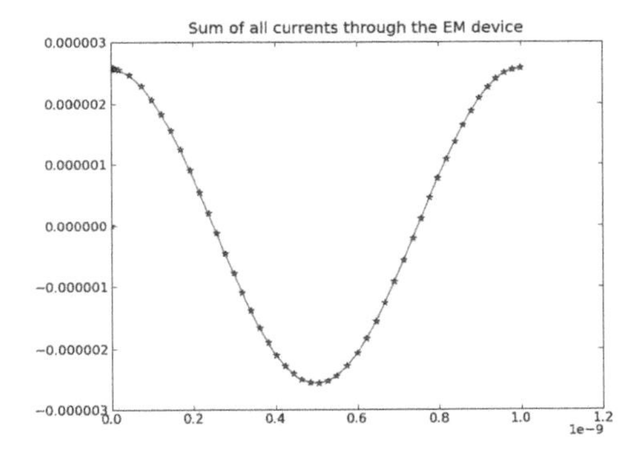

Figure 19.15 Sum of all simulated currents (including the substrate current) through the 8-shaped inductor after transient simulation with variable order and variable time step size.

Summary

In this chapter we presented a large-signal field solving approach that faithfully represents semiconducting material responses. The method exploits a Legendre transformation on the full-wave formulation such that we can apply standard backward Euler differential methods for adaptive time integration. Furthermore, we have incorporated Gauss' law as a constraint that should be respected at all time instances. The gauge condition results 'for free' once the solution is obtained. We demonstrated the correctness of the method by applying it to an industrial design problem, i.e. by computing the large-signal response of an integrated inductor above a semi-conducting substrate.

The transient information provides complementary insight in the behavior of complex designs in which electromagnetic interaction can jeopardize the ‘first-time-right’ EDA goal. In particular, RF modeling in the frequency domain is limited to the small-signal response superpositioned to a fixed operation point. Transient simulations are not limited to the perturbative nature of the stimuli and therefore are a valuable add-on to improve virtual prototyping.

20

Inclusion of Lorentz Force Effects in TCAD Simulations

Full-wave electromagnetic field solvers are commonly based on solving the electric field $\mathbf{E}$ and the magnetic field $\mathbf{B}$ using linear constitutive relations in the metallic regions[16]. Even in the static regime it is of interest to be able to compute both the electric as well as the magnetic field and to study their induced effects in devices. For example, the Hall sensor's operation depends essentially on the fact that the magnetic field distorts the flow of currents due to the Lorentz force (Ricobene et al. [1991]). Steady-state self-induced Lorentz force effects are equally realistic but are in general considered to be of less importance due to the fact that they are a few orders of magnitude less pronounced. However, the steady-state solvers may be very useful for estimating effects in the transient regime by extending their use to the quasi-static regime. In the latter case one may assume large currents being present for a short time. In Figure 20.1, an ESD structure is shown whose junctions are in normal operation reversely biased. In particular, switching currents need to stay below design rules thresholds and since they are large, the Lorentz force corrections are not negligible. Furthermore, even if one aims at a complete large-signal or transient simulation it is needed as a first step to include the Lorentz force in a steady-state description. Encouraged by these considerations we have addressed the inclusion of the Lorentz force into the static sector of a combined device simulator for semiconductors, metals and insulators.

The implementation not only considers the usual Hall effect originating from external magnetic fields but it is also capable of dealing with 'self-induced' Lorentz forces. This requires a complete self-consistent approach. The self-induced Lorentz force currents represent a non-linear correction to the current density constitutive relation and the artillery of computational

[16]This chapter is a reprint of Schoenmaker et al. [2011].

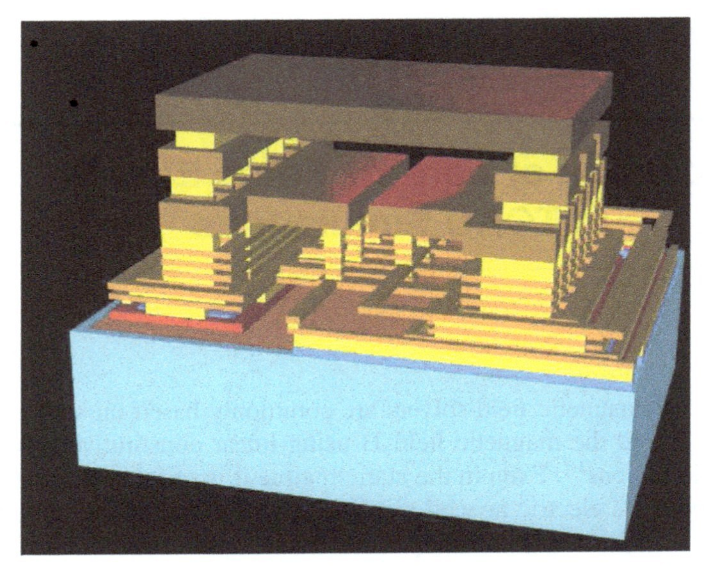

Figure 20.1 ESD protection structure whose position requires an accurate knowledge of substrate current flow.

techniques for semiconductor simulations are required to arrive at a solution *even* for metallic domains. It should be emphasized that the Lorentz force currents differ from the currents that are induced by time-varying electric fields. Eddy currents and proximity effects are the result of electric fields according to Ohm's law, but the fields are induced by Faraday's law of induction. On the other hand, the Lorentz force currents are induced by a magnetic force term and demands a modification of Ohm's law itself. In this paper we will present a new discretization of the inclusion of the Lorentz force term in Ohm's law. The method exploits the geometrical discretization techniques of the underlying finite-integration schemes for the vector potential (Schoenmaker and Meuris [2002a]).

20.1 Steady-State Equations

In this section we will summarize the equations that will serve as a starting point for the implementation of the Hall effect and the self-induced Lorentz forces. For semiconductors we refer to the usual classical drift-diffusion model at constant temperature. This model is sufficiently detailed to explain our approach. The current densities for holes and electrons in the drift-diffusion model *with* inclusion of the Lorentz force are

$$\begin{aligned}\mathbf{J}_n &= q\mu_n n\left(\mathbf{E} + \mathbf{v}_n \times \mathbf{B}\right) + kT\mu_n \nabla n \\ \mathbf{J}_p &= q\mu_p p\left(\mathbf{E} + \mathbf{v}_p \times \mathbf{B}\right) - kT\mu_p \nabla p\end{aligned} \tag{20.1}$$

where $\mathbf{J}_n = -qn\mathbf{v}_n$ and $\mathbf{J}_p = qp\mathbf{v}_p$ are implicitly defining the carrier velocities. The latter can be eliminated in favor of the current densities and both equations can be expressed as

$$\mathbf{J} + s\mu\mathbf{B} \times \mathbf{J} = \mathbf{K}_{\mathrm{DD}} \tag{20.2}$$

$$\mathbf{K}_{\mathrm{DD}} = q\mu c\mathbf{E} - skT\mu\nabla c \tag{20.3}$$

In here, $c = p$ and $s = +1$ (holes) or $c = n$ and $s = -1$ (electrons). Were it not for the second term at the left-hand side, Equation (20.2) would be the usual drift-diffusion expression for the current density. It is easily possible to obtain the current density in terms of the drift-diffusion current density $\mathbf{K}_{\mathrm{DD}}$. The results is:

$$\mathbf{J} = \frac{1}{(1 + \mu^2 B^2)}\left[\mathbf{K}_{\mathrm{DD}} + s\mu\,\mathbf{K}_{\mathrm{DD}} \times \mathbf{B} + \mu^2\left(\mathbf{K}_{\mathrm{DD}} \cdot \mathbf{B}\right)\mathbf{B}\right] \tag{20.4}$$

The parameter $|\mu B| << 1$ allows to write

$$\mathbf{J} = \mathbf{K}_{\mathrm{DD}} + s\mu\mathbf{K}_{\mathrm{DD}} \times \mathbf{B} \tag{20.5}$$

The current-continuity equations and the Poisson equations complete the description of the drift-diffusion model, i.e.

$$\nabla \cdot \mathbf{J} + s(R - G) = 0 \tag{20.6}$$

$$\nabla(-\epsilon\nabla V) = \rho(V, \phi_p, \phi_n) \tag{20.7}$$

Inside metallic regions, we use the following expression for the current density with μ_{H} the Hall mobility

$$\mathbf{J} = \sigma\mathbf{E} + \mu_{\mathrm{H}}\mathbf{J} \times \mathbf{B} \tag{20.8}$$

which implicitly determines $\mathbf{J}$ as a function of $\mathbf{E}$ and $\mathbf{B}$ and which leads after using $|\mu_{\mathrm{H}} B| << 1$ to

$$\mathbf{J} = \sigma\mathbf{E} + \mu_{\mathrm{H}}\sigma\mathbf{E} \times \mathbf{B} \tag{20.9}$$

The current density satisfies the current continuity equation $\nabla \cdot \mathbf{J} = 0$. The magnetic field consists of two terms of which one is representing the self-induced field and the other the external field $\mathbf{B}_{\text{ext}}$

$$\mathbf{B} = \nabla \times \mathbf{A} + \mathbf{B}_{\text{ext}} \tag{20.10}$$

$$\nabla \times \left(\frac{1}{\mu_{\text{perm}}} \nabla \times \mathbf{A} \right) = \mathbf{J} \tag{20.11}$$

As can be seen, we now end up with a system of equations that should be solved self-consistently, i.e. $\mathbf{J} = \mathbf{J}(V, \mathbf{A})$ since $\mathbf{B} = \nabla \times \mathbf{A}$ and $\mathbf{A} = \mathbf{A}(\mathbf{J})$.

20.2 Discretization of the Lorentz Current Densities

The Equations (20.5, 20.9) show that the total current density consists of two contributions. The usual finite-integration method will lead to the nodal current-balance equations

$$\sum_j \frac{d_{ij}}{h_{ij}} J_{ij}^{\text{standard}} + \sum_j \frac{d_{ij}}{h_{ij}} J_{ij}^{\text{Lorentz}} + s(R-G)_i \Delta v_i = 0 \tag{20.12}$$

where d_{ij} is the dual area and h_{ij} is the length of the link and J_{ij}^{standard} represents the current density expression as obtained without inclusion of the Lorentz force, e.g. for semiconductors it reads

$$J_{ij}^{\text{standard}} = K_{ij} = \frac{\mu_{ij}}{d_{ij}} \left(c_i B(sX) - c_j B(-sX) \right) \tag{20.13}$$

$$B(x) = \frac{x}{\exp(x) - 1}, \quad X = \frac{q}{kT}(V_i - V_j) \tag{20.14}$$

and J_{ij}^{Lorentz} represent the correction due to the Lorentz force, e.g. for semiconductors it is

$$J_{ij}^{\text{Lorentz}} = s\mu \left(\mathbf{K} \times \mathbf{B} \right) \cdot \mathbf{n}_{ij} \tag{20.15}$$

In here, $\mathbf{n}_{ij}$ is the unit vector along the link $\langle ij \rangle$. We will consider the discretization of (20.15). The current balance in each node is achieved by summing all contributions from each mesh element and its associated set of links that are attached to the node under consideration. In particular, a contributing link in some mesh element is an edge of two adjacent faces in the element. Figure 20.2 illustrates the situation for the link $\langle ij \rangle$ and the faces

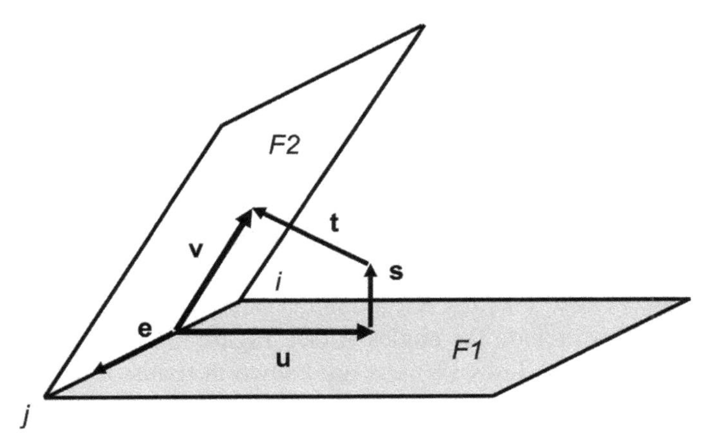

Figure 20.2 Mesh element ingredients that illustrate the decomposition of the Lorentz force vector product. d_{ij} is the (partial) dual area which is enclosed by the vectors $\mathbf{s}$, $\mathbf{t}$, $\mathbf{u}$ and $\mathbf{v}$. The distance between node i and j is $\langle h_{ij} \rangle$.

$F1$ and $F2$. Since we want the contribution of $\mathbf{K} \times \mathbf{B}$ in the direction $\langle ij \rangle$, we merely need the components of $\mathbf{K}$ and $\mathbf{B}$ in the plane of the dual area d_{ij}. To compute $(\mathbf{K} \times \mathbf{B}) \cdot \mathbf{n}_{ij}$ we exploit two local coordinate frames in this plane: one coordinate frame $\{\mathbf{s}, \mathbf{t}\}$ is used to perform a decomposition of $\mathbf{B}$ and another coordinate frame $\{\mathbf{u}, \mathbf{v}\}$, whose base vectors are perpendicular to the first ones, is used to compute $\mathbf{K}$ in the volume segment spanned by the link $\langle ij \rangle$ and its dual area d_{ij} in the element. Using these two frames, we obtain that

$$\begin{aligned} \mathbf{K} \times \mathbf{B} &= (K_u \mathbf{u} + K_v \mathbf{v}) \times (B_s \mathbf{s} + B_t \mathbf{t}) \\ &= K_u B_s (\mathbf{u} \times \mathbf{s}) + K_v B_t (\mathbf{v} \times \mathbf{t}) \\ &\quad + K_u B_t (\mathbf{u} \times \mathbf{t}) + K_v B_s (\mathbf{v} \times \mathbf{s}) \end{aligned} \tag{20.16}$$

As can be seen from Figure 20.2, the first two contribution will dominate the result whereas the last two terms vanish for orthogonal faces. For the last two terms it is observed that their contribution seems maximal for acute angles between the adjacent faces, but then the dual area diminishes which results again into a small correction. Therefore, we will implement the first two terms only which definitely suffices for structured grids and which will capture a large portion of the Lorentz force effects on unstructured grids (Schoenmaker et al. [2007d]). The magnetic fields perpendicular to the primary grid surfaces $F1$ and $F2$, i.e. B_t and B_s, are obtained from the circulation of the vector potential along these surfaces. The self-induced magnetic field is given by

$$B_{\perp} = \frac{1}{\Delta S} \sum_{k}^{N} A_k h_k \tag{20.17}$$

where $A_k \simeq \mathbf{A} \cdot \mathbf{e}_k$ and N is the number of links with lengths h_k around the surface and ΔS is the surface area. For completeness, we mention that the electric field is obtained in the finite-integration method from the voltage differences of the nodes of the discretization grid. Finally, in order to obtain the variables K_u and K_v, a weighted sum is taken from the current-density projections along $\langle ik \rangle$ and $\langle jk \rangle$ for K_u and in the same way for K_v, from $\langle il \rangle$ and $\langle jl \rangle$. The weights include the angles α (see Figure 20.3). Note that the projections of $\mathbf{K}$ along the links $\langle ij \rangle$ etc. are known in terms of the degrees of freedom $\{V_i, \phi_i^p, \phi_i^n, A_i\}$. We use

$$K_u = \frac{1}{2} \left(sin(\alpha_{jik}) K_{ik} + sin(\alpha_{ijk}) K_{jk} \right) \tag{20.18}$$

and a similar expression for K_v. The motivation for this procedure is found in the requirement that the final current density along the link $\langle ij \rangle$ should be anti-symmetric in its indices, i.e. $K_{ji} = -K_{ij}$. This can be achieved by using the same K expressions when assembling the contribution of the link $\langle ij \rangle$ to the nodes i and j. The requirement itself is needed to guarantee current balance.

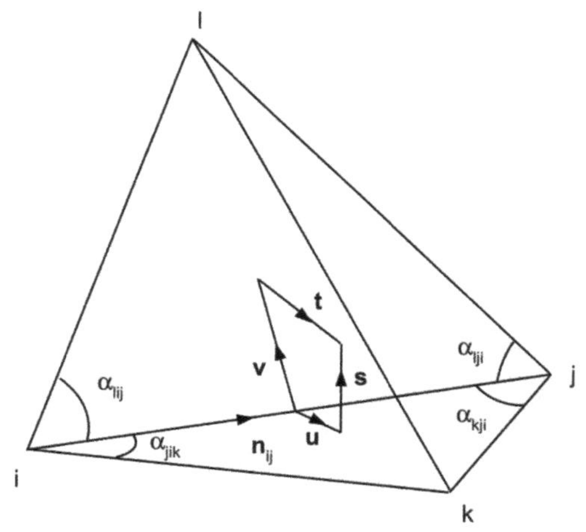

Figure 20.3 Locations of the angles α, that participate in the computation of the components K_u and K_v.

20.3 Static Skin Effects in Conducting Wires

The Hall effect induces a skin effect even for static currents. This can be understood by elementary considerations of the directions of the electric and magnetic fields. The electric field is in the direction of the current flow. The magnetic field circulates around the wire; moreover, it grows linearly inside the metallic region from the axial center of the wire. Therefore, the Lorentz force acts in the radial direction and the corresponding Lorentz current is compensated by a surface charge distribution that counteracts the Lorentz current (see Figure 20.4).

Of course, the net current in the radial direction is vanishing because there are no contacts in this direction to sustain the flow. Qualitatively, one can already conclude that that compensating current is induced by a voltage from the center to the edge that grows quadratically with the radius, r. This is because inside the wire, the magnetic field is proportional to the radius and therefore the Lorentz force grows linearly with r. The counter force is derived from a potential which is then increasing or decreasing quadratically. A detailed derivation for metals yields

$$V_L(r) = V_0 - k\left(\frac{r}{r_{max}}\right)^2, \quad k = \frac{\mu_0 R_H}{4\pi^2}\frac{I_{tot}^2}{r_{max}^2} \tag{20.19}$$

Here μ_0 is the permeability of vacuum, r_{max} the radius of the wire, R_H the Hall coefficient, I_{tot} the current in the wire and V_0 the value of the potential at the center of the wire. This value depends on the position along the wire, i.e.

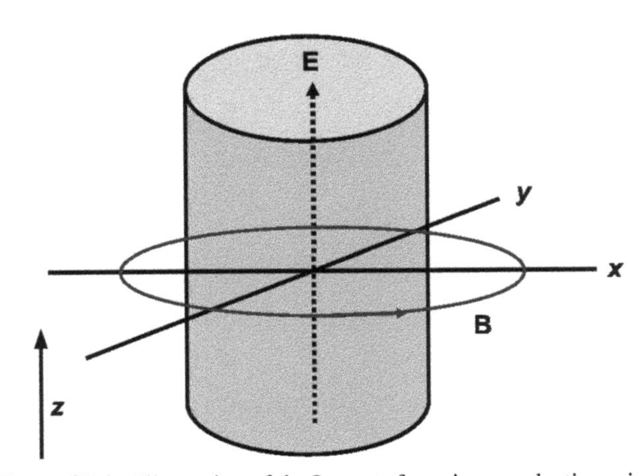

Figure 20.4 Illustration of the Lorentz force in a conducting wire.

the location with respect to the begin and end point of the wire. The maximum voltage difference ΔV is obtained by setting $r = r_{max}$. Since the parameter k depends on the total current or the applied voltage and the resistivity of the wire material, several equivalent ways exist to represent this parameter. For example, for semiconductor material one obtains

$$k = \mu_{holes} \frac{V_{bias}^2}{R_{wire} \cdot L_{wire}} \times 10^{-7} \text{ [V]} \tag{20.20}$$

$$k = -\mu_{elec} \frac{V_{bias}^2}{R_{wire} \cdot L_{wire}} \times 10^{-7} \text{ [V]} \tag{20.21}$$

Note the sign difference between electrons and holes. This reflects the positive or negative value of the Hall coefficient.

20.4 Self-Induced Lorentz Force Effects in Metallic Wires

In order to test the numerical implementation, we consider a wire with radius of 15 micron, a length of 100 micron, and a conductance $\sigma = 10^8$ S/m. The Hall coefficient is -10^{-10} Vm/AWb. The applied bias is 1 Volt. In Figure 20.5, the voltage is shown along an axis at height of 50 microns (in the middle between to top and bottom of the wire). The center of the wire is located at x = 50 micron. The maximum voltage difference between the voltage at the center and the voltage at the edge of the wire (x = 35, 65 micron) is 0.00699 Volt. The analytic result is obtained by inserting the current,

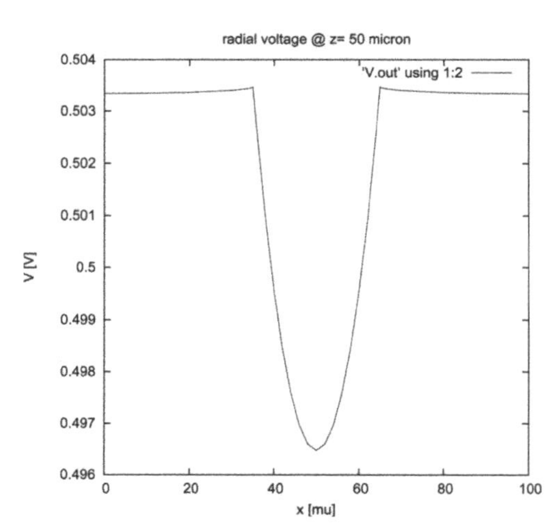

Figure 20.5 Radial voltage at mid cut of the conducting wire.

$I_{tot} = 7.0155 \times 10^2$ A into Equation (20.19) which gives $k = -0.00696$ V. The analytic and numerical results are in excellent agreement. The maximum value of the magnetic field being located at the edge of the wire is 9.2 T.

20.5 Self-Induced Lorentz Force Effects in Silicon Wires

The implementation for silicon wires is based on the use of the Scharfetter-Gummel current densities as presented in Equation (20.13). Moreover, the mobility represents also the Hall coefficient. For a silicon wire of equal dimensions as above we use a p-type doping of 10^{26} m^{-3} and we use a hole mobility $\mu_{hole} = 0.045$ m^2/Vsec. The resistance of the wire 0.198551 Ω. Using (20.20) we find that $k = +2.26 \times 10^{-4}$ Volt, whereas the numerical result is 2.264×10^{-4} Volt. Again excellent agreement is found between the numerical and analytic results. It should be noted that the drop in voltage towards the edge implies also a drop in the hole density since the latter is given by $p = n_i \exp(V - \phi_p)$ and ϕ_p is constant at fixed z. The maximum value of the magnetic field is 0.067 T.

20.6 External Fields

The method is as discussed in Section 20.3, can be straightforwardly extended to include external (constant) magnetic fields. For that purpose the external field is projected on the local frame base vectors $\{\mathbf{s}, \mathbf{t}\}$. As a test case we consider the silicon wire and consider a magnetic field in the y-direction.

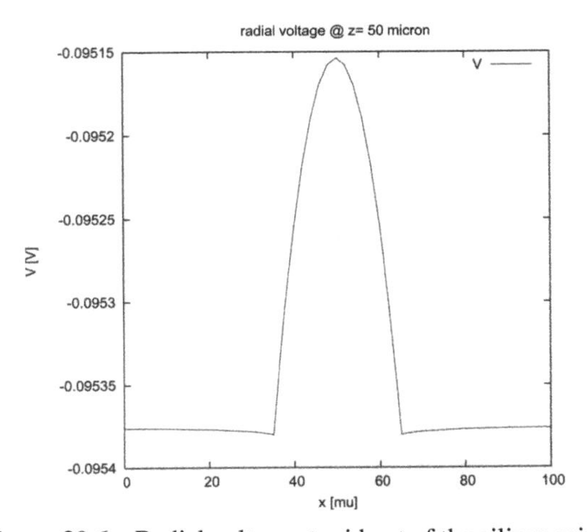

Figure 20.6 Radial voltage at mid cut of the silicon wire.

This field will break the rotational symmetry. At the left side of the central axis the magnetic field will decrease while at the right side, the field gets enhanced. This should be reflected in the compensating potential that is needed to cancel the Lorentz-force induced radial current. In Figure 20.7, the voltage is shown in the xy-plane at z = 50 micron. A cut along the x-axis is shown in Figure 20.6. As is seen in Figures 20.7 and 20.8, the radial symmetry is indeed distorted by the external field.

Figure 20.7 Radial voltage at mid cut of the silicon wire with an external field in the y-direction (arbitrary units).

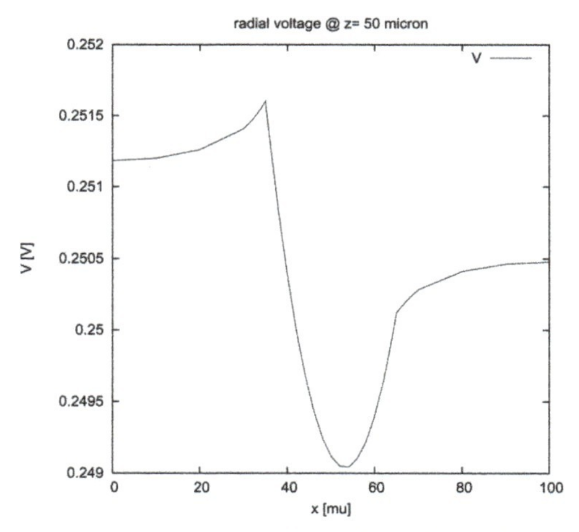

Figure 20.8 Radial voltage at mid cut of the silicon wire with an external field in the y-direction. The bias voltage is 0.5 V and the external field is taken 1 T.

Summary

We presented an implementation of the Lorentz force effects in a simulation program which incorporates both external field effects as well as self-induced fields for metallic materials as well as semiconducting materials. The implementation is rooted in the geometrical discretization of the vector potential which exclusively represents the vector potential by scalar variables assigned to the links of the computational grid. Although the method made use of some simplifying assumptions which depend on the use conditions, e.g. $|\mu B| << 1$ and on the obtuseness of the grid, these approximations can be easily elevated for completeness but their inclusion will lead to an enhanced filling of the Newton-Raphson matrices and larger CPU cost. The self-induced fields are sufficiently weak such that above approximations are sufficiently accurate to justify their use. Due to the non-linear nature and the self consistency of the self-induced Lorentz force, analytic results are scarce. However, for a simple conducting wire analytic results can be obtained and they serve as an excellent benchmark to test the correctness of the implementation as is illustrated by the comparison of numerical and analytic results. It should be emphasized that the implementation refers solely to local mesh element variables and therefore the corresponding code can be used to analyze the effects of the Lorentz force in arbitrary device layouts in three dimensions. Furthermore, the implementation is *not* a 'post-processing' scheme: the solver arrives at a solution by finding the nodal current balances of (20.12) in a fully self-consistent manner. Therefore, the sum of the currents of all contacts is equal zero within the numerical noise floor $\sim O(10^{-13})$. Finally we note that if $|\mu B| \geq 1$ then we must deal with all terms in (20.4) faithfully. In particular this implies that we must construct from the mesh ingredients a **B**-field component along each link. The available components are anchored in the plaquettes of the mesh. As a consequence the required component needs to be composed as a weighted sum of the **B**-field components of the plaquettes with the following selection rules: (1) collect all mesh cells that contain the link under consideration, (2) collect in each mesh cell the plaquettes that touch the link in just one node. The weights must lead to the anti-symmetry with the respect to the begin- and end nodes of the link.

21

Self-Induced Magnetic Field Effects, the Lorentz Force and Fast-Transient Phenomena

In this chapter We present a full physical simulation picture of the electromagnetic phenomena combining electromagnetic fields (EM) and carrier transport in semiconductor devices (TCAD) in the transient regime[17]. The simulation tool computes the electromagnetic fields in a self-consistent way and the resulting magnetic fields are incorporated in the computation of the current sources that get modified by the Lorentz force (LF).

The singled-out purpose of electrostatic discharge (ESD) devices is that these devices should protect electronic circuitry against fast-transient voltage/current spikes. Although the overall signal variation occurs within a nanosecond, the corresponding currents can ramp up to a multi-Ampere level. Fast varying current patterns give rise to equally fast varying induced magnetic fields being proportional to the time rate of change of the current. As a consequence a substantial part of the electric fields can be attributed to the variation of the induced magnetic fields and the full electromagnetic picture is required for understanding the effects of fast-transient input signals. A similar reasoning can be done for high-power switch devices. Continuing the reasoning along these lines: when the fields are varying sufficiently fast, both the induced magnetic fields and electric fields will ultimately have such large components such that the current flow is controlled by both fields. In particular, in semiconductors, the Hall coefficient is directly related to the carrier mobility and therefore *self-induced* Lorentz force modifications could become appreciable. So far, no simulation tool has been designed to address these concerns and strictly speaking, without actually computing these effects, we have no clue if it is justified to ignore these subtleties

[17]This chapter is a reprint of Schoenmaker et al. [2014].

all together or that these effects are really a concern. The purpose of the present work is to fill this gap. We present an implementation that allows the computation of these self-induced electromagnetic field effects for fast transient phenomena. In particular, the key ingredients are: (1) the semiconductor device equations (2) modifications there off to account for the Lorentz force (3) the Maxwell equations to compute the electromagnetic fields. All this is done in the time domain since in the frequency domain the "small-signal" analysis up front excludes high current/voltage signals at the ports. In the remainder of the paper, we refer to the Maxwell equations with the acronym EM (electromagnetism) and to the semiconductor equations with the acronym DD (drift-diffusion model). Finally, the modifications that are induced by the Lorentz force are referred to with the acronym LF. The EM-TCAD simulation refers to a concurrent solution of the Maxwell's equations that describe the EM dynamics and the transport models that describe the charge carrier dynamics in semiconductor devices. So far only a few frameworks have been developed to allow such coupling, see e.g., (Wang et al. [2002b] Bertazzi et al. [2006b] Willis et al. [2011]), based on different selections of EM and TCAD solvers. One experimentally verified co-simulation framework was firstly proposed in (Meuris et al. [2001b] Schoenmaker and Meuris [2002a]) in the frequency domain. The Maxwell's equations are formulated in terms of scalar potential V and vector potential $\mathbf{A}$ to obtain straightforward coupling with the drift and diffusion (DD) semiconductor model (called the A-V formulation henceforth). The spatial discretization scheme based on the Finite-Volume Method (FVM) and Finite-Surface Method (FSM) is fully compliant with the geometric meaning of the differential operators. The motivation for selecting the FVM is found in the desire to connect the field solver results directly to the variables that are commonly used in the electronic design environment, i.e. currents and voltages. In other words: the voltages (V) at the contacts are in one-to-one correspondence with node voltages at a SPICE net list and the currents entering/leaving the circuit fragment simulated by a field solver are in one-to-one correspondence with the end point(s) of some net list branch(es). A convenient link is established between the field solver and Kirchhoff's voltage and current laws. Successful simulations of semiconductor devices always have exploited charge conservation as an essential ingredient of the modeling that could be achieved using the FVM. It is noted that the FVM is mimetic (Lipnikov et al. [2011]) and relaxes the demand that for finite grid sizes, every variable should be calculable at every space-time point. In other words: only in the limit of zero grid-node distance ($\lim h \rightarrow 0$) such detailed information may be extracted. For example, the finite-integration method

(Weiland [1996]), which belongs to the class of FVM, puts the electric fields on the links of the grid and the magnetic field on the surfaces of the grid. No statement is made about these fields for space points that are neither on the links or surfaces. The benefit is that the geometrical character of the field variables are faithfully implemented and the constraints are obeyed without additional effort. Finally, the conservation of charge is built-in from start.

Besides these practical advantages the FVM can be justified from Noether's theorem (Noether [1918]) which states that to every conserved quantity there corresponds an invariance principle. Strictly speaking the relation is reverse, i.e. to every invariance principle there corresponds a conserved quantity. Thus in order to respect charge conservation we must build in the underlying invariance principle. For the electromagnetic fields the invariance is realized for the scalar potential and the vector potential which may be transformed without altering the physical content. Accepting voltages (V) as degrees of freedom in the modeling set up requires that for the inclusion of magnetic effects, the vector potential ($\mathbf{A}$) must be also part of the set of degrees of freedom.

A time-domain A-V simulator has been proposed in (Schoenmaker et al. [2012] Chen et al. [2013]) to meet the desire to handle large-signal response and to complement the small-signal analysis applied in its frequency-domain counterpart. The implicit Backward Euler (BE) approach was employed for time discretization and the Newton's method was used to solve the nonlinear system that arises from the semiconductor dynamics. Whereas (Chen et al. [2013]) deals with field transformations to capture ultra-fast time variations, this work does not exploit these transformations but aims at the inclusion of the Lorentz force. Due to the extended physical reality and its numerical treatment, five variables are used and solved simultaneously in the formulation. Three variables are located on nodes and two variables are defined on links. Therefore, the number of unknowns in each Newton iteration can be $\sim$ 9 times the number of nodes in the computational grid for large-scale problems. The discretization of the Lorentz force impact on currents was addressed in (Schoenmaker et al. [2010a, 2011]). The emphasis of this work was on finding a proper discretization of higher-order field expressions such as $\mathbf{E} \times \mathbf{B}$. Again, current balance is a key guideline in the discretization procedure and the mimetic methods that respect the geometrical meaning of the various observables are capable of arriving at very accurate results. In particular, comparing the outcome of the numerical approach with analytic results (for the rare case where such results can be found) demonstrates that the proposed discretization method is very reliable. Although it was anticipated in (Schoenmaker et al. [2010a, 2011]) that the self-induced Lorentz force will become only important, if it will be at all, in fast-transient regimes, the work in (Schoenmaker et al. [2010a, 2011]) is limited to the static regime. In this paper we will report about the full transient implementation of the proposed methods. This brings us to the following organization of this paper. Section 21.1 reviews the A-V formulation of EM-TCAD problem and the conventional solution scheme. Section 21.2 describes the extension to include the Lorentz force. Then we address the discretization of the Lorentz force as well as the numerical subtleties that are encountered when solving the full EM-TCAD-LF problem.

21.1 Time-Domain Formulation of EM-TCAD Problem

The starting point of the time-domain EM-TCAD formulation is the full-wave Maxwell's equations

$$\nabla \cdot \mathbf{D} = \rho, \quad \nabla \cdot \mathbf{B} = 0 \tag{21.1a}$$

$$\nabla \times \mathbf{E} = -\partial_t \mathbf{B}, \quad \nabla \times \mathbf{H} = \mathbf{J} + \partial_t \mathbf{D} \tag{21.1b}$$

$$\mathbf{D} = \epsilon \mathbf{E}, \quad \mathbf{B} = \mu \mathbf{H} \tag{21.1c}$$

where $\mathbf{D}, \mathbf{E}, \mathbf{B}, \mathbf{H}, \mathbf{J}$ and ρ are the displacement field, electric field, magnetic induction, magnetic field, free current density and charge density, respectively.

In the semiconductor region the Maxwell equations are complemented with the current continuity of electrons and holes

$$\frac{1}{q} \nabla \cdot \mathbf{J}_{\mathrm{n}} - \partial_t n - R(n, p) = 0 \tag{21.2a}$$

$$\frac{1}{q} \nabla \cdot \mathbf{J}_{\mathrm{p}} + \partial_t p + R(n, p) = 0 \tag{21.2b}$$

where n and p are the electron and hole densities, and $R(n, p)$ denotes the net generation/recombination rate of carriers and q is the elementary charge. The particle current densities in semiconductor are described by the DD model

$$\mathbf{J}_{\mathrm{n}} = q \mu_{\mathrm{n}} n \mathbf{E} + kT \mu_{\mathrm{n}} \nabla n \tag{21.3a}$$

$$\mathbf{J}_{\mathrm{p}} = q \mu_{\mathrm{p}} p \mathbf{E} - kT \mu_{\mathrm{p}} \nabla p \tag{21.3b}$$

where ϵ is the permittivity, μ is the magnetic permeability, μ_{p}, μ_{n}, k and T denote the carrier mobilities, Boltzmann constant and temperature, respectively.

To facilitate the coupling between the EM and TCAD solvers, the Maxwell equations are written in potential form using the variables V and $\mathbf{A}$ with the result that $\mathbf{B} = \nabla \times \mathbf{A}$ and $\mathbf{E} = -\nabla V - \partial_t \mathbf{A}$ (Meuris et al. [2001b] Schoenmaker and Meuris [2002a]). A new variable, the pseudo-canonical momentum

$$\mathbf{\Pi} = \partial_t \mathbf{A} \tag{21.4}$$

is also introduced to avoid the second-order time derivative (Schoenmaker et al. [2012]). The complete system of equations is then laid out in (21.5) utilizing a generalized "de Mari" scaling scheme (Schoenmaker and Meuris [2002a] Chen et al. [2013]).

$$\left\{ \begin{array}{c} \frac{1}{\nu}\nabla \cdot [\varepsilon_r (-\nabla V - \mathbf{\Pi})] - \rho = 0, \quad \rho = p - n + N_D \\ \frac{1}{\nu}\nabla \cdot [\varepsilon_r (-\nabla \partial_t V - \partial_t \mathbf{\Pi})] + \left\{ \begin{array}{ll} \nabla \cdot [\sigma (-\nabla V - \mathbf{\Pi})] & \text{metal} \\ \nabla \cdot \mathbf{J}_{\mathrm{sd}} & \text{semi/metal} \end{array} \right\} \\ = 0 \end{array} \right. \tag{21.5a}$$

$$\nabla \cdot \mathbf{J}_{\mathrm{n}} - \partial_t n - R(n,p) = 0 \tag{21.5b}$$

$$\nabla \cdot \mathbf{J}_{\mathrm{p}} + \partial_t p + R(n,p) = 0 \tag{21.5c}$$

$$\partial_t \mathbf{A} - \mathbf{\Pi} = 0 \tag{21.5d}$$

$$\begin{array}{l} -K\varepsilon_r (-\partial_t \mathbf{\Pi} - \nabla \partial_t V) - K\nabla (\varepsilon_r \partial_t V) + [\nabla \times (\nabla \times \mathbf{A}) - \nabla (\nabla \cdot \mathbf{A})] \\ -K\nu \left\{ \begin{array}{ll} \mathbf{J}_{\mathrm{c}} & \text{metal} \\ \mathbf{J}_{\mathrm{sd}} & \text{semi} \end{array} \right\} = 0 \end{array} \tag{21.5e}$$

where $\mathbf{J}_{\mathrm{c}}$ is the conduction current in metal and $\mathbf{J}_{\mathrm{sd}} = \mathbf{J}_{\mathrm{n}} + \mathbf{J}_{\mathrm{p}}$ is the total semiconductor current (see section 21.2). K and ν are two dimensionless constants of the scaling method (Chen et al. [2013]). In the scaling scenario where λ_{L} and λ_{T} are the natural units of length and time we find that $K = \mu_0 \epsilon_0 \lambda_{\mathrm{L}}^2 / \lambda_{\mathrm{T}}^2$. For example: for a small-size microelectronic device with a typical length of about one micron and a switching frequency around five GHz, we may set $\lambda_{\mathrm{L}} = 10^{-6}$ meter and $\lambda_{\mathrm{T}} = 10^{-10}$ second. The dimensionless parameter ν corresponds to the characteristic relaxation time of conductive materials. In particular, if λ_{COND} is the 'natural unit' of conductance (for example $\lambda_{\mathrm{COND}} = 10^7$ S/m for Aluminum) we find that $\nu = \lambda_{\mathrm{COND}} \times \lambda_{\mathrm{T}} / \epsilon_0$. The meaning of the parameter ν becomes clear when combining Gauss' law with the current continuity and eliminating the electric field, i.e. $\frac{\sigma}{\epsilon}\rho + \partial_t \rho = 0$ or $\rho \sim \exp\left(-\frac{\sigma t}{\epsilon}\right)$. We also note that 'natural units' can be chosen within a range of applicability. In other words: an equally valid choice is $\lambda_{\mathrm{L}} = 10^{-5}$ meter or $\lambda_{\mathrm{L}} = 10^{-7}$ meter in above discussion. An 'unnatural' choice would be to describe microelectronic devices using 1 km as a basis for expressing lengths. The scaling of the system can be further improved by using quasi-Fermi potentials, ϕ^p and ϕ^n. The relation is

$$n = n_i \exp\left(\frac{q}{kT}(V - \phi^{\mathrm{n}})\right), \quad p = n_i \exp\left(\frac{q}{kT}(\phi^{\mathrm{p}} - V)\right) \qquad (21.6)$$

and by scaling the semi-conductor current continuity equations with the nodal carrier densities. The upper equation in (21.5a) is the common Gauss law, and the lower one is the current continuity equation that is applied to nodes for which there is attached at least one grid cell consisting of a conductor (metal). For these nodes Gauss law is used to recover the charge densities which are not explicit unknowns. Note that a node may be located at a metal-insulator or metal/semiconductor or a metal/semiconductor/insulatior (triple point) interface. As a consequence the discretization or equation assembling receives contribution from both metallic and semi-conductor cells. Therefore both current types are found in (21.5a). The scaled current-continuity equations are given in (21.5b) and (21.5c) whereas (21.5d), equal in appearance as (21.4) is the scaled version of the pseudo-canonical momentum definition. The Maxwell-Ampere equation is given in (21.5e). It includes the subtraction of the divergence of Lorentz gauge condition

$$\nabla \cdot \mathbf{A} + K\varepsilon_r \partial_t V = 0 \qquad (21.7)$$

to eliminate the intrinsic singularity of the curl-curl operator (Chen et al. [2011b]) and obtaining a Laplacian-like operator $\nabla^2 = \nabla(\nabla\cdot) - \nabla \times (\nabla\times)$. The numerical subtleties in the various transient regimes have been addressed in (citeChen:13). In particular, the fact that the $\mathbf{A}V$-formulation ultimately breaks down if the transient times get too small (10^{-14} seconds). However, it is assumed that for such short time signals the drift-diffusion model becomes questionable (citeLaux:99) and the relevance the full EM-TCAD set up is at stake. Therefore, we limit the fast-transient signals in a range of $O(10^{-14}) - O(10^{-9})$ seconds.

21.2 Inclusion of the Lorentz Force

In this section we will present the equations that will serve as a starting point for the implementation of the Hall effect and the self-induced Lorentz forces. First of all, we emphasize that the Lorentz force, although having a magnetic origin is not related to skin effects and proximity effects. The latter correspond to altering the current flow due to the presence of time varying magnetic fields thereby altering the electric field. The currents are related to the electric field according to Ohm's law in metallic domains and according to the drift-diffusion current-continuity equations in semiconducting domains. However, the Lorentz force impacts Ohm's law and the drift-diffusion model itself. The starting point of the discussion of the Lorentz force is the well-known Lorentz force law:

$$\mathbf{F}(t) = q\mathbf{E}(t) + q\mathbf{v}(t) \times \mathbf{B}(t) \tag{21.8}$$

This force can be inserted into the Boltzmann transport equation that requires the description of external forces applied to the point particles moving in configuration space and by adapting the moment expansion, the upgraded equations of Ohm's law and the drift-diffusion model are obtained. In metallic regimes one finds

$$\mathbf{J}(t) = \sigma\mathbf{E}(t) + \mu_{\mathrm{H}}\mathbf{J}(t) \times \mathbf{B}(t) \tag{21.9}$$

where μ_{H} is the Hall coefficient[18] In order to keep track of the transient modifications, we explicitly denote the time dependence. In semiconducting regions the current densities for holes and electrons in the drift-diffusion model *with* inclusion of the Lorentz force are

$$\begin{aligned} \mathbf{J}_{\mathrm{n}}(t) &= q\mu_{\mathrm{n}}n(t)\left(\mathbf{E}(t) + \mathbf{v}_{\mathrm{n}}(t) \times \mathbf{B}(t)\right) + kT\mu_{\mathrm{n}}\nabla n(t) \\ \mathbf{J}_{\mathrm{p}}(t) &= q\mu_{\mathrm{p}}p(t)\left(\mathbf{E}(t) + \mathbf{v}_{\mathrm{p}}(t) \times \mathbf{B}(t)\right) - kT\mu_{\mathrm{p}}\nabla p(t) \end{aligned} \tag{21.10}$$

where $\mathbf{J}_{\mathrm{n}}(t) = -qn(t)\mathbf{v}_{\mathrm{n}}(t)$ and $\mathbf{J}_{\mathrm{p}}(t) = qp(t)\mathbf{v}_{\mathrm{p}}(t)$ are implicitly defining the mean carrier velocities. The latter can be eliminated in favor of the current densities and both equations can be expressed as (Rudan et al. [2006])

$$\mathbf{J}_{\mathrm{c}}(t) + s\mu_c\mathbf{B}(t) \times \mathbf{J}_{\mathrm{c}}(t) = \mathbf{K}_{\mathrm{c}}^{\mathrm{DD}}(t) \tag{21.11}$$

$$\mathbf{K}_{\mathrm{c}}^{\mathrm{DD}}(t) = q\mu_c c(t)\mathbf{E}(\mathbf{t}) - skT\mu_c\nabla c(t) \tag{21.12}$$

In here, $c(t) = p(t)$ and $s = +1$ (holes) or $c(t) = n(t)$ and $s = -1$ (electrons), $\mathbf{J}_{\mathrm{c}}$ is $\mathbf{J}_{\mathrm{p}}$ or $\mathbf{J}_{\mathrm{n}}$, $\mathbf{K}_{\mathrm{c}}$ is $\mathbf{K}_{\mathrm{p}}$ or $\mathbf{K}_{\mathrm{n}}$ and μ_{c} is μ_{p} or μ_{n}. The latter are taken here as scalars. Were it not for the second term at the left-hand side,

[18]The symbol μ is used ubiquitously in the present paper, but its meaning is clear from the context.

Equation (21.11) would be the usual drift-diffusion expression for the current density. It is easily possible to perform a full inversion of the current density in terms of the drift-diffusion current density $\mathbf{K}_c^{DD}(t)$:

$$\mathbf{J}_c(t) = \frac{1}{1+\mu^2 B^2}\left(\mathbf{K}_c^{DD}(t) + s\mu\mathbf{K}_c^{DD}(t)\times\mathbf{B}(t)\right.$$
$$\left. + \mu^2\left(\mathbf{K}_c^{DD}(t)\cdot\mathbf{B}(t)\right)\mathbf{B}(t)\right) \quad (21.13)$$

The parameter $|\mu B| << 1$ allows to write

$$\mathbf{J}_c(t) = \mathbf{K}_c^{DD}(t) + s\mu_c\mathbf{K}_c^{DD}(t)\times\mathbf{B}(t) \quad (21.14)$$

The current-continuity equations and the Poisson equations complete the description of the drift-diffusion model, i.e.

$$\nabla\cdot\mathbf{J}_c(t) + sR(t) + s\partial_t c(t) = 0 \quad (21.15)$$
$$\nabla\left(-\epsilon\nabla V(t)\right) = \rho\left(V(t),\phi_p(t),\phi_n(t)\right) \quad (21.16)$$

Inside metallic regions, the use of the approximation $|\mu_H B| << 1$ leads to the following expression for the current density

$$\mathbf{J}(t) = \sigma\mathbf{E}(t) + \mu_H\sigma\mathbf{E}(t)\times\mathbf{B}(t) \quad (21.17)$$

The current density satisfies the current continuity equation $\nabla\cdot\mathbf{J} + \partial_t\rho = 0$. The magnetic field consists of two terms of which one is representing the self-induced field and the other the external field $\mathbf{B}_{ext}$

$$\mathbf{B}(t) = \nabla\times\mathbf{A}(t) + \mathbf{B}_{ext}(t) \quad (21.18)$$

As can be seen from (21.5e), we now end up with a system of equations that should be solved self-consistently, i.e. $\mathbf{J}(t) = \mathbf{J}(V(t),\mathbf{\Pi}(t),\mathbf{A}(t))$ since $\mathbf{A}(t) = \mathbf{A}(\mathbf{\Pi}(t))$ and $\mathbf{\Pi}(t) = \mathbf{\Pi}(V(t),\mathbf{A}(t),\mathbf{J}(\mathbf{t}))$. The self-consistency requirement sustains into the static regime as was already observed in (Schoenmaker et al. [2011]).

21.3 Discretization of the Lorentz Current Densities

In order to prepare the discretization process, we slightly modify Equation (21.17) and cast it in the following form:

$$\mathbf{J}(t) = \mathbf{J}^{EM}(t) + \mu_H\mathbf{J}^{EM}\times\mathbf{B}(t) \quad (21.19)$$
$$= \mathbf{J}^{EM}(t) + \mathbf{J}^{LF}(t) \quad (21.20)$$

with

$$\mathbf{J}^{\mathrm{EM}}(t) = \sigma \mathbf{E}(t) \tag{21.21}$$
$$\mathbf{J}^{\mathrm{LF}}(t) = \mu_{\mathrm{H}} \mathbf{J}^{\mathrm{EM}}(t) \times \mathbf{B}(t)$$

The Equations (21.14, 21.17) and (21.19) show that the total current density consists of two contributions. The usual finite-integration method will lead to the nodal current-balance equations

$$\sum_j d_{ij} J_{ij}^{\mathrm{EM}}(t) + \sum_j d_{ij} J_{ij}^{\mathrm{LF}}(t) + sR_i(t)\Delta v_i + s\partial_t c_i(t)\Delta v_i = 0 \tag{21.22}$$

where d_{ij} is the dual area and Δv_i is nodal volume of node i and J_{ij}^{EM} represents the current density expression as obtained without inclusion of the Lorentz force, e.g. for semiconductors it reads[19]

$$J_{\mathrm{c},ij}^{\mathrm{EM}} = K_{\mathrm{c},ij}^{\mathrm{DD}} = s\frac{\mu_{\mathrm{c},ij} T_{ij}}{h_{ij}} \left(c_i B(sX_{ij}) - c_j B(-sX_{ij})\right) \tag{21.23}$$

$$B(x) = \frac{x}{\exp(x) - 1}, \quad X_{ij} = \frac{q}{kT_{ij}}(V_i - V_j + \mathrm{sgn}_{ij}\Pi_{ij}) \tag{21.24}$$

$$T_{ij} = (T_i + T_j)/2\,, \qquad \Pi_{ij} = \mathbf{e} \cdot \mathbf{\Pi} \tag{21.25}$$

In here, $\mathbf{e}$ is an *intrinsic* unit vector along the direction of the grid link and $\mathrm{sgn}_{ij} = +1$ if $\mathbf{e}$ points from node i to node neighbor j. If $\mathbf{e}$ points from node j to node neighbor i, we have $\mathrm{sgn}_{ij} = -1$. The variable h_{ij} is the length of the link. Furthermore, J_{ij}^{LF} represent the correction due to the Lorentz force, e.g. for semiconductors it is

$$J_{\mathrm{c},ij}^{\mathrm{LF}}(t) = s\mu_{\mathrm{c}} \left(\mathbf{K}_{\mathrm{c}}^{\mathrm{DD}}(t) \times \mathbf{B}(t) \right) \cdot \mathbf{n}_{ij} \tag{21.26}$$

The vector $\mathbf{n}_{ij}$ is the unit vector along the link $\langle ij \rangle$ pointing from node i to node j and is parallel or anti-parallel to $\mathbf{e}$. We will consider the discretization of (21.26). The current balance in each node is achieved by summing all contributions from each mesh element and its associated set of links that are attached to the node under consideration. In particular, a contributing link in some mesh element is a boundary segment of two adjacent faces in the element. The situation is illustrated in Figure 21.1 for the link $\langle ij \rangle$ and the faces $F1 = \langle ijk \rangle$ and $F2 = \langle ijl \rangle$. Since we need the contribution of $\mathbf{K} \times \mathbf{B}$

[19] We drop the explicit time dependence in the notation but it is tacitly assumed.

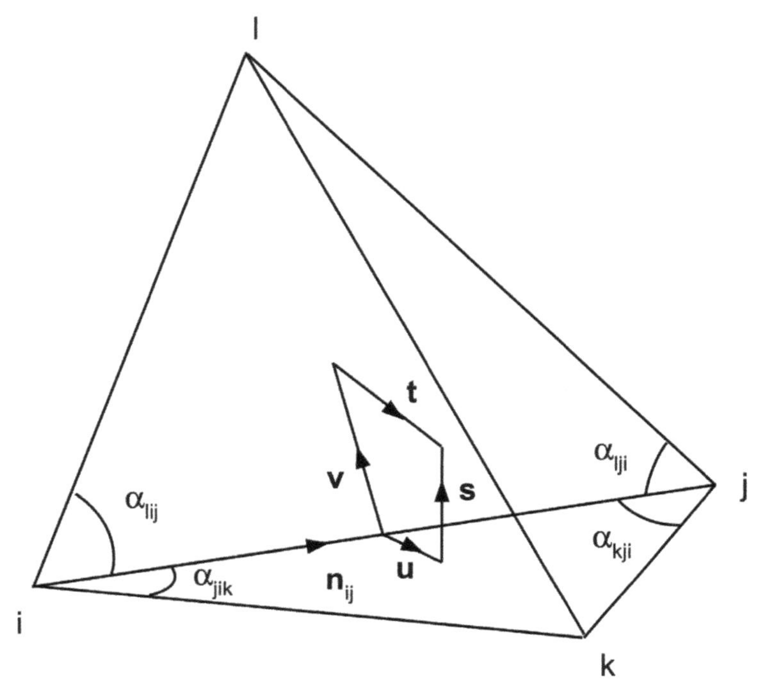

Figure 21.1 Mesh element illustrating the ingredients of the decomposition of the Lorentz force vector product. Part of the dual area, d_{ij} (see text) of the link $\langle ij\rangle$ is enclosed by the vectors $\mathbf{u}$, $\mathbf{v}$, $\mathbf{s}$, $\mathbf{t}$. The distance between node i and j is h_{ij}.

in the direction $\langle ij\rangle$, we merely need the components of $\mathbf{K}$ and $\mathbf{B}$ in the plane of the dual area d_{ij}. The dual area of a link is defined as the surface element of the dual grid and is perpendicular to the link. All mesh cells of the primary grid that have the link as a boundary segment contribute to the dual area of the link. To compute $(\mathbf{K} \times \mathbf{B}) \cdot \mathbf{n}_{ij}$ we exploit two local coordinate frames in this plane: one coordinate frame $\{\mathbf{s}, \mathbf{t}\}$ is used to perform a decomposition of $\mathbf{B}$ and another coordinate frame $\{\mathbf{u}, \mathbf{v}\}$, whose base vectors are perpendicular to the first ones, is used to compute $\mathbf{K}$ in the volume segment spanned by the link $\langle ij\rangle$ and its dual area d_{ij} in the element. Using these two frames, we obtain that (see Figure 21.2)

$$\begin{aligned}\mathbf{K} \times \mathbf{B} &= (K_u\mathbf{u} + K_v\mathbf{v}) \times (B_s\mathbf{s} + B_t\mathbf{t}) \\ &= K_uB_s(\mathbf{u} \times \mathbf{s}) + K_vB_t(\mathbf{v} \times \mathbf{t}) \\ &\quad + K_uB_t(\mathbf{u} \times \mathbf{t}) + K_vB_s(\mathbf{v} \times \mathbf{s})\end{aligned} \tag{21.27}$$

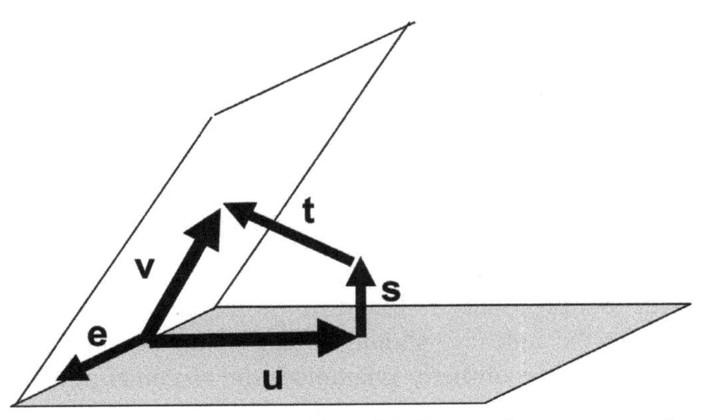

Figure 21.2 Illustration of the decomposition of the Lorentz force vector product in the local coordinate bases.

As can be seen from Figure 21.2, the first two contribution will dominate the result whereas the last two terms vanish for orthogonal faces. For the last two terms it is observed that their contribution seems maximal for acute angles between the adjacent faces, but then the dual area diminishes which results again into a small correction. Specifically, for the meshes that are used in this work, the last two terms do not contribute. Therefore, we will implement the first two terms only which definitely suffices for structured grids and which will capture a large portion of the Lorentz force effects on unstructured grids Schoenmaker et al. [2007c]. The magnetic fields perpendicular to the primary grid surfaces $F1$ and $F2$, i.e. B_t and B_s, are obtained from the circulation of the vector potential along these surfaces. The self-induced magnetic field is given by

$$B_{\perp} = \frac{1}{\Delta S} \sum_{k}^{N} A_k h_k \qquad (21.28)$$

where $A_k \simeq \mathbf{A} \cdot \mathbf{e}_k$ and N is the number of links with lengths h_k around the surface and ΔS is the surface area. For completeness, we mention that the electric field is obtained in the finite-integration method from the voltage differences of the nodes of the discretization grid and the link variable representing the first-order time derivative of the vector potential as is seen in Equation (21.24).

Finally, in order to obtain the variables K_u and K_v, a weighted sum is taken from the current-density projection along $\langle ik \rangle$ and $\langle jk \rangle$ for K_u and in the same way for K_v, from $\langle il \rangle$ and $\langle jl \rangle$. The weights include the angles α

and the lengths of the sides of the surface.The details of the K-calculation are found in (21.29).

$$
\begin{aligned}
K_u &= \frac{h_{ik}}{h_{ik}+h_{jk}}\sin(\alpha_{jik})K_{ik} + \frac{h_{jk}}{h_{ik}+h_{jk}}\sin(\alpha_{ijk})K_{jk} \\
K_v &= \frac{h_{il}}{h_{il}+h_{jl}}\sin(\alpha_{jil})K_{il} + \frac{h_{jl}}{h_{il}+h_{jl}}\sin(\alpha_{ijl})K_{jl}
\end{aligned} \tag{21.29}
$$

The motivation for this procedure is found in the requirement that the final current density along the link $\langle ij\rangle$ should be anti-symmetric in its indices, i.e. $K_{ji} = -K_{ij}$. This anti-symmetry guarantees the aforementioned charge conservation. The Lorentz current is a vector and the magnetic induction is a pseudo-vector. Therefore, **K** needs to be a vector. It can be achieved by using the same K expressions when assembling the contribution to the link $\langle ij\rangle$ from the node i to the node j as when assembling the contribution to the link $\langle ji\rangle$ from the node j to the node i. The requirement itself is needed to guarantee current balance. Let us now return to (21.26). Referring to Equation (21.18), we observe that

$$
\mathbf{K}(t) = \mathbf{K}(V(t), \phi^p(t), \phi^n(t), \mathbf{A}(t), \mathbf{\Pi}(t), \mathbf{B}_{\text{ext}}(t)) \tag{21.30}
$$

and find that the Lorentz contribution to the total current depends only on the instantaneous time for which the current is evaluated. In other words: the Lorentz force does not generate terms containing explicit time derivatives. Remember that the time derivative of **A** is given by **Π**.

To compute the time derivatives of the degrees of freedom, we apply a BDF rule. In general the BDF discretization performs the following substitution:

$$
\frac{\partial f}{\partial t} = \frac{1}{h_0}\sum_{i=0}\alpha_{-i} f(t_{-i}) \tag{21.31}
$$

The parameters α depend on the time step sizes and t_0 corresponds to the latest time instance. Our implementation is limited to the second order BDF. In this case we have

$$
\begin{aligned}
\alpha_{-1} &= -\frac{t_0 - t_{-2}}{t_{-1} - t_{-2}} \\
\alpha_{-2} &= -\frac{t_0 - t_{-1}}{t_0 - t_2} \times \frac{t_0 - t_{-1}}{t_{-2} - t{-1}} \\
\alpha_{-0} &= -\alpha_{-1} - \alpha_{-2}
\end{aligned} \tag{21.32}
$$

For equal time step sizes the values are $\alpha_{-1} = -2$ respectively $\alpha_{-2} = \frac{1}{2}$ and $\alpha_0 = \frac{3}{2}$,

We have now available all ingredients to build the discretized equations. Using the implicit BDF scheme, the solution at the latest time instance is found by applying a Newton-Raphson solver to the full set of equations. The degrees of freedom are the temporal nodal variables $V_i(t)$, $\phi_i^{\mathrm{p}}(t)$, $\phi_i^{\mathrm{n}}(t)$ and the link variables $A_i(t)$ and $\Pi_i(t)$. If we denote the Poisson equation or the current-continuity equation in metallic domains, i.e. (21.5a) as $\mathcal{P}(V, \phi^{\mathrm{p}}, \phi^{\mathrm{n}}, \Pi, A) = 0$, the hole current-continuity (21.5c) as $\mathcal{J}^{\mathrm{p}}(V, \phi^{\mathrm{p}}, \phi^{\mathrm{n}}, \Pi, A) = 0$, the electron current continuity (21.5b) as $\mathcal{J}^{\mathrm{n}}(V, \phi^{\mathrm{p}}, \phi^{\mathrm{n}}, \Pi, A) = 0$, the Maxwell-Ampere Equation (21.5e) as $\mathcal{M}(V, \phi^{\mathrm{p}}, \phi^{\mathrm{n}}, \Pi, A) = 0$ and the Π defining relation (21.5d) as $\mathcal{Q}(A) = 0$, then the Newton-Raphson updates ΔV, $\Delta\phi^{\mathrm{p}}$, $\Delta\phi^{\mathrm{n}}$, $\Delta\Pi$, ΔA are computed from the following linear system:

$$\begin{bmatrix} \frac{\partial \mathcal{P}}{\partial V} & \frac{\partial \mathcal{P}}{\partial \phi^{\mathrm{p}}} & \frac{\partial \mathcal{P}}{\partial \phi^{\mathrm{n}}} & \frac{\partial \mathcal{P}}{\partial \Pi} & \frac{\partial \mathcal{P}}{\partial A} \\ \frac{\partial \mathcal{J}^{\mathrm{p}}}{\partial V} & \frac{\partial \mathcal{J}^{\mathrm{p}}}{\partial \phi^{\mathrm{p}}} & \frac{\partial \mathcal{J}^{\mathrm{p}}}{\partial \phi^{\mathrm{n}}} & \frac{\partial \mathcal{J}^{\mathrm{p}}}{\partial \Pi} & \frac{\partial \mathcal{J}^{\mathrm{p}}}{\partial A} \\ \frac{\partial \mathcal{J}^{\mathrm{n}}}{\partial V} & \frac{\partial \mathcal{J}^{\mathrm{n}}}{\partial \phi^{\mathrm{p}}} & \frac{\partial \mathcal{J}^{\mathrm{n}}}{\partial \phi^{\mathrm{n}}} & \frac{\partial \mathcal{J}^{\mathrm{n}}}{\partial \Pi} & \frac{\partial \mathcal{J}^{\mathrm{n}}}{\partial A} \\ \frac{\partial \mathcal{M}}{\partial V} & \frac{\partial \mathcal{M}}{\partial \phi^{\mathrm{p}}} & \frac{\partial \mathcal{M}}{\partial \phi^{\mathrm{n}}} & \frac{\partial \mathcal{M}}{\partial \Pi} & \frac{\partial \mathcal{M}}{\partial A} \\ \frac{\partial \mathcal{Q}}{\partial V} & \frac{\partial \mathcal{Q}}{\partial \phi^{\mathrm{p}}} & \frac{\partial \mathcal{Q}}{\partial \phi^{\mathrm{n}}} & \frac{\partial \mathcal{Q}}{\partial \Pi} & \frac{\partial \mathcal{Q}}{\partial A} \end{bmatrix} \cdot \begin{bmatrix} \Delta V \\ \Delta\phi^{\mathrm{p}} \\ \Delta\phi^{\mathrm{n}} \\ \Delta\Pi \\ \Delta A \end{bmatrix} = - \begin{bmatrix} \mathcal{P} \\ \mathcal{J}^{\mathrm{p}} \\ \mathcal{J}^{\mathrm{n}} \\ \mathcal{M} \\ \mathcal{Q} \end{bmatrix}$$

The fragments $\frac{\partial \mathcal{P}}{\partial A}$, $\frac{\partial \mathcal{J}^{\mathrm{p}}}{\partial A}$, and $\frac{\partial \mathcal{J}^{\mathrm{n}}}{\partial A}$ are induced by the Lorentz force whereas other parts get additional contributions from the Lorentz force. Furthermore, we note that $\frac{\partial \mathcal{Q}}{\partial V} = \frac{\partial \mathcal{Q}}{\partial \phi^{\mathrm{p}}} = \frac{\partial \mathcal{Q}}{\partial \phi^{\mathrm{n}}} = 0$. Another interesting observation is that $\frac{\partial \mathcal{Q}}{\partial \Pi}$ and $\frac{\partial \mathcal{Q}}{\partial A}$ are diagonal matrices. Therefore it is straightforward to solve for ΔA in terms of $\Delta\Pi$. The solution is

$$\Delta A = -\left(\frac{\partial \mathcal{Q}}{\partial A}\right)^{-1} \left[\mathcal{Q} + \frac{\partial \mathcal{Q}}{\partial \Pi}\Delta\Pi\right] \tag{21.33}$$

In this way, we can eliminate the last column of matrix system.

$$\begin{bmatrix} \frac{\partial \mathcal{P}}{\partial V} & \frac{\partial \mathcal{P}}{\partial \phi^{\mathrm{p}}} & \frac{\partial \mathcal{P}}{\partial \phi^{\mathrm{n}}} & \frac{\partial \mathcal{P}}{\partial \Pi} - \frac{\partial \mathcal{P}}{\partial A}\left(\frac{\partial \mathcal{Q}}{\partial A}\right)^{-1}\frac{\partial \mathcal{Q}}{\partial \Pi} \\ \frac{\partial \mathcal{J}^{\mathrm{p}}}{\partial V} & \frac{\partial \mathcal{J}^{\mathrm{p}}}{\partial \phi^{\mathrm{p}}} & \frac{\partial \mathcal{J}^{\mathrm{p}}}{\partial \phi^{\mathrm{n}}} & \frac{\partial \mathcal{J}^{\mathrm{p}}}{\partial \Pi} - \frac{\partial \mathcal{J}^{\mathrm{p}}}{\partial A}\left(\frac{\partial \mathcal{Q}}{\partial A}\right)^{-1}\frac{\partial \mathcal{Q}}{\partial \Pi} \\ \frac{\partial \mathcal{J}^{\mathrm{n}}}{\partial V} & \frac{\partial \mathcal{J}^{\mathrm{n}}}{\partial \phi^{\mathrm{p}}} & \frac{\partial \mathcal{J}^{\mathrm{n}}}{\partial \phi^{\mathrm{n}}} & \frac{\partial \mathcal{J}^{\mathrm{n}}}{\partial \Pi} - \frac{\partial \mathcal{J}^{\mathrm{n}}}{\partial A}\left(\frac{\partial \mathcal{Q}}{\partial A}\right)^{-1}\frac{\partial \mathcal{Q}}{\partial \Pi} \\ \frac{\partial \mathcal{M}}{\partial V} & \frac{\partial \mathcal{M}}{\partial \phi^{\mathrm{p}}} & \frac{\partial \mathcal{M}}{\partial \phi^{\mathrm{n}}} & \frac{\partial \mathcal{M}}{\partial \Pi} - \frac{\partial \mathcal{M}}{\partial A}\left(\frac{\partial \mathcal{Q}}{\partial A}\right)^{-1}\frac{\partial \mathcal{Q}}{\partial \Pi} \end{bmatrix} \cdot \begin{bmatrix} \Delta V \\ \Delta\phi^{\mathrm{p}} \\ \Delta\phi^{\mathrm{n}} \\ \Delta\Pi \end{bmatrix}$$

$$= \begin{bmatrix} -\mathcal{P} + \frac{\partial \mathcal{P}}{\partial A}\left(\frac{\partial \mathcal{Q}}{\partial A}\right)^{-1} \mathcal{Q} \\ -\mathcal{J}^{\mathrm{p}} + \frac{\partial \mathcal{J}^{\mathrm{p}}}{\partial A}\left(\frac{\partial \mathcal{Q}}{\partial A}\right)^{-1} \mathcal{Q} \\ -\mathcal{J}^{\mathrm{n}} + \frac{\partial \mathcal{J}^{\mathrm{n}}}{\partial A}\left(\frac{\partial \mathcal{Q}}{\partial A}\right)^{-1} \mathcal{Q} \\ -\mathcal{M} + \frac{\partial \mathcal{M}}{\partial A}\left(\frac{\partial \mathcal{Q}}{\partial A}\right)^{-1} \mathcal{Q} \end{bmatrix}$$

We found that a submission of the matrix system without using this elimination to the indirect solver combo consisting of re-ordering, preconditioning and linear solving soon led to too much fill-in due to the fact that $\frac{\partial \mathcal{Q}}{\partial A}$ and $\frac{\partial \mathcal{Q}}{\partial \Pi}$ are of the same numerical size. Therefore, there will be a complete fill-in triggered in the lower-right corner of the matrix. Remember that the corresponding size is of the order of the number of links and is three times the order of number of nodes. As a consequence even such a partial fill-in is deteriorating the solving process. However, there are no issues anymore with fill-in at the pre-conditioning stage for the second version of the matrix system. As a post-processing step after solving we apply (21.33) for obtaining the update on the vector potential.

Another interesting observation is that the complete non-linear system of equations, i.e. $\mathbf{F}(\mathbf{X}) = 0$, where $\mathbf{X} = \{V, \phi^p, \phi^n, \mathbf{\Pi}, \mathbf{A}\}$ can be reformulated as

$$\mathbf{F}_0(\mathbf{X}) + \Delta\mathbf{F}(\mathbf{X}) = 0. \tag{21.34}$$

In here, $\mathbf{F}_0(\mathbf{X})$ consists of all terms that do not contain any reference to the Lorentz force while $\Delta\mathbf{F}(\mathbf{X})$ refers exclusively to terms induced by the Lorentz force. An iterative solution method can be realized that simplifies the Newton-Raphson matrix substantially. The pseudo code reads:

- *Compute the initial guess* $\mathbf{X} = \mathbf{X}_0$ *using* $\mathbf{F}_0(\mathbf{X}) = 0$,
- *while* $|\mathbf{X}_{n+1} - \mathbf{X}_n| \geq \epsilon$ *solve* $\mathbf{X}_{n+1}$ *using* $\mathbf{F}_0(\mathbf{X}_{n+1}) = -\Delta\mathbf{F}(\mathbf{X}_n)$.

An alternative approach performs the full Newton-Raphson matrix assembling but computes an intermediate solution based on the following pseudo code:

- *Compute an initial guess* $\mathbf{X} = \mathbf{X}_0$ *using* $\mathbf{F}_0(\mathbf{X}) = 0$,
- *Compute an update* $\Delta\mathbf{X}$ *from* $\frac{\partial \mathbf{F}}{\partial \mathbf{X}}|_{\mathbf{X}_0} \cdot \Delta\mathbf{X} = -\Delta\mathbf{F}$,
- *Solve* $\mathbf{F}(\mathbf{X}) = 0$ *using* $\mathbf{X}_0 + \Delta\mathbf{X}$ *as initial guess.*

21.4 Applications

Silicon wire Continuing with the example that was given in (Schoenmaker et al. [2011]) we first consider a silicon wire. The purpose is to demonstrate the convergence behavior of the solution scheme that is proposed in this paper and to obtain a first impression of the size of the modifications induced by the magnetic field effects as well as the Lorentz force in the transient regime. In Figure 21.3, a typical ESD current pulse is shown. The spike of the currents ramps up within a nanosecond. We consider a silicon wire with a square cross section of 40 x 40 μm^2. Contacts with areas of size 20 x 20 μm^2 are placed at the ends. A structured grid is used in this example although this is not essential. A two-dimensional cross-section of the wire that is embedded in a 100x100 μm^2 block of oxide is shown in Figure 21.4. The wire length is 100 μm. The doping of the silicon wire (N-type) is 10^{20}cm^{-3}. We apply a current boundary condition at one contact and a voltage boundary condition ($V = 0$) at the other end contact of the wire. The calculations are performed on a mesh of 4851 nodes.Since a current boundary condition is applied at the current injection contact, the contact voltage is computed. The result is shown in Figure 21.5. The two curves correspond to simulations with the magnetic field components switched on (curve EM) and with the magnetic field switched off (curve E). As can be seen there is a large difference at the rampup stage. This can be understood as follows: In order to push the current through the silicon wire an applied voltage must be set. Due to the induced electric field, that counteracts the applied field, a higher voltage must be applied. This is confirmed in the various plots of the field components. Figure 21.6 shows the current density in the silicon wire at the time instance 0.30×10^{-9} sec, Figures 21.7–21.9 show the electric field intensity, the contribution ∇V to the electric field and the induced contribution Π to this field. Both contributions are roughly equal in magnitude. For completeness, we also show the magnitude of the vector potential, A in Figure 21.10. The peak value is $\sim$ 10^{-7} Wb/m. Finally, Figure 21.11, shows the magnetic induction. All field plots are taken at 0.30×10^{-9} sec. At this time instance the fast-transient phenomena are very pronounced. Finally, Figure 21.12 gives the relative change in the voltage of the current-injection contact. The peak value is $\sim$ 0.07%, which is more or less inline with the observations for the static case (Schoenmaker et al. [2011]). All results were obtained with a relative current balance being of the order of $\sim$ 0.0001%. This accuracy is sufficiently high in order to detect the impacts of the various mechanisms.

ESD protection A second illustration deals with a stacked diode pair that serves as a prototype for a circuit topology corresponding to an electrostatic discharge (ESD) protection structure (Galy et al. [2011]).

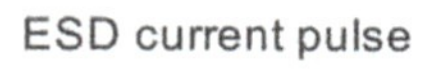

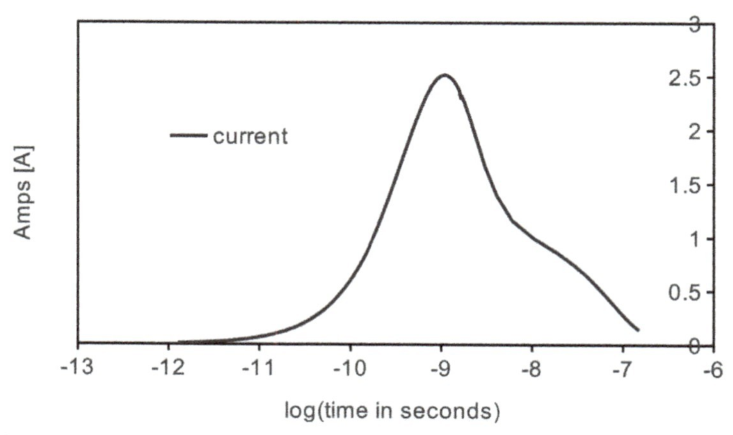

Figure 21.3 Typical ESD current pulse used in the simulation of the fast-transient signals.

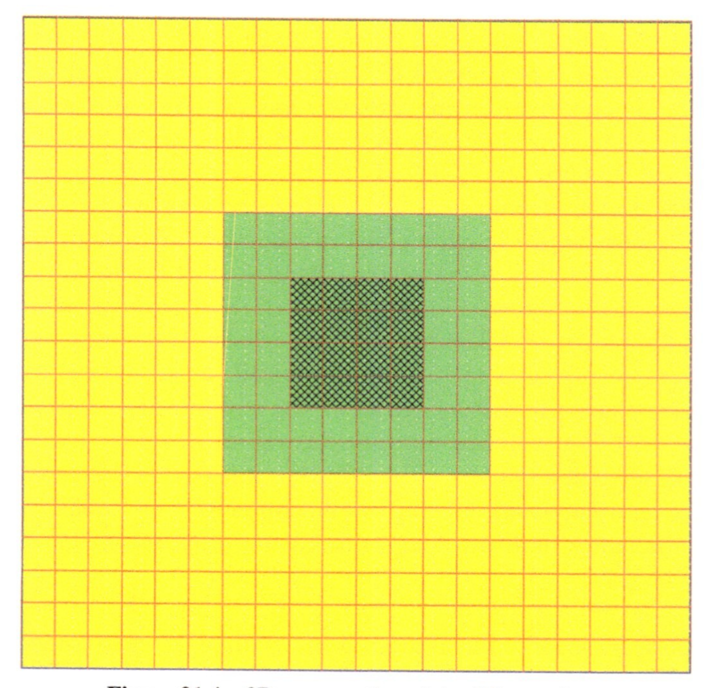

Figure 21.4 2D cross section of the Silicon wire.

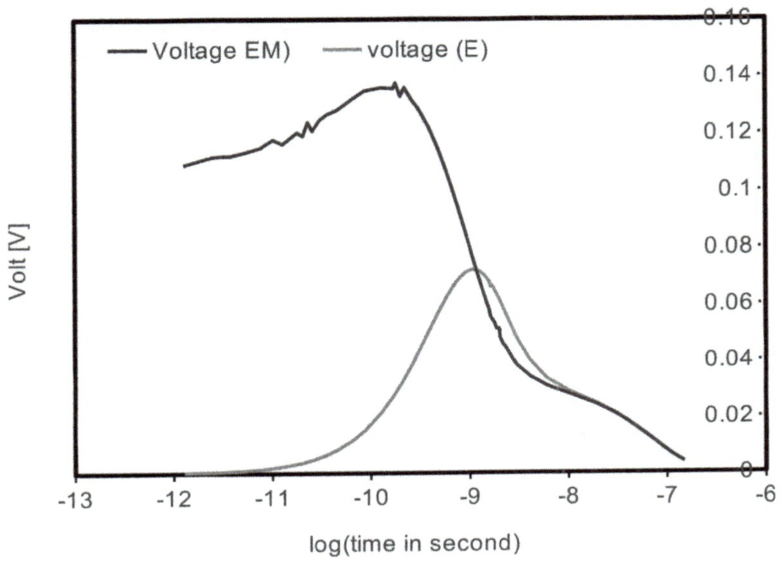

Figure 21.5 ESD voltage pulse computed in the simulation of the fast-transient signals at the injection contact.

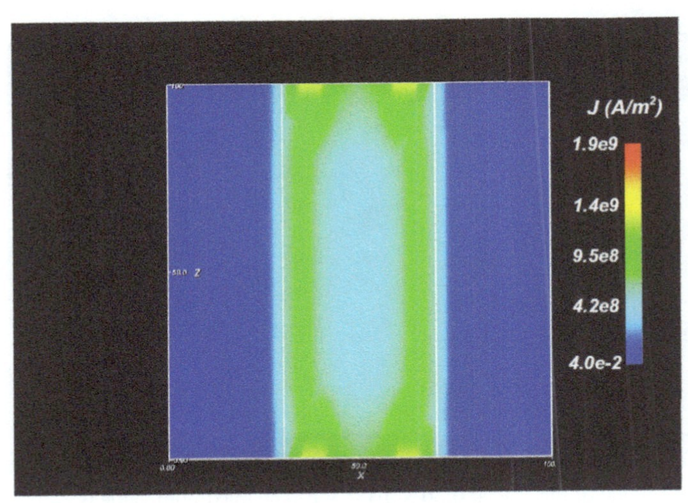

Figure 21.6 Current density at 0.30×10^{-9} sec. Some reduction is observed in the center due to the skin effect.

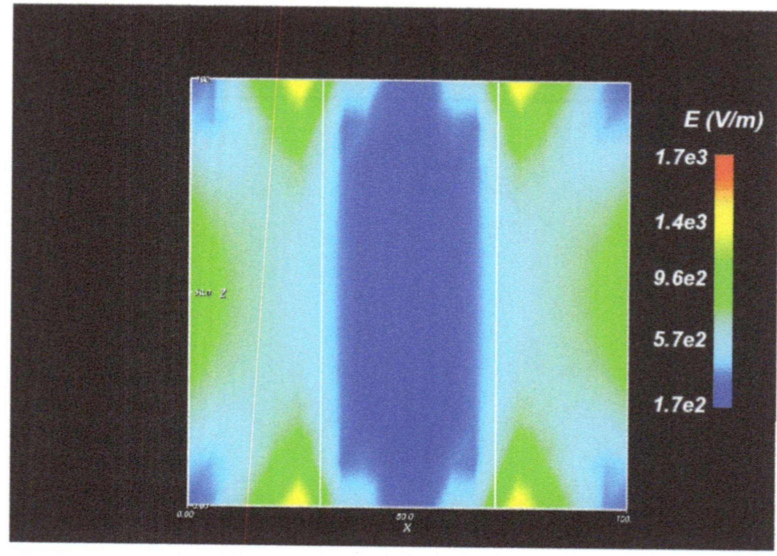

Figure 21.7 Electric field intensity at 0.30×10^{-9} sec. Some reduction in the center is observed due to the skin effect and the value is $\sim 10^3$ V/m.

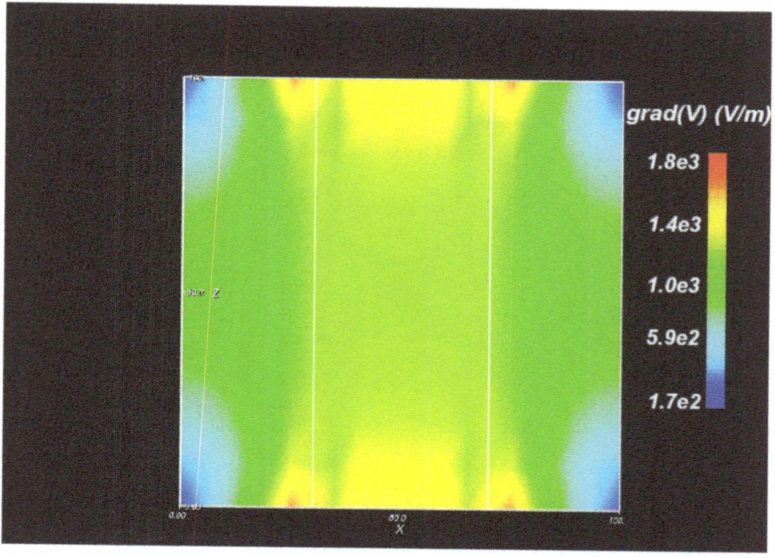

Figure 21.8 Magnitude of the ∇V at 0.30×10^{-9} sec. The maximum value is $\sim 10^3$ V/m.

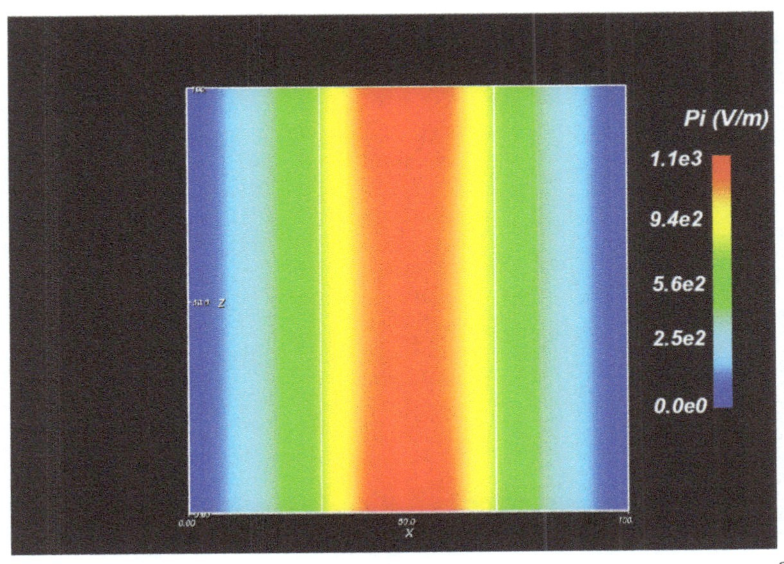

Figure 21.9 Magnitude of the pseudo-canonical momentum $\mathbf{\Pi} = \partial_t \mathbf{A}$ at 0.30×10^{-9}. The maximum value is $\sim 10^3$ V/m.

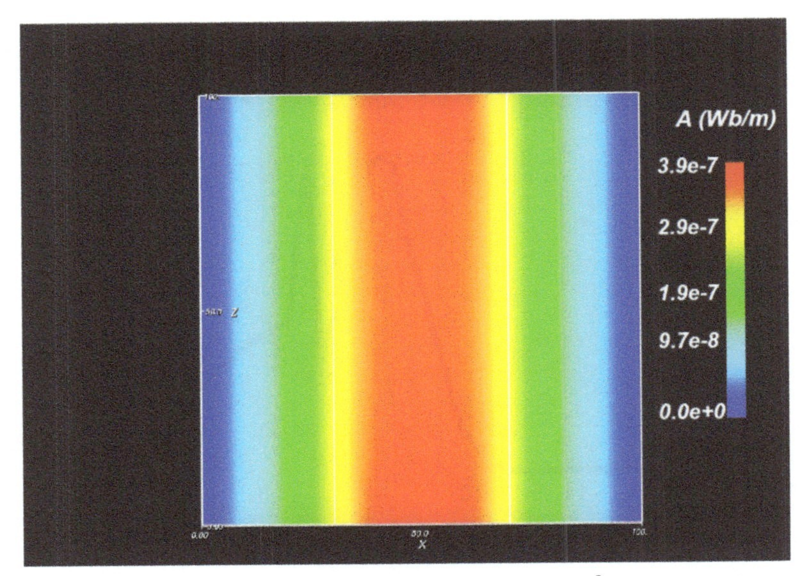

Figure 21.10 Magnitude of the vector potential **A** at 0.30×10^{-9} sec. The maximum value is $3 \sim \times 10^{-7}$ Vsec/m.

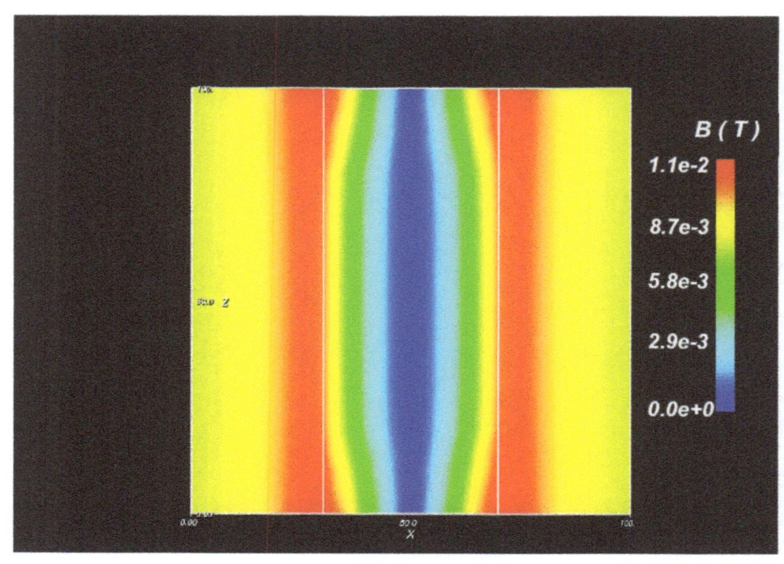

Figure 21.11 Magnitude of the magnetic induction, **B** at 0.30×10^{-9} sec. The maximum value is $\sim 10^{-2}$ T.

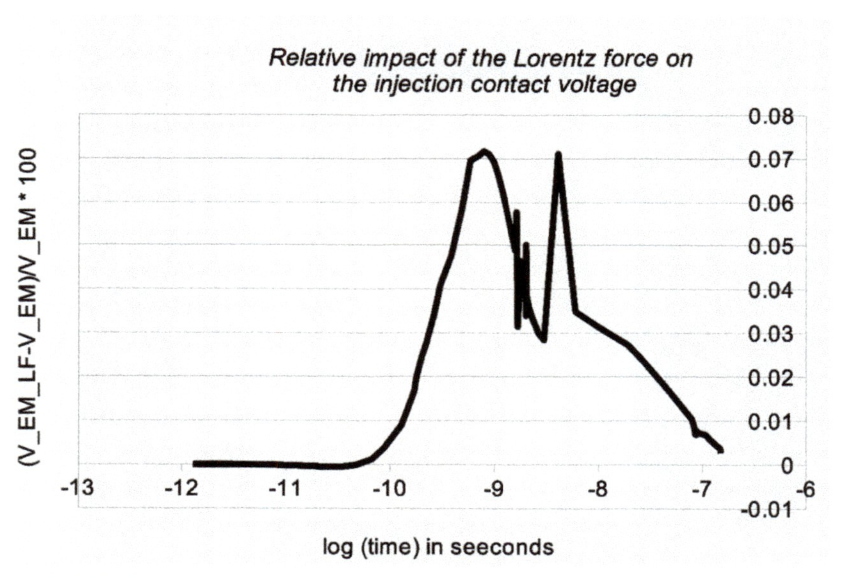

Figure 21.12 Relative change in the voltage of the current injection contact due to the Lorentz force.

The circuit lay-out is found in Figure 21.13 and a 'physical' realization is shown in Figures 21.14 and 21.15. The latter shows a stretched view of the Silicon-Controlled Rectifier (SCR) by stacking N-doped and

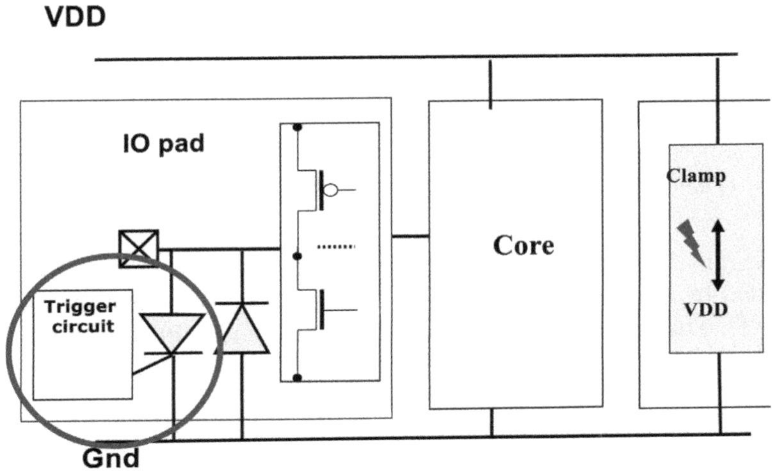

Figure 21.13 Circuit layout for use of a Silicon-Controlled Rectifier (SCR). The location of the SCR is encircled and is presented in more detail in Figure 21.14.

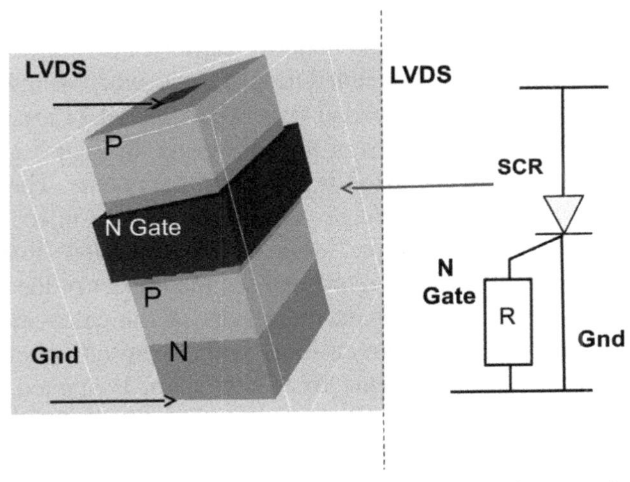

Figure 21.14 Device implementation the SCR. The left picture is a stretched view of the actual implementation that is presented in Figure 21.15.

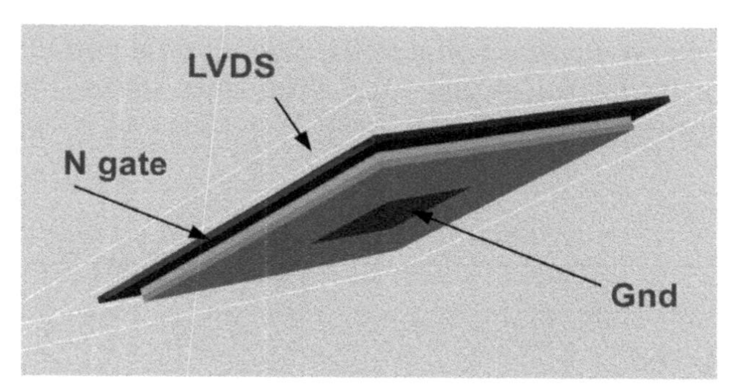

Figure 21.15 Actual implementation of the SCR structure.

P-doped regions. The N-gate contact, the third layer, counted from below, is surrounded by a metallic ring at which the 'gate' contact is placed with a load of 50 Ω. The current is as given in Figure 21.3 and is injected through the top contact LVDS. A ground contact (Gnd) is inserted at the bottom of the stack.

The results of the simulation of the voltage at the LVDS contact are found in Figure 21.16. The effect of the magnetic field on the gate current is shown in Figure 21.17. Both results clearly illustrate that the electromagnetic response has an important effect: There is a significant signal *delay* more than a *change* in amplitude of the signals. Finally, we consider the inclusion of the Lorentz force in the calculation. This is done in two stages. First we switch on the Lorentz force corrections for the majority carriers, i.e. in the N-doped regions, the Lorentz force effect is included for electrons, whereas in P-doped regions it is included for holes. At a second stage we include the Lorentz force for both types of carriers in both types of doped regions. In Figure 21.18 the impact of the majority-carrier inclusion is shown on the results. The dashed line shows the comparison of the voltage value at the current injection with and without inclusion of Lorentz force, whereas the continuous line shows the comparison for the gate current. In order to see the impact of the Lorentz force for minority carriers, we compare the results of the calculation with inclusion of the majority-carrier Lorentz-force with the results obtained by switching on all terms. i.e. the Lorentz force correction is applied to both majorities and majorities. Figure 21.19 shows the difference of the voltage value at the current injection (dashed line) and the difference for the gate current (continuous line) with and /without inclusion of the minority-carrier Lorentz-force term. As is seen in Figure 21.19, the minority carriers induce a change of the injection voltage value around 1.82 Volt, which corresponds to a relative change of 3.5 % as is seen in Figure 21.20.

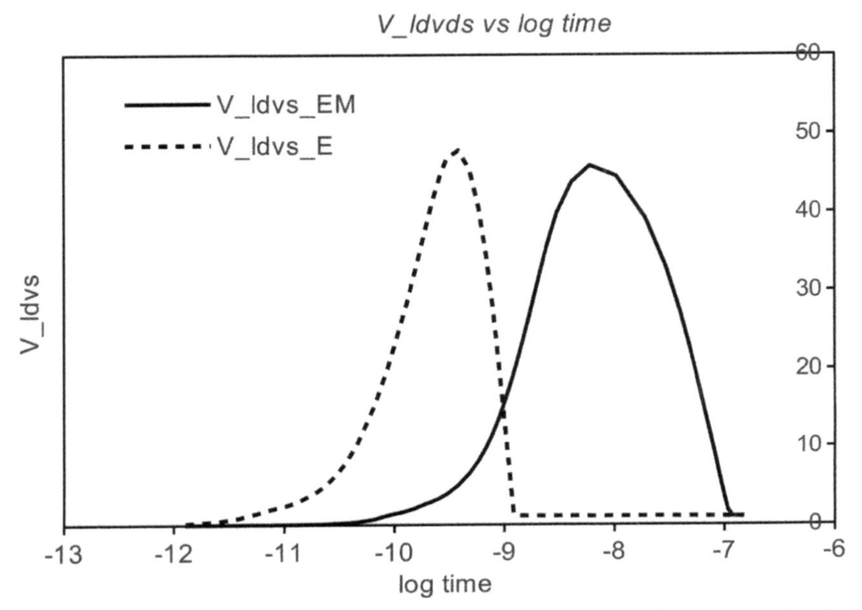

Figure 21.16 Change in the voltage of the current injection contact due to magnetic effects.

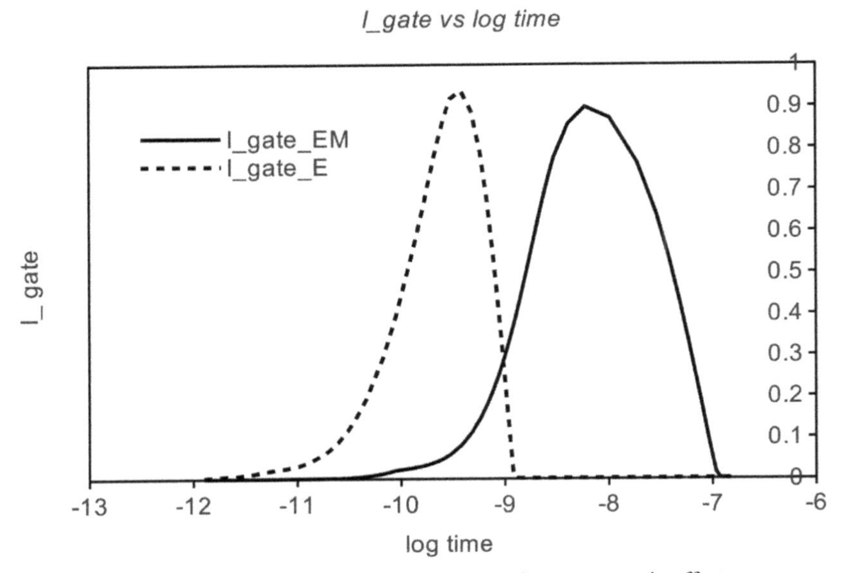

Figure 21.17 Change in the gate current due to magnetic effects.

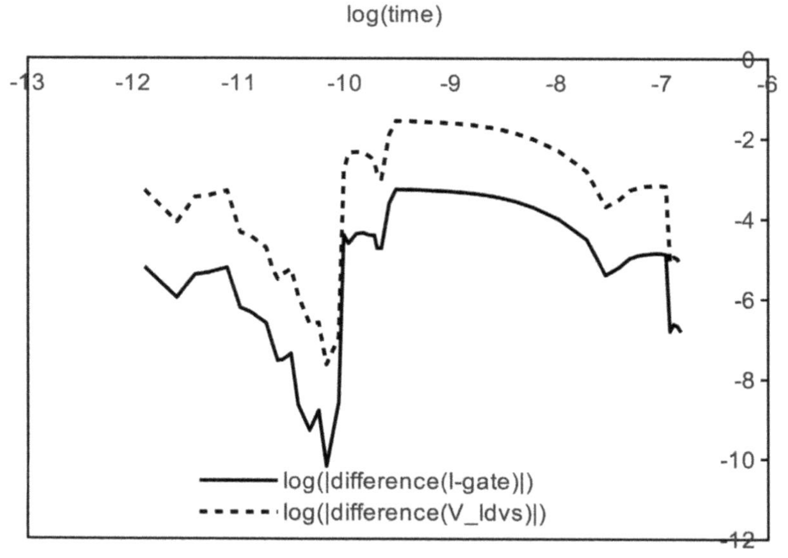

Figure 21.18 Change in the gate current and change in the voltage of the current injection contact (LVDS) due to switch-on of the Lorentz force for the majority carriers.

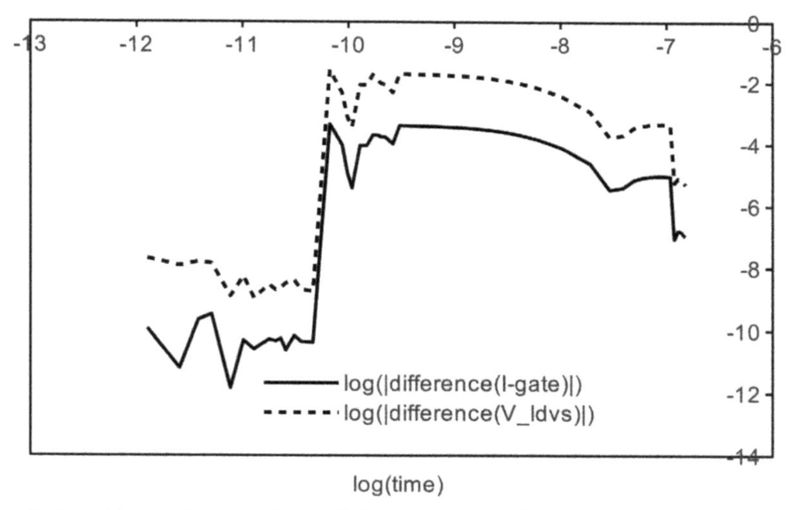

Figure 21.19 Change in the voltage of the current injection contact (LVDS, dashed line) and change in the gate current (continuous line) due to an additional switch-on of the Lorentz force for the minority carriers.

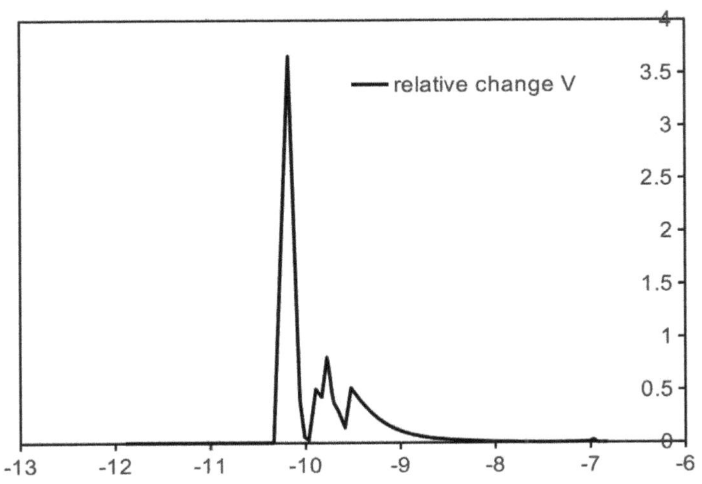

Figure 21.20 Relative change in the voltage of the current injection contact (LVDS) due to an additional switch-on of the Lorentz force for the minority carriers.

Summary

We presented a full simulation picture of fast-transient phenomena. Besides the electromagnetic effects, we provided a method to compute the impact of the Lorentz force. We applied the proposed method to a silicon wire. Furthermore, we applied the method to a simplified implementation of a Silicon-Controlled Rectifier that serves as a local protection structure for forward stress. There is convincing evidence that the magnetic contributions are of the same order of magnitude as the electric contributions. On the other hand, the Lorentz force effects are very minor and can be ignored for most practical purposes (despite the fact that it is an arduous task to compute these effects). In the fast-transient applications that we have done, their impact is of the order of $\sim 0.07\%$ in normal operation but minorities subjected to the Lorentz force and fast transient conditions lead to an appreciable deviation ($\sim 3\%$) for the voltage values that are needed to realize the large transient currents. The proposed method for calculating the Lorentz-force induced modifications may be applied to other fields of engineering and physics in which the Lorentz force plays a more dominant role. For example, the numerical scheme can be applied to simulate Hall sensors.

22

EM Analysis of ESD Protection for Advanced CMOS Technology

The purpose of this chapter is to present the main results of an Electro-Static Discharge (ESD) protection for advanced CMOS technology with Electro-Magnetic (EM) field effect and Lorentz Force (LF) contributions during fast transient and high-current surge[20]. The goal is to know the generated EM field during an ESD event and the impact within the structure where the LF could impact the final response. To address this goal the first step is building a tool to simulate fast transient conditions with all participating physical mechanisms included. The relevant equations describing these mechanisms are: the charge transport equations, the Maxwell equations to describe the electromagnetic fields. The Lorentz force is also included using an extended formulation of the current-continuity equations. An integrated approach is followed to simulate the full structure (metal connections + silicon device) during the ESD surge and to compare the results between EMLF simulations and transmission-line pulse TLP measurements. Obviously, in general, this work and tool can be used to address Electro Magnetic Compatibility (EMC) topics and more.

It is well known that advanced CMOS technology is very sensitive to ESD stress due to the fact that the down-scaling of transistors leads to a drastic reduction of its intrinsic robustness. It means that all input/output (IO) pads are protected by ESD protection. In this study, the ESD protection is local and uses a Silicon Control Rectifier (SCR) for the forward stress and a diode for the negative stress. Figure 22.1 gives an overview of the ESD structure to protect the IO. An SCR structure is a PNPN doping sequence where the first P region is the anode connected to the IO terminal and the last N region is the cathode of the SCR connected to the ground. A trigger circuit is connected to the N gate which is the second doping region. For this study the trigger

[20]This chapter is a reprint of Galy and Schoenmaker [2014].

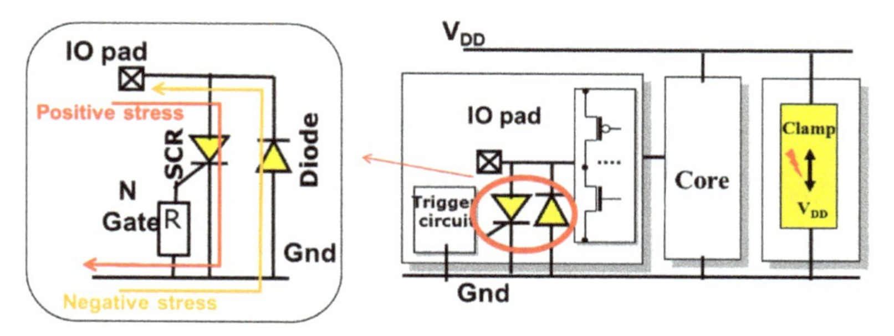

Figure 22.1 Basic electrical schematic of the local protection with SCR and diode.

circuit is a simple resistance $R_{\text{trig}} = 100\Omega$ and equivalent to a trigger value of 10Ω. This value should beadjusted for an efficient protection according to the load.

Figure 22.2 shows the concentric SCR with its diode layout top view and the 3D TCAD structure including the Cu metal stack and the silicon region with all doping regions.

22.1 Simulation of a Metallic Wire

For contacts, we apply for the scalar potential V and the quasi-Fermi levels, Dirichlets boundary conditions and Neumanns boundary conditions elsewhere at the boundary of the simulation domain. The boundary condition for the vector potential is chosen such that the magnetic field is tangential to the boundary of the simulation domain. Moreover, all simulations are done with isothermal condition at room temperature. The numerical discretization and the numerical kernel is superficially summarized here. A mesh is created for the simulation domain of the structure and the finite volume method is

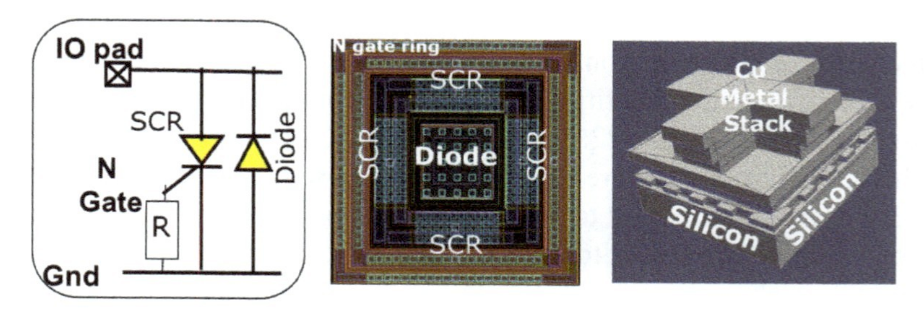

Figure 22.2 SCR schematic with its top view layout and 3D TCAD structure.

exploited. The current density and the charge density are discretized such that the charge conservation is respected faithfully. At each finite volume a nodal current balance equation is solved. Other variables are discretized using a generalized perspective of the finite-volume method. In particular, the Maxwell-Ampere equation is discretized using finite surfaces. The temporal evolution is discretized using a backward-difference formulation (BDF) up the second order. As a consequence, at each time step a Newton-Raphson cycle is launched that requires a multiple solving of linear systems. The complete system of variables to solve consists of the scalar potential V, the electrons and holes current densities or the quasi-Fermi levels and the vector potential A and it first-order time derivative, $\partial_t A$ i.e. $(V, \phi_n, \phi_p, A, \partial_t A)$.

The purpose of this section is to know if the solution method for the coupled equations converges for an elementary test case. An Al wire metal with a length of 100 μm is selected. The wire has a square cross section of S0 = 40 x 40 $\mu\mathrm{m}^2$. The contacts areas of size 20 x 20 $\mu\mathrm{m}^2$ are placed at the top and bottom of conductor. The total box of simulation is 100 x 100 $\mu\mathrm{m}^2$. Thus the computations are with E, EM and EMLF. A current stress pulse with duration of 100 nsec is applied on this structure and the electric field E, the magnetic field B and the current density J are extracted. (see Table 22.1).

Table 22.1 3D TCAD extraction of E, B and J in E, EM and EMLF simulation cases

	E simulation	EM simulation	EMLF simulation
Electric Field Extraction **E**	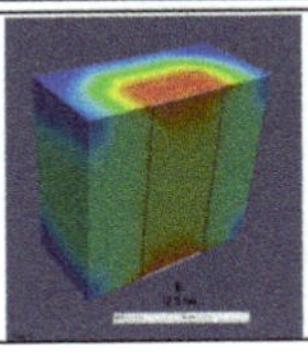	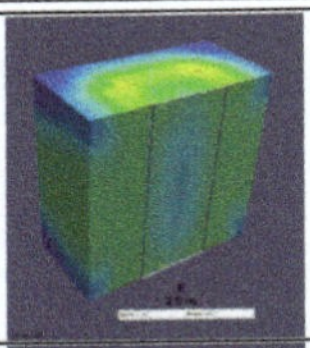	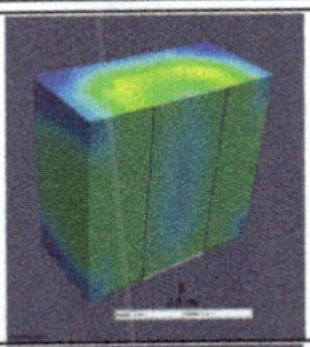
Magnetic Field Extraction bf B	Not Applicable	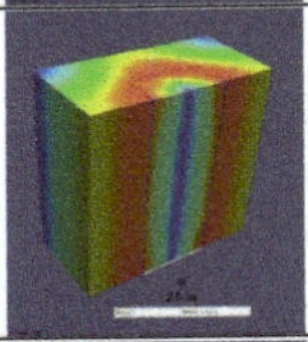	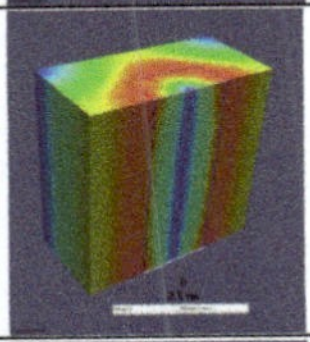
Current density extraction **J**	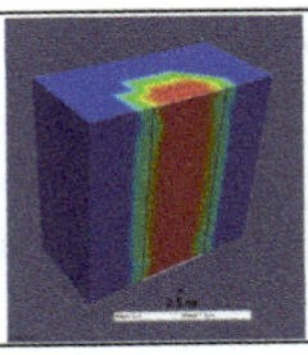	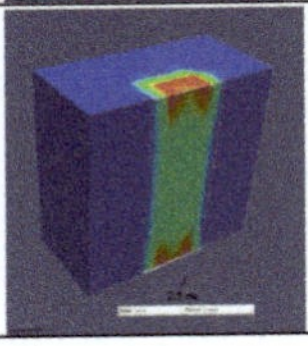	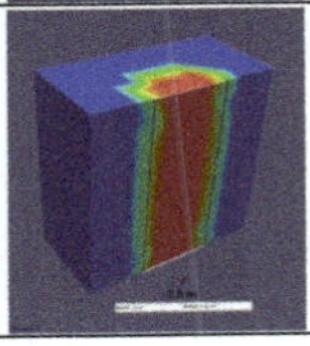

Good numerical convergence is achieved with efficient run time for all cases. Moreover, it is shown that a shift is found between E and (EM, EMLF) simulations. As expected, the dynamic responses changes during the fast transient event. No major difference is detected between EM and EMLF for this test case. We also note that the EM field distribution is computed all-around the conductor which is of interest to study the impact of EMC at the environment. Another way to compare the results is to extract the IV curves of the wire in a time interval and to calculate its dynamic resistance $r(t)$. Figure 22.3 extracts and compares the data. For the E-simulation, i.e. with the magnetic field effects switched off, the IV curve (in red) is linear as demanded by Ohms law and the calculated resistance is a constant equal to $r_E = 8m\Omega$. The classical calculation $R_0 = \Omega L/S0 = 7.5m\Omega$. The discrepancy is attributed to the coarseness of the computational grid and the size of the contact areas. An interesting point is that for the EM and EMLF simulations a strong impact is shown and the IV curves are totally different from the E-simulation case. This is due to self-induction. It is also the case for $r(t)$ where at the beginning of the current stress the resistance is 6 times higher than as for the E case. It only converges to the same value at the end of peak current (see Figure 22.3, part 2). Table 22.2 gives the initial resistance

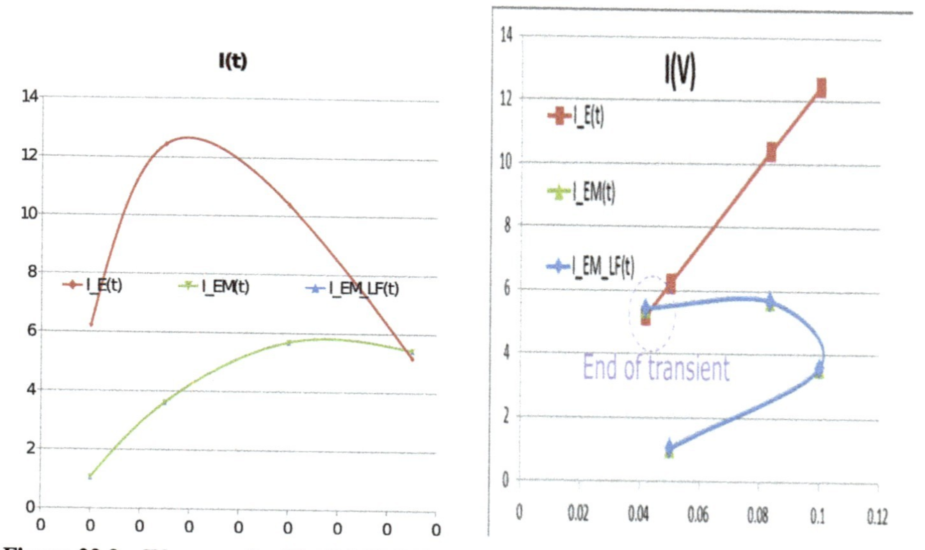

Figure 22.3 IV curves for (E, EM, EMLF) simulations and calculation of $r(t)$ for E, EM and EMLF.

Table 22.2 Resistance values

R0	7.50E-03 Ω
R0 50%	3.00E-02 Ω
R0 15%	5.00E-02 Ω
$\langle r_E \rangle$	8.03E-03 Ω
$\langle r_{EM} \rangle = \langle r_{EMLF} \rangle$	2.43E-02 Ω

R0 with two pondered conductor surfaces (50% & 15%) and the average $\langle r \rangle$of all r(t). These values are also reported in Figure 22.3 and show the variation of the resistance.

It should be emphasized that for a full ESD protection structure the impact of the Back-End Of Line (BEOL) is analyzed in conjunction with the Front-End Of Line (FEOL). Moreover, the interesting variables are E, B and J inside the structure and V(t) and I(t) for all terminal of this ESD protection. The next section addresses these variables for the full ESD protection structure.

22.2 In-depth Simulation of the Full ESD Structure

The purpose of this section is to perform a study of the full structure (FEOL + BEOL) with EMLF simulation for positive and negative ESD events. The main focus will be on EMLF simulations due to the fact that many physical parameters are calculated. The analysis is done with Rtrig=100 and for the high current pulse. It is reminded here that the positive stress involves the SCR and the negative surge is driven by the reverse ESD diode (see Figure 22.1).

Positive stress with SCR in action. For a positive ESD stress, Figure 22.4 reports the $I(t) - V(t)$ responses for the full structure with significant points (A, B, C, D) and the four-vector $A_\nu = (V, \mathbf{A})$ extractions on typical points (triggering time and maximal current drive) are also reported as inset.

At the beginning of the stress, the SCR is blocked and the drop voltage across the device increases (path #1) tills the break point A. The four vector is extracted at this point and the drop voltage reaches 15.7 V. Afterwards, the SCR is triggered and goes in snapback (path #2) to finally reach the point B. At this stage, the device drives around 500 mA in 100 psec of duration. The current increases to 1 A in point C and the maximal current is 2.5 A in point D. For the maximum stress current the four-vector is also extracted. Path #3 gives the followed IV curve with a quasi-constant voltage drop on the SCR. Path #4 corresponds to the current decrease all along of the end of the stress.

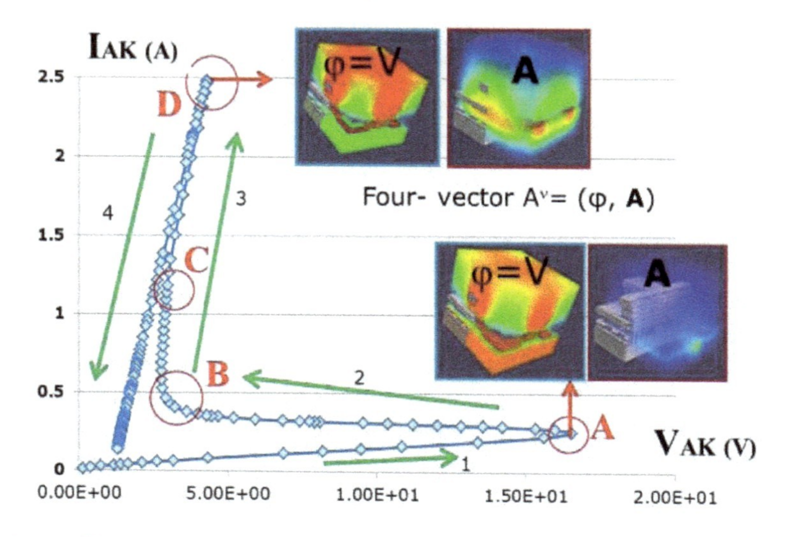

Figure 22.4 IV curves of SCR during ESD stress and typical point identification.

The next part of text focuses on several extractions of B, E, J at the points A, B, C, D. Table 22.3 gives the main calculated values on each significant point (A, B, C, D). Table 22.4 presents the magnetic field distribution during the stress. At the beginning of the event a dI/dt occurs in less than 40 psec and generates a magnetic field all around the structure where the N gate SCR is active. No major impact is observed in the silicon region. During the points B, C, D the current increases drastically up to 2.5 A in a duration of 1nsec. At this stage, the magnetic field is maximal and is distributed mainly inside the device where the anode and the cathode are located. An interesting observation is that now the silicon device is impacted by the magnetic field with a particular shape due to the specific silicon device and metal topology. Another observation is that the magnetic field generated during the ESD stress is confined inside the device thanks to the selected $3D$ topology.

Table 22.3 Extraction of main significant point on IV curve response

Point	Time (sec)	Current (A)	Voltage (V)
A	40 psec	258 mA	16.5 V
B	108 psec	637 mA	2.76 V
C	240 psec	1.26 A	3.01 V
D	1 nsec	2.50 A	4.35 V

Table 22.4 Extraction of magnetic field B in the full structure and in silicon region for A, B, C, D points

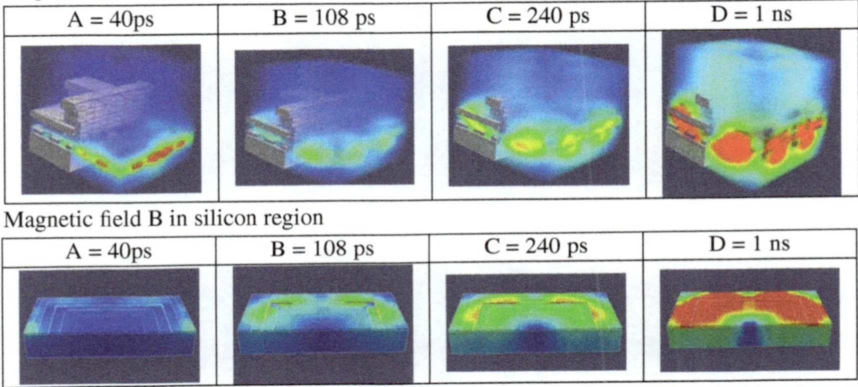

Magnetic field B in full structure

A = 40ps	B = 108 ps	C = 240 ps	D = 1 ns

Magnetic field B in silicon region

A = 40ps	B = 108 ps	C = 240 ps	D = 1 ns

Table 22.5 presents the electric field distribution. The maximum field is observed at the point A and obviously all around the N gate of the SCR. It is the same in the silicon area. Moreover, at the center of the structure the diode is blocked in reverse bias (red area) as indicated in the schematic of Figure 22.1.

We next focus on the current density **J** within the device and moreover a silicon inspection is done through Jn and Jp. Table 22.6 reports all extractions. For the point A the current is located to the N gate of the SCR with a low level

Table 22.5 Extraction of electric field E in the full structure and in silicon region for A, B, C, D points

Electric field E in full structure

A = 40ps	B = 108 ps	C = 240 ps	D = 1 ns

Electric field E in silicon region

A = 40ps	B = 108 ps	C = 240 ps	D = 1 ns

Table 22.6 Extraction of current density $\mathbf{J} \sim; 8 \times 10^{10} A/m^2$, $\mathbf{J}_{\mathrm{n}} \sim 1 \times 10^{11} A/m^2$, $\mathbf{J}_{\mathrm{p}} \sim 2 \times 10^{10} A/m^2$ in the full structure and in the silicon for A, B, C, D points

Current density $\mathbf{J}$ in full structure

A = 40ps	B = 108 ps	C = 240 ps	D = 1 ns
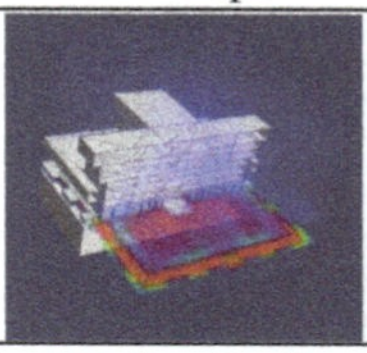	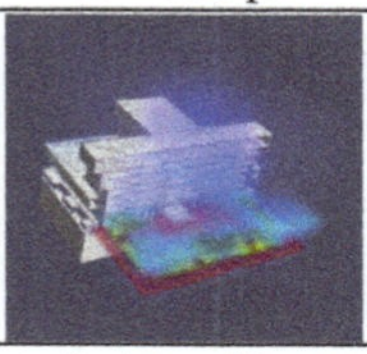	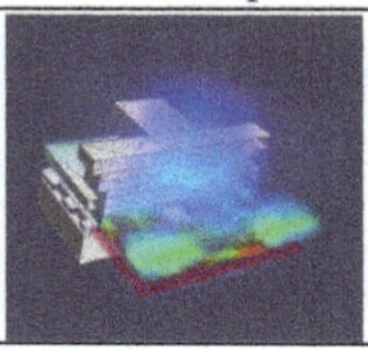	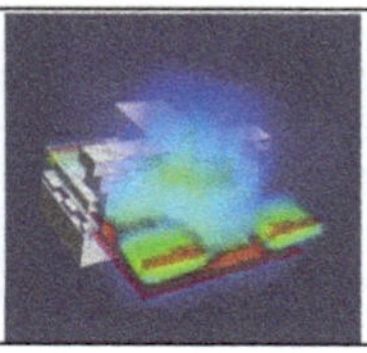

Current density $\mathbf{J} = \mathbf{J}_{\mathrm{n}} + \mathbf{J}_{\mathrm{p}}$ into silicon region

A = 40ps	B = 108 ps	C = 240 ps	D = 1 ns
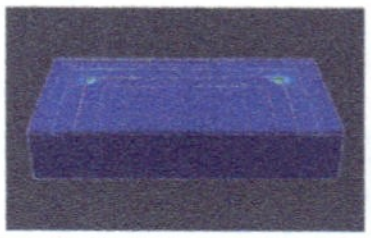	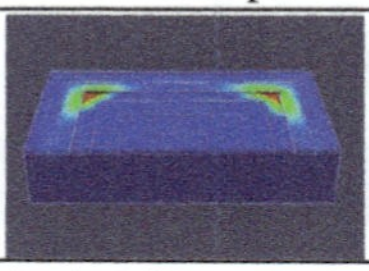	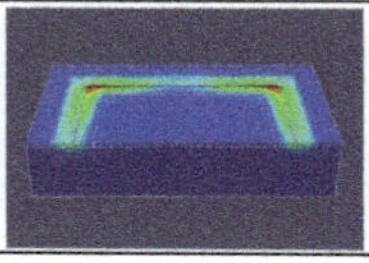	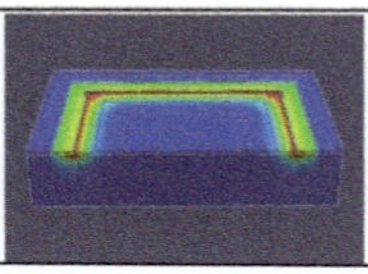

Current density $\mathbf{J} = \mathbf{J}_{\mathrm{n}}$ in the silicon region

A = 40 ps	B = 108 ps	C = 240 ps	D = 1 ns
	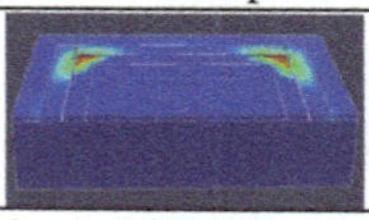	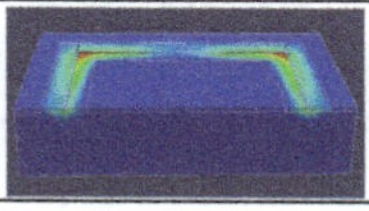	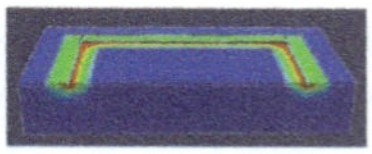

Current density $\mathbf{J} = \mathbf{J}_{\mathrm{p}}$ in the silicon region

A = 40ps	B = 108 ps	C = 240 ps	D = 1 ns
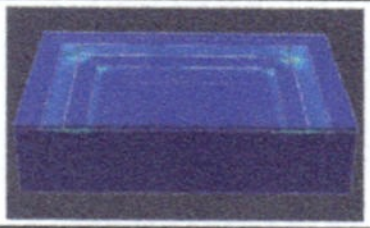	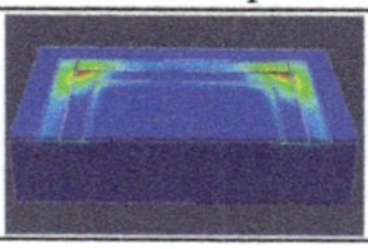	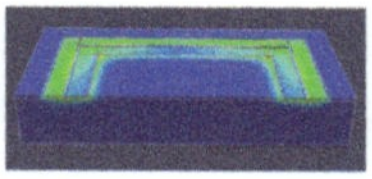	

Table 22.7 Main values for physical parameters extracted on points A, B, C, D

SCR	Unit	A = 40 ps	B = 108 ps	C = 240 ps	D = 1 ns
A max	Vsec/m	4.42E-08	2.41E-08	5.97E-08	1.25E-07
B max	T	0.0563947	0.0894767	0.0606908	0.082165
V max	V	17.1756	3.40176	3.63505	4.93343
E max	V/m	1.31E+08	4.17E+07	4.34E+07	4.46E+07
J max	A/m^2	3.30E+12	3.38E+12	3.28E+12	3.25E+12
$\mathbf{J}_{\mathrm{n}}$ max	A/m^2	5.02E+11	2.38E+12	1.38E+12	8.27E+11
$\mathbf{J}_{\mathrm{p}}$ max	A/m^2	8.20E+10	2.96E+11	1.25E+11	1.41E+11

of current (100 mA). After triggering, the current starts at the anode/cathode corners (see point B) and spreads along the anode to the cathode as indicated in points C and D. At point D the full structure is involved and drives 2.5A.

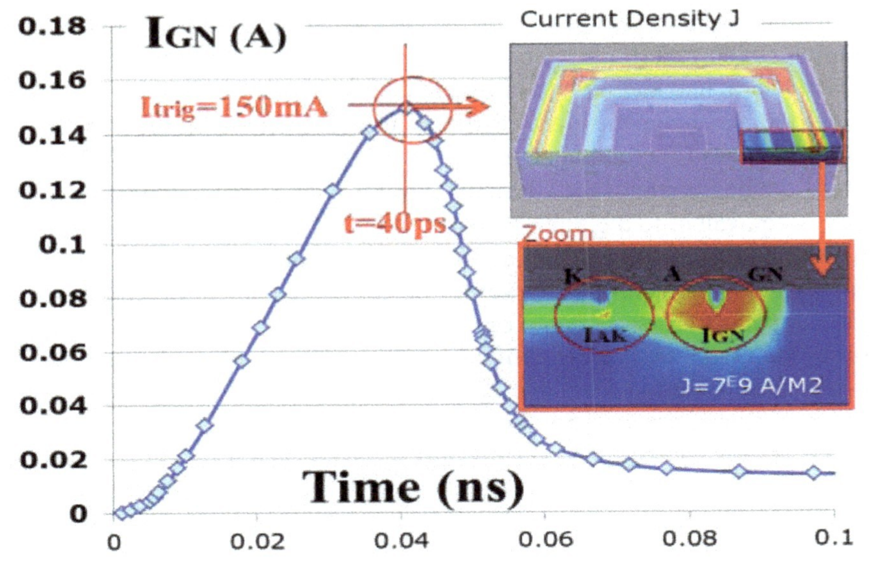

Figure 22.5 IGN(t) curves of N gate current during surge and current density extraction at A.

Electrons, Jn, are the main contributor for the A/K current and holes Jp are the main contributors to the gate current.

For the N gate the current extraction is done resulting into a low level current density. The result for the gate current is shown in Figure 22.5. It is shown that 150 mA is needed to trigger the SCR in a 40 psec duration. The current path from the anode to the N gate is well identified. Moreover, the AK current is also located with a filament shape and with a hot spot (see inset of Figure 22.5). This shape of the current density could be a weak spot and will be investigated in a future study. Moreover, the corners are also triggered and drive part of the total current. Thanks to the previous extractions, we conclude that the N gate current is 150 mA and for the SCR current I_{AK} is 108 mA.

22.3 Negative Stress with Active Diode

For a negative stress (Gnd vs. IO) the ESD diode is involved and drives the current stress. Basically, the simulated IV curve is the diode response that drives 2.5 A the maximum of the stress and is in agreement with the expectation. Three extractions are performed on this curve at the points, and, all along the rise time. The point is on the threshold voltage of the power diode, the point is in the middle of the rise up of the event and the point is on the maximum of current (no extraction during the relaxation). The four-vector A_ν is extracted on and and plotted in Figure 22.6.

As before, the magnetic field distribution is extracted for these three points. It is observed that a magnetic field is at the external boundary of the structure at the beginning of the stress (0.07 nsec). This is due to the fact that the diode is equivalent to a bipolar transistor where the N gate of the SCR is the collector. In this condition, the fast collector current provides a magnetic pulse all around the structure. Afterwards, the current is mainly driven by the diode (BE bipolar transistor) in the center of the structure. At the maximum of current, the magnetic field is confined with a strong penetration into the silicon region as is seen in see Table 22.8.

It is interesting to compare for the various stresses the distribution and the penetration of the magnetic field B at the top and bottom of the silicon region for the maximum current. Figure 22.7 shows the results for the diode and for the SCR. For the diode, the magnetic field distribution is more localized in the structure with strong penetration into silicon and reached a maximum of 66 mT. In the SCR case, the magnetic field distribution is more spread all around the structure with a low silicon penetration and reached a maximum of 82 mT. Obviously, this information is available thanks to the integrated simulation tool and is not accessible through measurements.

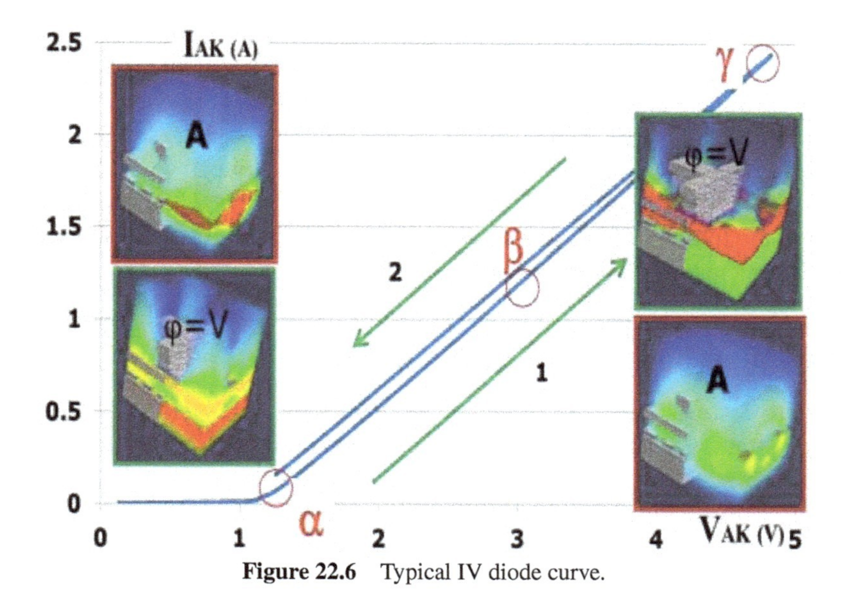

Figure 22.6 Typical IV diode curve.

Table 22.8 Extraction of Magnetic field B in the full structure and in silicon region for the α, β, γ points

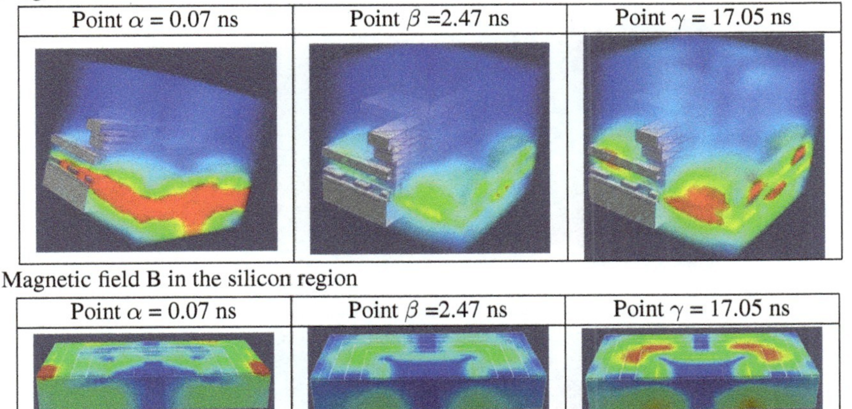

Magnetic field B in the full structure

Point α = 0.07 ns	Point β =2.47 ns	Point γ = 17.05 ns

Magnetic field B in the silicon region

Point α = 0.07 ns	Point β =2.47 ns	Point γ = 17.05 ns

Table 22.9 Extraction of current density $\mathbf{J}$, $\mathbf{J}_{\mathrm{n}}$ and $\mathbf{J}_{\mathrm{p}}$ in the structure and in silicon for α, β, γ points

$\mathbf{J} = \mathbf{J}_{\mathrm{n}} + \mathbf{J}_{\mathrm{p}}$ in the silicon region

Point α = 0.07 ns	Point β = 3.07 ns	Point γ = 17.05 ns

$\mathbf{J} = \mathbf{J}_{\mathrm{n}}$ in the silicon region

Point α = 0.07 ns	Point β = 3.07 ns	Point γ = 17.05 ns

$\mathbf{J} = \mathbf{J}_{\mathrm{p}}$ in the silicon region

Point α = 0.07 ns	Point β = 3.07 ns	Point γ = 17.05 ns

22.4 Diode SCR

The last extractions are focusing on the current density distribution J and its carrier contributions Jn and Jp. It is shown that the diode is acting as base-emitter transistor junction driving 2.5 A and the collector current around 200 mA.

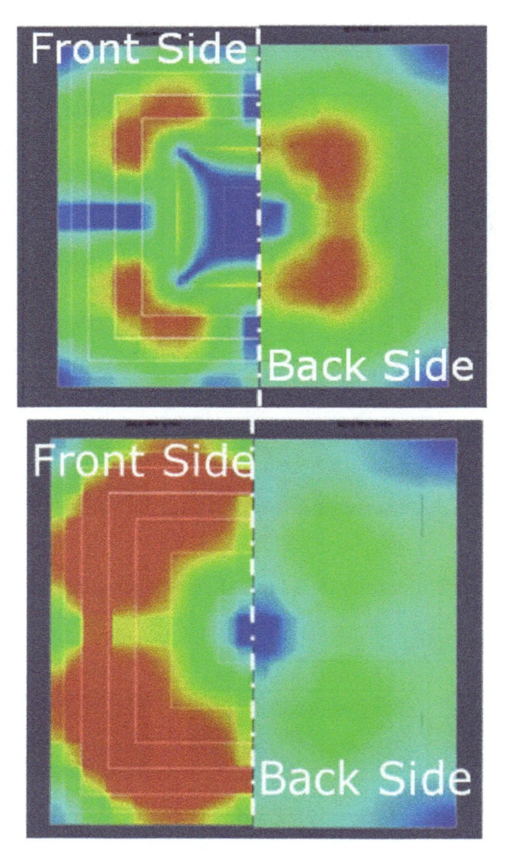

Figure 22.7 Front/Back side distribution magnetic field for the diode and SCR for 2.5A.

Following the previous remark, Figure 22.8 reports the temporal response of the N-gate which is the collector of the transistor where the BE junction is the ESD power diode. The current-density maps are reported on α and β points. For the γ point the scale is changed to observe the collector current. Moreover, comparing Figure 22.5 to Figure 22.8 of the two modes of protection, the N-gate current is totally different but with almost the same current magnitude (160 mA compared to 200 mA).

Table 22.10 gives the main physical values extracted on the three identified points during the ESD event. The maximum of the magnetic field B is around 66 mT at 2.5A for 5.3V drop voltage across the device in 17 nsec time range.

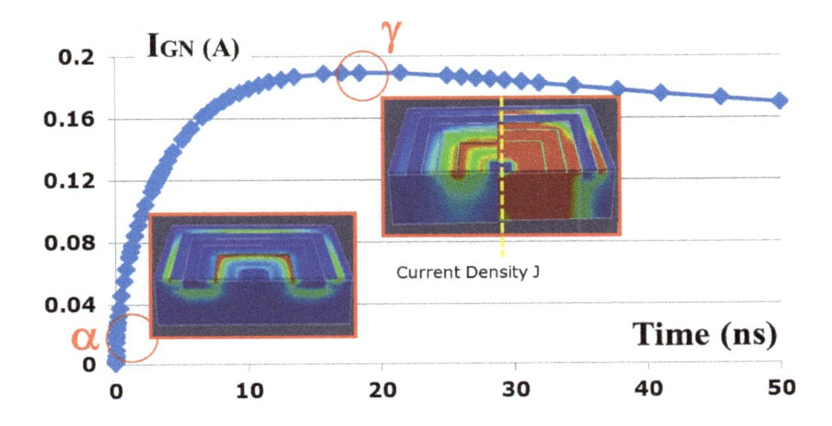

Figure 22.8 IGN(t) curves as collector current during surge and current density extraction at α & γ.

Table 22.10 Main values for extracted variables at the points α, β and γ

Diode	Unit	α = 0.07 ns IAK=37mA VAK=1.16V	β =3.07 ns IAK=1.22A VAK=3.04V	γ = 17.05 ns IAK=2.5A VAK=4.81V
A max	Vsec/m	5.27E-09	5.85E-08	1.13E-07
B max	T	0.007006	0.041654	0.066761
V max	V	1.77704	3.62909	5.35686
E max	V/m	5.45E+07	6.35E+07	7.13E+07
J max	A/m2	3.24E+12	3.25E+12	3.26E+12
Jn max	A/m2	5.44E+10	7.11E+11	1.03E+12
Jp max	A/m2	1.01E+10	1.01E+11	1.24E+11

22.5 Comparison with TLP Measurements

This last section is on the comparison between simulations and experimental measurements. The measurements are performed on a test chip with the standalone device with its associated trigger resistance and a complete trigger circuit. For the IV characterisation, a Transmission Line Pulse (TLP 50 ohms) is used with 10ns rise time and 100ns width. The comparison points are Vt1, Vhold and it is reminded here that the quasi-static behaviour is captured by TLP. That means that it is not possible to perform a direct temporal comparison between simulations and measurements. Moreover, for the robustness the It2 point is a sizing consideration and not discussed here. Table 22.6 summarizes the comparison.

Table 22.11 Comparison between measurements and simulation for Vt1 and Vhold on full ESD structure

Three Cases	TLP (10ns 100ms)	3D (BEOL+FEOL) EMLF Simulation
Case #1 : Rtrig = R0	Vt1= 13.7V Vhold =1.9V	Vt1= 16.5V Vhold =2V
Case #2 : Rtrig = R1 ¡ R0	Vt1= 4V Vhold =1.9V	Vt1= 4.5V Vhold =2V
Case #3 : Rtrig = R2 ¡ R1	Vt1= 1.9V Vhold =1.9V	Vt1= 2.4V Vhold =1.9V

At first order, the simulations are compliant with the measurements done at room temperature to characterize Vt1 and Vhold. The IV curve shapes are not the same for simulation and TLP due to the fact that the first one is time dependent and the TLP is a quasi-static behaviour ([80 ns-90 ns] capture window with average). Anyway, at first order the set of results is in good agreement and a fine assessment is required with more accurate models in the simulation, for example the thermal effect.

Summary

This chapter presents the main numerical and experimental results on an ultra-compact bi-directional ESD protection for advanced CMOS technology. To evaluate the numerical tool, a simple wire was selected as a first test case and demonstrator. It clearly shows that self-induction plays an important role during a fast-transient event. Next, an in-depth study was performed of the electromagnetic flash provided by a fast and high ESD surge in a complete structure (FEOL + BEOL). The structure is an SCR with an embedded diode. For positive stress, the SCR is active. It was shown that the magnetic field is confined in the SCR area and there is only a low silicon penetration. For the negative ESD event, the diode shows a magnetic field penetration in silicon with a typical signature. As the magnetic field B impacts the full structure. The simulation takes into account also the skin effect for metallic connections.

In order to provide a complete electromagnetic picture, we included the Lorentz force in the silicon transport equations. It is interesting to evaluate the electromagnetic distribution by solving the local equations in back-end-of-line (BEOL) and front-end-of-line (FEOL) simultaneously since this information is not yet accessible by near field measurements. Thanks to this study it is possible to give quantitative answers for the EM impact in its near environment at device level. Also, the calculated IV curves are compared to TLP for two typical points as $V)t1$ and V_{hold}. Good agreement is observed. Further assessment is needed to compare point-to-point the dynamic response. The numerical robustness of an integrated simulation approach was demonstrated and the tool was applied to demonstrate the good ESD response of a local ESD protection. This work and tool can be applied to Electro Magnetic Compatibility (EMC) topics. For example it is possible to calculate the EM field for a through silicon via (TSV) case or the generated EM field during high power switch. Other subjects of interest that become within reach are the magneto-resistance evaluation in transistors and the evaluation of the Hall effect in active/passive devices.

23

Coupled Electromagnetic-TCAD Simulation for Fast-Transient Systems

In this chapter we proceed with developing a numerically efficient formulation for electromagnetic-TCAD co-simulation for fast-transient computations[21]. The difficulties underlying the presently existing transient formulation stemming from the vector potential-scalar potential (A-V) framework are analyzed. A time-domain electric field-scalar potential (E-V) framework is then developed via equation and variable transformations. It results in better-conditioned systems that are friendly to iterative solution at fast switching times. Numerical examples show the proposed E-V solver renders a useful tool for addressing multi-domain simulation.

In recent years, there is a growing demand towards combining standalone electromagnetic (EM) solvers and technology computer-aided design (TCAD) semiconductor device simulators in mixed-signal, RF and multi-domain simulation. This is because the simplification of semiconductors (to conductors with equivalent conductivity) in linear EM analysis and the neglect of magnetic effects in TCAD simulation have become insufficient to capture the field-carrier interactions that are getting stronger as a direct consequence of the increasing operational frequency and the decreasing signal level. For instance, the nonlinear interplay between fast-varying EM fields and carrier flows in non-uniformly doped substrates is a subtle but important problem for the allocation of electrostatic discharge (ESD) structures (Galy et al. [2011]).

The EM-TCAD coupled simulation essentially refers to a concurrent solution of the Maxwell's equations that describe the ubiquitous EM dynamics, and the transport equations describing the charge carrier dynamics in semiconductors. A widely tested co-simulation framework in the frequency domain was proposed in (Meuris et al. [2001a] Schoenmaker and Meuris

[21]This chapter is a reprint of Chen et al. [2013].

[2002a]). The Maxwell's equations are formulated in terms of scalar potential V and vector potential $\mathbf{A}$ to obtain a straightforward coupling with the drift and diffusion (DD) semiconductor model. The A-V formulation has been translated into a series of tools (devEM [2003]), and verified against measurements with a number of industrial examples (CODESTAR [2003] ICESTARS [2008–2011]). Meanwhile, the A-V formulation has also been coupled with a quantum mechanical (QM) model to enable multiscale simulation (called QM/EM method) for emerging nano-electronic devices (Yam et al. [2011] Meng et al. [2012]).

Whereas the frequency-domain A-V solver has been developed in a rather advanced stage, the transient counterpart so far has been less explored. Unlike the objective of obtaining small-signal response in the frequency domain, the need of transient EM-TCAD simulation comes from the desire to handle large-signal response. A first implementation of time-domain A-V solver was reported recently in (Schoenmaker et al. [2012]) and (Meng et al. [2012]). The implicit backward Euler (BE) approach was employed for time discretization and the Newton's method was used to solve the nonlinear system thereby generated. Due to the extended physical reality and its numerical treatment, five variables are used and solved simultaneously in the formulation. The number of unknowns in each Newton iteration is thus $5 \sim 6$ times the number of nodes in the computational grid. As a consequence, the solution process for industrial problems represented in large linearized systems becomes a key concern.

Given the memory bottleneck and the lack of parallelizability for direct solvers, iterative methods are generally the most feasible tools for solving large sparse linear systems. However, the time-domain A-V formulation in (Schoenmaker et al. [2012]) inherits a numerical shortcoming from the frequency-domain version. As was shown in (Chen et al. [2011b]), the iterative solution of the linearized system tends to be increasingly inefficient as the frequency grows beyond 50GHz. The problem can be traced to the significant off-diagonal blocks in the Jacobian matrix arising from the high frequency and metal conductivity. The off-diagonal dominance induces an undesirable eigenvalue distribution in the preconditioned Jacobian matrices and compromises the performance of iterative solvers based on Krylov subspace methods such as GMRES. As a remedy, an E-V formulation was developed in the frequency-domain (Chen et al. [2011b]) relying on the variable transform (vector potential $\mathbf{A} \rightarrow$ electric field $\mathbf{E}$) and an equation transform for the nodes attached with metals (current continuity $\rightarrow$ gauge condition). The transformations substantially reduce magnitudes of the off-diagonal blocks at high frequencies and a remarkable performance boost of GMRES solver has been demonstrated.

In this chapter, an E-V based framework is developed in the time-domain for the fast-transient EM-TCAD simulation. The rationale is again to reduce the off-diagonal coupling and, more importantly, to prevent the potential singularity of the Jacobian matrix occurred in the metallic region at fast transients, which is also the root cause of slow convergence of iterative methods with the existing time-domain A-V formulation. In the rest of the paper, the time-domain A-V solver is briefly reviewed in the next section The fast-transient breakdown of A-V formula is analyzed next. The time-domain E-V solver is presented and numerical results are shown.

23.1 Time-Domain A-V formulation

The starting point of the time-domain EM-TCAD formulation are the full-wave Maxwell's equations

$$\nabla \cdot \mathbf{D} = \rho, \quad \nabla \cdot \mathbf{B} = 0 \tag{23.1a}$$

$$\nabla \times \mathbf{E} = -\frac{\partial \mathbf{B}}{\partial t}, \quad \nabla \times \mathbf{H} = \mathbf{J} + \frac{\partial \mathbf{D}}{\partial t} \tag{23.1b}$$

where $\mathbf{D}, \mathbf{E}, \mathbf{B}, \mathbf{H}, \mathbf{J}$ and ρ are the displacement field, electric field, magnetic induction, magnetic field, free current density and charge density, respectively. In semiconducting regions the source terms in the Maxwell's equations are submitted to the current continuity of electrons and holes

$$\nabla \cdot \mathbf{J}_n - \frac{\partial}{\partial t} n - R(n, p) = 0 \tag{23.2a}$$

$$\nabla \cdot \mathbf{J}_p + \frac{\partial}{\partial t} p + R(n, p) = 0 \tag{23.2b}$$

where n and p are the electron and hole densities, and $R(n, p)$ denotes the net generation/recombination rate of carriers. The particle current densities in semiconductor are described by the drift diffusion model

$$\mathbf{J}_n = q\mu_n n\mathbf{E} + qkT\mu_n \nabla n \tag{23.3a}$$

$$\mathbf{J}_p = q\mu_p p\mathbf{E} - qkT\mu_p \nabla p \tag{23.3b}$$

where μ, k and T are the carrier mobility, Boltzmann constant and temperature, respectively. (23.3) (Scharfetter and Gummel [1969]).

To facilitate the coupling between the EM and TCAD systems, the Maxwell's equations are reformulated by introducing the potential variables V and $\mathbf{A}$ such that $\mathbf{B} = \nabla \times \mathbf{A}$ and $\mathbf{E} = -\nabla V - \frac{\partial \mathbf{A}}{\partial t}$ (Meuris et al. [2001b] Schoenmaker and Meuris [2002a]). Furthermore, a new variable, called the pseudo-canonical momentum $\mathbf{\Pi} = \frac{\partial \mathbf{A}}{\partial t}$, is introduced with the purpose to avoid the second-order time derivative acting on the vector potential (Schoenmaker et al. [2012]). The complete system of equations is laid out in (23.4) (with scaling)

$$\begin{cases} \nabla \cdot [\varepsilon_r (-\nabla V - \mathbf{\Pi})] - \rho = 0, \ \rho = p - n + N_D \\ \nabla \cdot \left[\varepsilon_r \left(-\nabla \frac{\partial}{\partial t} V - \frac{\partial}{\partial t}\mathbf{\Pi}\right)\right] + \nabla \cdot [\sigma (-\nabla V - \mathbf{\Pi})] + \nabla \cdot \mathbf{J}_{sd} = 0 \end{cases} \tag{23.4a}$$

$$\nabla \cdot \mathbf{J}_n - \frac{\partial}{\partial t} n - R(n,p) = 0 \tag{23.4b}$$

$$\nabla \cdot \mathbf{J}_p + \frac{\partial}{\partial t} p + R(n,p) = 0 \tag{23.4c}$$

$$\frac{\partial}{\partial t}\mathbf{A} - \mathbf{\Pi} = 0 \tag{23.4d}$$

$$- K\varepsilon_r \left(-\frac{\partial}{\partial t}\mathbf{\Pi} - \nabla \frac{\partial}{\partial t} V\right) + K\nabla \left(-\varepsilon_r \frac{\partial}{\partial t} V\right) + [\nabla \times (\nabla \times \mathbf{A}) - \nabla (\nabla \cdot \mathbf{A})] - K\sigma (-\nabla V - \mathbf{\Pi}) - K\mathbf{J}_{sd} = 0 \tag{23.4e}$$

where $\mathbf{J}_{sd} = \mathbf{J}_n + \mathbf{J}_p$ is the total semiconductor current and K is a dimensionless constant which will be discussed later in the scaling scheme. The upper equation in (23.4a) is the conventional Gauss law, and the lower one represents the current continuity for the nodes attached with metals, where the Gauss law will be used to recover the charge densities that are not explicit unknowns. Equation (23.4e) represents a "modified" Maxwell-Ampere equation that includes the subtraction of the divergence of Lorentz gauge condition

$$\nabla \cdot \mathbf{A} + K\varepsilon_r \frac{\partial V}{\partial t} = 0 \tag{23.5}$$

to eliminate the intrinsic singularity of the curl-curl operator (Chen et al. [2011b]) by conversion to a "Laplacian" form $\nabla^2 = \nabla \times (\nabla \times) - \nabla(\nabla \cdot)$.

The generalized "de Mari" scaling is applied to (23.4) to improve the numerical range of the physical quantities that may differ by orders of magnitude. Four independent scaling parameters, λ for lengths, n_i for doping and carrier concentrations, V_T for scalar potentials and s_D for diffusion constants, are pre-determined, from which the remaining scaling parameters are derived from their physical relation. The length scaling λ is usually chosen to be the characteristic grid size to make the matrix entries generated by the differential operators of magnitude $O(1)$. The detailed scaling factors are presented in (Schoenmaker and Meuris [2002a]).

The system after spatial discretization can be assembled in a matrix format as shown in (21.6) on top of next page. V_1 and V_2 are partitioned according to the different governing equations in (23.4a). Similarly, $\mathbf{\Pi}$ is divided into $\mathbf{\Pi}_1$ and $\mathbf{\Pi}_2$ for the links attached without and with metallic volumes, respectively. I denotes the identity matrix with the appropriate dimension.

Equation (21.6) can be further condensed to a nonlinear differential equation

$$C\dot{x} = -Gx - F(x) - b \tag{23.6}$$

where the constant matrices C and G collect the linear dynamics in the system, F collects the nonlinear dynamics (generally the nonlinear semiconductor currents), and b the source term.

In temporal regime, (23.6) is generally discretized by backward differential formula (BDF). For simplicity, we employ here the first-order BDF, or BE, discretization. The Newton's method is then applied to handle the nonlinear algebraic system generated by BE. The linearized equation to be solved at the $k + 1$th Newton iteration of the $n + 1$ time step is therefore

$$\begin{aligned} &\left(J_F + \frac{C}{h} + G\right)\Delta x_{n+1}^{k+1} \\ &= -\left(F(x_{n+1}^k) + \left(\frac{C}{h} + G\right)x_{n+1}^k - \frac{C}{h}x_n + b_{n+1}\right) \end{aligned} \tag{23.7}$$

$$\left[\begin{array}{ccccc} -\begin{bmatrix} 0 & 0 \\ \nabla\cdot(\varepsilon_r\nabla) & 0 \end{bmatrix} & & & & -\begin{bmatrix} 0 & 0 \\ \nabla\cdot\varepsilon_r & 0 \end{bmatrix} \\ & -I & & & \\ & & I & & \\ & & & I & \\ \begin{bmatrix} K(\varepsilon_r\nabla - \nabla\varepsilon_r) & 0 \\ 0 & K(\varepsilon_r\nabla - \nabla\varepsilon_r) \end{bmatrix} & & & & \begin{bmatrix} K\varepsilon_r & 0 \\ 0 & K\varepsilon_r \end{bmatrix} \end{array}\right] \begin{bmatrix} \begin{bmatrix} \dot{V}_1 \\ \dot{V}_2 \end{bmatrix} \\ \dot{n} \\ \dot{p} \\ \dot{A} \\ \begin{bmatrix} \dot{\Pi}_1 \\ \dot{\Pi}_2 \end{bmatrix} \end{bmatrix} =$$

$$-\begin{bmatrix} -\begin{bmatrix} \nabla\cdot(\varepsilon_r\nabla) & \nabla\cdot(\varepsilon_r\nabla) \\ 0 & \nabla\cdot(\sigma\nabla) \end{bmatrix} & \begin{bmatrix} I_n \\ 0 \end{bmatrix} & \begin{bmatrix} -I_p \\ 0 \end{bmatrix} & & -\begin{bmatrix} \nabla\cdot\varepsilon_r & 0 \\ 0 & \nabla\cdot\sigma \end{bmatrix} \\ & 0 & & & \\ & & 0 & & \\ & & & 0 & -I \\ \begin{bmatrix} 0 & 0 \\ 0 & K\sigma\nabla \end{bmatrix} & & & \begin{bmatrix} \nabla^2 \\ \nabla^2 \end{bmatrix} & \begin{bmatrix} 0 & 0 \\ 0 & K\sigma \end{bmatrix} \end{bmatrix} \begin{bmatrix} \begin{bmatrix} V_1 \\ V_2 \end{bmatrix} \\ n \\ p \\ A \\ \begin{bmatrix} \Pi_1 \\ \Pi_2 \end{bmatrix} \end{bmatrix}$$

$$-\begin{bmatrix} \begin{bmatrix} 0 \\ \nabla\cdot\bar{J}_{sd} \end{bmatrix} \\ \nabla\cdot\bar{J}_n - R \\ \nabla\cdot\bar{J}_p + R \\ 0 \\ \begin{bmatrix} -K\bar{J}_{sd} \\ -K\bar{J}_{sd} \end{bmatrix} \end{bmatrix} - \begin{bmatrix} \begin{bmatrix} V_S \\ 0 \end{bmatrix} \\ 0 \\ 0 \\ 0 \\ \begin{bmatrix} 0 \\ 0 \end{bmatrix} \end{bmatrix}$$

where $J_F = \frac{\partial F}{\partial x}\Big|_{x^k_{n+1}}$ is the partial derivatives of nonlinear semiconductor currents and h is the time step size.

23.2 Analysis of Fast-Transient Breakdown

Performance of the iterative solution of (23.7) depends on the eigenvalue distribution of the Jacobian matrix $J = \left(\frac{\partial F}{\partial x}\Big|_{x^k_{n+1}} + \frac{C}{h} + G\right)$. More precisely, since preconditioners are inevitably needed, the spectrum of the preconditioned matrix will play a crucial role in determining the convergence properties of most existing iterative methods (Kharchenko and Yeremin [1995]). To further facilitate the analysis, we focus on the contribution of the linear components (e.g., back-end structures) to the Jacobian matrix by forming an M matrix as in (23.8), which includes only the blocks related to $(V, \mathbf{A}, \mathbf{\Pi})$. Note that the displacement currents for metal nodes/links are omitted due to the small values compared with the conduction currents. Semiconductors involved in typical EM-TCAD simulation have light to medium dopings, thus from a numerical point of view they can be regarded as equivalent conductors with low conductivity. Their existence in the Jacobian matrix, as will be evident in later discussions, does not affect results of the following analysis.

$$M = \begin{bmatrix} -\begin{bmatrix} \nabla\cdot\varepsilon_r\nabla & \nabla\cdot(\varepsilon_r\nabla) \\ 0 & \nabla\cdot(\sigma\nabla) \end{bmatrix} & 0 & -\begin{bmatrix} \nabla\cdot\varepsilon_r & 0 \\ 0 & \nabla\cdot\sigma \end{bmatrix} \\ 0 & I/h & -I \\ \begin{bmatrix} K\left(\varepsilon_r\nabla - \nabla\varepsilon_r\right)/h & 0 \\ 0 & K\sigma\nabla \end{bmatrix} & \begin{bmatrix} \nabla^2 \\ \nabla^2 \end{bmatrix} & \begin{bmatrix} K\varepsilon_r/h & 0 \\ 0 & K\sigma \end{bmatrix} \end{bmatrix} \tag{23.8}$$

Here the conductivity $\sigma = \sigma_0/s_\sigma$ has been scaled by $s_\sigma = \varepsilon_0/\tau$, where σ_0 is the real metal conductivity and ε_0 is the permittivity in vacuum. The step size h has also been scaled by τ. For general materials in integrated circuits, $\sigma \gg \varepsilon_r$. The K factor is defined by $\mu_0\varepsilon_0(\lambda/\tau)^2$ where μ_0 is the vacuum permeability.

Similar to the frequency-domain counterpart (Chen et al. [2011b]), the magnitudes of off-diagonal blocks in the time-domain formula (23.8) depends on the time stepping under consideration (h must be anti-proportional to the shortest timestep to resolve the transient signal behavior) as well as the metal conductivity. Spectral analysis in (Chen et al. [2011b]) indicated that the large off-diagonal blocks in time-domain simulation with metallic conductors would induce eigenvalues spread away from the unity in the complex plane in the preconditioned Jacobian matrix and significantly slow down the convergence of GMRES.

The numerical difficulty associated with (23.8) can be alleviated by scaling the second row by σ and the last row by $K\sigma$. In addition, one can force $h = O(1)$ by setting τ to the typical time step size to eliminate the timestep-dependence of most matrix blocks. This amounts to making τ an independent scaling parameter instead of s_D. The M matrix after balancing reads

$$M = \begin{bmatrix} -\begin{bmatrix} \nabla\cdot\varepsilon_r\nabla & \nabla\cdot(\varepsilon_r\nabla) \\ 0 & \nabla\cdot(\nabla) \end{bmatrix} & 0 & -\begin{bmatrix} \nabla\cdot\varepsilon_r & 0 \\ 0 & \nabla\cdot \end{bmatrix} \\ 0 & I & -I \\ \begin{bmatrix} K(\varepsilon_r\nabla - \nabla\varepsilon_r) & 0 \\ 0 & \nabla \end{bmatrix} & \begin{bmatrix} \nabla^2 \\ \nabla^2/(K\sigma) \end{bmatrix} & \begin{bmatrix} K\varepsilon_r & 0 \\ 0 & I \end{bmatrix} \end{bmatrix} \tag{23.9}$$

It is evident from (23.9) that M has generally balanced blocks except the the block $\nabla^2/(K\sigma)$, which decreases linearly with the reciprocal of τ, or the frequency of interest

$$K\sigma = \mu_0\varepsilon_0\frac{\lambda^2}{\tau^2}\frac{\sigma_0\tau}{\varepsilon_0} = \mu_0\sigma_0\frac{\lambda^2}{\tau} \propto \frac{1}{\tau} \tag{23.10}$$

As the $\nabla^2/(K\sigma)$ term becomes smaller at fast transients, the last row and the 2nd row in (23.9) become increasingly close to each other, rendering an ill-conditioned Jacobian matrix. The matrix will even become singular when the term $\nabla^2/(K\sigma)$ vanishes. This intuitively explains the difficulty in iterative solution the A-V solver encounters when simulating systems with rapid-varying dynamics.

An in-depth analysis can be carried out by looking into the modified Maxwell-Ampere equation (23.4e). It has been elucidated in (Chen et al. [2011b]) that the combination of the current-continuity and the original Maxwell-Ampere equation would lead to a redundant system. To eliminate the redundancy, the divergence of the gauge condition is subtracted from the original Maxwell-Ampere equation, resulting in (23.4e). Yet the two terms in the divergence of the gauge condition have magnitudes of $O(1)$ and $O(K\varepsilon_r)$ under the scaling. The large conductivity in metallic region will easily become so dominant that it "turns" the modified Maxwell-Ampere equation into nothing but another current continuity equation when $K\sigma$ is large, which induces ill-conditioned or even singular systems again.

A question follows immediately is when the conduction current term in (23.4e) will dominate the equation, or when $K\sigma$ will become large. One can derive from (23.10) that the real definition of "fast-transient" depends on the ratio of spatial and temporal scales of the simulation. Letting $K\sigma = 1$, and noting that $\mu_0\sigma_0 \sim O(1)$ for typical conducting materials, the following relation describes that when

$$\tau \approx \lambda^2 \tag{23.11}$$

the fast-transient breakdown starts to occur with the A-V formulation (recall that τ and λ are generally compliant with the time and length scales of the simulation). For micron-range structures where $\lambda \approx 10^{-6}$, $\tau \approx 10^{-12}$, which roughly corresponds to 50GHz ($\tau \sim 1/20$ of period). For nano-scale problems with $\lambda \approx 10^{-9}$, τ will be 10^{-18}, which suggests that the A-V solver will generally not encounter severe problem until 10^{17}Hz. In addition, if no metals are included in the structures to be simulated, the redundancy problem can also be largely avoided.

23.3 Time-Domain E-V Formulation

The numerical inefficiency associated with the A-V solver for fast-transient systems mainly results from the concurrent solution of the current continuity and the modified Maxwell-Ampere equation in the metallic region. A natural remedy will hence be replacing the current continuity in the metals (the lower one in (23.4a)) by the gauge condition (23.5) as in (Chen et al. [2011b]). It is a valid equation transformation as long as it is only applied on the nodes attached with metallic volumes, while keeping the original Gauss's law for the remaining nodes. This way, redundancy is avoided even when the modified Maxwell-Ampere equation "collapses" to the current continuity equation for fast transients.

Meanwhile, the comparable diagonal and off-diagonal blocks in (23.9) are not desirable for fast convergence of iterative methods, which is attributed to the A-V potential formulation of Maxwell's equations where the electric field $\mathbf{E}$ is separated into V and $\mathbf{\Pi}$ components but still used as an entity in the expression of displacement current. Therefore, a variable transformation $\mathbf{E} = -\nabla V - \mathbf{\Pi}$ is employed to minimize the cross-coupling between V and Π in the relevant equations. The complete E-V equation system is laid out in (23.12)

$$\begin{cases} \nabla \cdot (\varepsilon_r \mathbf{E}) - \rho = 0, \ \rho = p - n + N_D \\ K\varepsilon_r \frac{\partial}{\partial t} V + \nabla \cdot \mathbf{A} = 0 \end{cases} \tag{23.12a}$$

$$\nabla \cdot \mathbf{J}_n - \frac{\partial}{\partial t} n - R(n, p) = 0 \tag{23.12b}$$

$$\nabla \cdot \mathbf{J}_p + \frac{\partial}{\partial t} p + R(n, p) = 0 \tag{23.12c}$$

$$\nabla V + \frac{\partial}{\partial t} \mathbf{A} + \mathbf{\Pi} = 0 \tag{23.12d}$$

$$\begin{aligned} &- K\varepsilon_r \frac{\partial}{\partial t} \mathbf{E} + K\nabla \left(-\varepsilon \frac{\partial}{\partial t} V \right) + [\nabla \times (\nabla \times \mathbf{A}) - \nabla (\nabla \cdot \mathbf{A})] \\ &- K\sigma \mathbf{E} - K\mathbf{J}_{sd} = 0 \end{aligned} \tag{23.12e}$$

Analogous to the analysis of A-V solver, the M matrix in the E-V solver is constructed for $(V, \mathbf{A}, \mathbf{E})$ as

$$M = \begin{bmatrix} \begin{bmatrix} 0 & 0 \\ 0 & K\varepsilon_r \end{bmatrix} & \begin{bmatrix} 0 \\ \nabla\cdot \end{bmatrix} & \begin{bmatrix} \nabla \cdot \varepsilon_r & 0 \\ 0 & 0 \end{bmatrix} \\ \nabla & I & I \\ \begin{bmatrix} K\nabla\varepsilon_r & 0 \\ 0 & 0 \end{bmatrix} & \begin{bmatrix} -\nabla^2 \\ -\nabla^2 \end{bmatrix} & \begin{bmatrix} K\varepsilon_r & 0 \\ 0 & K\sigma \end{bmatrix} \end{bmatrix} \tag{23.13}$$

The merit of the E-V formulation lies in that the diagonal dominance in the last row of (23.13) is largely improved. The large metal conductivity only appears at diagonals, and the diagonal term increases linearly with respect to the frequency as $K\sigma \sim 1/\tau$. The negative side is that the diagonal term $K\varepsilon_r$ in the gauge condition is small when the frequency is low, rendering high off-diagonal dominance that affects the performance of iterative methods. Nevertheless, this term increases quadratically with the reciprocal of the time step, $K \sim 1/\tau^2$. Therefore, the E-V solver will have better performance in high-frequency simulations, similar to its counterpart in the frequency-domain.

Note that there is a limit that the E-V solver can gain from the increase of switching rate, which occurs when the last two rows of (23.13) starts to become off-diagonal dominant, i.e., $K\varepsilon_r \approx K\sigma$. The corresponding h is $h_{min} = \varepsilon_r\varepsilon_0/\sigma_0$, which is around 10^{-18} for typical materials. Beyond this step size the performance of E-V solver may decrease.

23.4 Numerical Results

The time-domain A-V and E-V solvers are both implemented in Matlab. Three structures with the specifications detailed in Table 23.1 were simulated by the two solvers. The iterative solutions are obtained by GMRES, along with the threshold-based incomplete LU preconditioner (ILUT), and the column approximate minimum degree (COLAMD) permutation (Davis et al. [2004b]) to reduce fill-ins in the preconditioner. This combination represents the state-of-the-art in iterative solution of large sparse linear systems. All tests were conducted on a 3.2GHz 16Gb-RAM Linux server.

The validity of the E-V formula is verified in Figure 23.1. The transient current through one conductor of the XWR structure is calculated by the A-V and the E-V solvers when a sinusoidal voltage of 10GHz is applied at one end of the upper conductor. The step size h is 10^{-11}s and the direct linear solver (backslash in Matlab) is applied. The A-V and E-V curves overlap on top of each other, which is expected since no approximation is introduced with the variable and equation transformations.

The GMRES efficiencies of the A-V and the E-V solvers with different step sizes are compared in Table 23.2 for the three test cases. The corresponding frequency range is roughly 100MHz$\sim$100THz. The same COLAMD+ILUT(10^{-4})+GMRES combo is applied in all cases. t_{pre} denotes the time for constructing the ILUT preconditioner. t_{GMRES} and N_{it} are the

Table 23.1 Specifications of test structures

Case	Description	# of Nodes	# of Unknowns
XWR	Cross metal wires embedded in insulator (Chen et al. [2011b])	1,400	6,520
SUB	Substrate noise isolation structure (Chen et al. [2011b])	6,384	41,368
VCO	8-shaped inductor in voltage controlled oscillator (Schoenmaker et al. [2012])	21,312	149,898

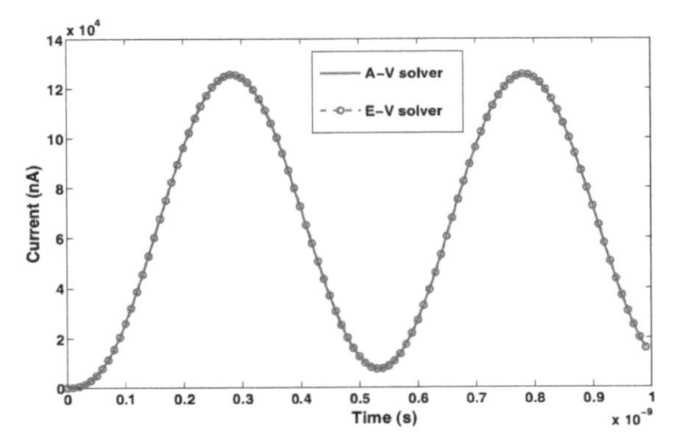

Figure 23.1 Conductor current in the XWR structure obtained by A-V and E-V solvers.

Table 23.2 Performance of iterative solution in A-V and E-V solvers

		A-V			E-V		
Case	h	t_{pre}	t_{GMRES}	N_{it}	t_{pre}	t_{GMRES}	N_{it}
XWR	10^{-9}	0.89	0.13	5	0.99	2.19	58
	10^{-11}	1.86	0.28	6	0.65	0.22	11
	10^{-13}	1.94	1.23	23	0.76	0.2	9
	10^{-15}	1.12	5.66	164	0.24	0.08	9
SUB	10^{-9}	10.61	4.21	27	17.85	23.98	81
	10^{-11}	15.31	5.34	27	17.59	3.64	14
	10^{-13}	49.71	13.65	37	30.92	3.96	12
	10^{-15}	55.45	153.71	262	21.41	5.93	20
VCO	10^{-9}	285	25.48	17	466.86	–	–
	10^{-11}	412.58	30.9	17	425.92	70.58	60
	10^{-13}	1274.9	–	–	210.25	51.33	42
	10^{-15}	–	–	–	87.41	18.74	34

total time of GMRES and the number of iterations required to achieve the tolerance of 10^{-6}, respectively. Symbol – indicates failure in computing preconditioner or getting convergence within 300 iterations

As demonstrated in Table 23.2, the numerical systems arising from the A-V formulation are favorable for iterative solution for slow-transient systems. As the necessary time step decreases, the convergence rate of GMRES continuously slows down and failures occur for some small step sizes. The E-V solver has an opposite performance trend. The GMRES convergence is slow at lower frequencies. For systems with fast-varying dynamics, however, the E-V solver significantly outperforms the A-V solver. This suggests that the E-V solver will be an valuable asset for EM-TCAD co-analysis for electronic devices operating in RF or even higher frequency range. The convenient switch between A-V and E-V solvers (Chen et al. [2011b]) facilitates a wide-band simulation across the preferred regions of both solvers. It is also shown that the inclusion of semiconductors (the SUB and VCO examples) does not change the general conclusion.

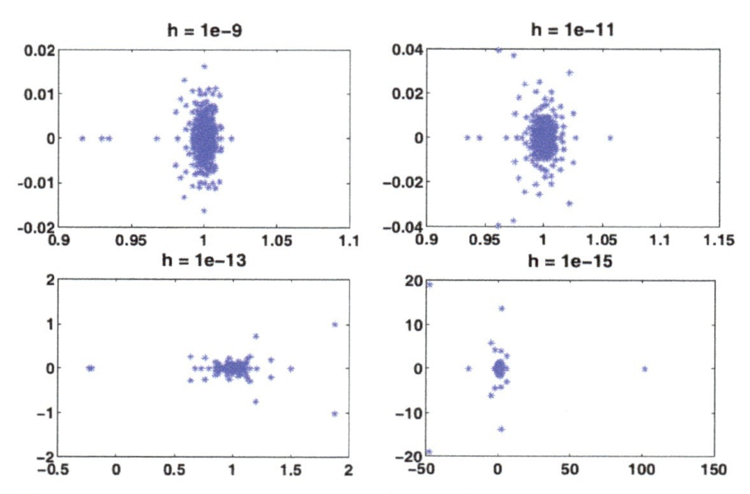

Figure 23.2 Eigenvalue distribution of A-V solver with different step sizes (preconditioned Jacobian matrix from the XWR case).

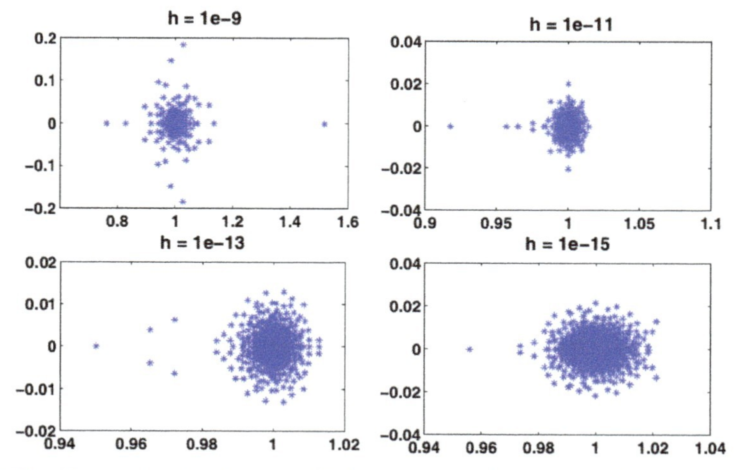

Figure 23.3 Eigenvalue distribution of E-V solver with different step sizes (preconditioned Jacobian matrix from the XWR case).

Spectral analysis is further conducted to explain the frequency-dependent behaviors of A-V and E-V solvers. The eigenvalue distributions of the Jacobian matrix preconditioned by the computed LU factor, i.e., $U^{-1}L^{-1}M$, are visualized for four step sizes in Figure 23.2 and Figure 23.3. It is seen that the eigenvalues of A-V solver at slow transients cluster tightly in the unity, but spread away at fast transients. The smaller is the step size, the more outlying eigenvalues appear, indicating a slower GMRES convergence. The spectrum of E-V solver follows a reversed trend. The eigenvalues are much more clustered for small time steps than for large time steps, which is consistent with the observations in Table 23.2.

Summary

We have revealed that the time-domain EM-TCAD co-simulation based upon the existing A-V formula will encounter numerical difficulty when simulating metallic structures at fast switching times. An E-V formulation is developed by equation and variable transformations to improve the conditioning of the numerical systems and thus enhance the efficiency of iterative solvers. It has been demonstrated that the E-V solver leads to a significant speedup for fast-transient systems, and would therefore be an indispensable tool for multi-domain simulation involving fast transients.

24

A Fast Time-Domain EM-TCAD Coupled Simulation Framework via Matrix Exponential with Stiffness Reduction

We present a fast time-domain multiphysics simulation framework that combines full-wave electromagnetism (EM) and carrier transport in semiconductor devices (TCAD)[22]. The framework features a division of linear and nonlinear components in the EM-TCAD coupled system. The former is extracted and handled independently with high efficiency by a matrix exponential approach assisted with Krylov subspace method. The latter is treated by ordinary Newton's method yet with a much sparser Jacobian matrix that leads to substantial speedup in solving the linear system of equations. More convenient error management and adaptive control are also available through the linear and nonlinear decoupling. Furthermore, a system formulation is developed to further enhance the efficiency of the proposed framework.

In recent years, there is a rising interest in combining stand-alone electromagnetic (EM) solvers and technology computer-aided design (TCAD) semiconductor device simulators in mixed-signal, RF and multi-domain simulation. This is because the simplifications made to decouple the EM and TCAD dynamics, i.e., equivalent conductivity model for semiconductors in linear EM analysis and the neglect of magnetic effects and finite conductivity of metals in TCAD simulation, fall short to characterize the ever-growing field-carrier interactions caused by the increasing operational frequency and the decreasing signal level.

EM-TCAD co-simulation refers to a concurrent solution of the Maxwell's equations that describe the EM dynamics and transport models that describe the charge carrier dynamics in semiconductor devices. So far only a few

[22]This chapter is a reprint of Chen et al. [2015].

frameworks have been developed to allow such coupling, see e.g., (Wang et al. [2002b] Bertazzi et al. [2006b] Willis et al. [2011]), based on different selections of EM and TCAD solvers for combination. One experimentally verified co-simulation framework was firstly proposed in (Meuris et al. [2001b] Schoenmaker and Meuris [2002a]) in the frequency domain. The Maxwell's equations are formulated in terms of scalar potential V and vector potential $\mathbf{A}$ to obtain straightforward coupling with the drift and diffusion (DD) semiconductor model (called the A-V formulation henceforth). The spatial discretization scheme based on the finite volume method (FVM) and finite surface method (FSM) is fully compliant with the geometric meaning of the differential operators. The A-V co-simulator has been translated into a series of commercial tools (devEM [2003]), and verified against measurements with a number of industrial examples (CODESTAR [2003] ICESTARS [2008–2011]). Meanwhile, the A-V formulation has also been coupled with a quantum-mechanical model to enable a multiscale simulation framework (called QM/EM method) for emerging nano-electronic devices (Yam et al. [2011] Meng et al. [2012]).

The time-domain A-V simulator has been proposed in (Chen et al. [2012]) to meet the desire to handle large-signal response and to complement the small-signal analysis applied in its frequency-domain counterpart. The latter ignores all non-linear effects originating from the signal amplitude. The implicit backward Euler (BE) approach was employed for time discretization and the Newton's method was used to solve the nonlinear system arising from semiconductor dynamics. Due to the extended physical reality and its numerical treatment, five variables are used and solved simultaneously in the formulation. Three variables are located on nodes and two variables are defined on links. Therefore, the number of unknowns in each Newton iteration can be $\sim$ 9 times the number of nodes in the computational grid for large-scale problems. Such rapid growth of problem size severely limits applicability of the co-simulation framework.

To overcome the computational hurdle, a new solution technique based on the matrix exponential (Chen et al. [2012]) method (MEXP). was proposed. The idea is to decompose an EM-TCAD system into linear and non-linear subsystems corresponding roughly to the back-end and front-end structures in the problem of interest. The transient response of the linear sector is described by the matrix exponential formula, which is computed efficiently by the Krylov subspace method only one time in each time step. The nonlinear subsystem is solved iteratively as usual by the Newton's method, but with a much sparser Jacobian matrix enabling faster linear solution than in traditional schemes, where the linear components are mixed into the Jacobian. This way, one can largely alleviate the bottleneck of solving large-scale linear systems in EM-TCAD and other nonlinear transient simulation problems. The MEXP method used in the linear numerical integration has also been shown to be a more scalable approach for million-scale problems than conventional linear multi-step methods (LMS) such as backward Euler and trapezoidal methods (Weng et al. [2011, 2012a]).

In this paper, we provide a proof for the validity of regularizing the differential-algebraic equation (DAE) system via differentiating the Gauss's law in our EM-TCAD framework, which is a crucial step to make the matrix exponential formula applicable through reducing the DAE into an ordinary differential equation (ODE). In addition, we present an alternative formulation of the EM-TCAD problem (called the E-V formulation) via variable and equation transformation to reduce the stiffness of the linear subsystem and accelerate the Krylov subspace computation. The numerical section is also largely enriched with more data and analysis, which in turn confirm the superiority of the proposed transient EM-TCAD framework.

24.1 Time-Domain Formulation of EM-TCAD Problem

The starting point of the time-domain EM-TCAD formulation are the common full-wave Maxwell's equations

$$\nabla \cdot \mathbf{D} = \rho, \quad \nabla \cdot \mathbf{B} = 0 \tag{24.1a}$$

$$\nabla \times \mathbf{E} = -\frac{\partial \mathbf{B}}{\partial t}, \quad \nabla \times \mathbf{H} = \mathbf{J} + \frac{\partial \mathbf{D}}{\partial t} \tag{24.1b}$$

where $\mathbf{D}, \mathbf{E}, \mathbf{B}, \mathbf{H}, \mathbf{J}$ and ρ are the displacement field, electric field, magnetic induction, magnetic field, free current density and charge density, respectively.

In the semiconductor region the Maxwell's equations are complemented with the current continuity of electrons and holes

$$\frac{1}{q}\nabla \cdot \mathbf{J}_n - \frac{\partial}{\partial t}n - R\,(n,p) = 0 \tag{24.2a}$$

$$\frac{1}{q}\nabla \cdot \mathbf{J}_p + \frac{\partial}{\partial t}p + R\,(n,p) = 0 \tag{24.2b}$$

where n and p are the electron and hole densities, and $R(n,p)$ denotes the net generation/recombination rate of carriers and q is the elementary charge.

The particle current densities in semiconductor are described by the drift-diffusion model

$$\mathbf{J}_n = q\mu_n n\mathbf{E} + kT\mu_n \nabla n \tag{24.3a}$$

$$\mathbf{J}_p = q\mu_p p\mathbf{E} - kT\mu_p \nabla p \tag{24.3b}$$

where μ, k and T denote the carrier mobility, Boltzmann constant and temperature, respectively.

To facilitate the coupling between the EM and TCAD solvers, the Maxwell's equations are recasted into the potential form using the variables V and $\mathbf{A}$ with the result that $\mathbf{B} = \nabla \times \mathbf{A}$ and $\mathbf{E} = -\nabla V - \frac{\partial \mathbf{A}}{\partial t}$ (Meuris et al. [2001b] Schoenmaker and Meuris [2002a]). A new variable the pseudo-canonical momentum $\mathbf{\Pi} = \frac{\partial \mathbf{A}}{\partial t}$ is also introduced to avoid the second-order time derivative (Schoenmaker et al. [2012]). The complete system of equations is then laid out in (24.4) utilizing a generalized "de Mari" scaling scheme (Schoenmaker and Meuris [2002a]))

$$\begin{cases} \frac{1}{\nu}\nabla \cdot [\varepsilon_r(-\nabla V - \mathbf{\Pi})] - \rho = 0, \ \rho = p - n + N_D \\ \frac{1}{\nu}\nabla \cdot \left[\varepsilon_r\left(-\nabla \frac{\partial}{\partial t}V - \frac{\partial}{\partial t}\mathbf{\Pi}\right)\right] + \nabla \cdot [\sigma(-\nabla V - \mathbf{\Pi})] + \nabla \cdot \mathbf{J}_{sd} = 0 \end{cases} \tag{24.4a}$$

$$\nabla \cdot \mathbf{J}_n - \frac{\partial}{\partial t}n - R(n,p) = 0 \tag{24.4b}$$

$$\nabla \cdot \mathbf{J}_p + \frac{\partial}{\partial t}p + R(n,p) = 0 \tag{24.4c}$$

$$\frac{\partial}{\partial t}\mathbf{A} - \mathbf{\Pi} = 0 \tag{24.4d}$$

$$\begin{aligned} &-K\varepsilon_r\left(-\frac{\partial}{\partial t}\mathbf{\Pi} - \nabla \frac{\partial}{\partial t}V\right) - K\nabla\left(\varepsilon_r \frac{\partial}{\partial t}V\right) + \\ &[\nabla \times (\nabla \times \mathbf{A}) - \nabla(\nabla \cdot \mathbf{A})] - K\nu\sigma(-\nabla V - \mathbf{\Pi}) - K\nu \mathbf{J}_{sd} = 0 \end{aligned} \tag{24.4e}$$

where $\mathbf{J}_{sd} = \mathbf{J}_n + \mathbf{J}_p$ is the total semiconductor current. K and ν are two dimensionless constants of the scaling method (Chen et al. [2013]). In the scaling scenario where λ_{L} and λ_{T} are the natural units of length and time we find that $K = \mu_0\epsilon_0\frac{\lambda_{\mathrm{L}}^2}{\lambda_{\mathrm{T}}^2}$. For example: for a small-size microelectronic device with a typical length of about one micron and a switching frequency around five GHz, we may set $\lambda_{\mathrm{L}} = 10^{-6}$ meter and $\lambda_{\mathrm{T}} = 10^{-10}$ second. The dimensionless parameter ν corresponds to the characteristic relaxation time of conductive materials. In particular, if $\lambda_{\rm COND}$ is the 'natural unit' of conductance (for example $\lambda_{\rm COND} = 10^7$ S/m for Alumium) we find that $\nu = \lambda_{\rm COND} \times \lambda_{\rm T}/\epsilon_0$. The meaning of the parameter ν becomes clear when combining Gauss' law with the current continuity and eliminating the electric field, i.e. $\frac{\sigma}{\epsilon}\rho + \frac{\partial \rho}{\partial t} = 0$ or $\rho \sim \exp\left(-\frac{\sigma t}{\epsilon}\right)$.

The upper equation in (24.4a) is the common Gauss law, and the lower one is the current continuity equation that is applied to for which are attached to at least one grid cell consisting of a conductor (metal). For these nodes Gauss law is used to recover the charge densities which are not explicit unknowns. Note that a node may be located at a metal-insulator interface. As a consequence the discretezation or equation assembling receives contribution from both metallic and semi-conductor cells. Therefore both currrent types are found in (24.4a). The Maxwell-Ampere equation is given in (24.4e). It includes the subtraction of the divergence of Lorentz gauge condition

$$\nabla \cdot \mathbf{A} + K\varepsilon_r \frac{\partial V}{\partial t} = 0 \tag{24.5}$$

to eliminate the intrinsic singularity of the curl-curl operator (Chen et al. [2011b]) and obtaining a Laplacian-like operator $\nabla^2 = \nabla(\nabla\cdot) - \nabla \times (\nabla \times)$.

After spatial discretization the system can be assembled in a matrix format as shown in (24.7) on top of next page. The voltage V is partioned into V_1 and V_2 according to the different governing equations in (24.4a). Similarly, $\mathbf{\Pi}$ is divided into $\mathbf{\Pi}_1$ and $\mathbf{\Pi}_2$ for the links attached to volumes that are metallic or not, i.e. insulating or semiconducting respectively. I denotes an identity matrix of appropriate size. Equation (24.7) can be further condensed to a nonlinear differential equation

$$C\dot{x} = -Gx - F(x) - b \tag{24.6}$$

$$\begin{bmatrix} \begin{bmatrix} 0 \\ -\frac{1}{\nu}\nabla\cdot(\varepsilon_r\nabla) \end{bmatrix} & & & & \begin{bmatrix} 0 \\ -\frac{1}{\nu}\nabla\cdot\varepsilon_r \end{bmatrix} \\ & I & & & \\ & & I & & \\ & & & I & \\ \begin{bmatrix} K(\varepsilon_r\nabla - \nabla\varepsilon_r) \\ K(\varepsilon_r\nabla - \nabla\varepsilon_r) \end{bmatrix} & & & & \begin{bmatrix} K\varepsilon_r \\ K\varepsilon_r \end{bmatrix} \end{bmatrix} \begin{bmatrix} \begin{bmatrix} \dot{V}_1 \\ \dot{V}_2 \end{bmatrix} \\ \dot{n} \\ \dot{p} \\ \dot{A} \\ \begin{bmatrix} \dot{\Pi}_1 \\ \dot{\Pi}_2 \end{bmatrix} \end{bmatrix} =$$

$$
-\begin{bmatrix}
\begin{bmatrix} -\frac{1}{\nu}\nabla\cdot(\varepsilon_r\nabla) \\ -\nabla\cdot(\sigma\nabla) \end{bmatrix} & \begin{bmatrix} I_n \\ 0 \end{bmatrix} & \begin{bmatrix} -I_p \\ 0 \end{bmatrix} & & \begin{bmatrix} -\frac{1}{\nu}\nabla\cdot\varepsilon_r \\ -\nabla\cdot\sigma \end{bmatrix} \\
 & 0 & & & \\
 & & 0 & & \\
 & & & 0 & -I \\
\begin{bmatrix} 0 \\ K\nu\sigma\nabla \end{bmatrix} & & & \begin{bmatrix} \nabla^2 \\ \nabla^2 \end{bmatrix} & \begin{bmatrix} 0 \\ K\nu\sigma \end{bmatrix}
\end{bmatrix}
\begin{bmatrix} \begin{bmatrix} V_1 \\ V_2 \end{bmatrix} \\ n \\ p \\ A \\ \begin{bmatrix} \Pi_1 \\ \Pi_2 \end{bmatrix} \end{bmatrix}
$$

$$
-\begin{bmatrix} \begin{bmatrix} 0 \\ \nabla\cdot\bar{J}_{sd} \end{bmatrix} \\ -\nabla\cdot\bar{J}_n + R \\ \nabla\cdot\bar{J}_p + R \\ 0 \\ \begin{bmatrix} -K\nu\bar{J}_{sd} \\ -K\nu\bar{J}_{sd} \end{bmatrix} \end{bmatrix}
-\begin{bmatrix} \begin{bmatrix} -N_D \\ 0 \end{bmatrix} \\ 0 \\ 0 \\ 0 \\ \begin{bmatrix} 0 \\ 0 \end{bmatrix} \end{bmatrix} \tag{24.7}
$$

where the constant matrices C and G collect the linear dynamics in the system, F collects the nonlinear dynamics (basically the nonlinear semiconductor currents) and b the excitation term.

Normally, a two-step approach is applied to solve (24.6). The time derivative is first approximated by a certain polynomial expansion leading to an algebraic non-linear equation which is then submitted to the Newton-Raphson method. For instance, the first-order BE approximation results in the following nonlinear equation for the solution of the $(n+1)$th step

$$
\begin{aligned}
C\frac{x_{n+1}-x_n}{h} &= -Gx_{n+1} - F(x_{n+1}) - b_{n+1} \\
F(x_{n+1}) + \left(\frac{C}{h}+G\right)x_{n+1} &- \frac{C}{h}x_n + b_{n+1} = 0
\end{aligned} \tag{24.8}
$$

in which h is the time step size. Then the Newton's method solves (24.8) by assembling and solving the Jacobian matrix in each iteration

$$
\begin{aligned}
&\left(\left.\frac{\partial F}{\partial x}\right|_{x_{n+1}^{k+1}} + \frac{C}{h} + G\right)\Delta x_{n+1}^{k+1} \\
&= -\left(F(x_{n+1}^{k}) + \left(\frac{C}{h}+G\right)x_{n+1}^{k} - \frac{C}{h}x_n + b_{n+1}\right)
\end{aligned} \tag{24.9}
$$

where the Jacobian $J = \left(\frac{\partial F}{\partial x}\Big|_{x_{n+1}^{k+1}} + \frac{C}{h} + G \right)$ and Δx_{n+1}^{k+1} denotes the update of x at the $(k+1)$th Newton iteration in the $(n+1)$th step.

24.2 Time-Domain Simulation with Matrix Exponential Method

Basic Matrix Exponential Method The matrix exponential method is essentially an ODE solution technique, i.e., it relies on converting the DAE (24.6) to an ODE by multiplying the inverse of C on both sides

$$\dot{x} = -C^{-1}Gx - C^{-1}F(x) - C^{-1}b \tag{24.10}$$

Note that at this stage, we assume the inverse of C can be constructed. The treatment for a singular C and the fast computation of C^{-1} will be presented in later subsections.

The analytical solution of (24.10) for x_{n+1} is given with the matrix exponential (Chua and Lin [1975])

$$x_{n+1} = e^{Mh}x_n + \int_0^h e^{M(h-\tau)} \left[-C^{-1}F(t_n+\tau) - C^{-1}b(t_n+\tau) \right] d\tau \tag{24.11}$$

in which $M = -C^{-1}G$.

Successful application of the above matrix exponential formula requires answers to the following three questions:

- Is C really invertible? How to deal with the systems with a singular C.
- How to compute the matrix exponential e^{Mh} (which involves C^{-1}) efficiently for large-scale problems?
- How to handle the nonlinear function F in the integral?

The following subsections will be devoted to answer these questions.

Regularization of C The coefficient matrix C is generally singular. The singularity arises from the Gauss law which is an algebraic constraint of the field degrees of freedom including no time derivatives. As is evident in (24.7), the nodes (collected in V_1) where the Gauss law is enforced generates empty rows in C. For the nodes on which current continuity is enforced (generally nodes attached to metals), the presence of displacement current allthough small compared with the conduction current, prevents such vacancy in C.

To overcome the singular C problem, we differentiate the Gauss law with respect to the time variable which gives

$$\nabla \cdot \left[\varepsilon \left(\nabla \frac{\partial V}{\partial t} + \frac{\partial \mathbf{\Pi}}{\partial t} \right) \right] + \frac{\partial \rho}{\partial t} = 0 \tag{24.12}$$

In the matrix Equation (24.7), this is equivalent to moving the first row of G to C and differentiating the b term accordingly, yielding a new equation system with a regular, invertable C (24.13) shown on top of next page. Note that G and M are allowed to be (they actually are) singular since the exponential function is well defined for a singular matrix.

$$\begin{bmatrix} \begin{bmatrix} -\frac{1}{\nu}\nabla\cdot(\varepsilon_r\nabla) \\ -\frac{1}{\nu}\nabla\cdot(\varepsilon_r\nabla) \end{bmatrix} & \begin{bmatrix} I_n \\ 0 \end{bmatrix} & \begin{bmatrix} -I_p \\ 0 \end{bmatrix} & & \begin{bmatrix} -\frac{1}{\nu}\nabla\cdot\varepsilon_r \\ -\frac{1}{\nu}\nabla\cdot\varepsilon_r \end{bmatrix} \\ & I & & & \\ & & I & & \\ & & & I & \\ \begin{bmatrix} K(\varepsilon_r\nabla - \nabla\varepsilon_r) \\ K(\varepsilon_r\nabla - \nabla\varepsilon_r) \end{bmatrix} & & & & \begin{bmatrix} K\varepsilon_r \\ K\varepsilon_r \end{bmatrix} \end{bmatrix} \begin{bmatrix} \begin{bmatrix} \dot{V}_1 \\ \dot{V}_2 \end{bmatrix} \\ \dot{n} \\ \dot{p} \\ \dot{A} \\ \begin{bmatrix} \dot{\Pi}_1 \\ \dot{\Pi}_2 \end{bmatrix} \end{bmatrix}$$

$$= -\begin{bmatrix} \begin{bmatrix} 0 \\ -\nabla\cdot(\sigma\nabla) \end{bmatrix} & & & & \begin{bmatrix} 0 \\ -\nabla\cdot\sigma \end{bmatrix} \\ & 0 & & & \\ & & 0 & & \\ & & & 0 & -I \\ \begin{bmatrix} 0 \\ K\nu\sigma\nabla \end{bmatrix} & & & \begin{bmatrix} \nabla^2 \\ \nabla^2 \end{bmatrix} & \begin{bmatrix} 0 \\ K\nu\sigma \end{bmatrix} \end{bmatrix} \begin{bmatrix} \begin{bmatrix} V_1 \\ V_2 \end{bmatrix} \\ n \\ p \\ A \\ \begin{bmatrix} \Pi_1 \\ \Pi_2 \end{bmatrix} \end{bmatrix} - \begin{bmatrix} \begin{bmatrix} 0 \\ \nabla\cdot\bar{J}_{sd} \end{bmatrix} \\ -\nabla\cdot\bar{J}_n + R \\ \nabla\cdot\bar{J}_p + R \\ 0 \\ \begin{bmatrix} -K\nu\bar{J}_{sd} \\ -K\nu\bar{J}_{sd} \end{bmatrix} \end{bmatrix} \tag{24.13}$$

One can establish the validity of such regularization from two perspectives. From a physical point of view, the time-differentiated Gauss law represents essentially a current-continuity equation in semiconductor and insulator regions for the displacement current with a source being the temporal change of charge density. The explicitly solved current continuity Equations (24.4b) and (24.4c), on the other hand, equate the particle current and the temporal change of charge density. These two continuity equations together recover the "true" current continuity in semiconductor (and insulator), viz. the displacement current = the particle current. Note that the generation/recombination terms for n- and p-type carriers cancel each other in the differentiated Gauss law.

From a numerical perspective, the regularization converts the original equation $f(t) = g(t)$ to an equation in the form of $\dot{f}(t) = \dot{g}(t)$. It should be noted that in the transition from $f(t) = g(t)$ to $\dot{f}(t) = \dot{g}(t)$ information is lost. This is because the latter may correspond to $0 = 0$ which is an empty statement. Therefore, it is important to guarantee the temporal evolution starts from a correct initial condition, i.e., ensure that $f(0) = g(0)$, which can be obtained by a prior static solution. The static Gauss' law is a non-empty statement and therefore the loss of information due to the time differentiation is restored by a careful selection of the initial conditions. A more detailed discussion of the use of the time-differentiated Gauss' law with focus on our specific EM-TCAD implementation is provided in Appendix 24.6.

In addition, the derivative of source excitation V_S is now required, which is trivial if the sources are defined by continuous differentiable functions. If the sources are defined in piece-wise-linear form (which is common in circuit simulation), the selection of time step must respect the turning points of the waveforms to ensure that within each step all the derivatives of sources are constant.

Fast Krylov Subspace Approximation of Matrix Exponential The matrix exponential vector product $e^{Mh}x$ is the main computation in the time marching scheme (24.11). It can be computed efficiently by projecting the pair of (M, x) onto an m-dimensional Krylov subspace defined by

$$\mathrm{K_m}(M, x) = span\{x, Mx, \ldots, M^{m-1}x\}$$

and computing the martrix exponential vector product on the Krylov subspace.

Specifically, an N-by-m orthonormal basis V_m ($N = dim(M)$) and an $(m + 1)$-by-m upper Hessenberg matrix H for the Krylov subspace $\mathrm{K_m}$ are first constructed by the Arnoldi process. Then $e^{Mh}x$ can be approximated by (Saad [1992])

$$e^{Mh}x \approx \|x\|_2 V_m e^{H_m h} e_1 \tag{24.14}$$

where $H_m = H(1 : m, 1 : m)$. Since m is small (< 100), the exponential of H_m can be easily computed. A posteriori error bound (Saad [1992]) is available to estimate the approximation error of (24.14)

$$err = \|x\|_2 H(m + 1, m)|e_m^{\mathsf{T}} e^{H_m h} e_1| \tag{24.15}$$

The main cost of the Krylov subspace approximation lies in the m matrix vector products Mx in the Arnoldi process. Recall that $M = C^{-1}G$, then each $Mx = -C^{-1}(Gx)$ involves one sparse matrix vector product and one sparse linear solve. Since, typically, C is much sparser than $C + G$ when dealing with linear structures, the MEXP method can be more scalable than LMS methods as reported in (Weng et al. [2011, 2012a]). Furthermore, the special block structure of the C and G matrices allows a fast computation of Mx requiring only one linear solution of a reduced-size matrix S, for which $dim(S) = \#$ of nodes $< \frac{1}{5}dim(M)$ (Chen et al. [2012]). Finally, convenient parallelization schemes are available for the matrix vector multiplications and the evaluation of e^{H_m} to further speed up the algorithm (Weng et al. [2012b]).

Solution of Nonlinear System The nonlinear function in (24.11) is approximated with a second-order implicit formula

$$F(\tau) = \left(F(x_n) + F(x_{n+1})\right)/2$$

which renders the scheme A-stable even for the nonlinear term (Beylkin et al. [1998]). Assuming further a constant derivative of source in each step i.e. constant b, the integral in (24.11) can be integrated explicitly, yielding

$$x_{n+1} = \frac{h}{2}\frac{e^{Mh} - I}{Mh}F\left(x_{n+1}\right) + e^{Mh}x_n + h\frac{e^{Mh} - I}{Mh}\left(\frac{F_n}{2} + b_n\right) \quad (24.16)$$

using the notation: $F_n = F(x_n)$. The operator $(e^{Mh} - 1)/Mh$ should be interpreted from its Taylor expansion. There is no real inverse of M, so M needs not to be invertible.

The function of the matrix exponential before $F(x_{n+1})$ is highly undesirable. In each Newton iteration the Jacobian matrix will be pre-multiplied with a matrix exponential. The computational burden is prohibitively long. Recall that we prefer computing only the product of a matrix exponential and a vector. A close analysis reveals that the first term in (24.16) represents a coupling between the linear dynamics in e^{Mh} and the nonlinear dynamics in $F(x_{n+1})$. Hence, we adopt a technique developed in (Nie et al. [2006]), which decouples the linear and the nonlinear terms by approximating the entire integrand in (24.11) with a Lagrange polynomial. The second-order approximation yields

$$\begin{aligned} x_{n+1} = &-\frac{h}{2}C^{-1}F\left(x_{n+1}\right) + e^{Mh}\left(x_n - \frac{h}{2}C^{-1}F_n\right) \\ &+ \frac{e^{Mh} - I}{M}(-C^{-1}b_{n+1}). \end{aligned} \quad (24.17)$$

Now $F(x_{n+1})$ has a constant coefficient and e^{Mh} is only multiplied with terms involving known quantities from the previous step. Further saving can be achieved by merging the last two terms in (24.17) into one single matrix exponential with a slightly larger matrix (Al-Mohy and Higham [2011])

$$\begin{aligned} x_{n+1} =& -\frac{h}{2}C^{-1}F\left(x_{n+1}\right) \\ &+ [I_N, 0]\exp\left\{\begin{bmatrix} M & -C^{-1}b_{n+1} \\ 0 & 0 \end{bmatrix} h\right\}\begin{bmatrix} x_n - \frac{h}{2}C^{-1}F_n \\ 1 \end{bmatrix} \\ =& -\frac{h}{2}C^{-1}F\left(x_{n+1}\right) + u_{n+1}. \end{aligned} \tag{24.18}$$

The vector u_{n+1}, denoted as the *linear update* term, is calculated only once at the beginning of each time step by the Krylov subspace method described in Subsection 24.2.

With (24.18), the Newton's method solves the following linear system in each iteration

$$\left(C + \frac{h}{2}\frac{\partial F}{\partial x}\bigg|_{x_{n+1}^{k+1}}\right)\Delta x_{n+1}^{k+1} = -\left(\frac{h}{2}F_{n+1}^{k} + Cx_{n+1}^{k} - Cu_{n+1}\right). \tag{24.19}$$

The local truncation error (LTE) of the approximation (24.17) is

$$-\frac{1}{12}((Mh)^2F_n + (Mh)\dot{F}_n + \ddot{F}_n). \tag{24.20}$$

One major advantage of the MEXP is that a sparser Jacobian matrix produced for the Newton iterations. Comparing (24.9) and (24.19), the Jacobian in BE consists of $\left(\frac{\partial F}{\partial x} + \frac{C}{h} + G\right)$, whereas in MEXP it is only $\left(C + \frac{h}{2}\frac{\partial F}{\partial x}\right)$. In large-scale computation, the improvement of the sparsity of the linear system can lead to significant computational savings.

The rationale behind this improved sparsity is that the linear components of a system generally do not change during the treatments for nonlinear components. As reflected in the composition of Jacobian matrix, the C and G matrices in the backward Euler context are both constant in the whole duration of Newton's method; only the Jacobian of F varies among iterations. In conventional approaches, the stable linear components are mixed into the Jacobian of each Newton iteration which do not provide new information while still consuming a substantial part of computation. In contrast, the MEXP approach separates the linear and the nonlinear parts of the system and represents the linear sub-system in a matrix exponential form. The linear sub-system is then handled by the efficient Krylov subspace technique in one run, leaving just the nonlinear sub-system participating in the subsequent Newton iterations.

24.3 Error Control and Adaptivity

Whereas constant step size is simple to implement, adaptive time-stepping commonly provides better accuracy and performance in simulation. The adaption requires certain error control to determine when to change a step size and re-evaluation scheme when the step size do change. The separation of linear and nonlinear components in MEXP, from this perspective, also enables convenient and efficient error control and adaptive scheme compared with that in traditional methods. The left flowchart in Figure 24.1 shows the flow of backward Euler methods, wherein the linear accuracy (i.e., the accuracy of the polynomial expansion of the time derivative) is tested only after the Newton iterations converge. Once the temporary solution cannot pass the test, the time step must reverse and a new round of Newton iterations will be restarted, incurring possibly a significant waste of computation.

On the contrary, the linear accuracy check (24.15) in MEXP, as shown in the right flowchart in Figure 24.1, is prior to the nonlinear solution. The accuracy in the matrix exponential approximation can be checked

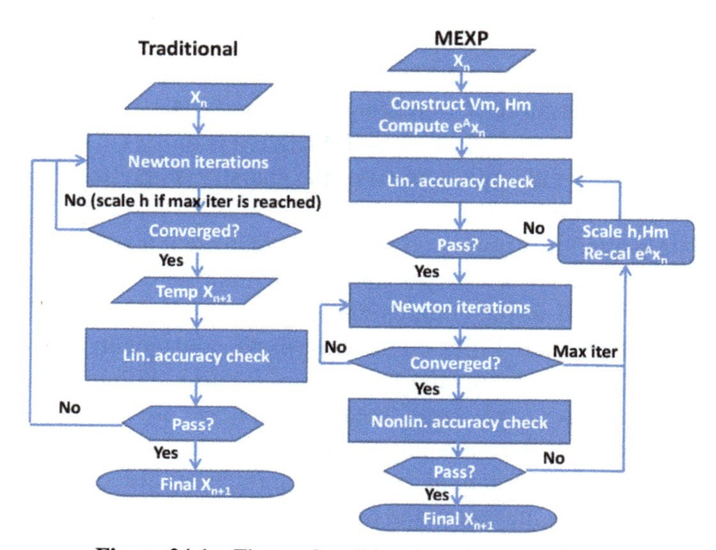

Figure 24.1 Flows of traditional methods and MEXP.

and tuned to be satisfactory before the algorithm enters the nonlinear phase. In addition, the Krylov subspace approximation (24.14) has a nice scaling invariant property that $\alpha M \rightarrow \alpha H_m$, which suggests that the calculated V_m and H can be reused (with corresponding scaling) to generate a new solution without demanding another Arnoldi process when h is changed, i.e.,

$$e^{Mh_1}x \approx \|x\|_2 V_m e^{\alpha H_m} e_1, \quad \alpha = h_1/h \tag{24.21}$$

Therefore, the step size can be adjusted multiple times at a negligible cost since each re-evaluation involves merely an exponential of H_m.

Although the nonlinear accuracy check (24.20) may also result in restart of Newton iterations in the MEXP flow, the chance is far lower than in the BE flow. This is because in general the majority of error in time-domain discretization comes from the linear components, e.g., the small parasitics in back-end structures that require small time step to capture. Hence, the separation of linear and nonlinear components effectively avoids the waste of Newton iterations when adaptive scheme is applied.

24.4 E-V Formulation of EM-TCAD for MEXP Method

Although MEXP possesses many advantages in handling the EM-TCAD problems, it has the similar problems as conventional LMS methods do when the DAE become stiff due to the coupling of physical dynamics with distinct time constants. A large dimension of Krylov subspace would be required by the Arnoldi process, which, as an eigenvalue algorithm, captures more accurately the eigenvalues with large magnitude than those with small magnitude. If the spectrum contains many large outlying eigenvalues, the Arnoldi process would "focus" on approximating these eigenvalues, leaving the spectrum near the origin, the more meaningful part in computing the matrix exponential, inadequately sampled. In the LMS context, stiff systems usually lead to ill-conditioned Jacobians in (24.9), causing a slow convergence when Krylov-subspace-based iterative solvers (such as GMRES) are applied. The difficulty can be alleviated by applying preconditioners to the Jacobian to push the eigenvalues away from the origin of the complex plane. The matrix to be solved becomes PJQ with P and Q being the left and right preconditioners. It is equivalent to applying row and/or column scaling (and permutation) to the original Equation (24.6). Nevertheless, the row/column scaling technique does not work for the MEXP method, because the DAE-ODE conversion renders the preconditioned matrix exponent $(PCQ)^{-1}(PGQ) = Q^{-1}(C^{-1}G)Q$ nothing but a similarity transformation of the original matrix which would not alter the eigenvalue distribution that determines the performance of Krylov subspace approximation.

The stiffness in the converted ODE mainly comes from the large contrast of electrical parameters of different materials. When metallic materials present, the high conducting currents give rise to a large number of fast modes (negative large eigenvalues in M) in the ODE via the current continuity equation. To reduce the stiffness in the converted ODE, and realizing that scaling cannot help, we propose to reformulate the EM-TCAD problem through a dedicated variable and equation transformations as done in (Chen et al. [2011b]), known as the time-domain E-V formulation. The underlying rationale is to replace the current continuity equation solved for metal nodes by the gauge condition (24.5) so that the metal conductivity would not affect the ODE properties as much as in the A-V formulation. In addition, the E-V solver introduces the variable transform $\mathbf{E} = -\nabla V - \mathbf{\Pi}$. The complete E-V system of equations to be solved is given in (24.22),

$$\begin{cases} \frac{1}{\nu}\nabla \cdot \left(\varepsilon_r \frac{\partial \mathbf{E}}{\partial t}\right) - \frac{\partial \rho}{\partial t} = 0, \ \rho = p - n + N_D \\ \alpha K \varepsilon_r \frac{\partial}{\partial t} V + \nabla \cdot \mathbf{A} = 0 \end{cases} \tag{24.22a}$$

$$\nabla \cdot \mathbf{J}_n - \frac{\partial}{\partial t} n - R(n,p) = 0 \tag{24.22b}$$

$$\nabla \cdot \mathbf{J}_p + \frac{\partial}{\partial t} p + R(n,p) = 0 \tag{24.22c}$$

$$\nabla V + \frac{\partial}{\partial t}\mathbf{A} + \mathbf{E} = 0 \tag{24.22d}$$

$$-K\varepsilon_r \frac{\partial}{\partial t}\mathbf{E} + \nabla\left(-\alpha K \varepsilon_r \frac{\partial}{\partial t} V\right) + \left[\nabla \times (\nabla \times \mathbf{A}) - \nabla(\nabla \cdot \mathbf{A})\right] - K\nu\sigma\mathbf{E} - K\nu\mathbf{J}_{sd} = 0 \tag{24.22e}$$

The E-V formulation used with the MEXP method differs from its use in the LMS context (Chen et al. [2013]) in the following perspectives. First, the differentiation of Gauss's law is again applied to enable the conversion to ODE, whereas in (Chen et al. [2013]) the original version is used. So, we need to prove that the A-V to E-V equivalence still holds in the regularization method which is done in Appendix 24.6.

Second, direct application of the gauge condition (24.5) can just slightly improve the spectrum of M, for $K\varepsilon_r$ remains a small constant requiring small step size to capture. One can fix this gauge-related trouble by going one step further to exploit the freedom in choosing a specific form of gauge condition, which by definition would not affect all observable physical quantities like currents and fields. For simplicity, we add a balancing parameter α in (24.22a) and (24.22e), where $\alpha = 1$ refers to the original Lorentz gauge and $\alpha = K^{-1}$ leads to a balanced gauge condition. The reduction in stiffness can improve the accuracy of the Krylov subspace method, either allowing a larger step size under the same m or a smaller m given the same step size.

Third, the E-V formulation improves the sparsity of the C matrix compared with the A-V formulation, which is more clear comparing the matrix form Equations (24.13) and (24.23). It makes no much difference in conventional LMS methods, for the numerical systems of interest are in the form of $C/h + G$, where the reduction of nonzeros in C is compensated by the increase in G. Nevertheless, in MEXP the improved sparsity in C can be more beneficial. As stated in Section 24.2, each Arnoldi step involves a linear solution with C (or its blocks in the fast computation), which takes up a considerable portion of computation in the Arnoldi process. A sparser matrix would therefore accelerate the process as well as the MEXP algorithm.

$$
\begin{bmatrix}
\begin{bmatrix} 0 & 0 \\ 0 & K\varepsilon \end{bmatrix} & \begin{bmatrix} I_n \\ 0 \end{bmatrix} & \begin{bmatrix} -I_p \\ 0 \end{bmatrix} & & \begin{bmatrix} \frac{1}{\nu}\nabla\cdot\varepsilon_r & 0 \\ 0 & 0 \end{bmatrix} \\
 & -I & & & \\
 & & I & & \\
 & & & I & \\
\begin{bmatrix} K\varepsilon_r\nabla & K\varepsilon_r\nabla \\ 0 & K\varepsilon_r\nabla \end{bmatrix} & & & & \begin{bmatrix} K\varepsilon_r & 0 \\ 0 & K\varepsilon_r \end{bmatrix}
\end{bmatrix}
\begin{bmatrix}
\begin{bmatrix} \dot{V}_1 \\ \dot{V}_2 \end{bmatrix} \\ \dot{n} \\ \dot{p} \\ \dot{A} \\ \begin{bmatrix} \dot{E}_1 \\ \dot{E}_2 \end{bmatrix}
\end{bmatrix}
$$

$$
= -\begin{bmatrix}
\begin{bmatrix} 0 & 0 \\ 0 & 0 \end{bmatrix} & & & \begin{bmatrix} 0 \\ \nabla \end{bmatrix} & \begin{bmatrix} 0 & 0 \\ 0 & 0 \end{bmatrix} \\
 & 0 & & & \\
 & & 0 & & \\
\nabla & & & 0 & I \\
\begin{bmatrix} 0 & 0 \\ 0 & 0 \end{bmatrix} & & & \begin{bmatrix} -\nabla^2 \\ -\nabla^2 \end{bmatrix} & \begin{bmatrix} 0 & 0 \\ 0 & K\nu\sigma \end{bmatrix}
\end{bmatrix}
\begin{bmatrix}
\begin{bmatrix} V_1 \\ V_2 \end{bmatrix} \\ n \\ p \\ A \\ \begin{bmatrix} E_1 \\ E_2 \end{bmatrix}
\end{bmatrix}
-
\begin{bmatrix}
\begin{bmatrix} 0 \\ 0 \end{bmatrix} \\ \nabla\cdot\bar{J}_n - R \\ \nabla\cdot\bar{J}_p + R \\ 0 \\ \begin{bmatrix} K\nu\bar{J}_{sd} \\ K\nu\bar{J}_{sd} \end{bmatrix}
\end{bmatrix}
$$

24.5 Numerical Results

The time-domain EM-TCAD simulation with MEXP solver is implemented in Matlab. The MEXP with the A-V formulation (24.13), the E-V formulation (24.23) with $\alpha = 1$ and the one with the balanced gauge ($\alpha = K^{-1}$) are named as MEXP-AV, MEXP-EV and MEXP-EVBG, respectively. For comparison we also implement the second-order Gear's (GR2) method known to be efficient for stiff systems. Three test structures with the specifications detailed in Table 24.1 are simulated. The linear/nonlinear nodes[23] refer to the nodes attached without/with semiconductor volumes, respectively. All tests are conducted on a 3.2GHz 32Gb-RAM server.

We begin by examining numerically the validity of the regularization technique in Section 24.2. Figure 24.2 shows the generalized eigenvalues of the $(C, -G)$ matrix pencils resulting from using the original Gauss law (see (23.6) and the differentiated one (see (24.13)). All the finite (not Inf) generalized eigenvalues are preserved after the regularization, whose effect is to turn all the infinite eigenvalues in the original system into zeros in the regularized system.

Next, we verify the accuracy of the MEXP-AV and MEXP-EV methods using the SWR example depicted in Figure 24.3 and a square wave input shown in Figure 24.4. The transient current through the left contact is shown in Figure 24.5 with a 50GHz excitation applied. GR2 with a small step size of $0.025ps$ is used to generate the benchmark result. The other three curves are from GR2, MEXP-AV and MEXP-EV using 10 times larger step size ($0.25ps$) respectively. The Krylov subspace dimension $m = 80$ is selected in MEXP to guarantee the (estimated) error throughout the simulation smaller

Table 24.1 Specifications of test structures

Name	Description	# of Nodes (Unknowns)	Breakdown of Lin./Nonlin. Nodes
SWR	Silicon wire with two metal leads	3,825 (27,294)	3,600/225
SUB	Substrate noise isolation structure (Chen et al. [2011b])	6,384 (41,368)	1,064/5,320
VCO	8-shaped inductor in voltage controlled oscillator (Schoenmaker et al. [2012])	21,312 (149,898)	4,736/16,576

[23]Nonlinear nodes can still generate nonzeros in the G matrix.

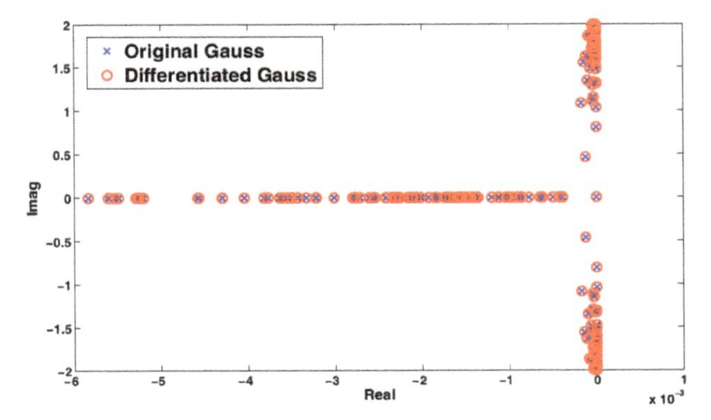

Figure 24.2 Generalized eigenvalues of the systems using the original and the differentiated Gauss law. Only the part near the origin is plotted.

Figure 24.3 3D view of the SWR case with one silicon wire in the middle and two copper leads having a cross section of $4 \times 4\mu m^2$, surrounded by an $8\mu m$ thick oxide layer (not shown). The lengths of the three parts are respectively $8\mu m$, $16\mu m$ and $8\mu m$. The silicon part has an n-type doping of $10^{21}m^{-3}$.

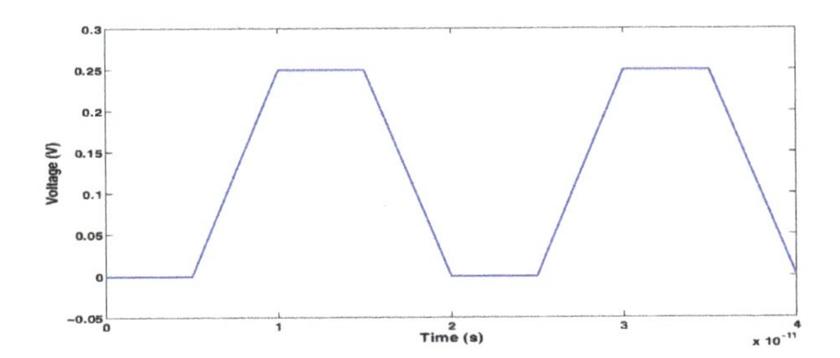

Figure 24.4 Square wave input.

than 10^{-6}. In Figure 24.3, a noticeable error is observed in GR2 due to the increase of step size. On the other hand, the MEXP curves match very well with the reference one, demonstrating a good accuracy of the method. The exact overlapping between MEXP-AV and MEXP-EV also confirms the validity of the E-V formulation.

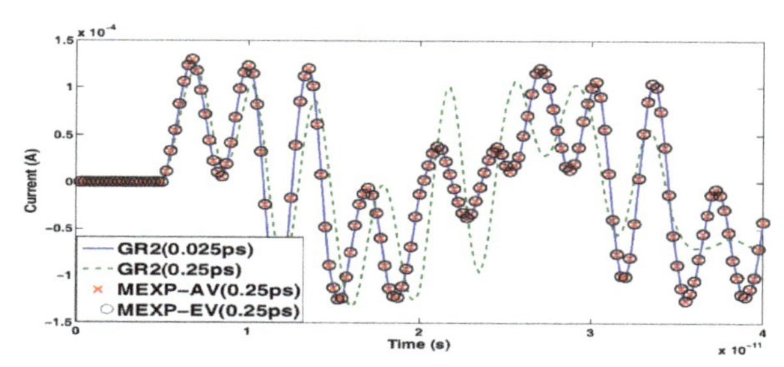

Figure 24.5 Current through the left contact obtained by GR and MEXP-AV and MEXP-EV for the SWR example.

The runtime data of the three methods, GR2, MEXP-AV and MEXP-EV for one step is collected in Table 24.2. MEXP-EVBG is similar to MEXP-EV and thus omitted. In the Krylov subspace approximation sector, the time taken by each matrix vector product Mx using the fast block matrix computation, and the total time for the Krylov subspace calculation of matrix exponential are reported. In the Newton's method sector, the runtime per iteration is measured both by using a direct solver (backslash in Matlab), and a iterative solver, which uses the combo of GMRES ($tol = 10^{-6}$), incomplete LU preconditioner ILU(10^{-3}) and column approximate minimum degree

Table 24.2 Runtime breakdown of GR2, MEXP-AV and MEXP-EV for the SWR case using $h = 0.25ps$ (time unit: sec)

		Krylov Approx. of Matrix Exp.			Newton's Method (Per Iteration)		
Cases	Method	m	Time Per Mx	Total Time	Direct	GMRES +ILU	nnz (Jacob)
SWR	GR2	–	–	–	1.68	4.31	200K
	MEXP-AV	80	0.048	4.50	0.07	1.21	80K
	MEXP-EV(BG)	80	0.051	4.81	0.06	1.20	59K
SUB	GR2	–	–	–	7.08	22.95	622K
	MEXP-AV	60	0.10	7.07	0.42	2.54	332K
	MEXP-EV(BG)	60	0.056	4.88	0.45	2.79	247K
VCO	GR2	–	–	–	50.60	360.90	1.8M
	MEXP-AV	80	0.50	41.07	2.58	16.90	1.2M
	MEXP-EV(BG)	80	0.32	30.01	2.61	17.11	0.9M

(COLAMD) permutation (Davis et al. [2004b]). The numbers of nonzeros in the Jacobian in Newton iteration are also shown.

The computation of matrix exponential, which is an extra cost for MEXP, is generally moderate because of the efficiency of Krylov subspace method. The E-V formulation enables a faster Mx multiplication and overall Krylov subspace approximation owing to the improved sparsity in the C matrix. For the runtime of Newton iteration, considerable speedup is achieved with the MEXP solver, up to $19X$ for the direct solver and up to $21X$ for the iterative solver. In terms of absolute numbers, the saving from one Newton iteration with MEXP is usually enough to pay off the extra cost required in the matrix exponential computation, and the more Newton iterations there are, the higher benefit one can gain from a sparser Jacobian. This improved sparsity is also evident from the reduction in the number of nonzeros.

Figure 24.6 visualizes the difference in sparsity of the Jacobian in GR2 and MEXP-AV. Because of the absence of the G component, the Jacobian in MEXP-AV is free of the ∇^2 term in the last row of (24.13) and the identity term for Π in the equation of A. In particular, the matrix block for the former tends to be the densest one in the Jacobian of BE, the elimination of which in MEXP-AV, therefore, represents rather significant saving in solving the sparse system.

After showing the advantages of MEXP-type frameworks over the LMS, we focus now on comparing the variants of MEXP based on different formulation. The error of the Krylov subspace approximation of MEXP (estimated by (24.15)) w.r.t. step size given the same $m = 80$ are plotted in Figure 24.7 for the three MEXP methods. One can see that the switch to the original E-V formulation can only improve slightly the accuracy over MEXP-AV given the same m and h, which is expected due to the small time constants induced by

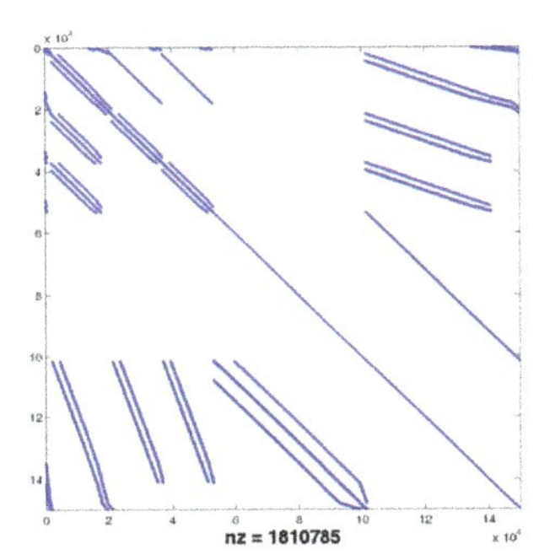

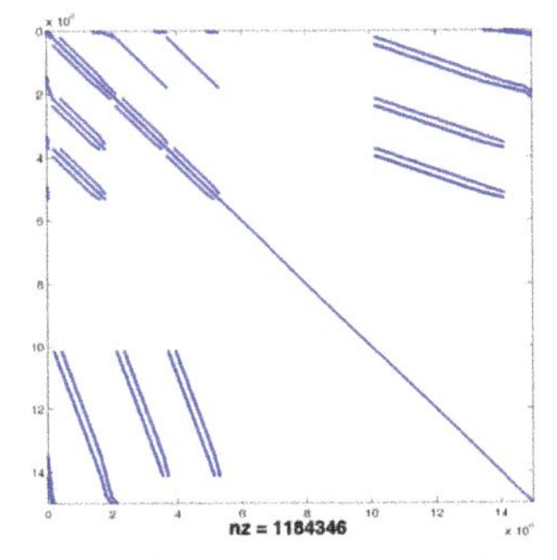

Figure 24.6 Sparsity pattern of the Jacobian matrices in GR2 and MEXP-AV solvers (VCO example).

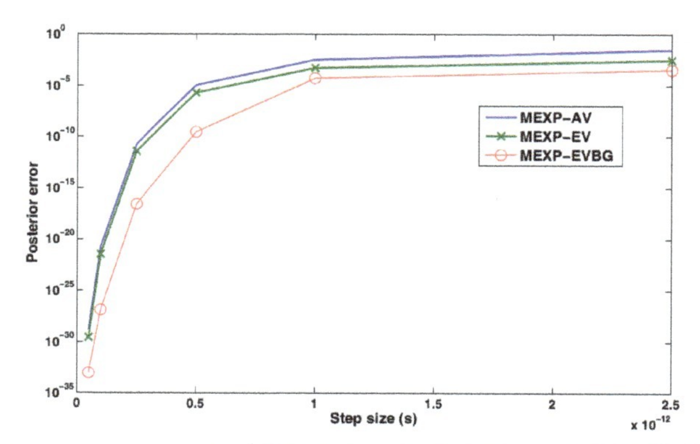

Figure 24.7 Error vs. step size for MEXP-AV, MEXP-EV and MEXP-EVBG (SWR case with fixed $m = 80, h = 0.25ps$.

the Lorentz gauge. On the other hand, MEXP-EVBG constantly outperforms MEXP-AV and MEXP-EV in the quality of Krylov subspace approximation under the same computational expense, owing to the use of a more balanced version of gauge condition.

In Table 24.3, we compare the performance of the four methods using adaptive time-stepping. The SWR structure is simulated with a total time span of $40ps$ and an initial h of $0.25ps$. We adopt the common SPICE estimation of LTE for GR2 and selection scheme of step size described in (Palusinski [2005]). The target accuracy is 10^{-6}. The same linear error control is applied in MEXP except that the LTE is replaced by the posterior error (24.15) and the step size can be tuned within the current step. The nonlinear error check with (24.20 is employed as well in MEXP. The m is fixed to be 80 and the direct solver is applied.

First, the adaptive time-stepping is not an efficient practice in GR2 (actually in all LMS methods), in that a new round of Newton iterations has to be started once a time step is rejected. In this case, over $1/5$ of the total Newton iterations are wasted. On the other hand, no Newton iteration is wasted in all the adaptive MEXP, since the linear accuracy has been guaranteed before the

Table 24.3 Performance using adaptive step size (time unit: sec)

Method	# of Time Step	Total NT	Wasted NT	Total Time
GR2	475	1263	316	2738.80
MEXP-AV	158	316	0	860.20
MEXP-EV	152	304	0	822.01
MEXP-EVBG	81	162	0	365.86

nonlinear solution and the time-domain error from the linear components is more dominant. Second, consistent with the above error vs. step size experiment, MEXP-EV does not improve much the adaptive performance compared with MEXP-AV; basically the same numbers of time steps are required. The MEXP-EVBG, on the other hand, provides a significant speedup by allowing larger step sizes, and only half number of time steps is needed to complete the simulation, which again confirms the efficiency of using the balanced gauge condition in the E-V solver. The specific time points selected are shown in Figure 24.8.

Lastly, we investigate the effect of the proportion of linear and nonlinear components on the efficiency of GR2 and MEXP in Table 24.4. The VCO structure is simulated with three different sizes of the semiconductor region (thus different breakdown of linear and nonlinear nodes). Normally, fewer linear nodes means sparser C and G matrices, and thus faster computation of matrix exponential. The MEXP-AV solver is more sensitive to the sparsity variations. On the other hand, more nonlinear nodes leads to denser Jacobian, and thus the slower Newton iteration in all solvers. However, even the structure contains (nearly) no linear nodes, the Jacobian in MEXP frameworks remains much sparser than that in GR2 because, as mentioned before, the nonlinear nodes can still generates entries in the linear matrix G, which will be a part of the Jacobian in GR2, but not in MEXP. There nonzeros are mainly related to the Maxwell-Ampere system, e.g., the ∇^2 term (c.f. Figure 24.6).

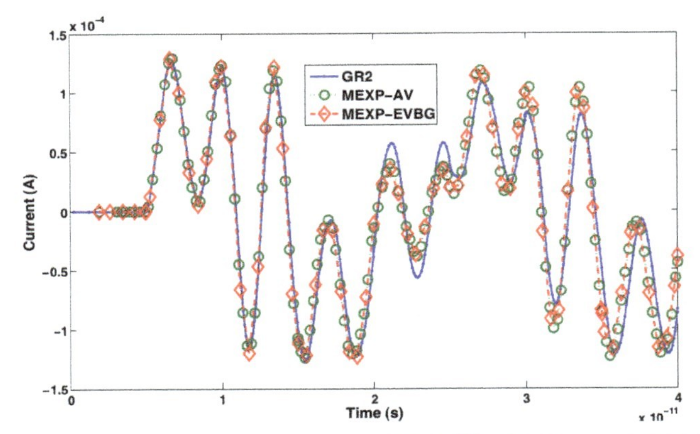

Figure 24.8 Left-contact current computed by GR2, MEXP-AV and MEXP-EVBG with adaptive time-stepping.

Table 24.4 Effect of breakdown of linear/nonlinear nodes (unit: sec)

	GR2		MEXP-AV		MEXP-EV(BG)	
LIN/NL nodes	EXP	Per NT	EXP	Per NT	EXP	Per NT
14,208/7,104	–	34.67	56.56	1.82	30.19	2.01
4,736/16,576	–	50.60	41.07	2.58	30.01	2.61
39/21,273	–	59.92	33.21	4.32	29.36	4.99

Some remarks are in order:

- Solving large linear system in Newton-like methods has long been a bottleneck in time-domain scientific computing involving nonlinearity. Workarounds have been investigated actively from various perspectives, including divide-and-conquer (e.g., domain decomposition) to reduce the problem size of individual solution, matrix-free nonlinear solvers (fixed point method and matrix-free Newton's method, etc.) to avoid solution of linear systems, and fast linear solvers (exploiting advanced preconditioners and parallelization).
- The MEXP framework proposed in this work pursues yet another direction. Since it is still desirable to keep Newton's method in the game for its better convergence and stability, we aim to minimize the portion of system that will participate in the nonlinearity treatment and handle the remaining (linear) portion with techniques that have better scalability than the methods requiring solution of linear systems. This way, the simulation capacity can be substantially elevated until the solution to the nonlinear fraction itself hits again the ceiling of computation resources. The division of linear and nonlinear components also facilitates error control and adaptive time-stepping.
- Slightly different from the judgment in (Chen et al. [2012]), the benefit of MEXP, as demonstrated in the experiments, depends on the proportions of linear and nonlinear components in a more complex manner. The more linear components, the higher gain one can expect from solving a sparser matrix in the Newton's method, but at the cost of a (potentially) more expensive computation of the matrix exponential. And even though the structure contains no linear component, the MEXP scheme can still enable a much faster nonlinear solution by removing a substantial part of nonzeros that would be constant during Newton iterations. Hence, the MEXP method is a promising technique, not only for EM-TCAD problems where front-end and back-end structures are commonly coexistent, but also for standard TCAD problems when magnetic dynamics are to be included.

24.6 Validity Proof of Regularization with Differentiated Gauss' Law

One specific aspect of the EM-TCAD formulation is the subtraction of the divergence of the gauge condition (24.5) from the original Maxwell-Ampere equation, as given in (24.1b). This treatment for one thing serves to regularize the singular curl-curl operator[24] and for the other thing avoids the redundancy with the current continuity equation enforced explicitly in metal and semiconductor (Chen et al. [2011b]) regions. The full EM-TCAD problem can be formulated in two ways: One formulation is given as the system consisting of the Maxwell-Ampere Equation (24.1b) along with either the current continuity and Gauss' law as done in (23.6). An equivalent formulation also exploits the Maxwell-Ampere equation but replaces the current continuity by the gauge condition, which is the foundation of the E-V formulation (Chen et al. [2011b]). Hence, it is necessary to show that the differentiated Gauss's law, when used as part of the equation system to be solved, still preserves the validity of the equivalent implementation.

The differentiated Gauss's law reads

$$\nabla \cdot \frac{\partial \mathbf{D}}{\partial t} = \frac{\partial \rho}{\partial t}, \tag{24.23}$$

In semiconductors we solve explicitly the current continuity $\nabla \cdot \mathbf{J} = \frac{\partial \rho}{\partial t}$, where $\mathbf{J}$ represents the current due to charge carriers, and in insulators we requires $\mathbf{J} = 0, \rho = 0$. Therefore, together with the current continuity equation solved in metals, the consequence of solving (24.23) in semiconductors and insulators is that we enforce in all domains

$$\nabla \cdot \mathbf{J} + \nabla \cdot \frac{\partial \mathbf{D}}{\partial t} = 0. \tag{24.24}$$

For the Maxwell-Ampere system, what we really solve is the following equation

$$\nabla \times (\frac{1}{\mu} \nabla \times \mathbf{A}) - \nabla \cdot \left[(\frac{1}{\mu} \nabla \cdot \mathbf{A}) + f(V, t) \right] - \mathbf{J} - \frac{\partial \mathbf{D}}{\partial t} = 0, \tag{24.25}$$

in which the gauge is not restricted to the Lorentz one. Taking the divergence at both sides, one obtains

$$\nabla^2 \left[\frac{1}{\mu} \nabla \cdot \mathbf{A} + f(V, t) \right] = \nabla \cdot \left(\mathbf{J} + \frac{\partial \mathbf{D}}{\partial t} \right). \tag{24.26}$$

[24] In the static regime or zero-frequency limit the singular character of the curl-curl operator is not elevated by the displacement current variation and therefore some regularizing action is required.

Together with (24.25), solving (24.24) implies that $\nabla^2[\frac{1}{\mu}\nabla \cdot \mathbf{A} + f(V,t)] = 0$ on all internal nodes. Enforcing additionally $\frac{1}{\mu}\nabla\cdot\mathbf{A}+f\,(V,t) = 0$ on the boundary of simulation domain via a special discretization scheme (Chen et al. [2011b]) one can recover implicitly the gauge condition everywhere which is the rationale underlying the A-V solver. Instead, if the gauge condition is solved explicitly on all internal nodes, the current continuity (24.24) will be satisfied implicitly as well, supporting the equation transform in the E-V solver. Hence, the use of the differentiated Gauss's law maintains the equivalence between the A-V and the E-V systems. Note that a correct initial condition $\nabla \cdot \mathbf{D}(0) = \rho(0)$ must be ensured via a static computation beforehand.

24.7 Fast Computation of *Mx* in E-V Formulation

The matrix vector product Mx can be computed by the following block formula

$$
\begin{aligned}
Mx &= -C^{-1}Gx \\
&= -\begin{bmatrix} C_{VV} & C_{Vn} & C_{Vp} & 0 & C_{VE} \\ & -I & & & \\ & & I & & \\ & & & I & \\ C_{EV} & 0 & 0 & 0 & C_{EE} \end{bmatrix}^{-1} \begin{bmatrix} 0 & 0 & 0 & G_{VA} & G_{VE} \\ & & 0 & & \\ & & 0 & & \\ G_{AV} & & & 0 & I \\ 0 & 0 & 0 & G_{EA} & G_{EE} \end{bmatrix} \begin{bmatrix} x_V \\ x_n \\ x_p \\ x_A \\ x_E \end{bmatrix} \\
&= -\begin{bmatrix} X & \tilde{C}_{Vn} & \tilde{C}_{Vp} & 0 & Y \\ & -I & & & \\ & & I & & \\ & & & I & \\ Z & \tilde{C}_{En} & \tilde{C}_{Ep} & 0 & U \end{bmatrix} \begin{bmatrix} 0 & 0 & 0 & G_{VA} & G_{VE} \\ & & 0 & & \\ & & 0 & & \\ G_{AV} & & & 0 & I \\ 0 & 0 & 0 & G_{EA} & G_{EE} \end{bmatrix} \begin{bmatrix} x_V \\ x_n \\ x_p \\ x_A \\ x_E \end{bmatrix} \qquad (24.27) \\
&= -\begin{bmatrix} (XG_{VA}+YG_{EA})\,x_A + (XG_{VE}+YG_{EE})\,x_E \\ 0 \\ 0 \\ G_{AV}x_V + x_E \\ (ZG_{VA}+UG_{EA})\,x_A + (ZG_{VE}+UG_{EE})\,x_E \end{bmatrix},
\end{aligned}
$$

where

$$
\begin{aligned}
S = C_{VV} - C_{VE}C_{EE}^{-1}C_{EV}, X = S^{-1},\ Y = -S^{-1}C_{V\Pi}C_{\Pi\Pi}^{-1},\\
Z = -C_{EE}^{-1}C_{EV}S^{-1},\ U = C_{EE}^{-1}(C_{EV}S^{-1}C_{VE}C_{EE}^{-1} + I).
\end{aligned}
\tag{24.28}
$$

Using the same data reuse strategy as in (Chen et al. [2012]), one can define

$$
\begin{aligned}
v_1 &= G_{AV}x_A + G_{VE}x_E,\\
v_2 &= C_{EE}^{-1}\left(G_{EA}x_A + G_{EE}x_E\right),\\
v_3 &= S^{-1}(v_1 - C_{VE}v_2),
\end{aligned}
\tag{24.29}
$$

by which the final computation boils down to

$$
Mx = -\begin{bmatrix} v_3 \\ 0 \\ 0 \\ G_{AV}x_V + x_E \\ -C_{EE}^{-1}C_{EV}v_3 + v_2 \end{bmatrix}
\tag{24.30}
$$

Summary

We have presented a fast solution technique for time-domain EM-TCAD co-simulation. The essence lies in the division of linear components and nonlinear components, and handling the two with different strategies. The sparsity of the Jacobian matrix in the nonlinear treatment is therefore largely improved, since the linear components will no longer participate in the treatment. Several techniques have been developed to guarantee the validity and scalability of the proposed method. In particular, a special E-V formulation is developed to reduce the stiffness of the ODE system and further improve the matrix sparsity. It has been demonstrated that the new MEXP method can improve substantially the performance of the coupled simulator and provides convenient and efficient control of error and adaptive stepping.

PART IV

Advanced Topics

The last part of the book deals with advanced issues which so far have not received a complete answer. The purpose is to promote the techniques that were developed in Part II outside the limited field of microelectronics, where the mainstream of applications is found. Furthermore our hope is that the content of this part inspires the interested reader to participate in contributing to the further development of accounting for the geometrical reasoning behind discretization methods. The style of presentation in Part IV deviates from the style of Part III. Whereas several chapters in Part III are in a style of peer-reviewed journal papers, the chapters in Part IV are more 'narrative'. Chapter 25 aims at applying the surface impedance modeling approximation using the scalar and vector potentials as dynamical variables. The surface impedance approximation is explored for formulations using the electric and magnetic field as underlying variables it is required to translate the surface impedance condition to the potentials. Chapter 26 is dealing with floating domains and illustrates to use of ghost fields in another setting. The ghost field approach served as a source of inspiration to deal with floating domain simulations.

Next, Chapter 27 addresses the problem of high-frequency computations where fields tend to adapt to the boundary conditions in an exponential manner. The skin effect is a well-known illustration of it. In 1985 Schilders et al. (Schilders et al. [1985]) observed that if equations contain a singular perturbation then the solutions account for its presence by fast correction at the boundary regions. Here we attempt to exploit this idea for the Maxwell-Ampere system but of course the problem is much more complicated due to the fact that there are three dimensions and three components of the vector potential.

Finally, in Chapter 28, we give a detailed account of the work that led us to the conclusion that the gauge approach of simulating electromagnetic field problems is stable. Instead of presenting the work in peer-reviewed paper style, the reader can witness step-after-step towards finally coming to above conclusion. The chapter demonstrates how a code bug may have a profound impact om the stability behavior or how there is an interplay with the meshing algorithm. To conclude, the chapter illustrates nicely how a scientific conclusion materializes by inventing test structures, code testing and physical reasoning.

25

Surface-Impedance Approximation to Solve RF Design Problems

We will discuss the use of the surface-impedance approximation to deal with the simulation of RF fields around integrated passive device structures. We have encountered situations that the solving of electromagnetic waves in metallic structures combined with dielectric regions can become very cumbersome due to the stiffness of the Newton-Raphson matrices. A substantial reduction of the stiffness can be obtained by avoiding the solving of the full wave system inside metals. There are situations that such detailed solving in metals may be avoided. This is the case when the frequency of the wave is sufficiently high such that the wave is only present in the surface region of the metal. The use of the surface impedance approximation is not new. It is a well-known method to set boundary conditions in wave guide calculations. The metal in wave guide modeling is solved by ignoring the wavy details inside the metal and to consider the metallic walls as boundaries of the simulation domain. When considering electronic devices, the exclusion of the metallic bulk regions requires a careful filtering of the degrees of freedom since the metallic structures are not just an encapsulating container for the EM waves but they are routed inside the simulation domain. In this document we demonstrate the validity of the surface impedance approximation for computing currents and fields for micro-electronic applications.

25.1 Surface Impedance Approximation

Consider the case of an electromagnetic wave bouncing on a flat metallic plane. First suppose that the conductivity in the metal is infinite (super conducting). In that case there is no electric field inside the conductor. As a consequence Gauss' law is

$$\nabla \cdot \mathbf{D} = 0 \tag{25.1}$$

where S is the surface charge density. Using the small volume element crossing the metal/insulator interface we find the perpendicular electric field is discontinuous with a step determined by the size of the surface charge density. Using the Faraday's law for the circulation of the electric field, e.g.

$$\oint \mathbf{E} \cdot \mathbf{dl} = \frac{d}{dt} \int \mathbf{dS} \cdot \mathbf{B} \tag{25.2}$$

over a small area element crossing the metal/insulator interface we find the tangential component of the electric field is continuous, since we may choose the surface area arbitrary small while keeping the tangential circumference finite. Let $\mathbf{n}$ be a the vector perpendicular to the metal/insulator.

$$\mathbf{n} \times (\mathbf{E}_{\text{cond}} - \mathbf{E}_{\text{insul}}) = 0 \tag{25.3}$$

Moreover, the infinite conductivity also implies that the charges are so mobile that a tangential magnetic field collapses to zero inside the conductor since a surface charge current bfK, will account for a discontinuous tangential magnetic field. interface.

$$\mathbf{n} \times \mathbf{H} = \mathbf{K} \tag{25.4}$$

For the perpendicular components we can refer to Biot-Savart's law (4) applied again to a small volume element crossing the insulator/metal interface.

$$\nabla \cdot \mathbf{B} = 0 \tag{25.5}$$

Therefore we find

$$\mathbf{n} \cdot (\mathbf{B}_{\text{cond}} - \mathbf{B}_{\text{insul}}) = 0 \tag{25.6}$$

In Figure 25.1 the various components at the interface are illustrated. In the non-ideal case, corresponding to large but finite conductance, the quantitative details are not valid anymore but the qualitative picture is still preserved. Inside the conductor there exists a small tangential component of the electric field that is responsible for the current flow $\mathbf{J}$ inside the conductor. Starting from the Maxwell-Ampere equation

$$\nabla \times \mathbf{H}_{\text{cond}} = \mathbf{J}_{\text{cond}} + \frac{\partial}{\partial t}\mathbf{D}_{\text{cond}} \tag{25.7}$$

In the frequency regime it leads to using $\mathbf{X}(t) = \mathbf{X}\exp(-\text{i}\omega t)$

$$\mathbf{E}_{\text{cond}} = \frac{1}{\sigma + \text{i}\omega\varepsilon} \nabla \times \mathbf{H}_{\text{cond}} \tag{25.8}$$

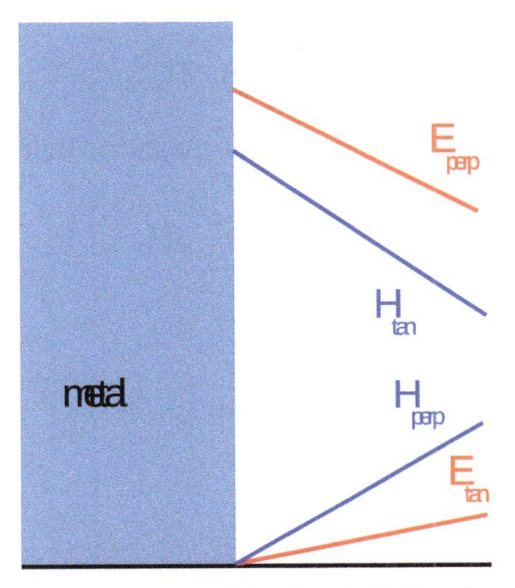

Figure 25.1 Various components of the electric field and magnetic induction.

From Faraday's law

$$\nabla \times \mathbf{E}_{\text{cond}} + \frac{\partial}{\partial t}\mathbf{B}_{\text{cond}} \tag{25.9}$$

we find

$$\mathbf{H}_{\text{cond}} = -\frac{\mathrm{i}}{\omega\mu_{\text{cond}}}\nabla \times \mathbf{E}_{\text{cond}} \tag{25.10}$$

Now suppose a planar wave is bouncing on the metal/insulator interface with constant component in the y and z direction. In this case

$$\mathbf{H}(x,y,z,t) = \mathbf{H}\exp\mathrm{i}(-k\xi + \omega t) \ \ , \ \ \mathbf{H}(x,y,z,t) = \mathbf{H}\exp\mathrm{i}(-k\xi + \omega t) \tag{25.11}$$

Then we may replace

$$\nabla \to \mathbf{n}\frac{\partial}{\partial \xi} \tag{25.12}$$

From (25.8) and (25.10) we obtain

$$\mathbf{H}_{\text{cond}} = \frac{\mathrm{i}}{\omega\mu}\nabla \times \left(\frac{1}{\sigma}\nabla \times \mathbf{H}_{\text{cond}}\right) \tag{25.13}$$

where we assumed that $\sigma >> \mathrm{i}\omega\varepsilon$. Equation (25.13) is can be elaborated to

$$\mathbf{H}_{\text{cond}} = \frac{\mathrm{i}}{\omega\mu_{\text{cond}}\sigma}\left[\mathbf{n}\frac{\partial}{\partial\xi}\left(\mathbf{n}\cdot\frac{\partial}{\partial\xi}\mathbf{H}_{\text{cond}}\right) - \frac{\partial}{\partial\xi}\mathbf{H}_{\text{cond}}\right] \tag{25.14}$$

Now

$$\mathbf{n} \cdot \frac{\partial}{\partial \xi} \mathbf{H}_{\text{cond}} = \mathbf{n} \cdot \frac{\partial^2}{\partial \xi^2} \mathbf{n} \cdot \mathbf{H}_{\text{cond}} \simeq 0 \tag{25.15}$$

since $\mathbf{n} \cdot \mathbf{H}$ corresponds to the perpendicular magnetic field. Thus we find

$$\mathbf{H}_{\text{cond}} = -\frac{\mathrm{i}}{\omega \mu_{\text{cond}} \sigma} \left(\frac{\partial^2}{\partial \xi^2} \mathbf{H}_{\text{cond}} \right) \tag{25.16}$$

This equation is solved by

$$\mathbf{H}_{\text{cond}}(\xi) = \mathbf{H}_{\text{cond}}(0) \exp \left[(1 + \mathrm{i}) \frac{\xi}{\delta} \right] \tag{25.17}$$

and δ is the skin depth.

$$\delta^2 = -\frac{\mathrm{i}}{\mu_{\text{cond}} \omega \sigma} (1 + \mathrm{i})^2 \rightarrow \delta = \sqrt{\frac{2}{\mu_{\text{cond}}\, \omega \sigma}} \tag{25.18}$$

For the electric field in the conductor we find by substitution into (25.8)

$$\mathbf{E}_{\text{cond}}(\xi) = (1 + \mathrm{i}) \sqrt{\frac{\omega \mu_{\text{cond}}}{2\sigma}} B_{\tan}(\xi) \tag{25.19}$$

The metal/insulator interface is characterized by the following condition:

$$E_{\tan} = (1 + \mathrm{i}) \sqrt{\frac{\omega \mu_{\text{cond}}}{2\sigma}} H_{\tan} \tag{25.20}$$

where both fields are tangential component at the interface and mutually orthogonal. Equation (25.17) is basically the surface impedance boundary condition (SIBC). We will use Equation (25.17) in the following way:

$$E_{\tan} = (1 + \mathrm{i}) \sqrt{\frac{\omega \mu_{\text{cond}}}{2\mu_{\text{insul}}^2 \sigma}} B_{\tan} \tag{25.21}$$

In Figure 25.2 we show the qualitative pattern of the various field components

25.2 Formulation of the BISC in Potentials

So far, we presented well-known results (Jackson [1975]). However our next task is to find a proper formulation in term of the Poisson potential and vector potential. Furthermore, we must have a method to computed the currents at

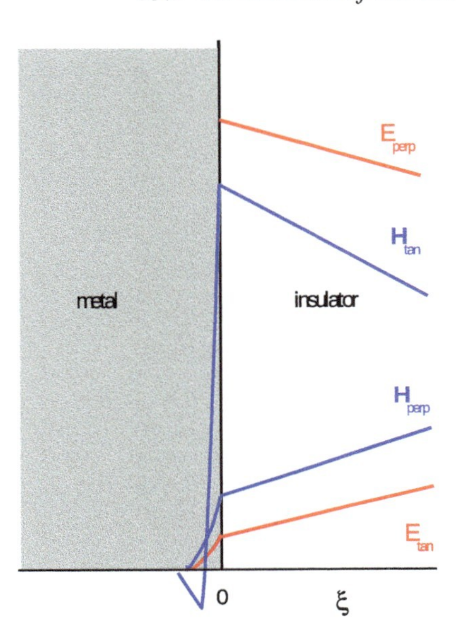

Figure 25.2 Various components of the electric field and magnetic induction at large but finite metallic conducutance.

the contacts. In order to benefit from the surface impedance boundary condition we want to avoid the calculation of the vector potential in the metallic regions. Therefore, in stead of computing the vector potential everywhere, we skip the calculation in the metallic regions and replace the discretized Maxwell-Ampere equation for the links at metal/insulator interfaces by a suitable version of Equation (25.17). Only links that are really inside the metal are skipped. Links at the metal/insulator interfaces are still participating in the generation of degrees of freedom. From the field potential relations

$$\mathbf{B} = \nabla \times \mathbf{A}, \quad \mathbf{E} = -\nabla V - \mathrm{i}\omega \mathbf{A} \tag{25.22}$$

we find that an appropriate translation of (25.17) is given by the following expression on a discretization grid

$$\frac{(V_j - V_i)}{h_{ij}} + \mathrm{i}\omega s_{ij} A_{ij} + (1+\mathrm{i})\sqrt{\frac{\omega_{\mathrm{cond}}}{2\sigma}}\frac{1}{\Delta S_{\mathrm{surf}}}\sum_{\mathrm{links}\in\mathrm{surf}} s_{\mathrm{link}\in\mathrm{surf}} A_{jk} h_{jk} = 0 \tag{25.23}$$

The meaning of the various various is illustrated in Figure 25.3. The variables s_{ij} and $s_{\mathrm{link}\in\mathrm{surf}}$ are 1 or -1 depending on the orientation of the links with

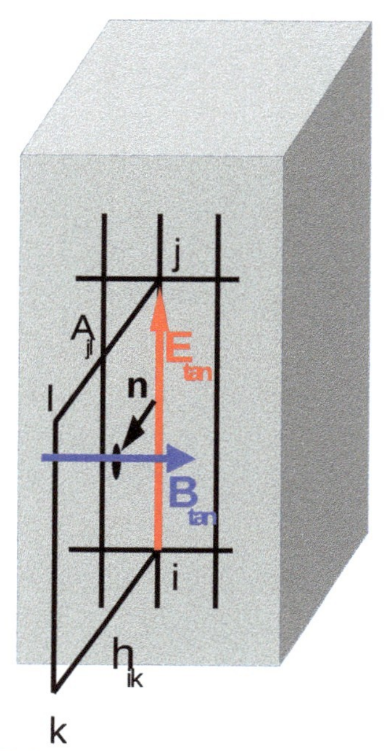

Figure 25.3 Variables in the assembling of the SIBC.

respect to the mesh and surfaces in the mesh. Δ_{surf} is the area of the surface perpendicular at the metal/insulator interface that contains the link $\langle ij \rangle$ for which the SIBC is applied.

25.3 Scaling Considerations

The discretization also requires that the scaling of the variables and equations is done. From Equation (25.17) we find that

$$E_{\text{tan}} = (1 + \mathrm{i}) \sqrt{\frac{\omega \mu_{con}^{r}}{2\mu_0 \left(\mu_r^{insul}\right)^2 \sigma}} B_{\text{tan}} \tag{25.24}$$

and

$$s_E E_{\text{tan}}^0 = (1 + \mathrm{i}) \sqrt{\frac{s_f}{s_\sigma} \frac{\omega^0 \mu_r^{\text{cond}}}{2\mu_0 \left(\mu_r^{\text{insul}}\right)^2 \sigma^0}} s_B B_{\text{tan}}^0 \tag{25.25}$$

where '0' indicates scaled variables and s is the dimensional scaling factor e.g. s_E scales the electric field, s_B scales the magnetic inductance, s_f scales the frequency, s_σ scales the conductance. Furthermore, let λ be the length scaling factor. Furthermore k is a dimensionless scaling parameter related to the speed of light and ν is the scaling parameter related to the relaxation of charge excess in conductors. Since

$$\begin{aligned} \frac{s_B}{s_E} &= \frac{1}{s_f} \\ \nu &= \frac{s_\sigma}{\varepsilon_0 s_f} \\ k &= \varepsilon_0 \mu_o \lambda^2 s_f^2 \end{aligned} \tag{25.26}$$

we find that the scaled equation is

$$E^0 = F\,(1+\mathrm{i})\sqrt{\frac{\omega^0 \mu_r^{\text{cond}}}{2\sigma^0}}\frac{B^0}{\mu_r^{\text{insul}}} \tag{25.27}$$

where

$$F = \frac{1}{\sqrt{k\nu}} \tag{25.28}$$

The variables k and ν are the dimensionless parameters defined in (25.26). It is also interesting to have a look at the scaling of the skin depth.

$$\delta = \sqrt{\frac{2}{\mu\omega\sigma}} = \frac{1}{\mu_0 s_f s_\sigma}\sqrt{\frac{2}{\mu_r^{\text{con}}\omega^0\sigma^0}} \tag{25.29}$$

From the definitions of k and ν we find

$$\frac{1}{\sqrt{\mu_0 s_f s_\sigma}} = \frac{1}{k\nu}\lambda \tag{25.30}$$

Therefore the scaled equation for the skin depth is with $\delta = \lambda\delta^0$ and

$$\delta^0 = \frac{1}{\sqrt{k\nu}}\sqrt{2}\mu_r\omega^0\sigma^0 \tag{25.31}$$

For example with a value for the frequency of $f = 100$ MHz and the conduction of Copper which corresponds to a value of $\sigma = 0.33\ 10^8$ S /m, the angular frequency $\omega = 2\pi 10^8 \text{sec}^{-1}$. The vacuum permeability is $\mu_0 = 4\pi 10^7 \text{mkg}/\text{C}^2$. Then the skin depth is $\delta = 8.76\ \mu$m.

25.4 One-Dimensional Test Example

We will test the validity of the surface-impedance model using a simple square pillar. The structure is shown in Figure 25.4 and Figure 25.5. The test case will also serve as a test vehicle to test further discretization steps. For example: What to do with the computation of the Poisson potential in metal? For the one-dimensional test example we find that in the standard approach the voltage drop from the top contact to the bottom contact is linear. This feature must be reproduced in the surface-impedance boundary conditions (SIBC) approach. The most convenient way to arrive at this result

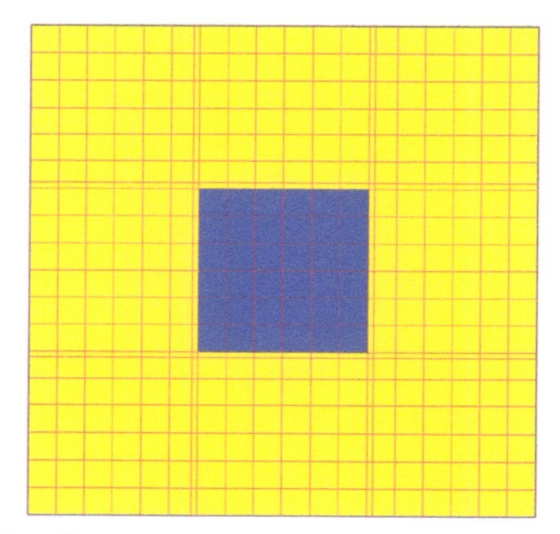

Figure 25.4 Two-dimensional cross section of the test structure with mesh.

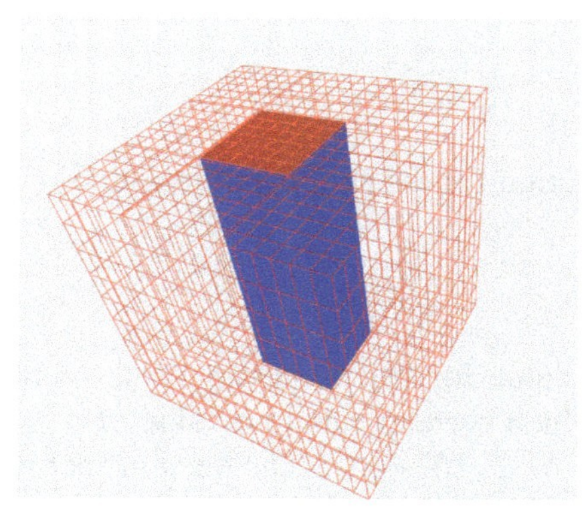

Figure 25.5 Three-dimensional test structure for the surface impedance approximation.

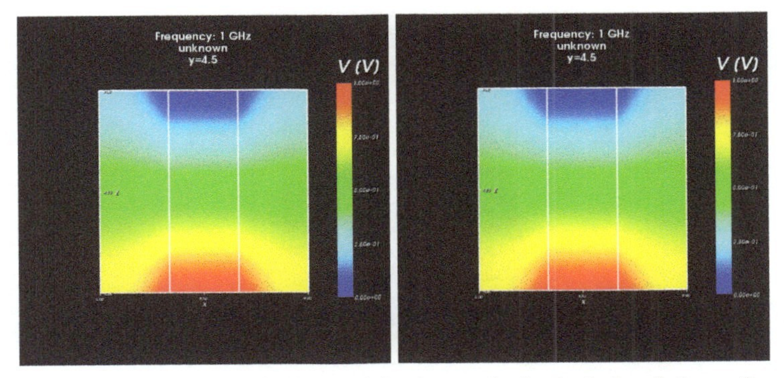

Figure 25.6 The Poisson potential in metal for the standard calculation (left panel) and SIBC (right panel).

is to keep the assembling of the current continuity in metal as in the standard approach. However the A field inside the metal does not contribute. This does not generate a big mismatch for the calculation of the Poisson potential but it does so for the calculation of the current density. In Figure 25.6 the plots are shown for the Poisson potential using the standard approach and the SIBC approach.

The structure is one-dimensional and its purpose is to test the implementation of the surface-impedance boundary conditions (SIBC). Since there is only a current in the z-direction we only need to compute the z-component of the vector potential. From the circulation of the vector potential we can extract the magnetic induction in the xy – directions. In the next paragraph, we focus on the magnetic induction at the insulator side of the metal/insulator interface. The size of the simulation domain is 9x9x9 μm^3 and the xy size of the metal is 3x3 μm^2.

Results in the range of 1–10 GHz In Figures 25.8, 25.9 and 25.10 we show the results for the magnetic in induction in the cells at the edge of the metal pillar in the insulator. To be more precise: the magnetic induction component tangential to the metal surface is shown (see Figure 25.7). Figure 25.8 compares the results for the real part of the magnetic induction and Figure 25.9 shows the imaginary part. As is seen in Figure 25.10, the imaginary part matches the default (standard) calculation quite well. However, the real part has a mismatch factor if size two. However, as we will see in the next paragraph the factor gets closer to one if the frequency gets larger. In other words, when the skin depth reduces and the surface-impedance model is more accurate. Note that in the frequency range [1–3] GHz, the SIBC overestimates

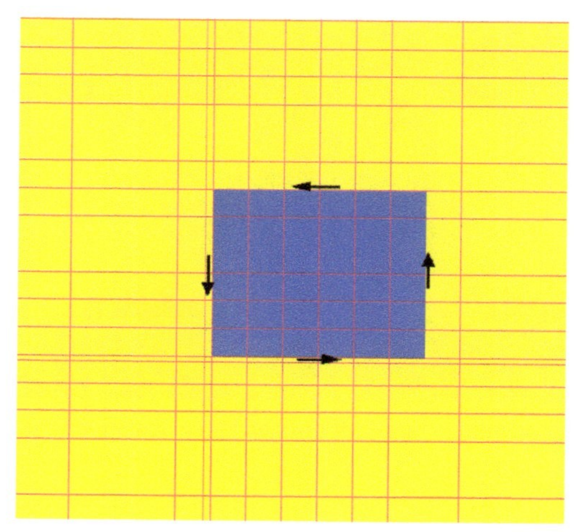

Figure 25.7 Component of the B-field that is reported in Figures 25.8–25.10.

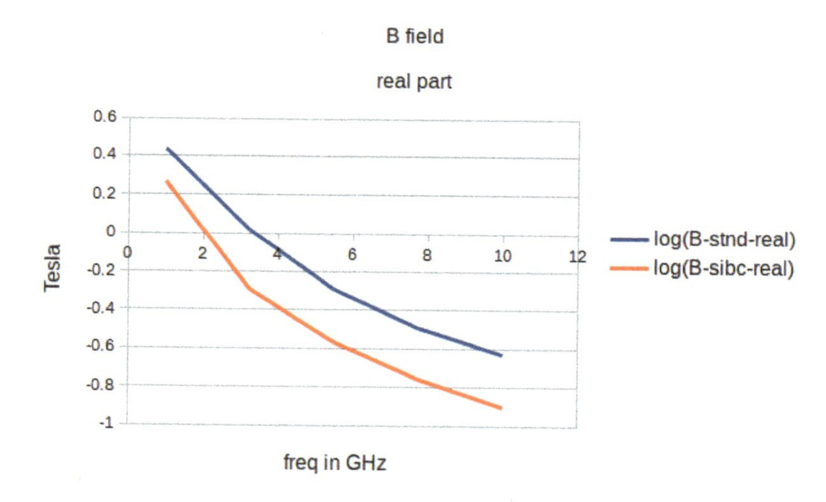

Figure 25.8 Real part of the magnetic induction tangential to the metal surface.

the imaginary part of the magnetic induction. In this range the skin depth is 2.7 micron. The SIBC model assumes that the metal can accommodate all skin current. However, the size of the cross section of the metal pillar is too small to do so. Therefore the computed skin current is larger than the current obtained in the non-approximated (standard) computation. As a consequence

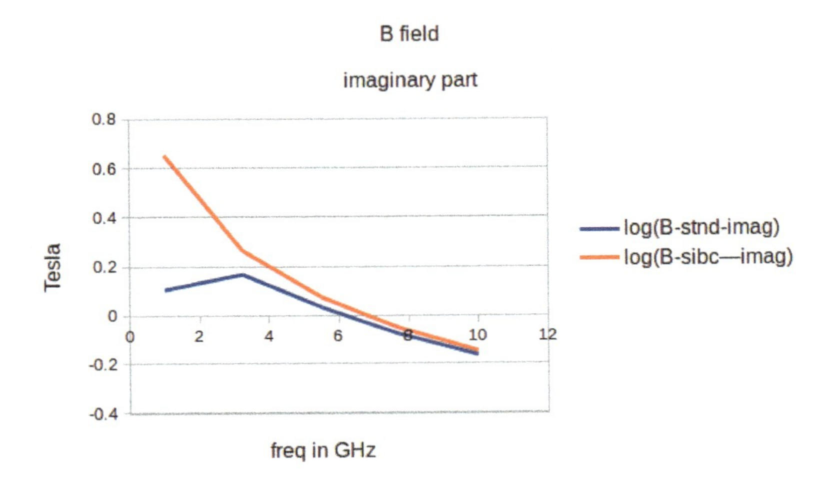

Figure 25.9 Imaginary part of the magnetic induction tangential to the metal surface.

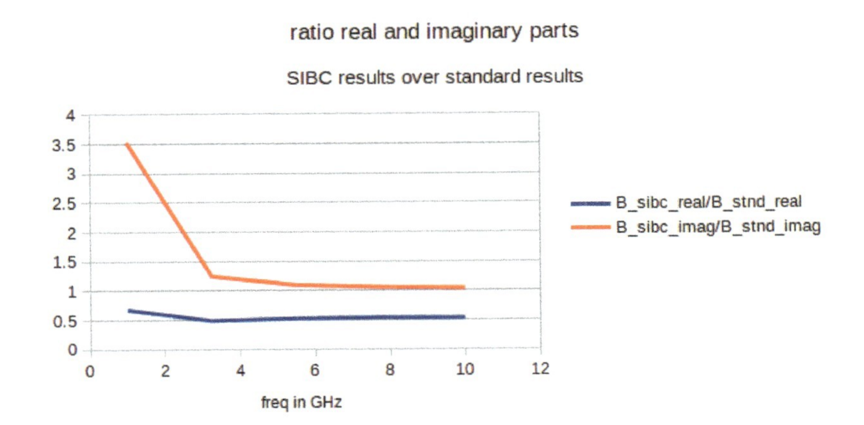

Figure 25.10 Ratio of the real and imaginary parts in the range 1–10 GHz.

the resulting magnetic induction will also be larger since the current in the metal pillar and the induction are related by

$$\oint \mathbf{H} \cdot \mathbf{dl} = I \tag{25.32}$$

As far as the real part is concerned we have a mismatch of a factor 2 that can not be explained by above arguments since the corresponding BISC current is too small.

Results over the range from 1GHz to 1THz We also did this calculation over a much wider range. The results are shown in Figures 25.11, 25.12 and 25.13.

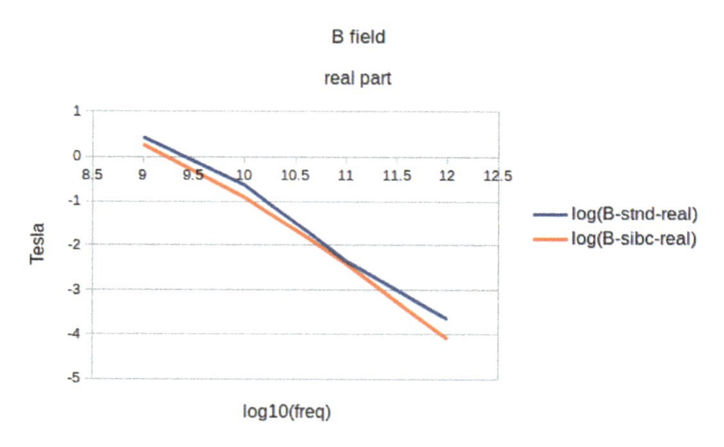

Figure 25.11 Real part of the magnetic induction tangential to the metal surface.

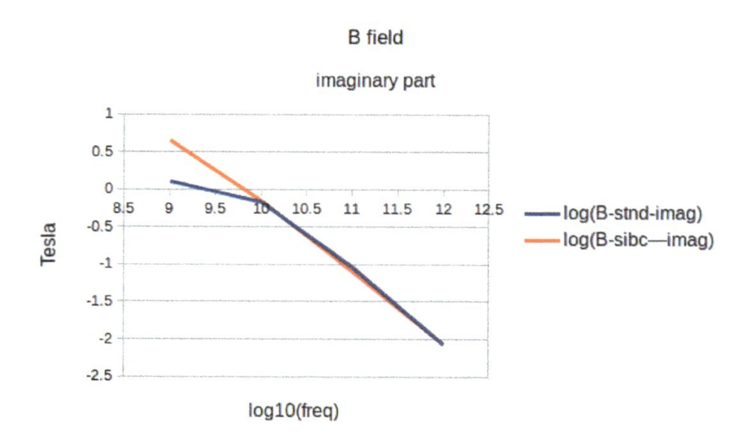

Figure 25.12 Imaginary part of the magnetic induction tangential to the metal surface.

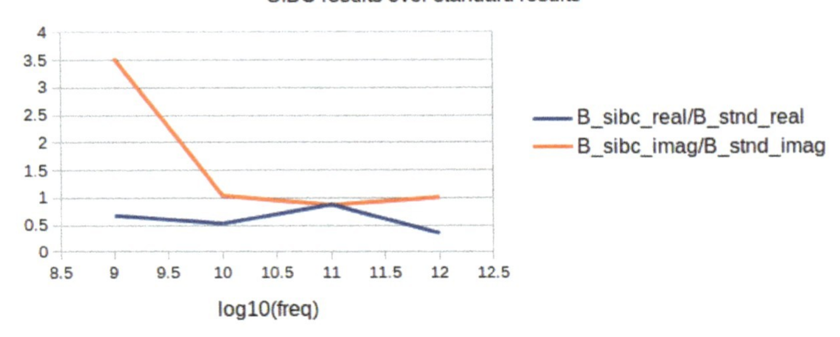

Figure 25.13 Ratio of the real and imaginary parts in the range 1GHz.

Current calculations As said, inside metal the current evaluation using local fields does not make sense since the vector potential is not computed anymore. However, we can compute the current by (25.33) e.g.

$$\oint \mathbf{B} \cdot \mathbf{dl} = \mu I \tag{25.33}$$

For that purpose we place consider all surfaces around the contact as depicted in Figure 25.14. The currents are compared in Figure 25.15 and Figure 25.16. The BISC model results are compared with the standard (no approximations) results. The numbers are surely of the right order of magnitude and can serve as a first order result that can be corrected with refined setting of conductance parameters.

Surface Impedance Boundary Conditions in the Transient Regime In order to start the SIBC in the transient regime (SIBC-TR) we start with Equation (25.21) e.g.

$$E_{\mathrm{tan}} = (1 + \mathrm{i}) \sqrt{\frac{\omega \mu_{\mathrm{cond}}}{2\mu_{\mathrm{insul}}^2 \sigma}} B_{\mathrm{tan}} \tag{25.34}$$

Squaring this expression gives

$$E_{tan}^2 = \mathrm{i} \frac{\omega \mu_{\mathrm{con}}}{(\mu_{ins})^2 \sigma} B_{tan}^2 \tag{25.35}$$

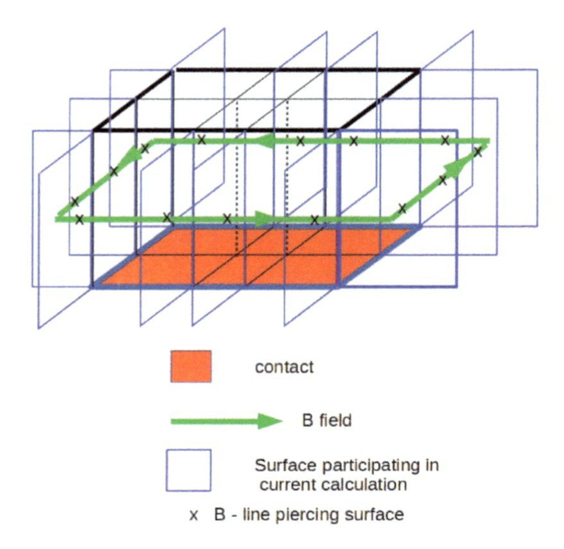

Figure 25.14 Illustration of contact current calculation.

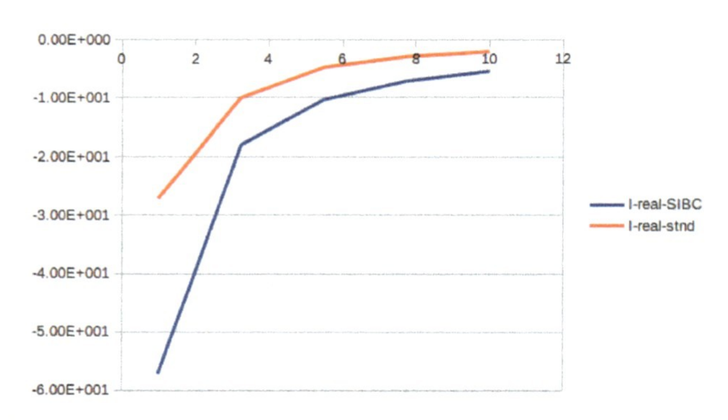

Figure 25.15 Comparison of the real parts of the contact currents from 1–10 GHz.

Now a factor $\mathrm{i}\omega$ is in one-to-one correspondence with a time differentiation. Beware that the fields appear in squared form. So (25.35) is in fact

$$E_{\tan}^{2}(\omega)\exp(2I\omega t) = \mathrm{i}\frac{\omega\mu_{\mathrm{con}}}{(\mu_{\mathrm{ins}})^{2}\sigma}B_{\tan}^{2}(\omega\mu_{\mathrm{con}})exp(2\mathrm{i}\omega t)$$

$$= \frac{\mu_{\mathrm{con}}}{2(\mu_{\mathrm{ins}})^{2}\sigma}\frac{\partial}{\partial t}B_{\tan}^{2}(\omega)\exp(2\mathrm{i}\omega t) \tag{25.36}$$

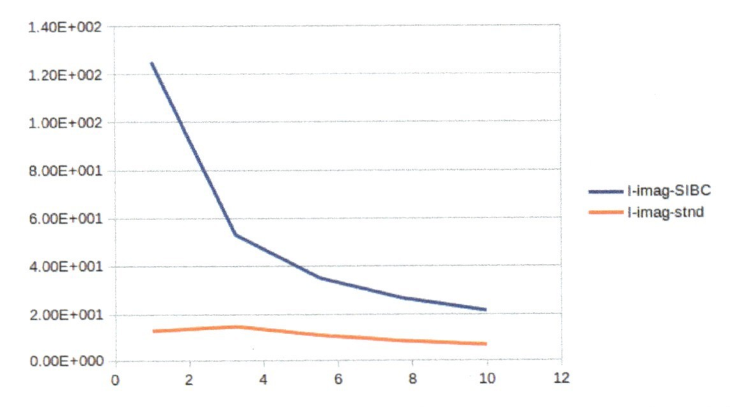

Figure 25.16 Comparison of the imaginary parts of the contact currents from 1–10 GHz.

Therefore we set

$$E_{\text{tan}}^2 = \frac{\mu_{\text{con}}}{2(\mu_{\text{ins}})^2\sigma}\frac{\partial}{\partial t}B_{\text{tan}}^2 \tag{25.37}$$

So at least for an harmonic wave we have a relation between the electric field and the magnetic induction at the metal/insulator interface.

We may attempt to come to this conclusion from the Maxwell equations directly. The local conservation of energy can be expressed as

$$\begin{aligned}\frac{\partial w}{\partial t} + \nabla \cdot \mathbf{S} &= -\mathbf{J}.\mathbf{E} \\ w &= \frac{1}{2}(\epsilon E^2 + \frac{B^2}{\mu}) \\ \mathbf{S} &= \frac{1}{\mu}\mathbf{E} \times \mathbf{B} \\ \mathbf{J} &= \sigma \mathbf{E}\end{aligned} \tag{25.38}$$

Since the conductance is very large we may keep the leading-order terms e.g. assume that

$$\frac{1}{2}\epsilon E^2 << \sigma E^2 \quad , \quad |E||B| << \mu\sigma E^2 \tag{25.39}$$

which differs from (25.37) with a factor -1 comparing left and right hand side.

Discretized form In order to take a discretized version of (25.37) we use

$$E_{ij} = -\frac{V_j - V_i}{h_{ij}} - s_{ij}\Pi_{ij} \tag{25.40}$$

and

$$H_{surf} = \frac{1}{\mu_{\text{insul}}} \left(\frac{1}{\Delta S_{\text{surf}}} \sum_{links \in surf} \sigma_{jk} A_{jk} h_{ij} \right) \tag{25.41}$$

Therefore, translation of (25.37) to

$$E_{\text{tan}}^2 = \frac{\mu_{\text{con}}}{2\sigma} \frac{\partial}{\partial t} H_{\text{tan}}^2 \tag{25.42}$$

gives

$$\frac{\mu_{con}}{(\mu_{\text{ins}})^2 \sigma} \left(\frac{1}{\Delta S_{\text{surf}}} \sum_{links \in surf} \sigma_{jk} A_{jk} h_{jk} \right) \left(\frac{1}{\Delta S_{\text{surf}}} \sum_{links \in surf} \sigma_{jk} \Pi_{jk} h_{jk} \right) - \left(\frac{(V_j - V_i)}{h_{ij}} - s_{ij} \Pi_{ij} \right)^2 = 0 \tag{25.43}$$

Equation (25.43) is the new result. A first implementation has been realized. If successful it allows us to perform fast transient simulations and avoid detailed calculations in metallic region.

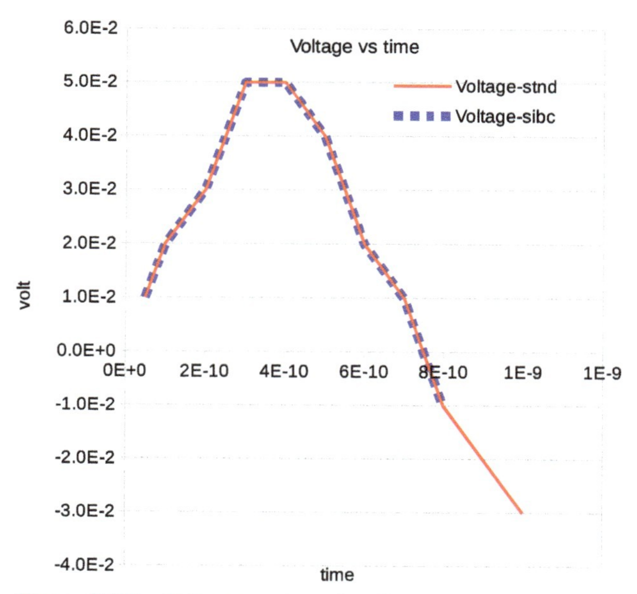

Figure 25.17 Voltage vs. time plot of the transient test set up.

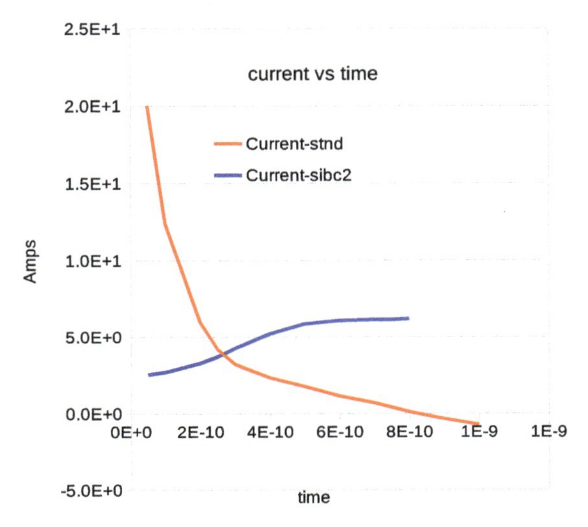

Figure 25.18 Current vs. time plot of the transient test set up.

One-dimensional Test Example in the Transient Regime We use the same structure as when testing the SIBC modeling in the frequency domain. Here we take a fixed voltage applied to one contact. The voltage plot is given in Figure 25.17. We compute the current through the pillar using the regular formulation, e.g. this is the standard set up of the equations. The results are marked with the label stnd. Next we compute the results using SIBC for metallic domains. The results are shown in Figure 25.18. The results are is not acceptable and needs further investigation.

26

Using the Ghost Method for Floating Domains in Electromagnetic Field Solvers

Floating regions in device simulation are a potential source of convergence problems due to the singular character of the matrices that need to be inverted during the iteration towards the solution in the Newton-Raphson method. Floating regions typically occur if there exist domains where charge can get trapped. Such domains can be n-type regions surrounded by p-type domains, p-type regions surrounded by n-type domains, n- or p-type regions surrounded by insulators and variations thereof. Physically, the origin of the singular nature of the Newton-Raphson matrix is quite obvious: if the value of the trapped charge is not taken into account, as is often the case, then the solution is not unique (a different value of the total trapped charge gives rise to a different potential level). The mathematical problem is ill-posed and the result is a numerical problem that gives rise to convergence problems such as norm oscillations (Edwards et al. [1988]). Several work-around methods have been suggested in the past. One option is to exploit transient simulations, thereby carefully controlling the amount of trapped charge. Another solution has been to introduce an artificial contact to the floating region and together with the constraint that no current is flowing into this contact. Finally, one can exploit the recombination-generation term in the drift-diffusion equations and elevate the isolating status of the floating region. Whereas, the latter option is feasible for floating regions that are only shielded from the contacts by pn junctions, e.g. SOI devices, such a work-around is not possible if the floating region is truly insulated. This is the case if the region is fully surrounded by insulators. Furthermore, if the floating region is metal surrounded by insulating materials, e.g. floating gates or dummy structures in the interconnect layout, then such an approach is not possible and alternative solutions must be found. Here, we will present such an alternative method that explicitly takes into account the trapped charge in the floating domain. We will concentrate here on floating metallic regions.

We introduced a method for solving electromagnetic field problems by including an additional scalar ghost field that needs to be obtained as part of the solution method. The solution for this additional field does not carry energy and can be viewed as being a mathematical aid that allows for the construction of a gauge-fixed, regular matrix representation of the curl-curl operator acting on edge elements (Meuris et al. [2001b]).

Instead of the curl-curl operator combined with the gauge condition

$$\mathcal{M}_{\text{old}} = \begin{bmatrix} \nabla \times \nabla \times \\ \nabla \cdot \end{bmatrix}, \tag{26.1}$$

leading to a sparse, well-posed, but non-square matrix $\mathcal{M}_{\text{old}}$ that acts only on the vector field $\mathbf{A}$, the operator

$$\mathcal{M}_{\text{new}} = \begin{bmatrix} \nabla \times \nabla \times & \gamma \nabla \\ \nabla \cdot & \nabla^2 \end{bmatrix}, \tag{26.2}$$

is considered that acts on the pair of variables $\mathbf{A}$ and a ghost field χ, according to

$$\mathcal{M}_{\text{new}} \star \begin{bmatrix} \mathbf{A} \\ \chi \end{bmatrix} = \mu \begin{bmatrix} \mathbf{J} - \epsilon \frac{\partial}{\partial t} \mathbf{E} \\ 0 \end{bmatrix}. \tag{26.3}$$

This operator leads to matrices $\mathcal{M}_{\text{new}}$ that are sparse, regular, square and semi-definite. Implicitly, we made an important conceptual step: by adding additional degrees of freedom, e.g. the ghost-field variables χ, we turned a numerically unattractive problem into an easy one: although the starting problem was well-defined it is problematic due to the fact that the matrices that need to be solved are not square. The extra degrees of freedom transformed the problem into one which generates well-defined and square matrices. Having noted this message, we will deal with the floating regions in a similar way: the inclusion of additional degrees of freedom will assist in turning an unsolvable (singular) problem into a solvable (regular) problem. In particular, we will address a long standing problem of dealing with floating domains in general-purpose electromagnetic field solvers.

26.1 Problem Description

In general-purpose electromagnetic field solvers the constitutive laws for the current densities are used in such a way that they reflect the material properties of the underlying domain. In particular, for metals Ohm's law is used in the form

$$\mathbf{J} = \sigma \mathbf{E} = -\sigma \nabla V, \tag{26.4}$$

together with the continuity equation:

$$\nabla \cdot \mathbf{J} = 0. \tag{26.5}$$

In insulating regions the Poisson equation

$$-\nabla(\epsilon \nabla V) = \rho \tag{26.6}$$

is solved. Continuity at the interfaces between metallic and insulating regions should be guaranteed. A finite-element implementation of both equations for the several domains automatically takes the continuity into account, but ignores the fact that *no* currents flow in the floating regions, since the finite-element method as well as the box-integration method refer to the *balance* of the current flow in the metallic nodes. Since the metal-insulator nodes do participate in the current balance, the latter ones are dealt with according to their metallic nature. All this works fine provided that the metal is not floating. However, for floating regions, this implementation leads to solutions that describe constant currents in the metallic regions, thereby respecting the balance in the metal nodes but at the same time putting arbitrary values of the potential on the interface nodes.

In order to illustrate above remarks we consider a simple one-dimensional static problem that is described with a grid of seven nodes. However, the idea can be equally applied to bigger problems at higher frequencies. A metallic domain is squeezed between two insulating regions as is illustrated in Figure 26.1. There are five variables $(V_2, V_3, V_4, V_5, V_6)$ for which the finite-element equations read:

$$\begin{bmatrix} -2\epsilon & \epsilon & 0 & 0 & 0 \\ 0 & \sigma & -\sigma & 0 & 0 \\ 0 & -\sigma & 2\sigma & -\sigma & 0 \\ 0 & 0 & \sigma & -\sigma & 0 \\ 0 & 0 & 0 & \epsilon & -2\epsilon \end{bmatrix} \star \begin{bmatrix} V_2 \\ V_3 \\ V_4 \\ V_5 \\ V_6 \end{bmatrix} = \begin{bmatrix} -\epsilon V_1 \\ 0 \\ 0 \\ 0 \\ -\epsilon V_7 \end{bmatrix} \tag{26.7}$$

It can be easily checked that the determinant of the matrix in (26.7) is zero. A basis for the null space is given by the vector $\mathbf{v}_0 = (1/2, 1, 1, 1, 1/2)$.

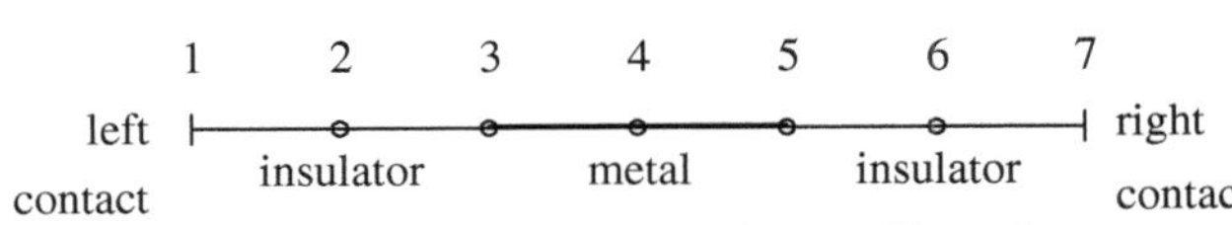

Figure 26.1 Grid of a one-dimensional structure of a metallic region squeezed between two insulating regions.

Therefore, any vector that obeys Equation (26.7) can be displaced by an arbitrary amount proportional to $\mathbf{v}_0$. For such a small system one can easily recognize the singularity of the matrix, however this is not longer the case for real world problems. For such problems, the matrix is huge and also the linear system can be solved to get *a solution*. With this treatment of the floating areas their potential value is undefined, and can become whatever value due to the ill-posedness of the problem and the singularity of the corresponding matrix.

Thus we find that the cause of the arbitrariness is the ill-posed formulation of above assignment of the various equations for the Poisson potential. For the potentials at the nodes, the discretized problem reads

$$\mathcal{M}_{\text{float}} \star [\mathbf{V}] = [\mathbf{b}], \tag{26.8}$$

where $\mathbf{V}$ is the column vector of the Poisson potential at the internal nodes and $\mathbf{b}$ describes the coupling to the contacts. $\mathcal{M}_{\text{float}}$ is a singular matrix. In the spirit of the ghosts, we will now transform this problem into a regular one by extending the size. In other words, we will construct a *regular* matrix and a larger vector for the unknown variables such that the floating region problem is described by the following equation:

$$\begin{bmatrix} \mathcal{M} & \mathcal{P} \\ \mathcal{Q} & \mathcal{N} \end{bmatrix} \star \begin{bmatrix} \mathbf{x} \\ \mathbf{y} \end{bmatrix} = \begin{bmatrix} \mathbf{b} \\ \mathbf{0} \end{bmatrix}, \tag{26.9}$$

where $\mathcal{P}$, $\mathcal{Q}$ and $\mathcal{N}$ are extra entries for making the complete matrix non-singular and keeping it square at the same instant.

26.2 Proposed Solution

In order to construct the extended matrix system we will use the fact that the total charge on each floating region is fixed. Moreover, since no current can flow in the floating regions, each region has a fixed value of the potential, V_{fl}. These two facts suffice to perform the construction. From Gauss' law we find that for the k-th floating region, we must add to the evaluation of the Poisson equation at the insulator side of the interface, a charge term

$$\rho_i = \sum \epsilon \Delta A_i \left(V_{\text{fl}}^k - V_i \right), \tag{26.10}$$

where the sum runs over all nodes in the insulating region that couple to the k-th floating region and ΔA_i is the interface area assigned to the i-th node. The variable V_{fl}^k can be added to the vector $\mathbf{y}$, thereby extending the set of unknowns. At the same time we obtain the entries for the matrix part $\mathcal{P}$, since they follow from the Newton-Raphson derivatives $\partial\rho/\partial V_{\text{fl}}^{k}$. The additional equations follow from setting the total charge that is trapped on each floating region: all that needs to be done is to sum the charge density that is stored at each node of the surface of the k-th floating region. Therefore, each floating region generates one additional equation. The sums determine the matrix parts $\mathcal{Q}$ and $\mathcal{N}$. Finally, we note that the node equations for the Poisson potentials at the interfaces are simply

$$V_{\text{int},i} - V_{\text{fl}}^{k} = 0. \tag{26.11}$$

Using the simple example of Figure 26.1, we arrive at the following non-singular matrix problem:

$$\begin{bmatrix} -2\epsilon & \epsilon & 0 & 0 & 0 & 0 \\ 0 & 1 & 0 & 0 & 0 & -1 \\ 0 & -\sigma & 2\sigma & -\sigma & 0 & 0 \\ 0 & 0 & 0 & 1 & 0 & -1 \\ 0 & 0 & 0 & \epsilon & -2\epsilon & 0 \\ -1 & 0 & 0 & 0 & -1 & 2 \end{bmatrix} \star \begin{bmatrix} V_2 \\ V_3 \\ V_4 \\ V_5 \\ V_6 \\ V_{\text{fl}}^{1} \end{bmatrix} = \begin{bmatrix} -\epsilon V_1 \\ 0 \\ 0 \\ 0 \\ -\epsilon V_7 \\ Q^1/\epsilon \end{bmatrix} \tag{26.12}$$

26.3 Example 1: Metal Blocks Embedded in Insulator

We have inserted above ideas in a general-purpose electromagnetic field solver. In Figure 26.2 a structure is depicted with two floating metallic regions and a applied bias of two Volts from the bottom contact to the top contact. In Figure 26.3, the potential along a cut line from the bottom contact to the

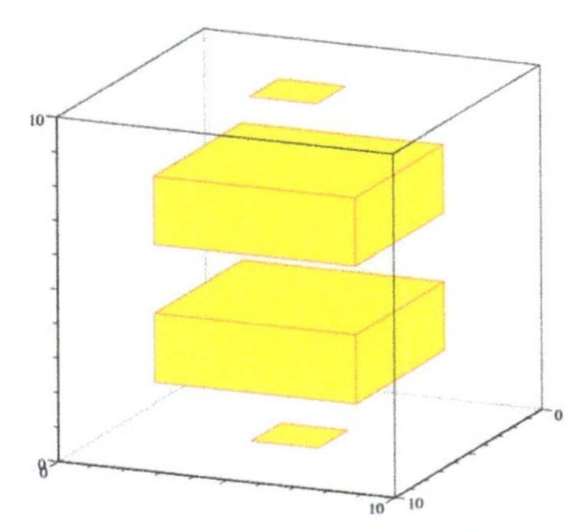

Figure 26.2 Two floating metallic regions embedded into an insulating volume and two contacts.

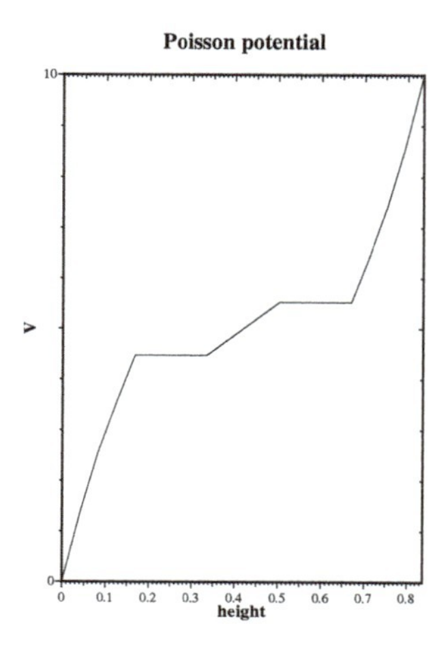

Figure 26.3 The Poisson potential along a line from the bottom to the top contact.

top contact is shown that is obtained by solving the problem according to above described method. Note that the potential in the metal regions is flat (equipotential) and that the level is between the values of the applied biases (no charge was put on the floating regions).

26.4 Example 2: A Transformer System

To show that the method also works on more complex application, we examined the system shown in Figure 26.4.

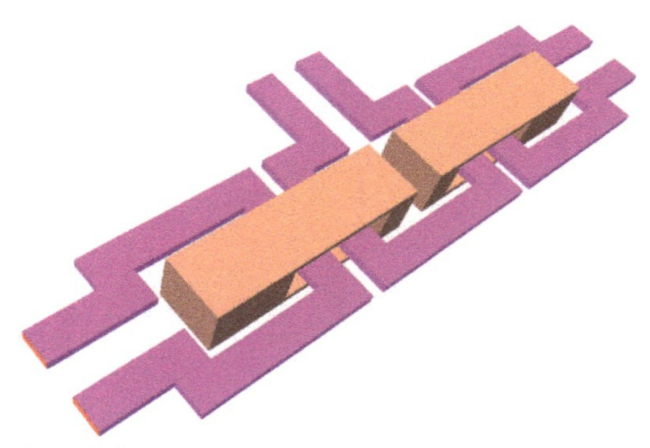

Figure 26.4 A 3D view of the geometry of a ring transformer used to show the validity of the method.

A three ring transformer system has been analyzed. Three metal rings in the horizontal plane, containing each two ports (or contacts) are connected by two metal floating regions (rings in the vertical plane). Although we restrict the analysis to the static case, also the high-frequent analysis can be carried out as will be shown later. The transformer is embedded in a dielectric.

All the ports of the rings have zero-voltage boundary conditions for the electric potential, except for one of port of one ring (Figure 26.5). The rings with zero-voltage boundary conditions are equipotential volumes at zero Volt as expected. The floating areas in between are also equipotential volumes. Their potential value is fixed by the extra condition (26.10).

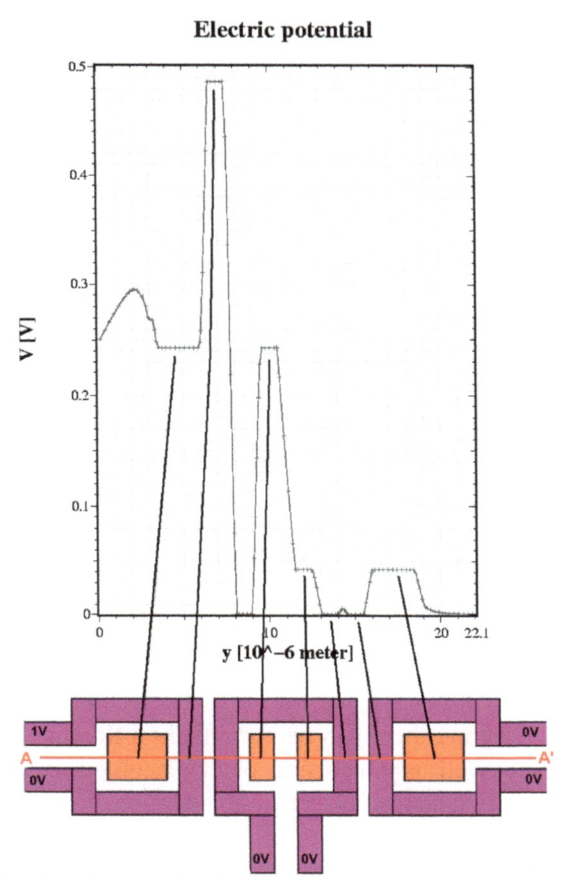

Figure 26.5 The electric potential along the line AA' in the cross-section. The applied potentials on the ports are also shown.

One of the advantages of this technique is the treatment of trapped charges. Indeed in Equation (26.12), the variable Q^i pops up. This charge stands for the trapped charge on the floating region i. The influence of the trapped charge on the solution of the Poisson problem, can be studied very easily using this technique. Additional positive (negative) charge will increase (decrease) the floating potentials.

26.5 Initial Guess

In realistic simulations the additional equation for each floating region involves the nodes of surface of the floating domain. Although this number of nodes is an order of magnitude less than the total number of nodes, it represents still a considerable amount of non-local coupling. As a consequence the matrix inversion (in practice: the iterative solving procedure) is hampered by such a degree of non-locality.

Moreover, the new matrix (26.12) is not symmetric and the conjugate gradient method for symmetric systems, that exist for solving this kind of systems fails.

In order to improve the iterative solution scheme, a good initial guess will usually substantially reduce the calculation time. For floating regions, one may obtain a good initial guess by realizing that the Poisson potential is constant in each floating domain. A negligible gradient of the Poisson potential can also be obtained by treating the metal as a high-K dielectric material, since the large permittivity will force the electric field to become small. Therefore, as an initial guess this approach will lead to almost flat Poisson potentials in the floating domains. The cusp in the potential at the interface nodes is fully determined by the the ratios of the permittivities and therefore in general this initial guess work fine, if the total charge on the floating domain vanishes.

26.6 High-Frequency Problems

In the high-frequency regime there can be currents in a floating domain due to inductance effects as well as conversion of displacement currents in the insulating domains to conductance currents at the surface of metallic floating domains. In general we must solve the full system of Maxwell equations to explain the observed effects. A difficulty is the limiting procedure corresponding to zero frequency, e.g, when do steady-state approximations apply?

Let us first summarize a line of reasoning which seems plausible but is actually wrong: The electric field is

$$\mathbf{E} = -\nabla V - \frac{\partial \mathbf{A}}{\partial t} \tag{26.13}$$

The first term corresponds to a conservative field and can therefore be dealt with as before in floating domains. In other words, we have that $\mathbf{E} = \mathbf{E}_{\mathrm{C}} + \mathbf{E}_{\mathrm{NC}}$, where the first term is the conservative contribution arising from the scalar potential and the second term is the electric field arising from inductive effects which is in general non-conservative. In floating regions we can put $\mathbf{E}_{\mathrm{C}} = 0$, therefore the electric field is *fully* inductive. Moreover, with the choice of the Coulomb gauge, the Poisson equation for the electric potential remains the same for the static and the dynamic treatment. This means that also Gauss' law must hold for the dynamic regime, and that we can write down (26.10) in the dynamic regime.

The flaw in above line of reasoning is that the conservative contribution $\mathbf{E}_{\mathrm{C}} \neq 0$. The charge density inside the metallic floating domain can still be zero but there are voltage differences possible over a floating domain.

In Figure 26.6 a structure is shown of a floating ring and a transmission line. The bottom plate is grounded. When sending a high-frequency signal over the the transmission line, the ring will resonate a some frequency. In Figures 26.7 and 26.8, the results for the s11 and s21 parameters are shown. Besides the experimental results, we also present the results of the simulation. The curves denoted with E_C non zero clearly reproduce the presence of the resonance. Here all Maxwell equations are solved 'as-is'. On the other hand, the curves denoted by E_C zero, have been obtained assumed that there Poisson potential is constant in the ring. These curves do not show any sign of the resonance. Therefore, the assumption that the conserved contribution to th electric field in floating domains can be neglected is wrong.

Figure 26.6 Design of a resonator structure. The floating ring (light-gray colored) is resonating with a narrow stripe transmission line (blue colored). The right view is a stretched version of the left unstretched view.

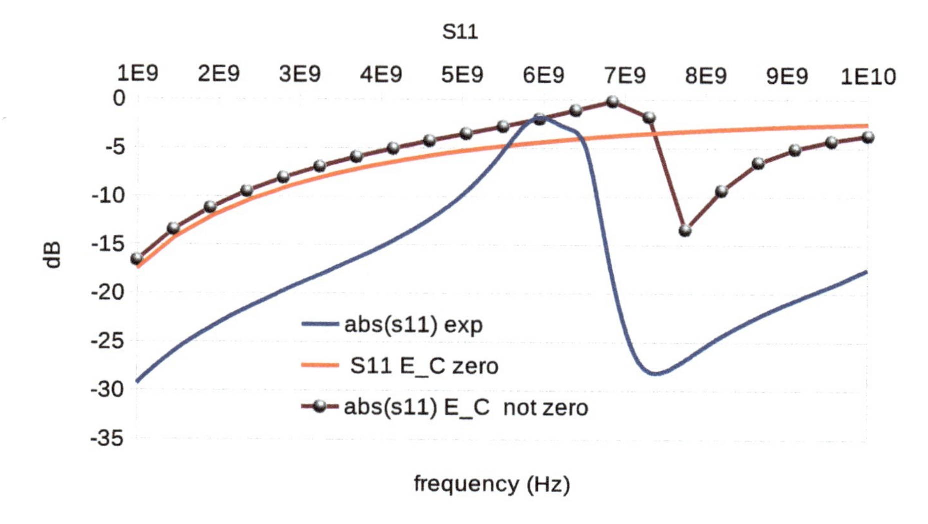

Figure 26.7 S11 parameter. The resonance is observed in the measurement and in the simulation set that does not put $E_C = 0$.

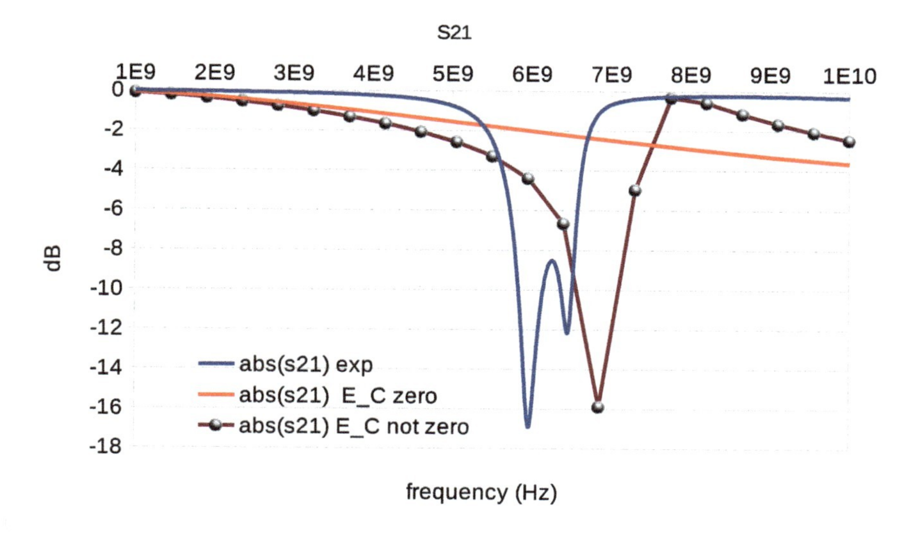

Figure 26.8 S21 parameter. The resonance is observed in the measurement and in the simulation set that does not put $E_C = 0$.

The resonance manifests itself clearly in a detailed view at the voltage and vector potential patterns over the ring. In Figures 26.9 and 26.10 the voltages over the ring are shown. It is clear that the voltage is not constant. Furthermore in Figures 26.11 and 26.12, the vector potential is shown over the ring.

Figure 26.9 Voltage over the resonating ring at 6.85 GHz.

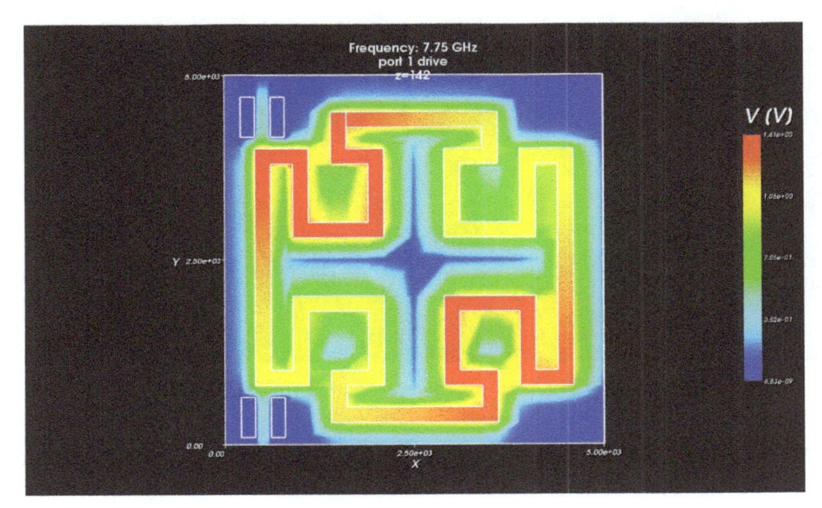

Figure 26.10 Voltage over the resonating ring at 7.75 GHz.

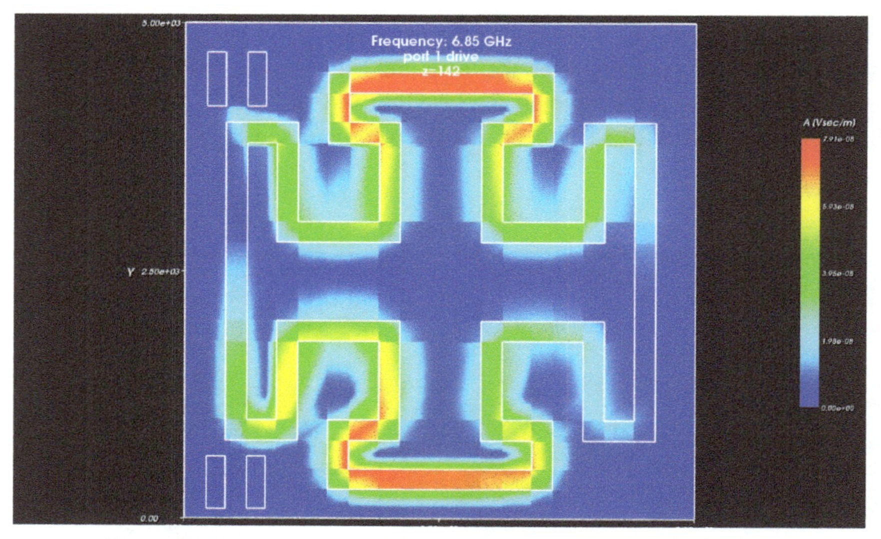

Figure 26.11 Vector potential over the resonating ring at 6.85 GHz.

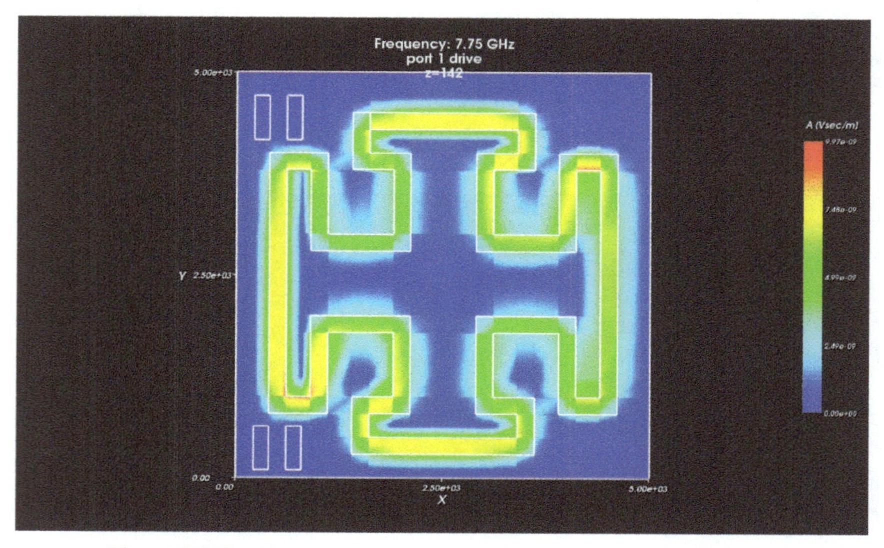

Figure 26.12 Vector potential over the resonating ring at 7.75 GHz.

So far, we followed the 'ghost' approach in order to deal with singular problems. The novelty of the method consists in the way of handling the metallic nodes inside the floating region: their degrees of freedom are not eliminated in favor of the floating potential. In stead, the degrees of freedom corresponding to the floating regions, are treated according to the local Ohm's law as is done for non-floating metallic domains. Results on test problems correspond with the expected values. If trapped charge is introduced, the corresponding floating region potential is shifted.

We also may exploit the knowledge that the electric field in floating metallic domains is zero. Therefore, assembling the current-continuity equation in floating domains is futile and an alternative approach must be pursued. The alternative approach can be summarized as follows: Since no current is flowing in the floating domain, there is only one potential value to be computed. Therefore, do not assemble the current continuity law in the floating domain and take the conservation of the total charge as a condition for determining the value of the floating domain potential. We implemented this approach and it leads also to regular linear system.

In Figure (26.13) we show a stack of two floating regions between two biased blocks of metal. With a bias of one volt at the top block and zero volt at the bottom block, the resulting voltage profile is shown in Figure (26.14).

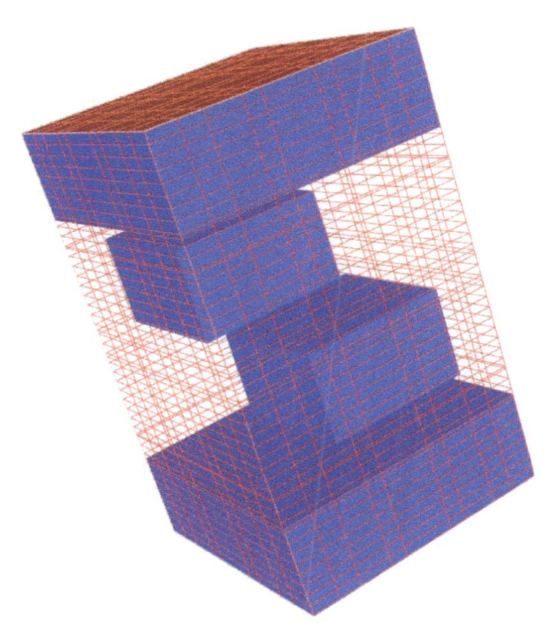

Figure 26.13 Structure with a stack of two floating regions and the used mesh.

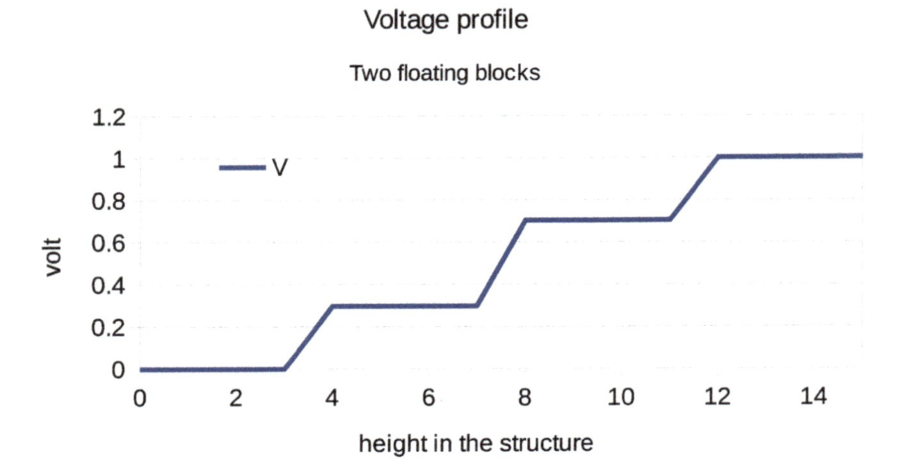

Figure 26.14 The resulting voltage along a vertical cut through the stack of Figure 26.13.

26.7 Floating Semiconductor Regions

(Floating) metal does not have bulk charges. All charge (if any) is found at the surface as a surface charge density and gives rise to a discontinuous electric field. Floating semiconductor does have bulk charge. This requires that the charge conservation can not be exploited as a surface integral. Gauss' law is still valid but does not provide new information on the floating potential. The trick that is successful for metals can not be simply repeated for semiconductors So what next? Remember that semiconductors are described by quasi Fermi levels ϕ_p and ϕ_n besides the Poisson potential V. The carrier densities are given by

$$p = n_i e^{q/k_B T(\phi_p - V)}, \quad n = n_i e^{q/k_B T(V - \phi_n)} \tag{26.14}$$

We also know that

$$\mathbf{J}_p = q\mu_p p\mathbf{E} - qD_p \nabla p, \quad \mathbf{J}_n = q\mu_n n\mathbf{E} + qD_n \nabla n \tag{26.15}$$

Current conservation is given by

$$\nabla.\mathbf{J}_n - q(R - G) = 0, \quad \nabla \mathbf{J}_p q(R - G) = 0 \tag{26.16}$$

$$R = R_{SRH} = \frac{np - n_1^2}{(\tau_p(n + n_1) + \tau_n(p + p_1))} \tag{26.17}$$

We now exploit the key observation that in the floating region the current is equal zero. Thus only one Fermi level ϕ_F needs to be determined. For its determination we apply charge conservation as a volume integral: Since $\mathbf{J}_p = \mathbf{J}_n = 0 \rightarrow R_{SRH} = 0 \rightarrow pn = n_i^2$. Therefore we obtain $\phi_p = \phi_n = \phi_F$ and

$$\int_{\text{floating region}} \mathrm{d}V(p - n + N_D) = 0 \tag{26.18}$$

The equations that set the values of V_i (one for each node) and ϕ_F are the Poisson's equation and the condition expressing charge conservation

$$\nabla(\varepsilon\nabla V) = n_i \mathrm{e}^{q/k_BT(\phi_F - V)} - n_i \mathrm{e}^{q/k_BT(V - \phi_F)} + N_D \tag{26.19}$$

$$\int_{\text{floating region}} \mathrm{d}V \left(n_i \mathrm{e}^{q/k_BT(\phi_F - V)} - n_i \mathrm{e}^{q/k_BT(V\phi_F)} + N_D\right) = 0 \tag{26.20}$$

The extension to several isolated floating semiconductor regions requires that each region is equipped with its Fermi potential ϕ_F^k. The Fermi level ϕ_F plays the same role as V_{fl} does for metallic floating regions. The current density in the floating region is zero since $\mathbf{J}_p = -q\mu_p p \nabla\phi_p, \quad D_p = k_BT/q\mu_p$ and $\mathbf{J}_n = -q\mu_n n \nabla\phi_n, \quad D_n = k_BT/q\mu_n$. To illustrate the proposed method we consider a block of p-type silicon with doping level $10^{16}/\mathrm{cm}^3$ between two metallic blocks as shown in Figure 26.15.

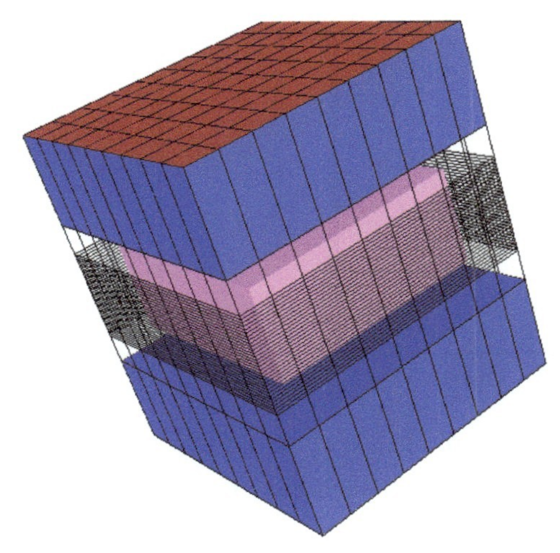

Figure 26.15 Structure with a semiconductor floating region between two metallic blocks.

In Figures 26.16 and 26.17 the hole and electron densities are shown for an applied bias of 0.5 volt at the top metallic block.

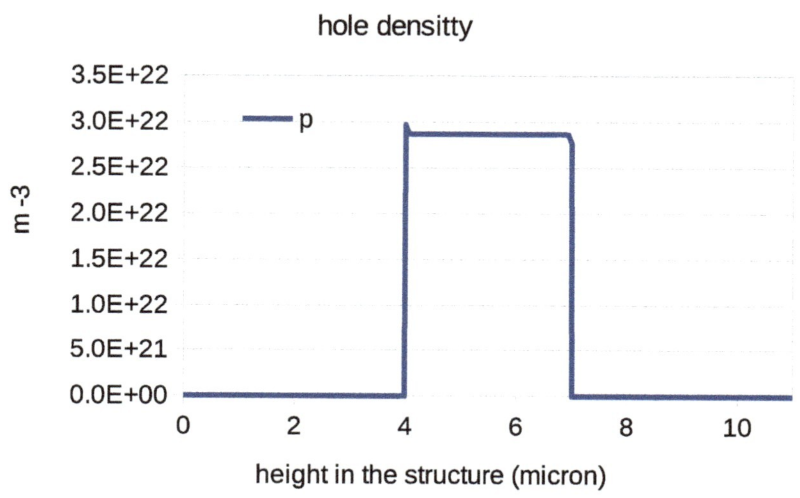

Figure 26.16 Hole density profile in the floating floating semiconductor regions. The applied bias is 0.5 volt.

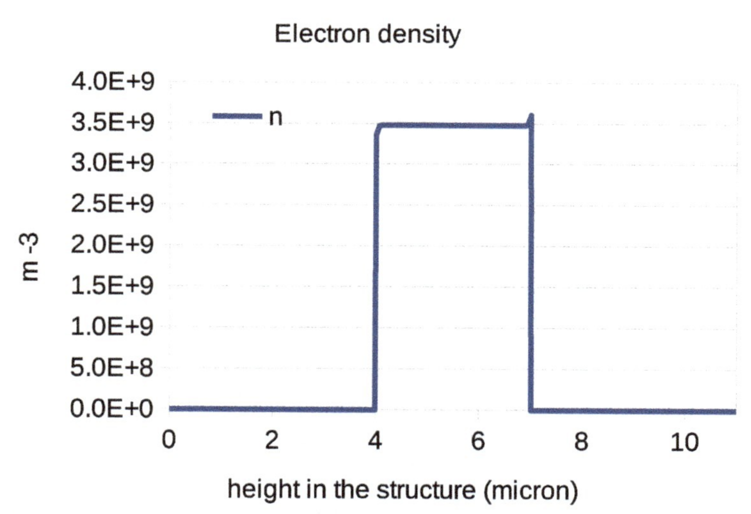

Figure 26.17 Electron density profile in the floating floating semiconductor regions. The applied bias is 0.5 volt.

In Figures 26.18, 26.19 and 26.20 the Poisson potential, the hole Fermi level and the electron Fermi level are shown for an applied bias of 0.5 volt at the top metallic block.

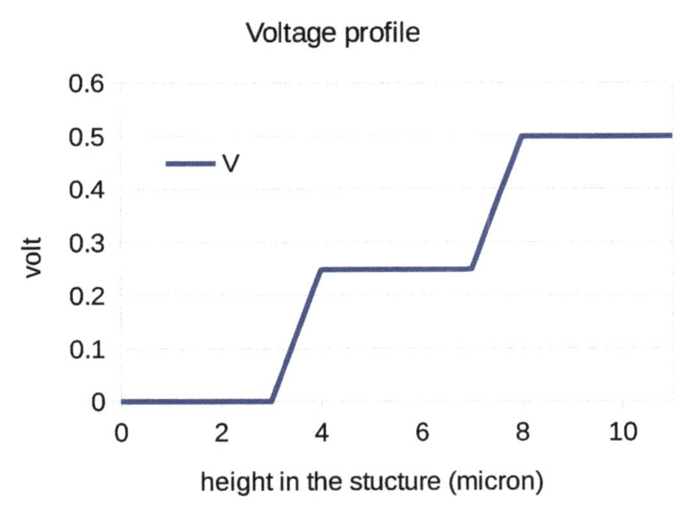

Figure 26.18 Poisson potential profile in the floating floating semiconductor regions. The applied bias is 0.5 volt.

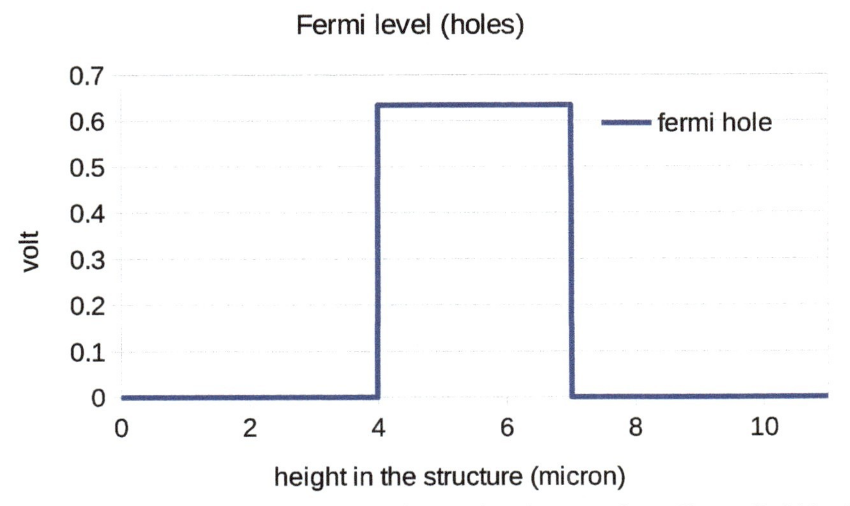

Figure 26.19 Hole Fermi level in the floating semiconductor regions. The applied bias is 0.5 volt.

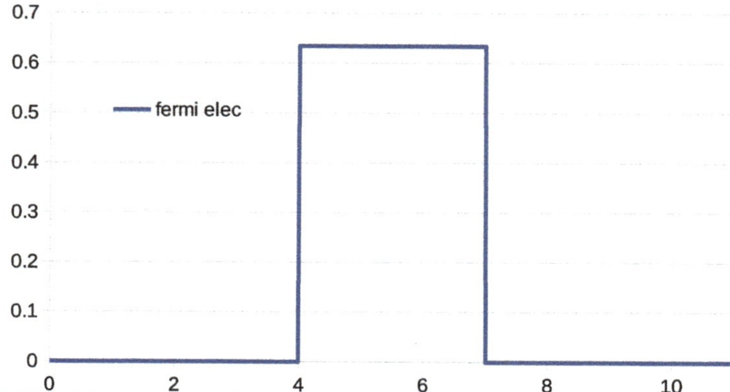

Figure 26.20 Electron Fermi level in the floating semiconductor regions. The applied bias is 0.5 volt.

When increasing the bias, the electric potential in the semiconducting region increases as well as the hole density. This is partially compensated by a higher Fermi level. For example, with an applied bias $V_{\text{applied}} = 0.7V$, we find $\phi_F = 0.7V$, $V_{\text{semi}} = 0.35V$ and $p = 6.43x10^{16}/\text{m}^3$.

For floating semiconductor regions attached to metallic parts whereas the combined configuration is floating requires that the combined charge integral is zero. There is one equation for one additional variable $\phi_F = V_{fl}$ which is the global floating potential. The charge integral can obtained as the sum of the surface integral over the metal and the volume integral over the semiconductor. At the metal/semi interface we have charge neutrality. We consider a test structure that combines three micron of semi (p-type $10^{16}/\text{cm}^3$) with three micron of Aluminum. The structure is shown in Figure 26.21.

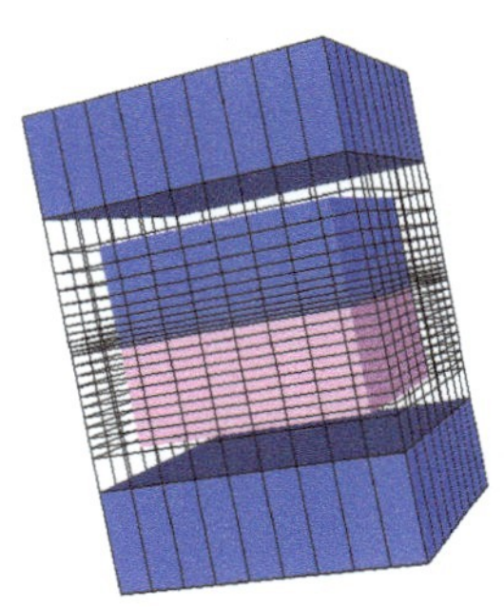

Figure 26.21 Structure of a floating metal/semiconductor stack . The lower part is semiconductor (pink).

In Figures 26.22, 26.23 and 26.24 the Poisson potential, the hole density and the electron density is given for an applied bias at the top metallic block of 0.5 volt. Furthermore, Figure 26.25 and 26.26 present the hole and electron Fermi levels.

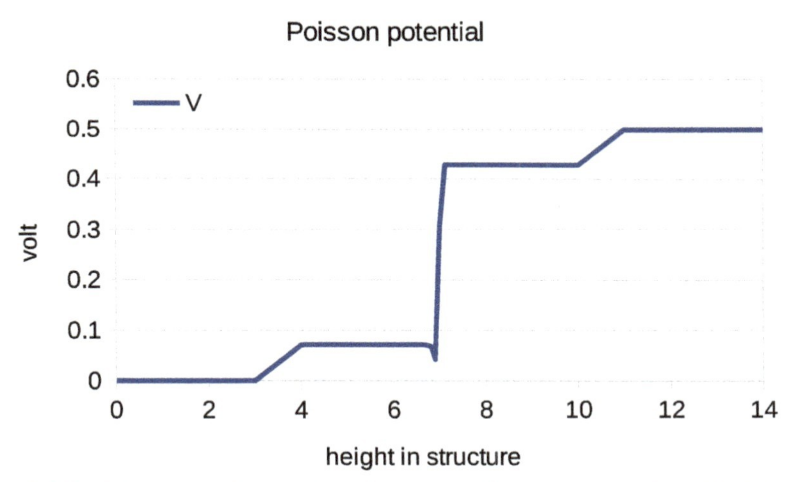

Figure 26.22 Poisson potential profile in the vertical direction through Semi/Metal floating stack. Applied bias at top is 0.5 volt.

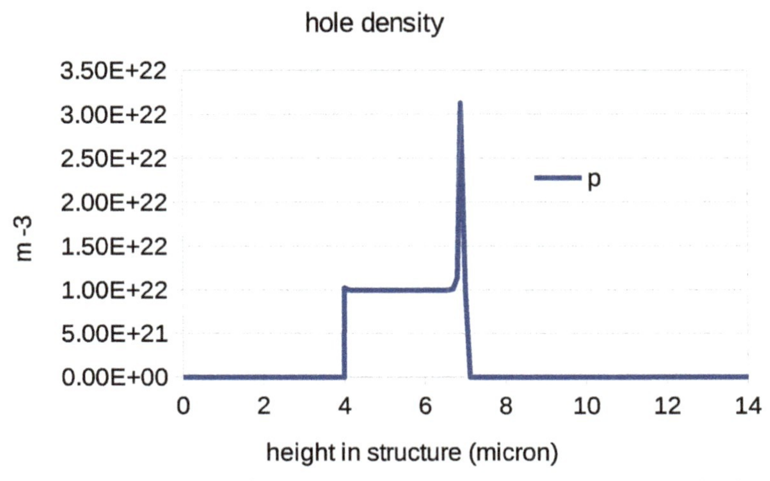

Figure 26.23 Hole density profile in the vertical direction through Semi/Metal floating stack. Applied bias at top is 0.5 volt.

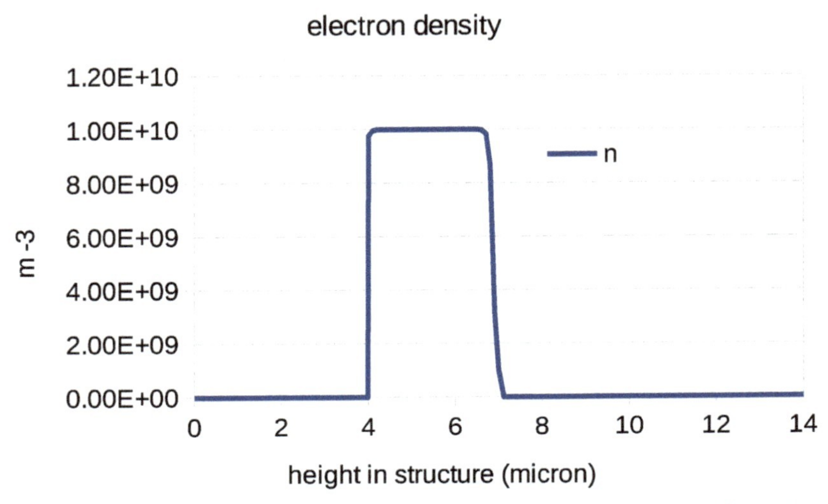

Figure 26.24 Electron density profile in the vertical direction through Semi/Metal floating stack. Applied bias at top is 0.5 volt.

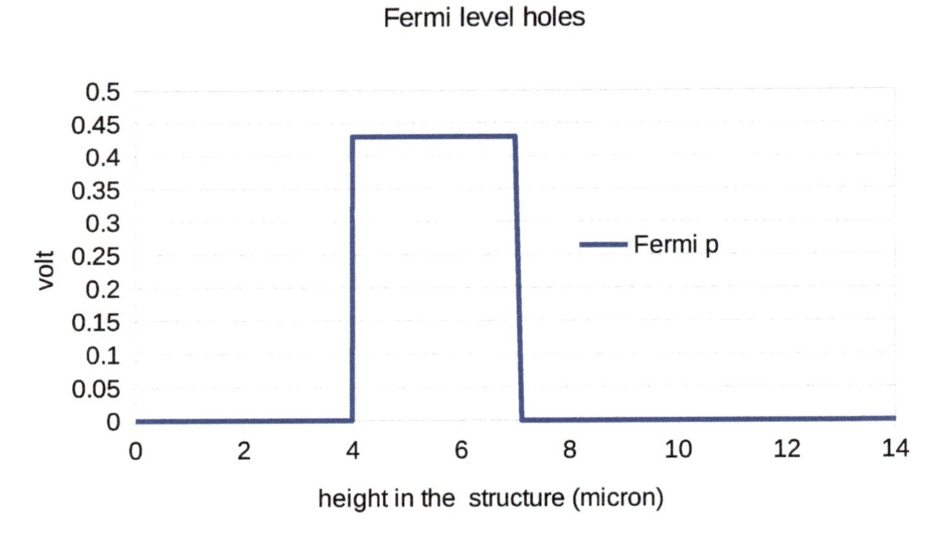

Figure 26.25 Hole Fermi level profile in the vertical direction through Semi/Metal floating stack. Applied bias at top is 0.5 volt.

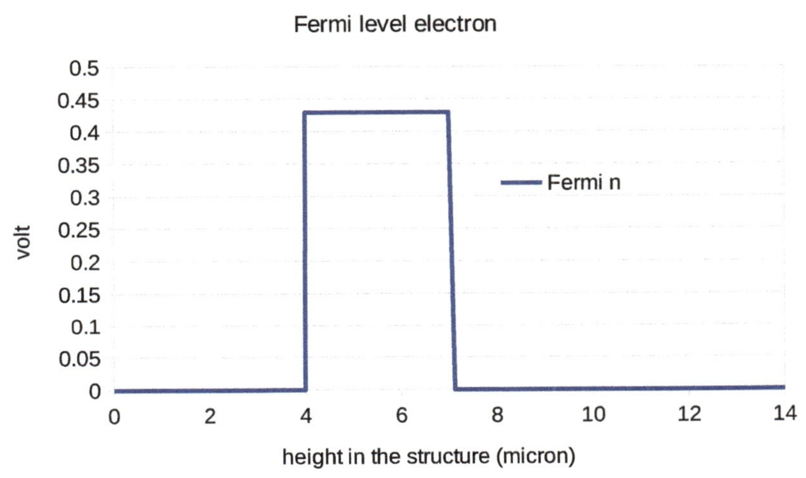

Figure 26.26 Electron Fermi level profile in the vertical direction through Semi/Metal floating stack. Applied bias at top is 0.5 volt.

Note that there is a subtle interplay between: Having one overall value for the Fermi potential The consistent determination of the Poisson potential. There is a push of holes to the metal/semi interface and attraction of electrons to the positive bias contact. The overall charge neutrality of the full floating system is reflected in the equality of dV/dz in the insulators.

27

Integrating Factors for Discretizing the Maxwell-Ampere Equation

Integrating factors play an important role for discretizing differential equations. This is also the case for partial differential equations despite the fact that is is not always clear how such factors should look like in higher-dimensional problems. We present the construction of the integrating factor that is needed for the Maxwell-Ampere equation in three dimensions. The construction is proposed by considering a simpler case from which we deduce a series of guidelines in order to make an educated guess for the higher dimensional problem. In particular, we will consider one of the most successful applications of integrating factors, i.e. the Scharfetter-Gummel Scharfetter-Gummel scheme approach for the simulation of semiconductor devices as a vehicle to identify the essential steps for the construction of the integration factor for the Maxwell-Ampere equation. In the next section, the Scharfetter-Gummel discretization is reviewed. The section ends with a summary of key observations that will help us to construct the integrating factor for electromagnetic problems. We proceed with the the discretization of the full-wave Maxwell equations in the potential formulation is reviewed on structured grids. The latter is not an essential limitation, but the review is a aid in visualizing the discretization. In fact, we show that the discretized equations can be formulated in a diagrammatic manner. The diagrammatic equations are of the type “picture (left hand side) = picture (right hand side)” or “picture = 0”. The diagrammatic equations pave the way for dealing with operators like $\nabla \times \nabla \times$, but working out the details shows that the generalization to unstructered grids is best done by respecting the diagrammatic meaning. After having presented the discretization scheme, we present the integrating factor for the Maxwell-Ampere equations. The generalization to unstructured grids is also addressed. given. Implementation details are given, and finally the impact is given for a structure borrowed from Part III.

27.1 Review of the Scharfetter-Gummel Discretization

We will consider a dielectric medium whose loss effects can be neglected. Therefore, no current continuity equations need to be solved in the dielectric materials. In the semiconducting regions, the current $\mathbf{J}$ consists of negatively and positively charged carrier currents obeying the current continuity equations.

$$\nabla \cdot \mathbf{J}_n - q\frac{\partial n}{\partial t} = U(n,p) \tag{27.1}$$

$$\nabla \cdot \mathbf{J}_p + q\frac{\partial p}{\partial t} = -U(n,p) \tag{27.2}$$

The charge and current densities are

$$\rho = q(p - n + N_D - N_A) \tag{27.3}$$

$$\mathbf{J}_n = q\mu_n n\mathbf{E} + kT\mu_n \nabla n \tag{27.4}$$

$$\mathbf{J}_p = q\mu_p p\mathbf{E} - kT\mu_p \nabla p, \tag{27.5}$$

and $U(n,p)$ is the generation/recombination rate of charge carriers. The current continuity equations provide the solution of the variables n and p. The permittivity ε is real for the applications envisaged, we may safely assume that the structure is non-magnetic, i.e. $\mu = \mu_0$.

The current densities can be collected as:

$$\mathbf{J} = q\mu c\mathbf{E} \pm kT\mu\nabla c, \tag{27.6}$$

where the plus (minus) sign refers to negatively (positively) charged particles and c denotes the corresponding carrier density. A 'physics-inspired' discretization of this equation argues that we first build a discretization grid. Next we consider the projections of the current density on the grid links. It is assumed that both the current $\mathbf{J}$ and the electric field $\mathbf{E}$ are constant along a link and that the potential V varies linearly along the link. Adopting a local coordinate axis u with $u = 0$ corresponding to node i, and $u = h_{ij}$ corresponding to node j, we may perform a projection of Equation (27.6) along the link ij to obtain

$$J_{ij} = q\mu_{ij} c\left(\frac{V_i - V_j}{h_{ij}}\right) + skT\mu_{ij}\frac{\mathrm{d}c}{\mathrm{d}u}, \tag{27.7}$$

which is a first-order differential equation in c and $s = \pm$. The latter is solved using the aforementioned boundary conditions and gives rise to a non-linear carrier profile. The current J_{ij} can then be rewritten as

$$\frac{J_{ij}}{\mu_{ij}} = -\frac{\alpha}{h_{ij}} B\left(\frac{-\beta_{ij}}{\alpha}\right) c_i + \frac{\alpha}{h_{ij}} B\left(\frac{\beta_{ij}}{\alpha}\right) c_j, \tag{27.8}$$

using the Bernoulli function

$$B(x) = \frac{x}{\mathrm{e}^x - 1}. \tag{27.9}$$

Furthermore, we used $\alpha = skT$ and $\beta_{ij} = q(V_i - V_j)$. The use of the factors $B(x)$ has been decisive for successful modeling of semiconductor devices. We will now make some essential observations on these Bernoulli expression.

27.2 Observations

First of all we note that that the current is expressed as a sum over the end-point concentration values:

$$J_{ij} = \sum_{l=1}^{2} c_l\, w_l(\alpha, \beta) \tag{27.10}$$

In here, w are weight functions assigned to each concentration. We observe that the exponential factor precisely correspond to the solutions of the *homogeneous* differential equation. We may rewrite (27.7) as

$$\frac{J_{ij} h_{ij}}{\mu_{ij} kT} = s h_{ij} \frac{\mathrm{d}c}{\mathrm{d}u} + \frac{q E_{ij} h_{ij}}{kT} c \tag{27.11}$$

using $E_{ij} = (V_i - V_j)/h_{ij}$. The homogeneous equation is

$$s h_{ij} \frac{\mathrm{d}c}{\mathrm{d}u} - X\, c = 0 \tag{27.12}$$

where $X = -\frac{qE_{ij}h_{ij}}{kT}$ and the homogeneous solution is $c_H(u) = k\ \mathrm{e}^{Xu}$, with k some constant. We observe that the weight factors are functions of the homogeneous solutions. We can turn this observation into a guiding principle. Without loss of generalization we will now focus on the simplified Scharfetter-Gummel equation

$$J = -\frac{\mathrm{d}c}{\mathrm{d}x} + Ec \tag{27.13}$$

where E is a singular perturbation parameter. By this it is understood that if E becomes very large, then the term $\frac{\mathrm{d}c}{\mathrm{d}x}$ is much weaker then the term Ec. The differential equation written as

$$-\frac{1}{E}\frac{\mathrm{d}c}{\mathrm{d}x}+c=J'=J/E \tag{27.14}$$

suggests the solution $c=J/E$. However, the boundary conditions also need to be taken into account. Indeed, the solution accounts for this by a fast 'correction' in the regions of the boundary points (Schilders et al. [1985]).

The naive discretization of this equation uses $\sigma = 1$ in the following discrete formulation:

$$J=-\sigma\frac{c_2-c_1}{h}+\frac{E}{2}(c_1+c_2) \tag{27.15}$$

An improved treatment exploits the fact that the homogeneous solution should perfectly fit into the discretized formulation. This can be achieved by demanding that $\sigma \neq 1$ and is obtained from

$$0=-\sigma\frac{c_H(x+h)-c_H(x)}{h}+\frac{E}{2}(c_H(x+h)+c_H(x)) \tag{27.16}$$

and $c_H(x)=\mathrm{e}^{Ex}$ is the homogeneous solution. This results into

$$\sigma=\frac{Eh}{2}\frac{E^{Eh}+1}{\mathrm{e}^{Eh}-1}. \tag{27.17}$$

Therefore

$$J=-\frac{Eh}{2}\left(\frac{E^{Eh}+1}{\mathrm{e}^{Eh}-1}\right)\frac{c_2-c_1}{h}+\frac{E}{2}(c_1+c_2) \tag{27.18}$$

This result is exactly equal to the result that is derived by using the physical arguments:

$$J=\frac{1}{h}B(Eh)c_1-\frac{1}{h}B(-Eh)c_2 \tag{27.19}$$

Thus we arrive at a series of guidelines for building integrating factors for discrete differential operators:

- A differential operator acquires a pre-factor, which is in general non-trivial to account for singular distortion
- The factor is determined by insertion of the homogeneous solution into the homogeneous equation.

The singular distortion can be viewed as follows. The derivative (diffusion) term is much smaller than the non-derivative (drift) term for large E. In order to match the boundary conditions the solution catches up with the boundary conditions by 'amplifying its curvature' towards the boundary points by the factor σ. In the next sections we review the Maxwell-Ampere system and its discretization on structured grids. This a good starting point for generalizing the scheme to unstructured grids.

27.3 Maxwell Equations

The Maxwell equation are summarized below:

$$\nabla \cdot \mathbf{D} = \rho \nabla \times \mathbf{E} + \partial \mathbf{B}/\partial t = 0$$
$$\nabla \cdot \mathbf{B} = 0 \nabla \times \mathbf{H} = \mathbf{J} + \partial \mathbf{D}/\partial t$$

From $\nabla \cdot \mathbf{B} = 0$, we obtain that $\mathbf{B} = \nabla \times \mathbf{A}$. Using Faraday's equation, we may write: $\mathbf{E} = -\nabla V - \partial \mathbf{A}/\partial t$. Using the constitutive relation $\mathbf{H} = \nu \mathbf{B}$ and $\mathbf{D} = \epsilon \mathbf{E}$ we obtain that the magnetic vector potential and the current density are related through a curl-curl equation that can be derived from Maxwell's equation

$$\nabla \times (\nu \nabla \times \mathbf{A}) = \mathbf{J}_{\mathrm{TOT}} \tag{27.20}$$

The current is the sum of the conduction and the displacement current $\mathbf{J}_{\mathrm{TOT}} = \mathbf{J}_{\mathrm{c}} + \mathbf{J}_{\mathrm{d}}$ and $\nu = 1/\mu$ is the reluctance. The curl-curl operator is singular in the sense that there does not exists a Green function (inverse) that allows a unique determination of $\mathbf{A}$ once that the current distribution is prescribed. Any solution $\mathbf{A}$ can be modulated by adding the gradient of an arbitrary scalar field. In order to elevate this arbitrariness an additional condition is imposed, e.g. a gauge condition. The Coulomb gauge is specified by the following additional constraint on the vector field $\mathbf{A}$:

$$\nabla \cdot \mathbf{A} = 0 \tag{27.21}$$

When the magnetic vector potential and its partial derivatives are continuous and the magnetic permeability is constant, the curl-curl operator can be replaced by the grad-div operator minus the Laplace operator:

$$\nabla \times \nabla \times \mathbf{A} = \nabla (\nabla \cdot \mathbf{A}) - \nabla^2 \mathbf{A} \tag{27.22}$$

The first term vanishes by virtue of the Coulomb gauge. The Laplace operator itself is regular and can be decomposed into the three components x, y and z that are decoupled and one obtains the following system of Poisson equations:

$$\begin{aligned} -\nabla^2 \mathbf{A}_x &= \mu \mathbf{J}_x \\ -\nabla^2 \mathbf{A}_y &= \mu \mathbf{J}_y \\ -\nabla^2 \mathbf{A}_z &= \mu \mathbf{J}_z \end{aligned} \tag{27.23}$$

Note that μ acts on $\mathbf{J}$ to generate a 1-form, therefore the Laplacian acting on a 1-form should result into a 1-form.
In the next section we will give the discretization of the curl-curl operator that respects the geometrical interpretation of this operator. We will argue that the result is only algebraically equivalent with the result that is obtained with the box-integration method of the three Laplace Equations (27.23), provided that

- the discretization mesh is Cartesian and equidistant,
- the permeability is constant,
- the boundary conditions are properly selected.

27.4 Discretization of the Curl-Curl Operator

For determining the discretized equation for a particular $\mathbf{A}$ on a link we integrate the Maxwell-Ampere equation over the surface of the dual mesh that the link intersects.

$$\int_S \nabla \times (\nu \nabla \times \mathbf{A}) \cdot \mathrm{d}\mathbf{S} = \int_S \mathbf{J} \cdot \mathrm{d}\mathbf{S} \approx |S| \mathbf{J}. \tag{27.24}$$

Applying Stokes' law this becomes

$$\oint \nu \nabla \times \mathbf{A} \cdot \mathrm{d}\mathbf{l} = \int_S \mathbf{J} \cdot \mathrm{d}\mathbf{a} \approx |S| \mathbf{J}. \tag{27.25}$$

where the integral is over the circumference of S. Replacing the integration by a finite summation we obtain

$$\sum_{i=1}^{4} \nu \mathbf{B}_i . l_i \approx |S| \mathbf{J}. \tag{27.26}$$

Where $\mathbf{B} = \nabla \times \mathbf{A}$ can be discretized in the same manner by integrating over the surfaces in the primary mesh and applying Stokes law again.

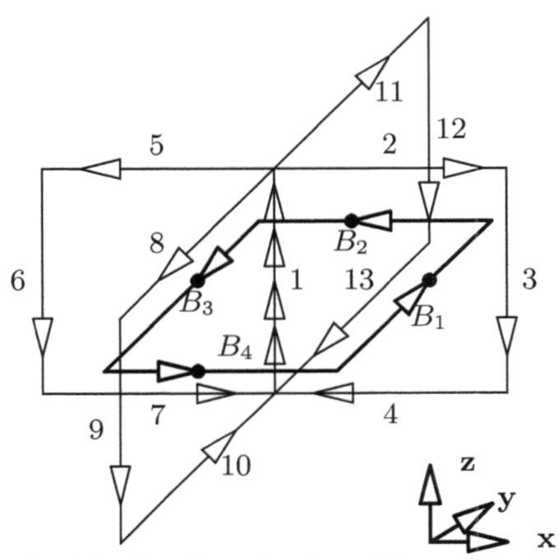

Figure 27.1 Links involved in the discretization of the curl-curl operator for the central link 1.

The method is illustrated in Figure 27.1, where the thick lines represent the **B** – fields piercing through the plaquettes at the black dots.

We will now give the detailed discretized equation for the middle link in z-direction in Figure 27.1. In the notation below the variables x_i and y_i represent the lengths of the links that point towards the positive ($i = +$) or negative ($i = -$) x and y directions. The variable z_0 represents the length of the link (1) under consideration. Equation (27.26) then becomes:

$$
\begin{aligned}
B_1\frac{(y_- + y_+)}{2} - B_2\frac{(x_- + x_+)}{2} - B_3\frac{(y_- + y_+)}{2} + B_4\frac{(x_- + x_+)}{2} \\
= \mu J_1\frac{(x_- + x_+)(y_- + y_+)}{4}
\end{aligned}
\tag{27.27}
$$

with

$$
\begin{aligned}
B_1 &= \frac{1}{x_+ z_0}\left(A_1 z_0 + A_2 x_+ - A_3 z_0 - A_4 x_+\right) \\
-B_2 &= \frac{1}{y_+ z_0}\left(A_1 z_0 + A_{11} y_+ - A_{12} z_0 - A_{13} y_+\right) \\
-B_3 &= \frac{1}{x_- z_0}\left(A_1 z_0 - A_5 x_- - A_6 z_0 + A_7 x_-\right) \\
B_4 &= \frac{1}{y_- z_0}\left(A_1 z_0 - A_8 y_- - A_9 z_0 + A_{10} y_-\right)
\end{aligned}
\tag{27.28}
$$

We may substitute (27.28) into (27.27) and obtain the algebraic relation between the vector field variables $(A_1, \ldots A_{13})$ and the current density J_1 on link 1. The results is:

$$\sum_{k=1}^{13} \Lambda_k A_k = \mu \Theta J_1 \tag{27.29}$$

where the weight factors Λ and Θ are:

$$\begin{aligned}
\Lambda_1 &= (\frac{1}{x_+} + \frac{1}{x_-})(\frac{y_- + y_+}{2}) + (\frac{1}{y_+} + \frac{1}{y_-})(\frac{x_- + x_+}{2}) \\
\Lambda_2 &= \frac{1}{z_0}(y_- + y_+), \qquad \Lambda_3 = -\frac{1}{x_+}(y_- + y_+) \\
\Lambda_4 &= -\Lambda_2, \qquad \Lambda_5 = -\frac{1}{z_0}(\frac{y_- + y_+}{2}) \\
\Lambda_6 &= -\frac{1}{x_-}(\frac{y_- + y_+}{2}), \qquad \Lambda_7 = -\Lambda_5 \\
\Lambda_8 &= -\frac{1}{z_0}(\frac{x_- + x_+}{2}), \qquad \Lambda_9 = -\frac{1}{y_-}(\frac{x_- + x_+}{2}) \\
\Lambda_{10} &= -\Lambda_8, \qquad \Lambda_{11} = \frac{1}{z_0}(\frac{x_- + x_+}{2}) \\
\Lambda_{12} &= -\frac{1}{y_+}(\frac{x_- + x_+}{2}), \qquad \Lambda_{13} = -\Lambda_{11} \\
\Theta &= \frac{1}{4}(x_- + x_+)(y_- + y_+)
\end{aligned}$$

This result is valid for constant μ. For variable μ we have to assign ν according to its occurrence in Equation (27.26). For later use, we note that Θ corresponds to the area of the dual surface of the link under consideration and Λ corresponds to the product of dual-link lengths and primary-link lengths divided by primary surface areas.

27.5 Discretization of the Divergence Operator

The discretization of the equation $\nabla \cdot \mathbf{A} = 0$ is performed by applying Gauss' law over a volume element on the dual mesh and replacing the surface integration over $\mathbf{A}$ by a finite summation. This method is illustrated Figure 27.2. In this figure, the lengths of the links that are involved in the finite summation are also denoted. By defining the variables $A_i = \mathbf{e}_i \cdot \mathbf{A}$, we obtain

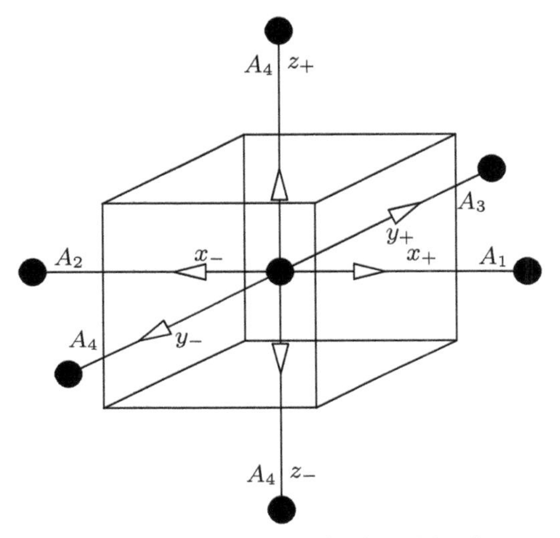

Figure 27.2 Links involved in the discretization of the divergence operator.

$$\begin{aligned}
&\frac{1}{4}A_1(y_- + y_+)(z_- + z_+) - \frac{1}{4}A_2(y_- + y_+)(z_- + z_+)\\
&\quad + \frac{1}{4}A_3(x_- + x_+)(z_- + z_+) - \frac{1}{4}A_4(x_- + x_+)(z_- + z_+)\\
&\quad + \frac{1}{4}A_5(x_- + x_+)(y_- + y_+) - \frac{1}{4}A_6(x_- + x_+)(z_- + z_+) = 0
\end{aligned} \tag{27.30}$$

It is instructive for later use to present this equation in a symbolic manner. This is done in Figure 27.3. This symbolic equation can be rewritten as is illustrated in Figure 27.4.

The four links that compose a cross in the xy-plane are also found in the the composition of the curl-curl operator as is illustrated in Figure 27.1. Indeed, we may apply the symbolic equation of Figure 27.4. For that purpose we consider the collection of links as is illustrated in Figure 27.5, where two new links (# 14, #15) are added.

By grouping the links in the left-hand side of Equation (27.29), it can be rewritten as

$$\sum_{l=1}^{3}\sum_{G_l} \Lambda_k A_k = \mu\ J_1 \frac{1}{4}(x_- + x_+)(y_- + y_+), \tag{27.31}$$

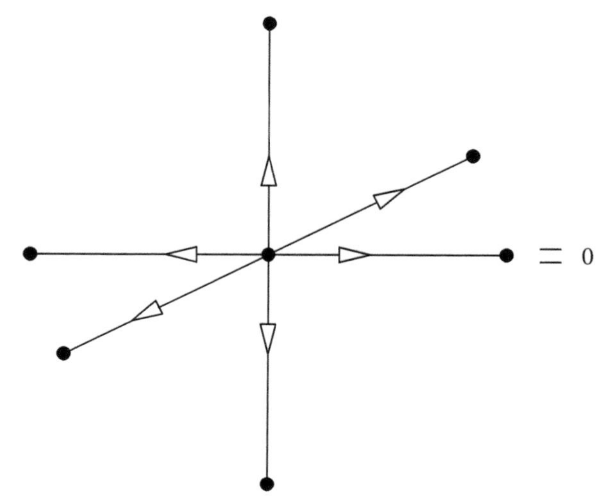

Figure 27.3 Symbolic representation of the equation $\nabla \cdot \mathbf{A} = 0$.

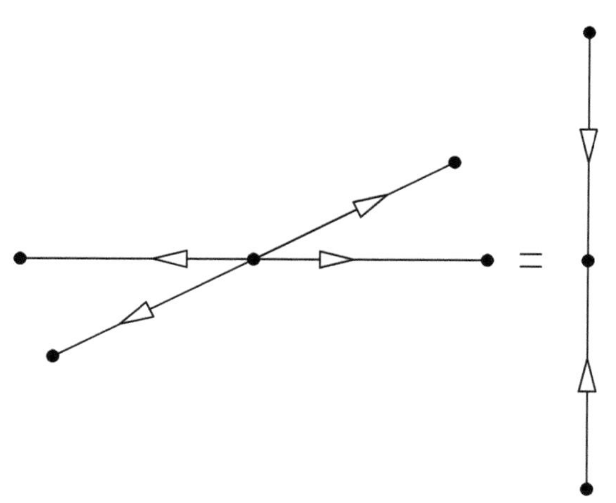

Figure 27.4 Symbolic equation extracted from the equation $\nabla \cdot \mathbf{A} = 0$.

where $G_1 = \{1, 3, 6, 9, 12\}$, $G_2 = \{2, 5, 8, 11\}$ and $G_3 = \{4, 7, 10, 13\}$. Using $\nabla \cdot \mathbf{A} = 0$ in its symbolic equivalence of Figure 27.4, we find that

$$\sum_{G_2} \Lambda_k A_k = \frac{(x_- + x_+)(y_- + y_+)}{2\, z_0(z_0 + z_+)}(A_1 - A_{15}) \tag{27.32}$$

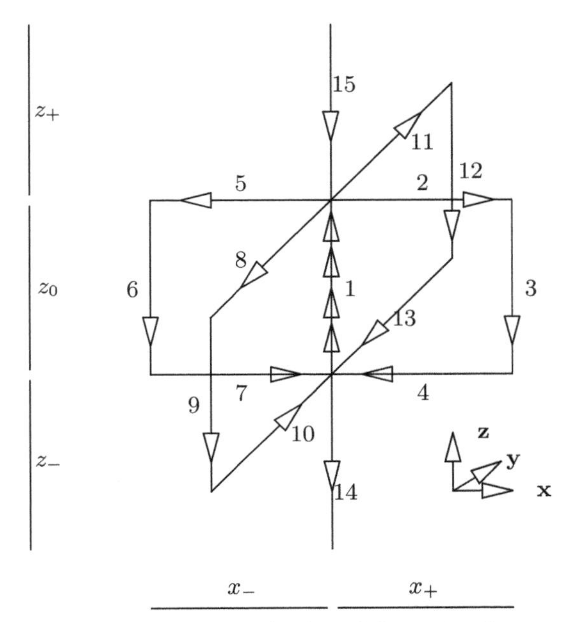

Figure 27.5 Links involved in the discretization of the curl-curl operator for the central link 1 and the transition to the Laplace operator for this link.

and

$$\sum_{G_3} \Lambda_k A_k = \frac{(x_- + x_+)(y_- + y_+)}{2\, z_0(z_0 + z_-)}(A_1 - A_{14}) \tag{27.33}$$

This finally leads to a representation of the curl-curl operator that does not contain the links from the sets G_2 and G_3 anymore. In other words: only links parallel to the starting link and being neighbor contribute to the discretization. The final result is illustrated in Figure 27.6. Equation (27.29) can be written as

$$\sum_{k=1'}^{7'} \lambda_k A_k = \frac{1}{4}\, \mu J_{1'}\, (x_- + x_+)(y_- + y_+) \tag{27.34}$$

where the summation goes over the primed labeling of the links in Figure 27.6 and

$$\lambda_{1'} = (\frac{1}{x_+} + \frac{1}{x_-})(\frac{y_- + y_+}{2}) + (\frac{1}{y_+} + \frac{1}{y_-})(\frac{x_- + x_+}{2}) + \frac{(x_- + x_+)(y_- + y_+)}{2\, z_0(z_0 + z_+)} + \frac{(x_- + x_+)(y_- + y_+)}{2\, z_0(z_0 + z_-)}$$

$$\begin{aligned}
\lambda_{2'} &= -\frac{1}{x_+}(\frac{y_- + y_+}{2}), \quad \lambda_{3'} = -\frac{1}{x_-}(\frac{y_- + y_+}{2}) \\
\lambda_{4'} &= -\frac{1}{y_+}(\frac{x_- + x_+}{2}), \quad \lambda_{5'} = -\frac{1}{y_-}(\frac{x_- + x_+}{2}) \\
\lambda_{6'} &= -\frac{(x_- + x_+)(y_- + y_+)}{2\ z_0(z_0 + z_+)} \\
\lambda_{7'} &= -\frac{(x_- + x_+)(y_- + y_+)}{2\ z_0(z_0 + z_-)}
\end{aligned} \tag{27.35}$$

This representation leads to an assembling of the discretized full set of equations with a considerably narrower band width. The width of the matrix stencil reduces from 13 to 7. Moreover, since the gauge condition is implicitly applied, the resulting matrix is also regular. In the next section we will compare this result with the result that will be obtained if a direct implementation of the Laplace operator is constructed.

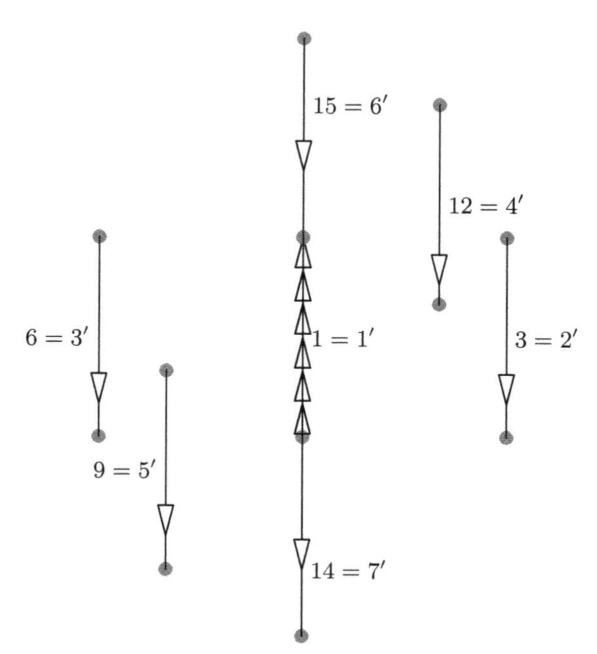

Figure 27.6 Links contributing to the Laplacian representation of the discretized curl-curl operator for the central link 1.

27.6 Discretization of Poisson-Type Operators

To discretize the Poisson-type equations for **A**, three new meshes need to be considered. These three meshes are obtained by taking the midpoints of the links in resp. x-, y- and z-direction. In each mid point is stored the projection of **A** in the direction of the link. These values are then considered as nodal values on these new meshes. The discretization is constructed by replacing the gradients in the Laplace operator by finite differences, integrating these over a volume-element, applying Gauss law, and replacing the surface integrals by finite summations. The volume element is selected in such a way that the link direction of the link is inscribed in its size and for the two other directions the surfaces are chosen to coincide with the surfaces of the dual mesh. In the example of Figure 27.7, the size of the volume element is $\frac{1}{4}(x_- + x_+)(y_- + y_+)z_0$.

For the link # 1 in Figure 27.6 we have to consider the mesh corresponding to the z-direction and according to above prescription the Poisson-type equation:

$$-\nabla^2 A_z = \mu J_z$$

results into

$$\begin{aligned} &\frac{1}{2}\frac{(y_- + y_+)z_0}{x_+}(A_1 - A_2) \\ &\quad + \frac{1}{2}\frac{(y_- + y_+)z_0}{x_-}(A_1 - A_3) \\ &\quad + \frac{1}{2}\frac{(x_- + x_+)z_0}{y_+}(A_1 - A_4) \end{aligned}$$

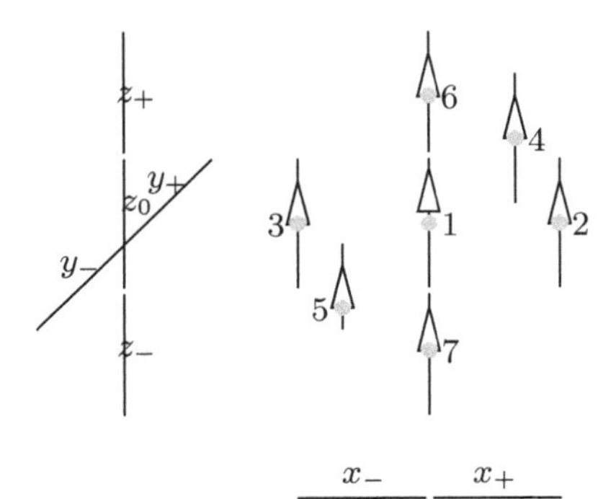

Figure 27.7 Laplace nodes and corresponding links central link and node 1.

$$
\begin{aligned}
&+ \frac{1}{2}\frac{(x_- + x_+)z_0}{y_-}(A_1 - A_5) \\
&+ \frac{1}{2}\frac{(x_- + x_+)(y_- + y_+)}{(z_0 + z_+)}(A_1 - A_6) \\
&+ \frac{1}{2}\frac{(x_- + x_+)(y_- + y_+)}{(z_- + z_0)}(A_1 - A_7) \\
&= \frac{1}{4}\mu(x_- + x_+)(y_- + y_+)(z_0)\ J_1
\end{aligned}
\tag{27.36}
$$

27.7 Equivalence

After multiplication of Equation (27.36) with $1/z_0$, we obtain

$$
\begin{aligned}
&\frac{(y_- + y_+)}{2\ x_+}(A_1 - A_2) + \frac{(y_- + y_+)}{2\ x_-}(A_1 - A_3) \\
&+ \frac{(x_- + x_+)}{2\ y_+}(A_1 - A_4) + \frac{(x_- + x_+)}{2\ y_-}(A_1 - A_5) \\
&+ \frac{(x_- + x_+)(y_- + y_+)}{2\ z_0(z_0 + z_+)}(A_1 - A_6) \\
&+ \frac{(x_- + x_+)(y_- + y_+)}{2\ z_0(z_0 + z_-)}(A_1 - A_7) \\
&= \frac{1}{4}\mu(x_- + x_+)(y_- + y_+)\ J_1
\end{aligned}
\tag{27.37}
$$

A comparison of (27.37) with (27.34) shows that equivalent results are obtained for the right-hand side. Moreover, the links and node entries also have identical weight factors. Therefore we conclude that above method of evaluating the Poisson operator for each component of a vector field is equivalent to computing the curl-curl operator extended with the Coulomb gauge condition. It should be noted that the equivalence is bi-directional. Starting from the Poisson operator and the Coulomb gauge condition the correct representation of the discretized curl-curl operator can be constructed. Since both the curl-curl operator extended with the Coulomb gauge condition as well as the Laplace operator are regular, their solutions are unique and identical.

27.8 High-Frequency Maxwell Equations

After collecting the terms proportional to $\mathbf{A}$ and applying the scaling procedure, the inhomogeneous Maxwell-Ampere equation takes the following form

$$\nabla \times \left(\frac{1}{\mu_r}\nabla \times \mathbf{A}\right) - \nabla\left(\nabla \cdot \mathbf{A}\right) + k\mathrm{i}\omega\left(\sigma + \mathrm{i}\omega\varepsilon_r\right)\mathbf{A}$$
$$= k\mathbf{J}_{\text{diff}} - k\left(\sigma + \mathrm{i}\omega\varepsilon_r\right)\nabla V \qquad (27.38)$$

When applying above discretization scheme we note that the final discretization of the spatial differential operators results into a sum over at most 15 links for a discretized version.

$$\nabla \times \left(\nu_r \nabla \times \mathbf{A}\right) - \nabla\left(\nabla \cdot \mathbf{A}\right) \rightarrow \sum_{k=1}^{15} \Lambda_k A_k \qquad (27.39)$$

and $\nu_r = 1/\mu_r$. The Λ-factors will now be adapted such that the solution of the homogeneous equation is faithfully matched. An arbitrary homogeneous solution takes the following form

$$\mathbf{A}(\mathbf{x}) = \int_{\mathbf{p}} \mathrm{d}\mathbf{p}\ \mathbf{a}_{\mathbf{p}} \mathrm{e}^{\mathrm{i}\mathbf{p}\cdot\mathbf{x}} \quad p^2 = k\mathrm{i}\omega\left(\sigma + \mathrm{i}\omega\varepsilon_r\right) \qquad (27.40)$$

In the Coulomb gauge, the $\mathbf{p}$-vector is orthogonal to $\mathbf{a}_{\mathbf{p}}$. There are now (at most) 15 numbers A_i that correspond to the line integrals of this solution projected on the the 15 links

$$A_i = \frac{1}{h_i}\int_{\mathbf{p}} \mathrm{d}\mathbf{p}\ \left(\mathbf{a}_{\mathbf{p}} \cdot \mathbf{n}_i\right) \int_{h_i} \mathrm{d}x\ \mathrm{e}^{\mathrm{i}\mathbf{p}\cdot\mathbf{x}} \qquad (27.41)$$

Next we consider the homogeneous equation. Our results are also applicable to the Lorentz gauge since the difference of the two gauges can be assigned to the inhomogeneous term.

Let us focus on the link for which we are assemble the equation, say link #1 in Figure 27.5. Then it makes sense to choose $\mathbf{a}_{\mathbf{p}}$ in the direction of this link. Since $\mathbf{p} \cdot \mathbf{a}_{\mathbf{p}} = 0$, we obtain that $\mathbf{p} \cdot \mathbf{n}_1 = 0$. As a consequence, the large collection of homogeneous solutions that could be used for constructed the corrected Λ-factors is substantially reduced. Since $\mathbf{p}$ is in a plane perpendicular to $\mathbf{n}_1$, it suffices to consider the two-dimensional homogeneous equation

$$-\frac{1}{\mu_r}\nabla^2_{(x,y)} A_z(x,y) + p^2 A_z(x,y) = 0 \qquad (27.42)$$

Thus, in order to construct the Λ-factors, we consider the vector potential field projected in the direction of the link under consideration. For the present Cartesian grid under consideration it implies that the links #2, #4, #5, #7, #8, #10, #11, #13 in Figure 27.5 do not obtain a correction to the weight factor. In this specific example we can argue that these links do not appear in the resulting equation for A_1, as is seen from Figure 27.6. Although in the Cartesian grid the orthogonal links on link, #1 are not participating at all, we extract from this that in unstructured grids the weight factor should account for the component of the vector field that is parallel to the direction of link #1. In general an arbitrary link, i, contributes with a correction that is proportional the $\mathbf{n}_1 \cdot \mathbf{n}_i$. To be more precise, the replacement of Equation (27.39) now becomes

$$\nabla \times (\nu \nabla \times \mathbf{A}) - \nabla (\nabla \cdot \mathbf{A}) \to \sum_{k=1}^{15} \sigma_G \Lambda_k A_k \tag{27.43}$$

where σ_G is determined by the insertion of a solution of Equation (27.42). The subscript G refers to the group ordering as discussed below Equation (27.31. Moreover, inspired by our second approach to the Scharfetter-Gummel discretization method we demand that

$$0 = \sum_{i=1}^{3} \sigma_{G_i} \sum_{k \in G_i} \Lambda_k \, \mathbf{n}_1 \cdot \mathbf{n}_k \, A_k^H + \Theta \, \mu \, p^2 \, A_1^H \tag{27.44}$$

and A_k^H and $k \in G_1$ is a solution of (27.42) evaluated according to (27.41). For $k \in G_2$ or $k \in G_3$, we demand that the gauge condition is respected. We have selected a rather standard gauge condition for which singular perturbed differential equations do not arise. Therefore, we will from now on set $\sigma_{G_2} = \sigma_{G_3} = 1$. The Θ and Λ factors represent the geometrical information. The solutions of this equation is most conveniently found in polar coordinates for which (27.42) becomes

$$-\frac{1}{r}\frac{\partial}{\partial r}\left(r\frac{\partial A_z}{\partial r}\right) - \frac{1}{r^2}\frac{\partial^2 A_z}{\partial \phi^2} + \mu_r p^2 A_z = 0 \tag{27.45}$$

This equation is conveniently solved by setting $A_z(r, \phi) = P(r)Q(\phi)$, leading to

$$\frac{\mathrm{d}^2 Q}{\mathrm{d}\phi^2} + m^2 \, Q = 0$$

$$\frac{\mathrm{d}^2 P}{\mathrm{d}r^2} + \frac{1}{r}\frac{\mathrm{d}P}{\mathrm{d}r} + \left(\mu_r p^2 + \frac{m^2}{r^2}\right) P = 0 \tag{27.46}$$

We can further shrink the number of options by inserting the restriction that only the rotational symmetric solution is of interest. Therefore, we consider $m = 0$. Furthermore setting the complex variable $z = \sqrt{\mu_r}\, pr$, we find that

$$\frac{\mathrm{d}^2 P}{\mathrm{d}z^2} + \frac{1}{z}\frac{\mathrm{d}P}{\mathrm{d}z} + P = 0 \tag{27.47}$$

Therefore the solution is the zeroth-order Bessel function $P(r) = J_0(z)$. Note that $J_0(0) = 1$. Using the notation of Figure 27.5, we find that

$$\begin{aligned}\sigma = \; & -\Theta\ \mu_r p^2\ \left[\Lambda_3 J_0(\sqrt{\mu_r}\ px_+) + \Lambda_6 J_0(\sqrt{\mu_r}\ px_-)\right. \\ & \left. +\Lambda_9 J_0(\sqrt{\mu_r}\ py_-) + \Lambda_{12} J_0(\sqrt{\mu_r}\ py_+) + \Lambda_1\right]^{-1}\end{aligned} \tag{27.48}$$

Note that $\Lambda_1 = -\Lambda_3 - \Lambda_6 - \Lambda_9 - \Lambda_{12}$. Therefore, for $p \to 0$ we obtain that $\sigma \to 1$.

For completeness, we give here the series representation of the zeroth Bessel function:

$$\begin{aligned} J_0(z) &= \sum_{k=0}^{\infty} (-)^k \frac{(\frac{1}{4}z^2)^k}{(k!)^2} \\ &= 1 - \frac{1}{4}z^2 + \frac{1}{64}z^4 + \ldots \end{aligned} \tag{27.49}$$

27.9 Integrating Factors for Unstructured Grids

The building of integrating factors in higher dimensions requires some steps that can not always easily be justified from a strict mathematical perspective. Some choices are made as being merely plausible. This is even more the case when constructing the integrating factors on unstructured grids. Collecting our experience in the foregoing sections, we propose to proceed as follows:

- (1) A non-trivial integrating factor will be exclusively assigned to the discretization of the curl-curl operator.
- (2) The grad-div operator has a trivial integrating factor because the gauge condition is of the Coulomb-Lorenz type.
- (3) The midpoint value of the homogeneous solution will be used in the evaluation of the non-trivial integrating factor. Therefore we do not have to take integrals over $J_0(z)$.
- (4) All links of the curl-curl operator contribute to the integrating factor with a weight proportional to $\mathbf{n}_1 \cdot \mathbf{n}_k$.

This then leads to the following expression as the integrating factor in front of the curl-curl discretized operator:

$$\sigma_{G_1} = \frac{\Theta \, \mu_r p^2}{\sum_k (\mathbf{n}_1 \cdot \mathbf{n}_k) \Lambda_k J_0(\sqrt{\mu_r} p d_{1k})} \tag{27.50}$$

and $p^2 = k\text{i}\omega \, \varepsilon_r \, (\sigma + \text{i}\omega\varepsilon_r)$ and d_{1k} is the distance between the midpoints of the links #1 and #k. In order to compute Θ we must guarantee the the limit $p \to 0$ gives $\sigma \to 1$. Since for $p \to 0$ we have that $J_0(z) \to 1$, we demand that

$$\sum_k (\mathbf{n}_1 \cdot \mathbf{n}_k) \Lambda_k = 0 \tag{27.51}$$

However, this is easily seen to be the case, since

$$\sum_k (\mathbf{n}_1 \cdot \mathbf{n}_k) \Lambda_k = \mathbf{n}_1 \cdot (\sum_k \mathbf{n}_k \Lambda_k) = 0 \tag{27.52}$$

Since $\sum_k \mathbf{n}_k h_k$ =0 for a closed loop, Equation (27.52) is valid provided that $\Lambda_k = \gamma_{\text{surf}} h_k$, where h_k is the length of link k. Indeed, the construction of Λ involves circulatons. The sum consists of circulations around each surface that is involved in constructing the curl-curl operator and each surface brings it weight factor γ_{surf}. Since the numerator and denominator both become zero for $p^2 \to 0$, it is desirable to apply 'l Hopital's rule. From

$$1 = \frac{\frac{\partial}{\partial(\mu_r p^2)} \left[\Theta \, \mu_r p^2\right]_{\mu_r p^2 = 0}}{\frac{\partial}{\partial(\mu_r p^2)} \left[\sum_k (\mathbf{n}_1 \cdot \mathbf{n}_k) \, \Lambda_k J_0(\sqrt{\mu_r} p d_{1k})\right]_{\mu_r p^2 = 0}}, \tag{27.53}$$

we deduce

$$\Theta = -\frac{1}{4} \sum_k (\mathbf{n}_1 \cdot \mathbf{n}_k) \Lambda_k d_{1k}^2. \tag{27.54}$$

27.10 Implementation Details

The implementation of the integrating factor as it was described above is valid for links in bulk locations in the simulation domain. For links that are part of interface surfaces, a complication arises. In particular, the value of p^2 depends on the conductance σ and therefore on the material under consideration. Here, we decouple the regions by applying the solution of the homogeneous equation piece-wise in regions of constant material. The detailed dependence is expressed by finding the correspondences $\sigma(\text{link}) \to \sigma(\text{surf}) \to \sigma(\text{cube})$.

27.11 Effect of the Inclusion of the Integrating Factor

This design considers an inductor that is shielded by a closed-loop grounded guard ring. The layout is shown in Figure 27.8. When an alternating current is injected into the inductor, an induced current in opposite phase is induced in the closed loop. Although the loop is grounded at both ends, the vector potential is still present and therefore the induced current exists despite the fact that the loop is grounded. This configuration is also instructive to obtain an in-depth understanding of the measurement set-up for S-parameters.

27.12 Simulation Set Up and Results

The structure is simulated using the Y-parameter extraction method. This design has a high-resistive substrate. Therefore, the decay of the field strength is rather slow and a full stack of 625 micron substrate is included in the simulation. The experimental data of the substrate is: $\sigma = 0.1$ S/m. The computed curves are obtained by using 1) the MAGWEL solver with calibrated substrate resistance, 2) the MAGWEL solver using an integrating factor method. Finally, the results obtained using Agilent Momentum (Fach et al. [1984]) are shown. The curves are identified with the labels EXP, OPTIM, INTFAC, MOM respectively. In Figures 27.9 the Y-parameters are shown. The impact of the integrating factor on the curves of Y11 and Y12 is very weak in the frequency range 0-10 GHz. In Figures 27.10, 27.11, 27.12 the S-parameters are shown. The Y-parameter information can be transformed into compact-model parameter information. For details, we refer to the next section. Here we plot the 2-port resistance, the 2-port inductance and the Q-factor. The results are shown in Figure 27.13.

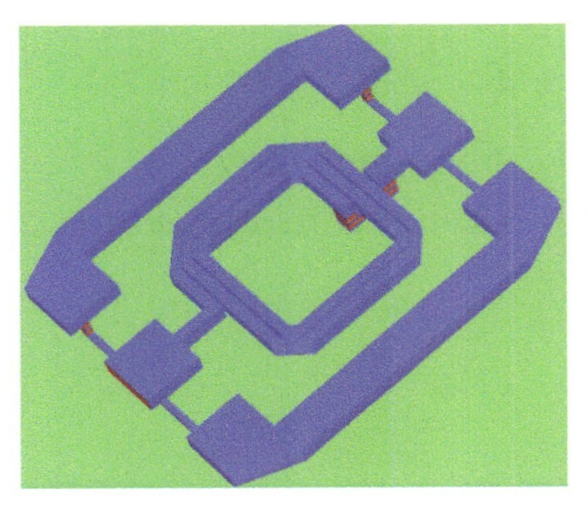

Figure 27.8 Inductor layout of an inductor with a closed guard ring.

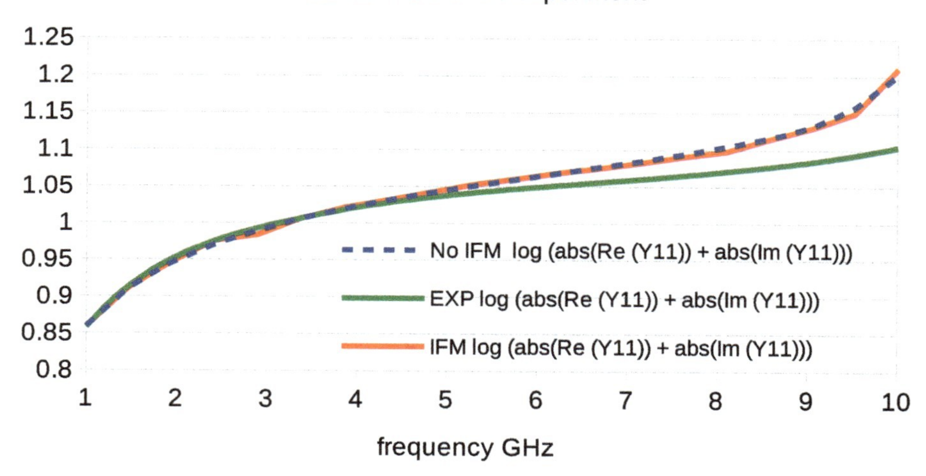

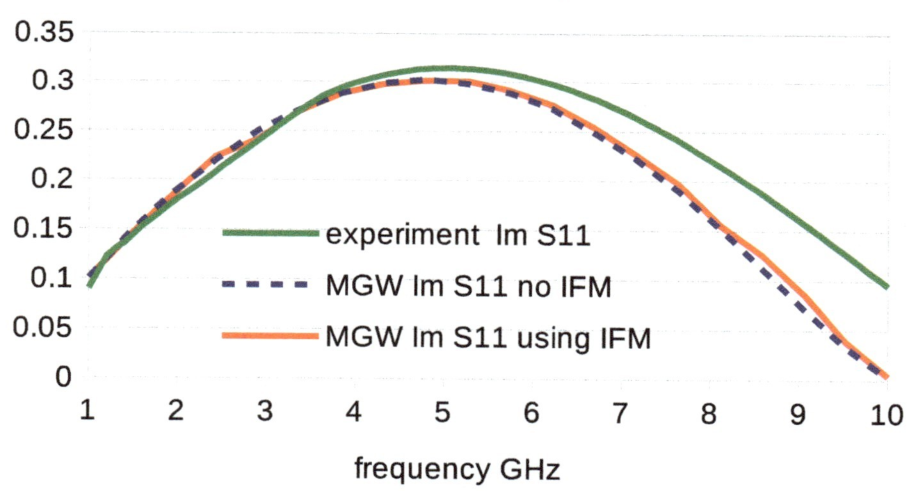

Figure 27.9 log(abs(Re(Y11))+abs(Im(Y11))) and log(abs(Re(Y12))+abs(Im(Y12))).

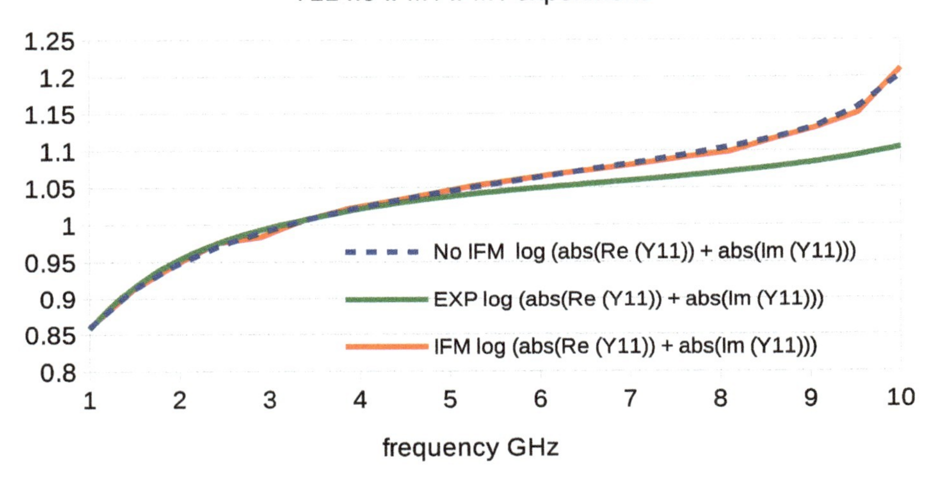

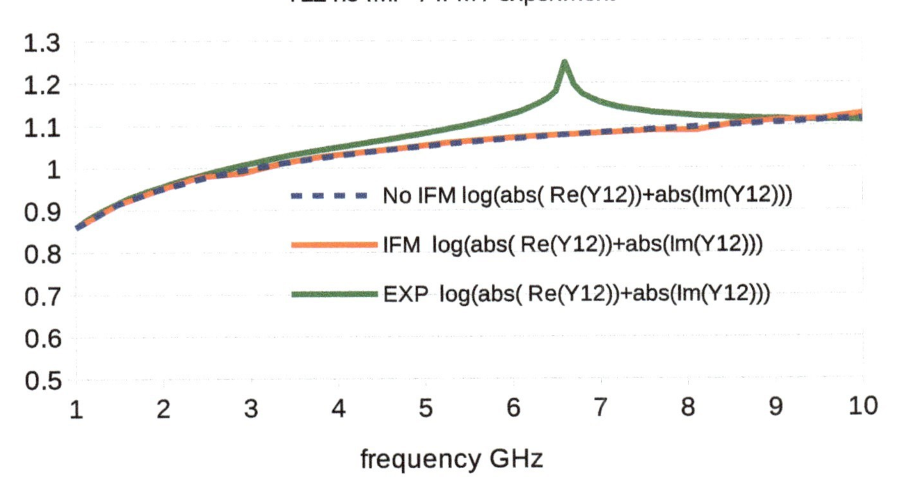

Figure 27.9 log(abs(Re(Y11))+abs(Im(Y11))) and log(abs(Re(Y12))+abs(Im(Y12))).

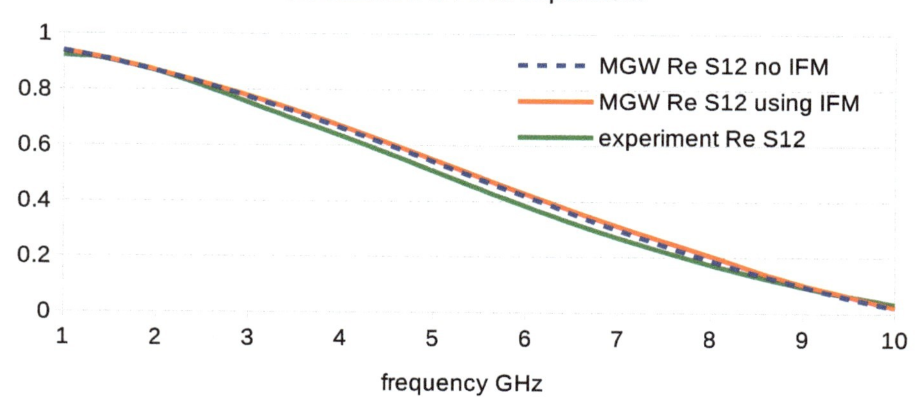

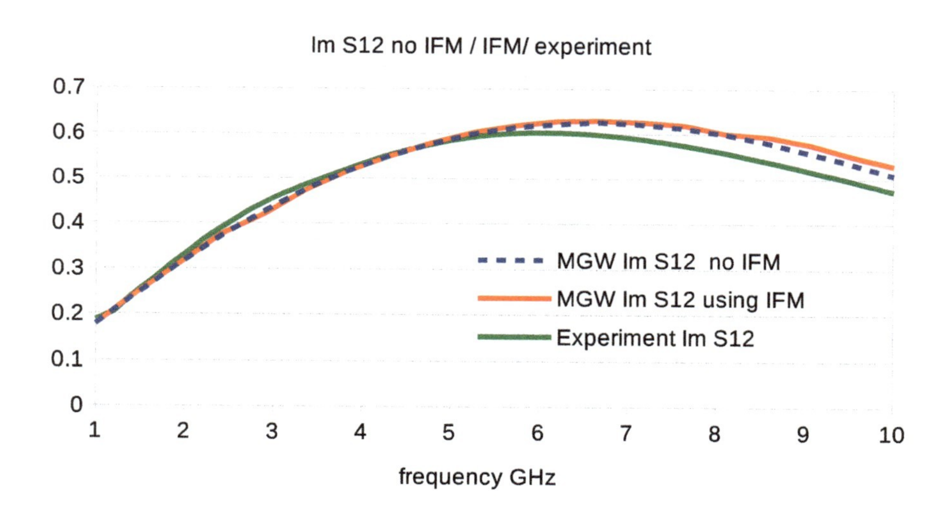

Figure 27.11 Re (S12) and Im (S12).

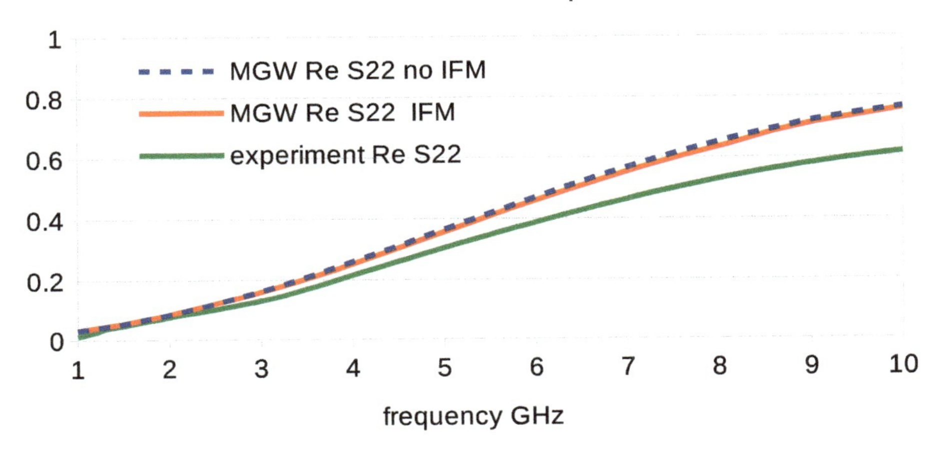

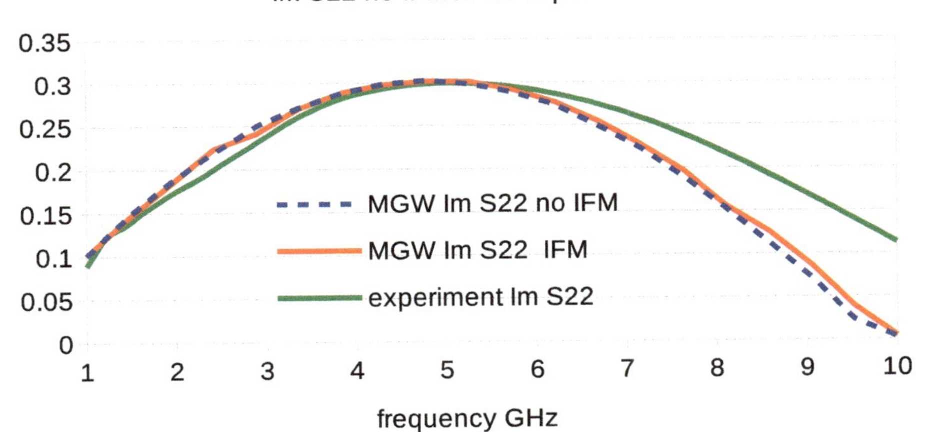

Figure 27.12 Re (22) and Im (S22).

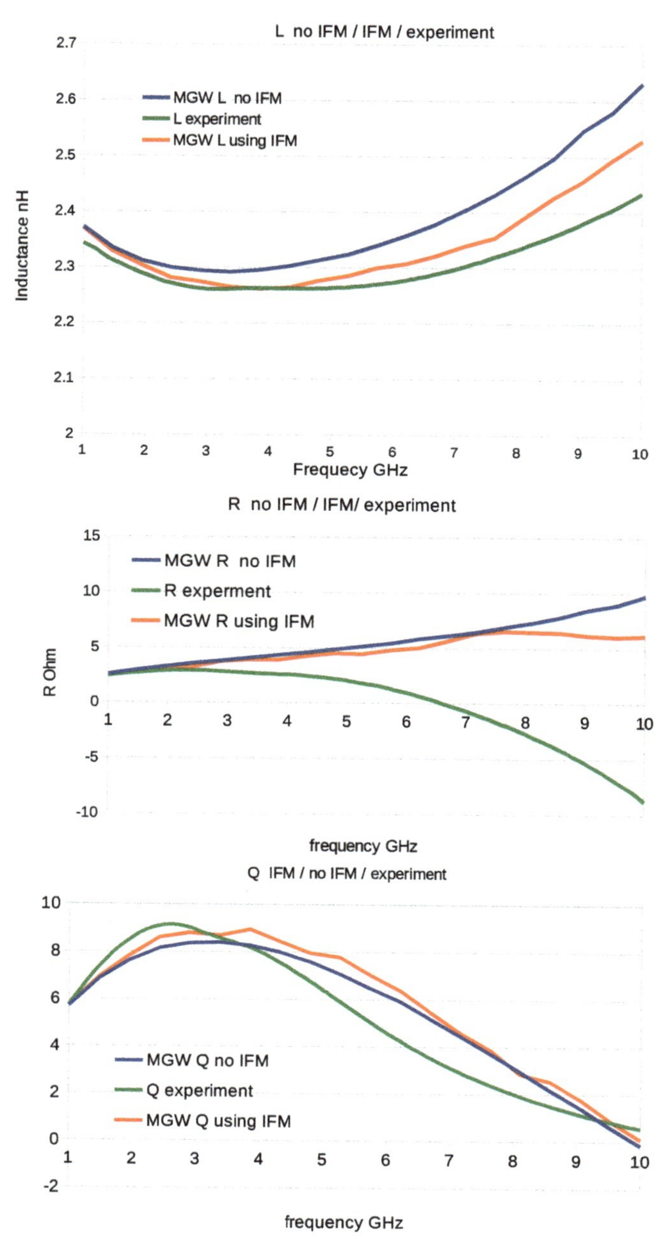

Figure 27.13 The inductance, resistance and Q factor of the inductor with grounded closed-guard ring.

Summary

From this simulation work we found that the implementation of the Hodge operators is important for getting results that are within 1% accuracy for the inductance. A naive implementation gives errors of 10%. These errors can be counteracted by using an enhancement factor for the permeability μ_r. This factor is tuned only once at zero frequency. Furthermore, in order to get results that are very accurate for inductors with wide windings, the integrating factor method improves the results at higher frequency. This is in agreement with expectations since the integrating factor becomes important for high frequency and large mesh elements.

28

Stability Analysis of the Transient Field Solver

The implementation of algorithms is a laborious task last but not least because it requires intensive testing before the corresponding code is sufficiently robust and reliable in its response to explore new territories of applications. This statements turned out to be also true for the integrated device/circuit simulation tool. This chapter deals with the stability problem that was finally solved after an in-depth analysis of the eigenvalue spectra of small-scaled devices.

We consider a twisted piece of metal in oxide. The structure is shown in Figure 28.1. The non-linear interface generates the matrices A and the jacobian J. They are connected to the transient equation in the following way.

$$A \cdot \frac{d\mathbf{x}}{dt} + J \cdot \mathbf{x} = 0 \tag{28.1}$$

We have reworked the assembling such that A is regular. Then we may write

$$\frac{d\mathbf{x}}{dt} = \left(-A^{-1}J\right) \cdot \mathbf{x} \tag{28.2}$$

This has a formal solution

$$\mathbf{x}(t) = \mathrm{e}^{\left(-A^{-1}J\right)} \cdot \mathbf{x}(t_0) \tag{28.3}$$

From this we conclude that all eigenvalues of the matrix in the operator must have semi-positive values. Our test structure consists of metal and oxide. We choose a 'small' value for the conductance. e.g. $\sigma = 1.75 \times 10^3$ S/m. If we choose the permittivities in metal and oxide equal 1 we obtain the eigenvalue plot of Figure 28.2.

Next we change the value of the relative permittivity of oxide to 4 and keep the the ALUM (metal) value equal one. Then we obtain the eigenvalue

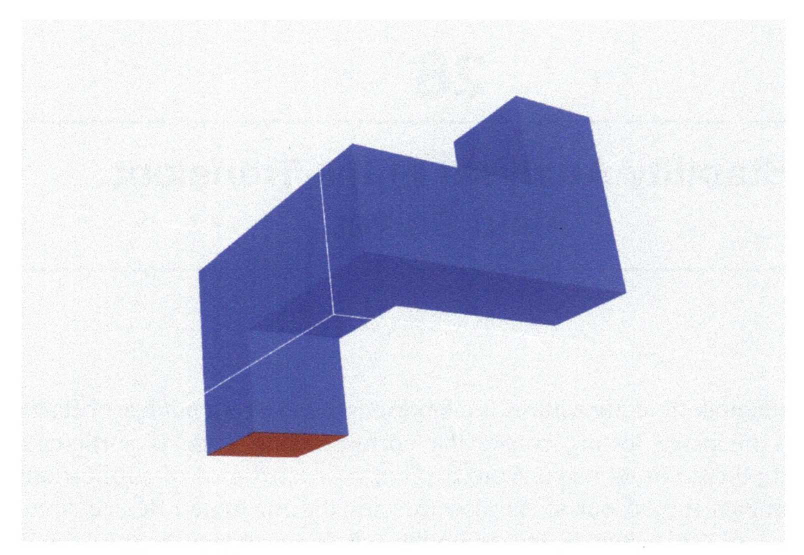

Figure 28.1 Test structure to study the stability problem.

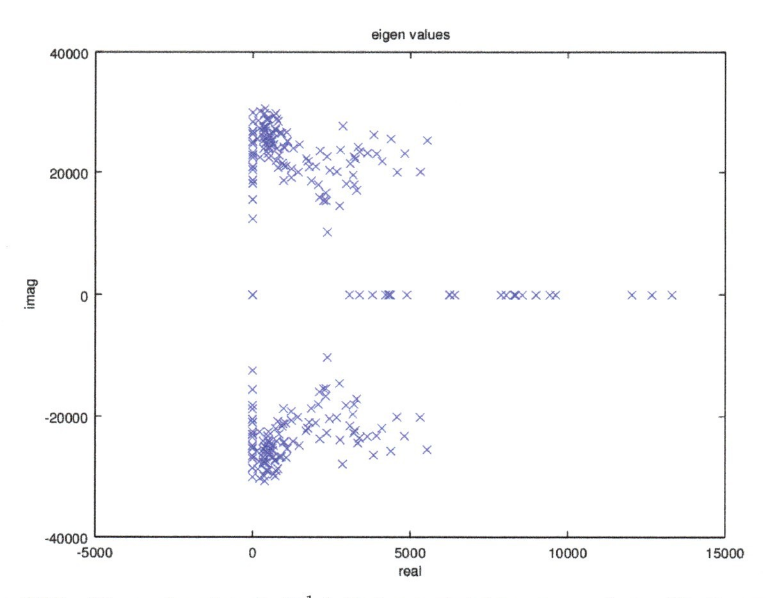

Figure 28.2 Eigenvalue plot of $A^{-1}J$. Both permittivities are equal one. The Lorenz gauge is used.

spectrum as depicted in Figure 28.3. As seen in Figure 28.3, there are eigenvalues with large negative real parts. It means the system is unstable which is not tolerable. Next we change the setting of permittivities by making the permittivity of metal also equal to 4. Then we obtain the spectrum as depicted in Figure 28.4.

From these observations we can draw the following conclusion. The instability problem is triggered not by the values of the permittivity but by the discontinuity in the permittivity. This triggers the question what happens if the conductivity has a discontinuity. To sort out this question we change the conductivity of the up and down pillar to 1.75×10^5 S/m and keeping the permittivities equal (to 4). The structure is shown in Figure 28.5 and the result is shown in Figure 28.6. The discretization of the equations needs modification if there is a discontinuity both in the conductance and the permittivity. The large conductance has clustered the spectrum. It is interesting to zoom in to some parts in the plot. This is done in Figure 28.7. We observe that the discontinuity in the conductance also leads to negative real parts of the some eigenvalues. When redoing the case with different permitivities (1 and 4 respectively for metal and insulator) but with higher conductance (1.75×10^5 S/m) for the metal, Figure 28.3 gets replaced by Figure 28.8.

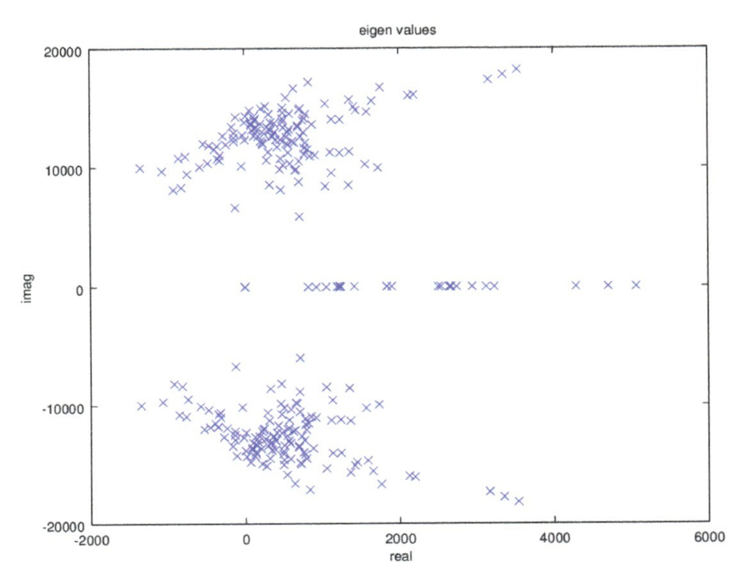

Figure 28.3 Eigenvalue plot of $A^{-1}J$. We have different permittivities and a moderately low conductance. (Lorenz gauge).

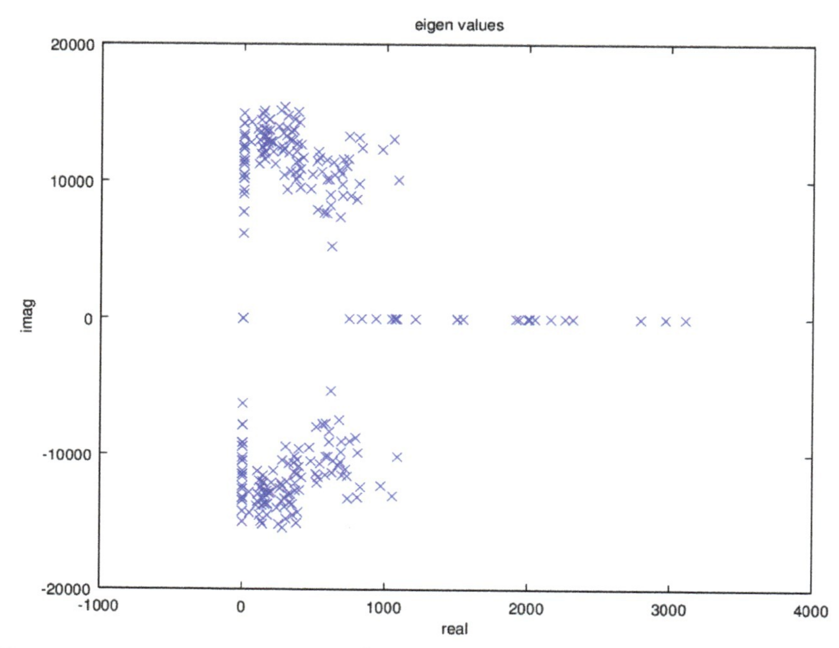

Figure 28.4 Eigenvalue plot of $A^{-1}J$. We have equal permittivities with value equal 4. (Lorenz gauge).

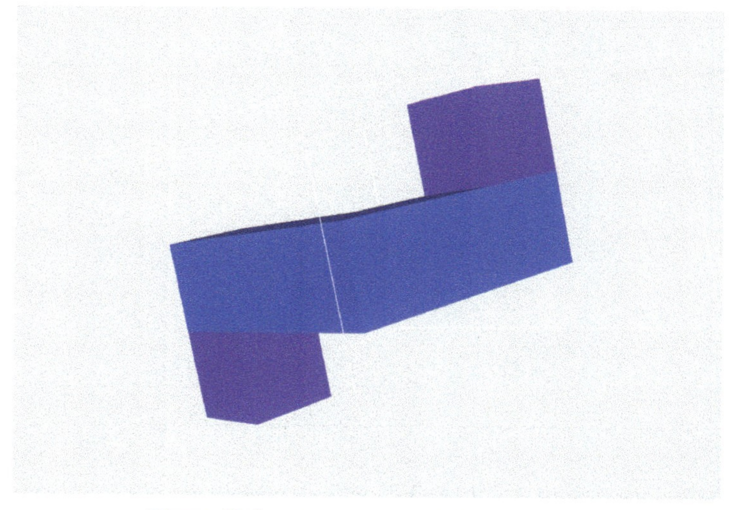

Figure 28.5 Structure with different metals.

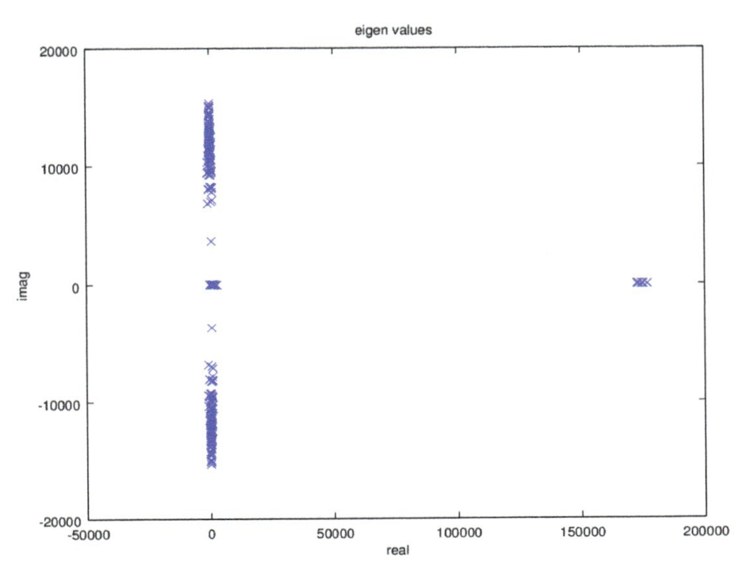

Figure 28.6 Eigenvalue plot of $A^{-1}J$. Using different conductances but same permittivities (Lorenz gauge).

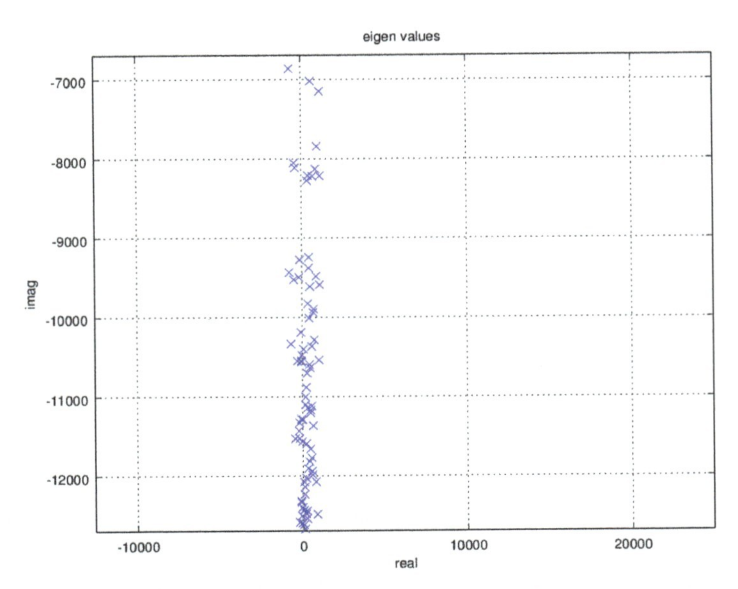

Figure 28.7 Zoom in to the eigenvalue spectrum for different conductances but equal permittivities. (Lorenz gauge).

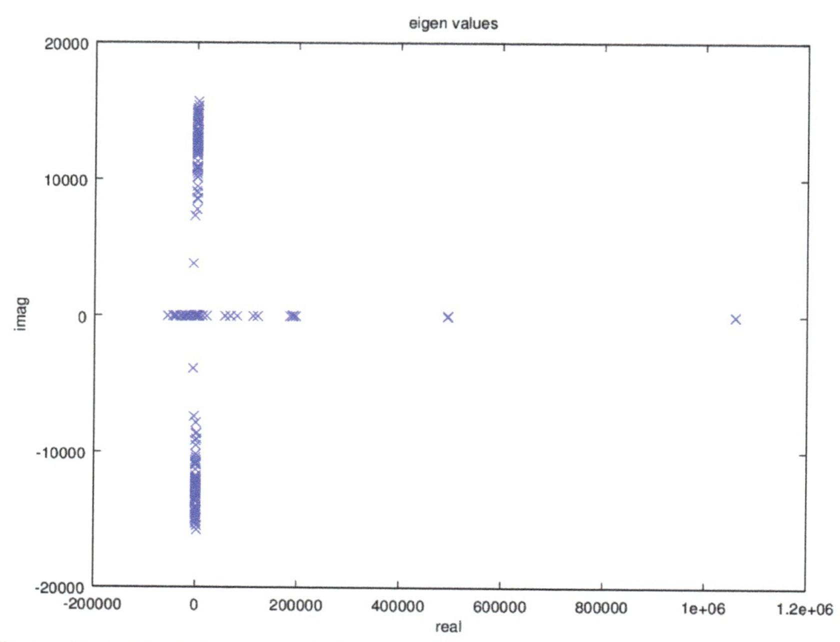

Figure 28.8 Zoom in to the eigenvalue spectrum for different conductances but equal permittivities. (Lorenz gauge).

Observation: Step-function changes in the conductance and/or the permittivity trigger unstable transient behaviour.

Our next experiment takes the permittivity in the metal equal to the permitivity in the oxide whereas the conductivity in metal is put back to realistic values (1.75×10^7 S/m). The idea is that it does not matter what the precise value of the permittivity in metal is since the conduction current will overrule the displacement current in metal. The result is shown in Figure 28.9. The eigenvalue spectrum seems fine. Looking into the numbers we find that numerical noise size negative real part appear. The most negative real part eigenvalues are:

- –2.1893e-06 + 6.2213e+03i
- –2.1893e-06 – 6.2213e+03i

Comparing the real-part size with the imaginary part size we consider this as computational noise. The stability problem is deeply rooted in the way how step-function interface material parameters are discretized and is not related to a bug in the present implementation.

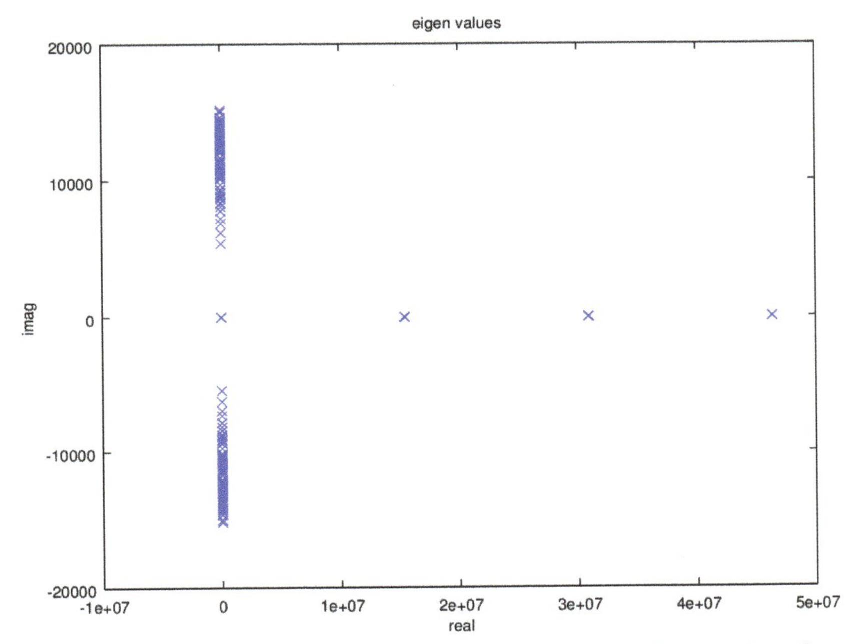

Figure 28.9 The eigenvalue spectrum for metal permittivity taken equal to the oxide permittivity. (Lorenz gauge).

Remedy: In order to avoid negative eigenvalues it is needed to revisit the coding of the conductance and permittivity at abrupt interfaces. Note that the cases that are considered above correspond to step functions in the material parameters. This suggests to soften the step function. Another way of dealing with this situation is to use refine the mesh around the mesh to make confine the unphysical interface results. For the time being we have a 'work-around' by replacing the permittivity in the metal by a value determined by the surrounding insulator material.

28.1 Impact of the Gauge Condition

An essential ingredient in the set up of the field dependence is the freedom to choose different gauge conditions. Therefore, if we make judgments about the location of the eigenvalues, the question arises if their locations are 'gauge' dependent. In order the answer this question we repeated the simulation corresponding to the set up described for Figure 28.3, but now the eigenvalues are calculated in the Coulomb gauge. The result is shown in Figure 28.10.

We observe that the eigenvalues do depend on the choice of gauge. So the stability question is a gauge-sensitive question. It is observed that eigenvalues with negative real part are much more sensitive to the choice of gauge than the eigenvalues with positive real part. At the next page we put together Figure 28.3 and Figure 28.10 on one page for comparison.

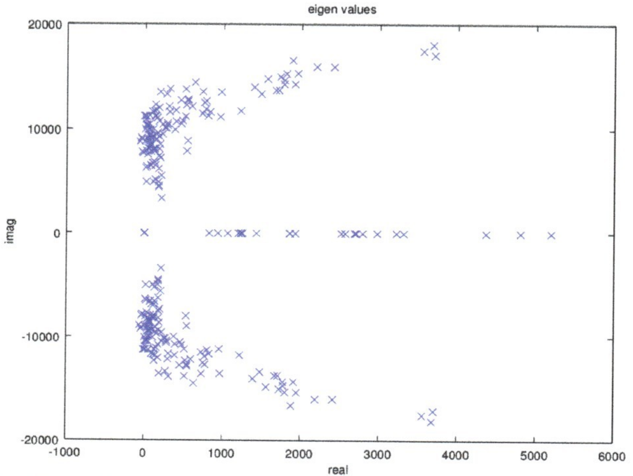

Figure 28.10 Eigenvalues for the set up belonging to Figure 28.3 but computed in the Coulomb gauge.

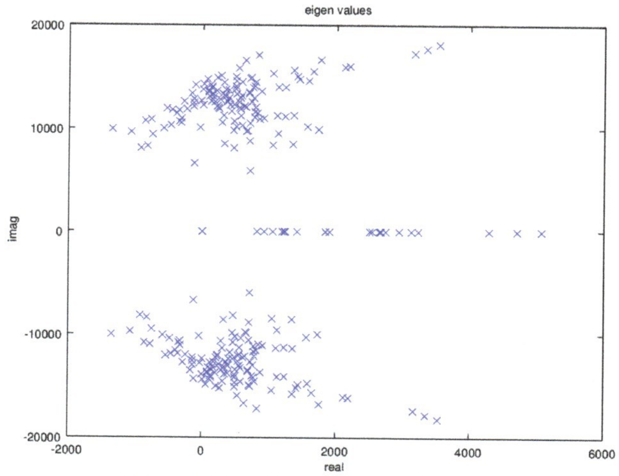

Figure 28.11 Lorenz gauge eigenvalues.

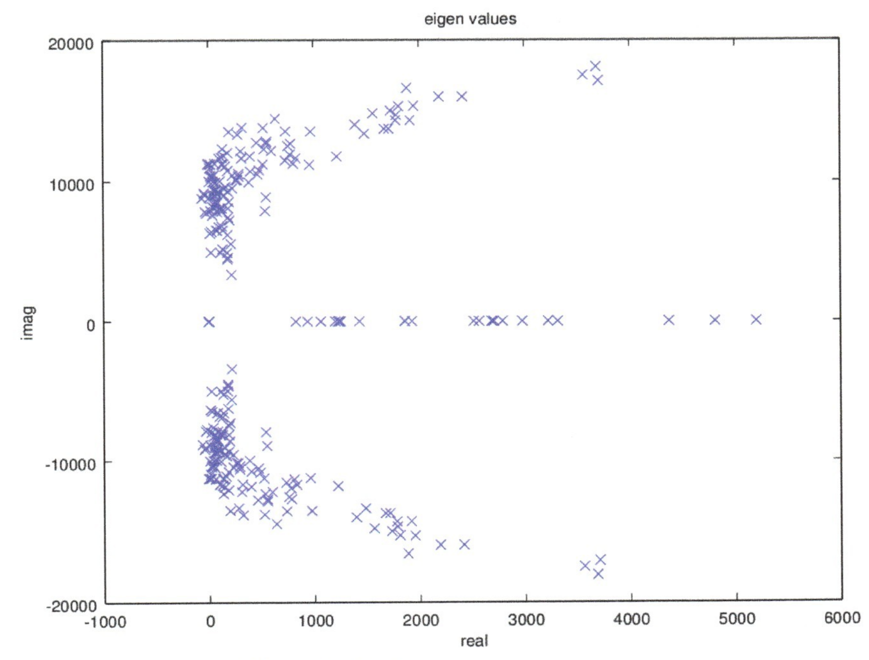

Figure 28.12 Coulomb gauge eigenvalues.

Of course the physical quantities should not depend on the choice of gauge. In Table 28.1 we compared the currents at different time steps at all contacts as well as the current balance.

The sensitivity to the gauge choice has triggered a careful re-analysis of the implementation of the Lorenz gauge. The subtlety of the implementation is found in the term:

$$k\epsilon\nabla\left(\frac{\partial V}{\partial t}\right) \tag{28.4}$$

The discretizaton requires that this term is integrated over the surface area corresponding to a node on the boundary of the simulation domain. The critical decision for getting correct results in the Lorentz gauge is that the permittivity must be averaged over *volume* elements and not over surface areas surrounding the nodes. Having incorporated this approach we obtain the following result for the Lorenz gauge we obtain Figure 28.13 for the distribution of eigenvalues.

Table 28.1 Comparison of Lorenz and Coulomb gauge results

Time	Contact	Lorenz Gauge	Coulomb Gauge
1e-9 sec	C1	−7.187088742165e-04	−7.187088793377e-04
	C2	7.187088742164e-04	7.187088793377e-04
	balance	7.934928358149e-12	7.708647076277e-12
2e-9 sec	C1	−1.437104467262e-03	−1.437104467260e-03
	C2	1.437104467262e-03	1.437104467260e-03
	balance	3.817442028574e-12	3.968329065281e-12
3e-9 sec	C1	−1.436791161378e-03	−1.436791158818e-03
	C2	1.436791161378e-03	1.436791158818e-03
	balance	1.509199390461e-14	6.036797572600e-14
4e-9 sec	C1	−1.177175531310e-12	−1.586168568470e-11
	C1	1.086623567363e-12	1.574094973277e-11
	balance	9.960716034159e-13	1.568058175680e-11
5e-9 sec	C1	8.904276454779e-13	1.557493779886e-11
	C2	7.998756815310e-13	1.548438583492e-11
	balance	7.244157115752e-13	1.537874187698e-11
5e-9 sec	C1	3.133058872722e-07	3.133084475212e-07
	C2	−3.133058872018e-07	−3.133084474496e-07
	balance	2.246627850880e-08	2.285308567642e-08

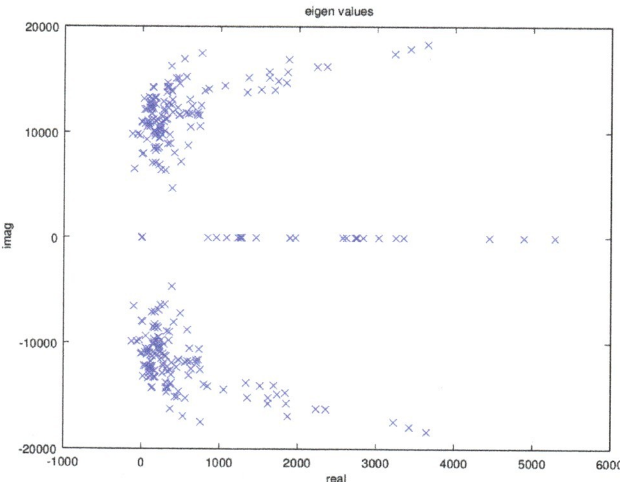

Figure 28.13 Eigenvalue distribution in the Lorenz gauge after different implementation of Equation (28.4).

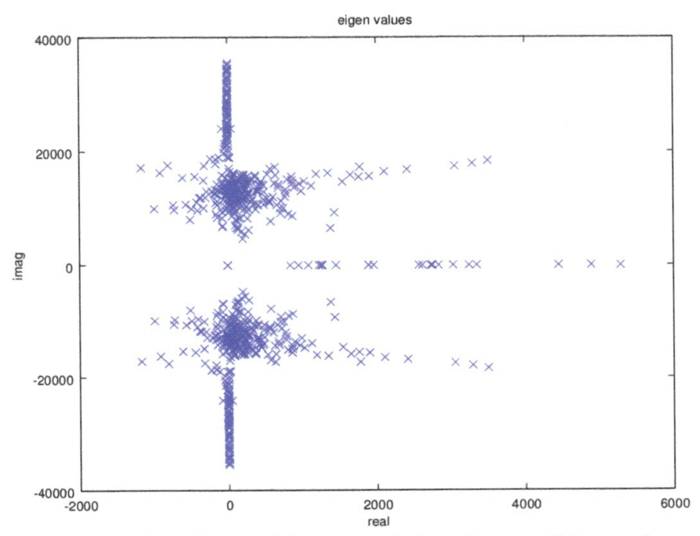

Figure 28.14 Eigenvalue when applying magnetic boundary conditions and $\varepsilon_{\text{oxide}} \neq \varepsilon_{\text{metal}}$.

28.2 Magnetic Neumann Boundary Conditions

Although it is not essential in the use of the simulator we may set magnetic Neumann boundary conditions. The first attempt to apply them on our test structure gives the eigenvalue spectrum as given in Figure 28.14. The spectrum indicates that there is still a flaw in the implementation. Therefore we recommend to refrain from using magnetic Neumann boundary conditions for the non-linear interface. With a modified setting of the permittivities $\varepsilon_{\text{oxide}} = 1$ and Neumann boundary conditions we find the results of Figure 28.15. When using the same set up but using the Coulomb gauge we find the results of Figure 28.16. For a simple straight line we obtain the results of Figure 28.17. This shows that also for Neumann boundary conditions we must expect a stable spectrum.

28.3 Results for Larger Values of the Conductance

We consider the Lorenz gauge normal conductance of the metal (10^7 S/m) and take $\varepsilon_{\text{metal}} = 0$. We take $\varepsilon_{\text{oxide}} = 10$ and the structure of Figure 28.1. All experiments are in Lorenz gauge. Then we obtain the following eigenvalue spectrum of Figure 28.18. At first sight everything looks fine.

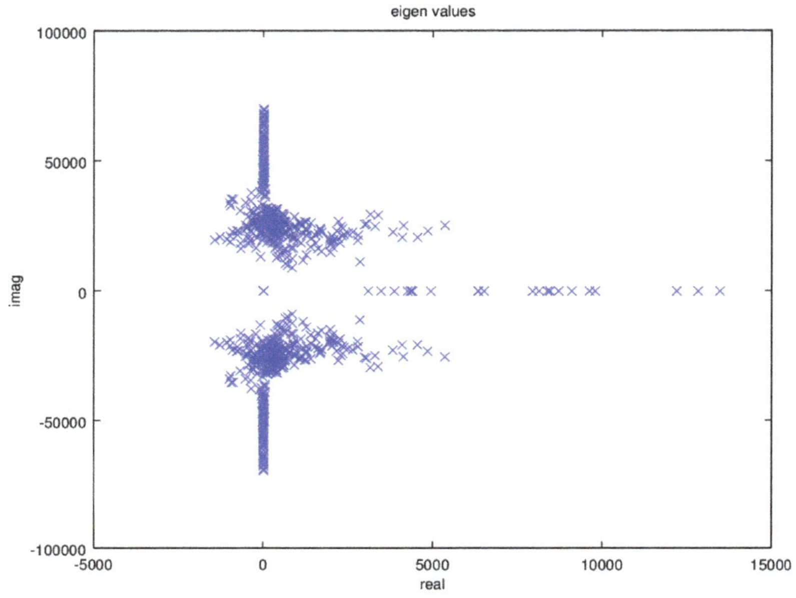

Figure 28.15 $\varepsilon_{\text{oxide}} = \varepsilon_{\text{metal}}$ and Neumann boundary conditions and Lorenz gauge.

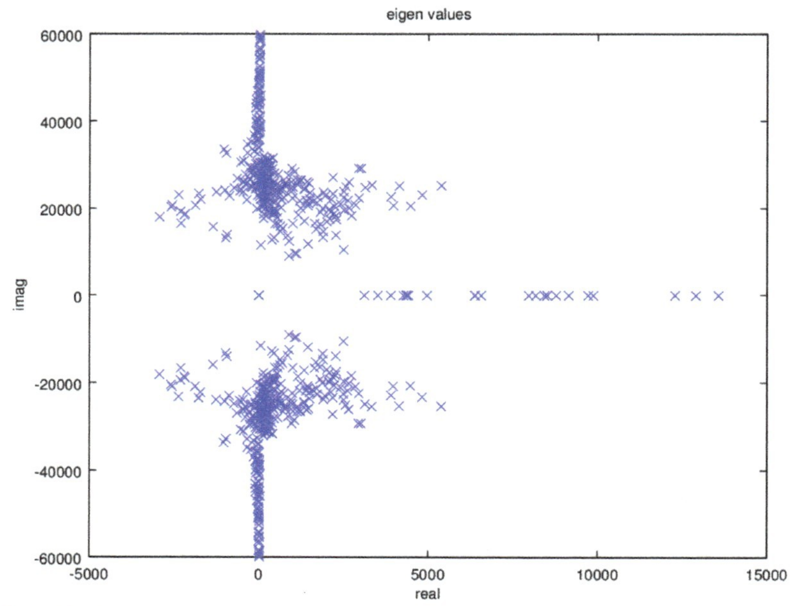

Figure 28.16 $\varepsilon_{\text{oxide}} = \varepsilon_{\text{metal}}$ and Neumann boundary conditions and Coulomb gauge.

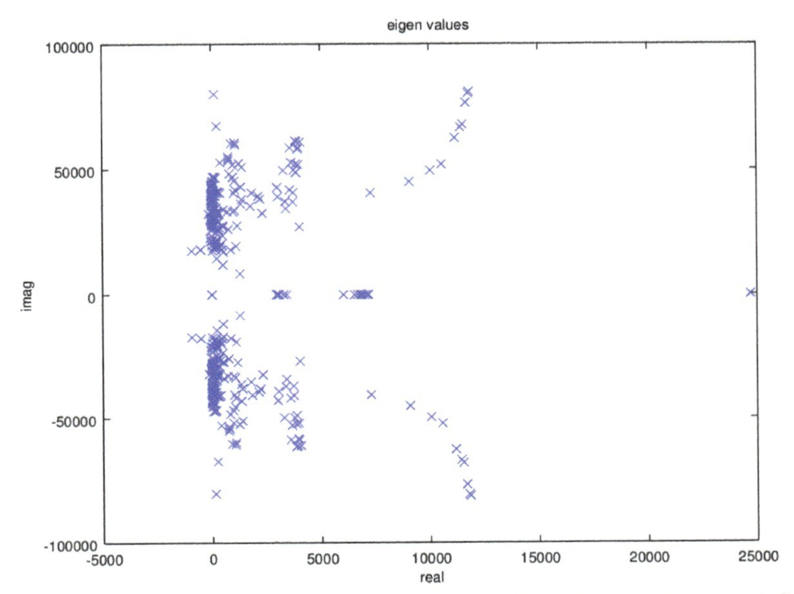

Figure 28.17 A straight line structure and Neumann boundary conditions and Coulomb gauge.

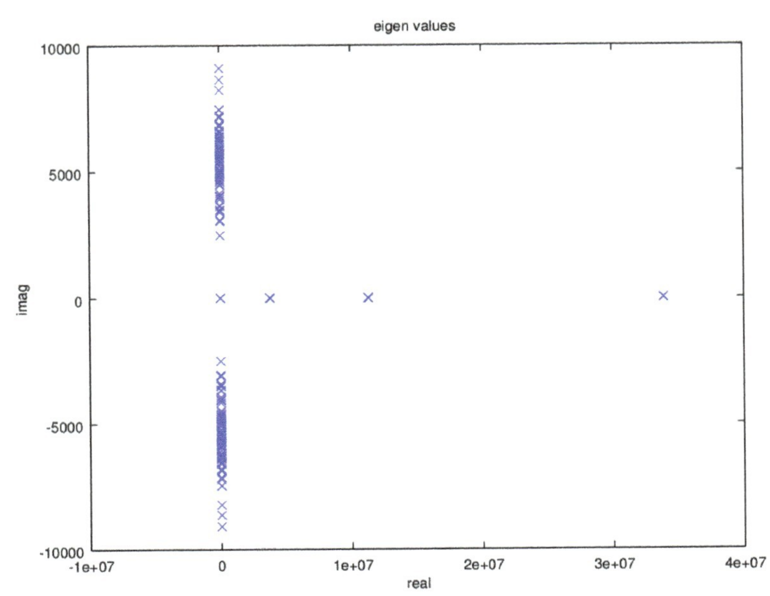

Figure 28.18 Global view at the eigenvalue spectrum $\varepsilon_{\mathrm{oxide}} = 10,\ \varepsilon_{\mathrm{metal}} = 0$ and Lorenz gauge.

In the plot (Figure 28.19) we zoom in at the values around zero. The calculation is repeated with $\varepsilon_{\text{metal}} = 5$ and the rest is unchanged giving the result of Figure 28.20. Next we repeat the exercise for $\varepsilon_{\text{metal}} = 8$ and for $\varepsilon_{\text{metal}} = 10$. The results are shown in Figure 28.21 and Figure 28.22. The real parts of the eigenvalues get closer to zero.

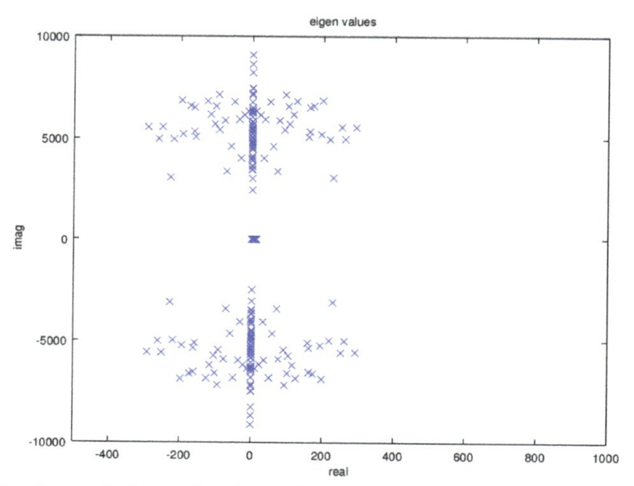

Figure 28.19 Zoomed view at the eigenvalue spectrum $\varepsilon_{\text{oxide}} = 10,\ \varepsilon_{\text{metal}} = 0$ and Lorenz gauge.

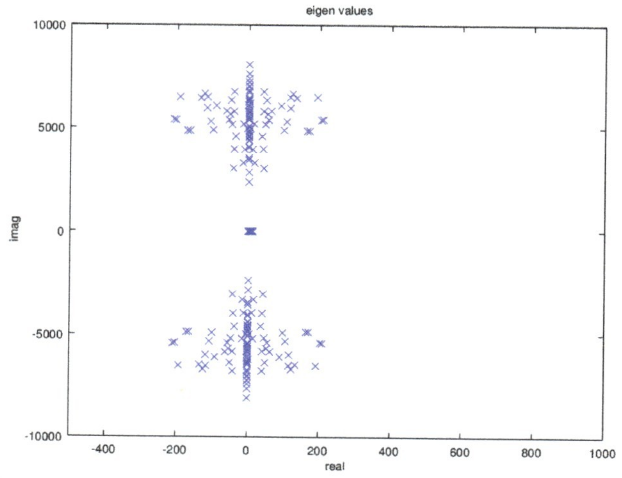

Figure 28.20 Zoomed view at the eigenvalue spectrum $\varepsilon_{\text{oxide}} = 10,\ \varepsilon_{\text{metal}} = 5$ and Lorenz gauge.

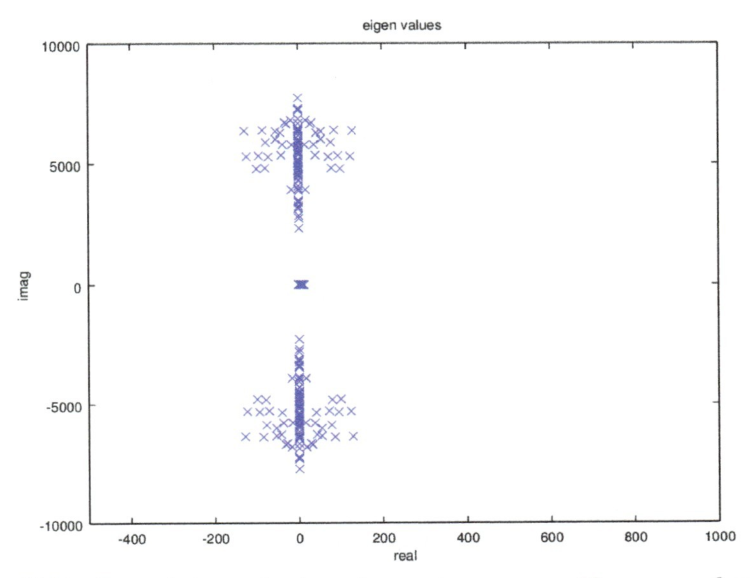

Figure 28.21 Zoomed view at the eigenvalue spectrum $\varepsilon_{\text{oxide}} = 10,\ \varepsilon_{\text{metal}} = 8$ and Lorenz gauge.

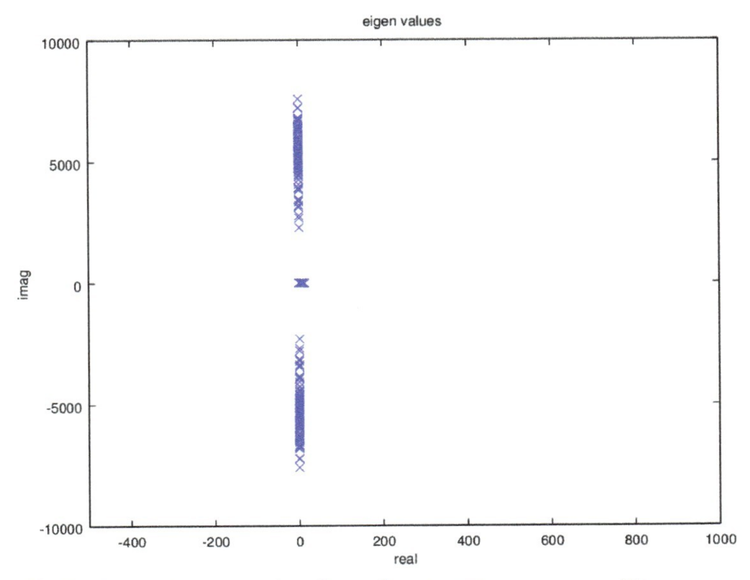

Figure 28.22 Zoomed view at the eigenvalue spectrum $\varepsilon_{\text{oxide}} = 10,\ \varepsilon_{\text{metal}} = 10$ and Lorenz gauge.

28.4 Yet Another Experiment

Inspired by the symmetric positioning of the eigenvalues we suspect something real is seen that needs further explanation. In order to confirm the trend we next set up the experiment and set the permittivity in the metal larger than the permittivity in oxide. The result is shown below in Figure 28.23. Keep on zooming in we find the plots as given in Figure 28.24 and Figure 28.25. We also did an experiment with negative permittivity. Then the spectrum is unphysical since large negative real values of the eigenvalues are found.

A good conductor is identified as having $\frac{\sigma}{\omega\varepsilon} >> 1$, or with $\varepsilon_{\rm im} \simeq \sigma$. Therefore it is looks that stability issues may be avoided by tuning the parameters σ. and $\varepsilon_{\rm re}$ such that the highest $\varepsilon_{\rm re}$ is found in the metal.

28.5 Inductor Experiments

The prototype inductor is somewhat different from the twisted bar because the structure has 'internal' contacts. Therefore we did an analysis of the eigenvalue spectrum by setting the relative permittivity of the metal at 10

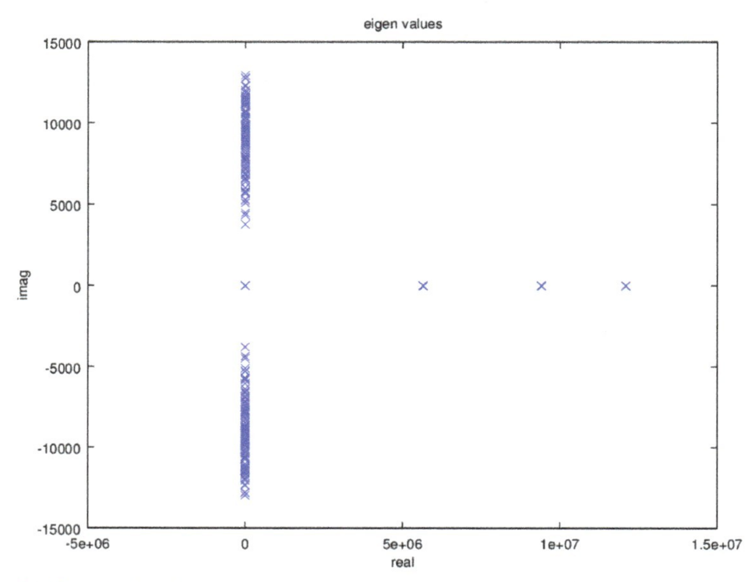

Figure 28.23 Global view at the eigenvalue spectrum $\varepsilon_{\rm oxide} = 5$, $\varepsilon_{\rm metal} = 10$ and Lorenz gauge.

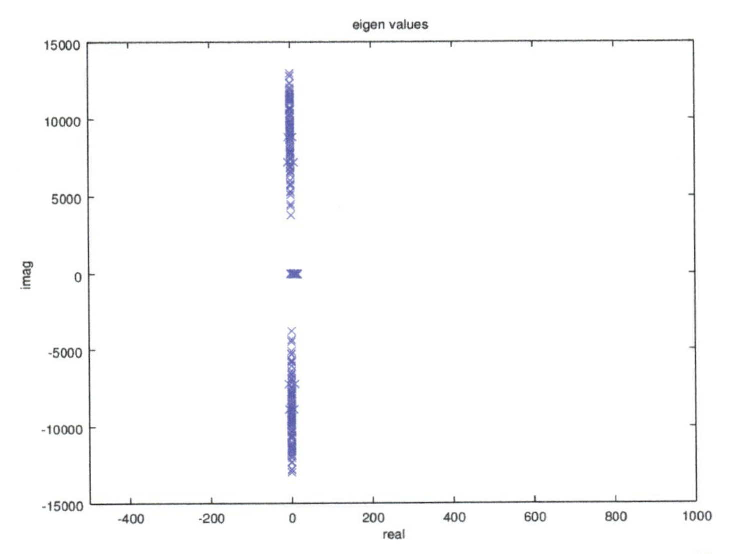

Figure 28.24 Zoomed view at the eigenvalue spectrum $\varepsilon_{\text{oxide}} = 5,\ \varepsilon_{\text{metal}} = 10$ and Lorenz gauge.

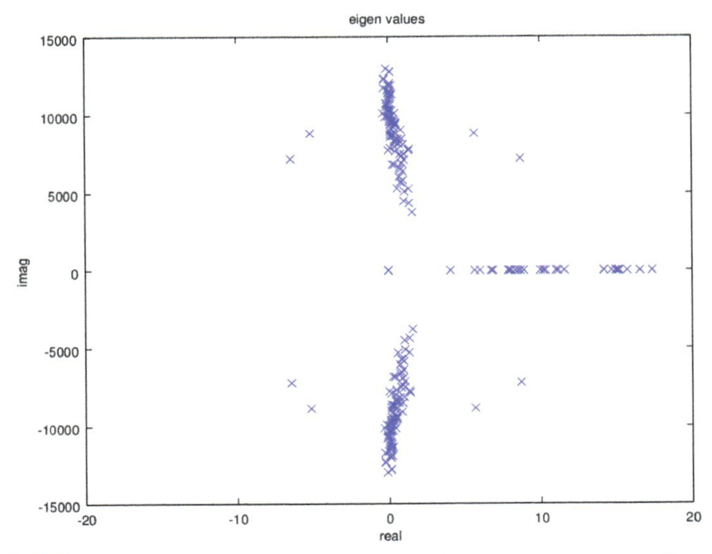

Figure 28.25 Further zoomed view at the eigenvalue spectrum $\varepsilon_{\text{oxide}} = 5,\ \varepsilon_{\text{metal}} = 10$ and Lorenz gauge.

and the relative permittivity of the oxide at 3.96. The design is shown in Figure 28.26. The eigenvalue spectrum is shown in Figure 28.27. A zoomed-in view is given in Figure 28.28. The smallest negative real part eigenvalues are (–1.9554e-01 + 2.0830e+03i) and (–1.9554e-01 – 2.0830e+03i). We take a look very close to real part equal zero and get the result as shown in Figure 28.29.

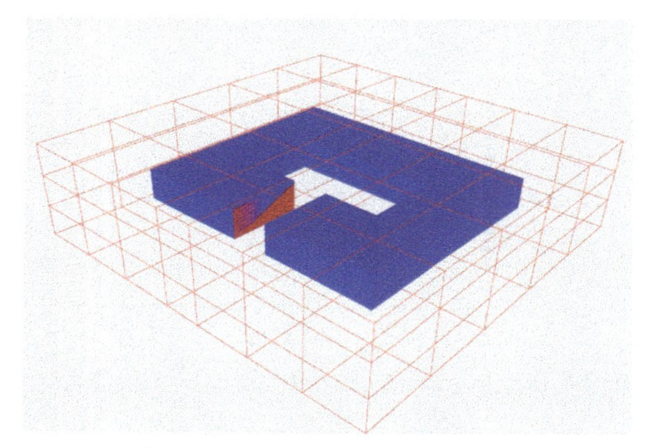

Figure 28.26 Inductor design.

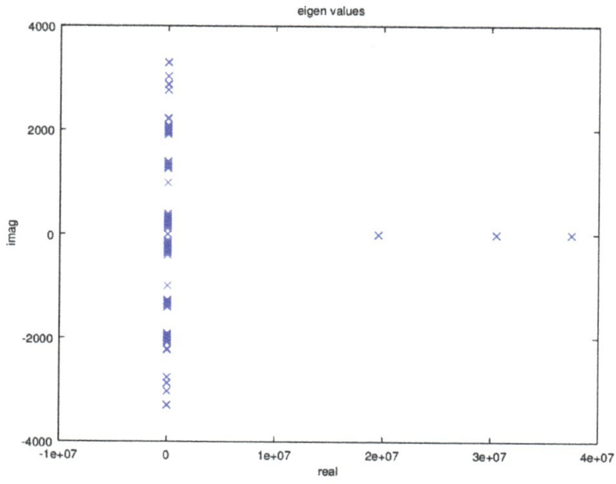

Figure 28.27 Global view at the eigenvalue spectrum $\varepsilon_{\text{oxide}} = 3.9$, $\varepsilon_{\text{metal}} = 10$ and Lorenz gauge.

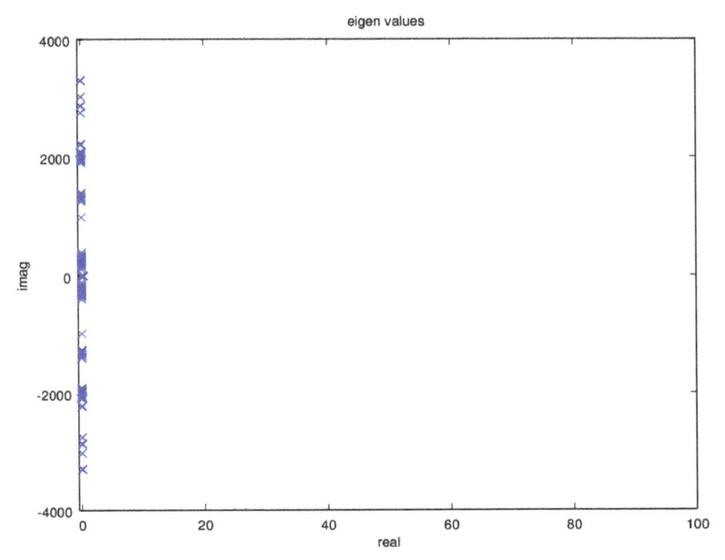

Figure 28.28 Global view at the eigenvalue spectrum $\varepsilon_{\mathrm{oxide}} = 3.9$, $\varepsilon_{\mathrm{metal}} = 10$ and Lorenz gauge.

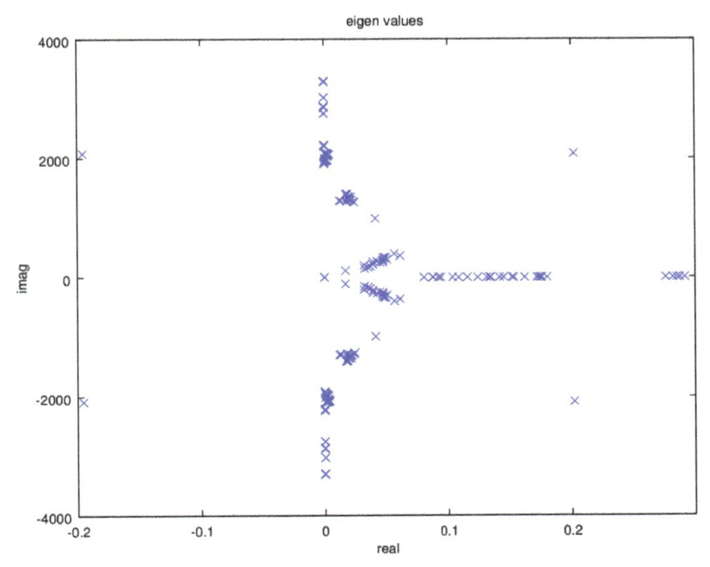

Figure 28.29 Global view at the eigenvalue spectrum $\varepsilon_{\mathrm{oxide}} = 3.9$, $\varepsilon_{\mathrm{metal}} = 10$ and Lorenz gauge.

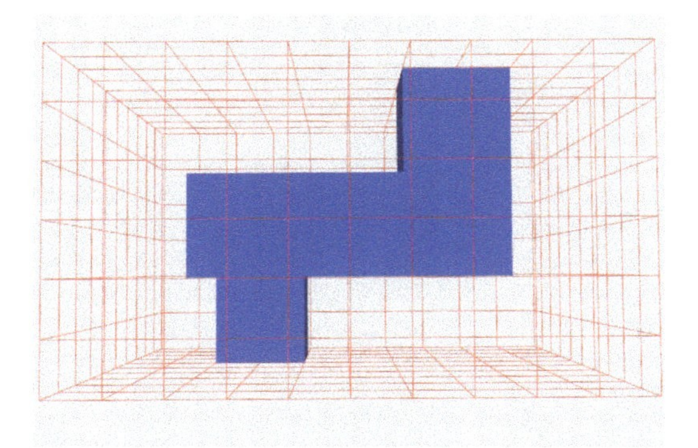

Figure 28.30 Mesh and structure of the 3D folded conductor.

Again this result suggests to adapt the permittivity of metal in order to enhance stability. The presence of the symmetric pairing (quadrupling) of eigenvalues does not appear to be a bug nor a coincidence but a signal that the code tries to inform us something about EM simulation that so far is not yet fully grasped. In order to gain further insight into this issue we need to find out how persistent our results are if the mesh size is increased. We next consider the structure with a refined mesh. The structure and mesh are shown in Figures 28.30 and 28.31. We set the permitivity in metal $\varepsilon_{\text{metal}} = 10$ and in oxide $\varepsilon_{\text{oxide}} = 5$. The spectrum is shown in Figure 28.32 and Figure 28.33.

Next we set the values of the permittivities equal to eachother. Then we obtain the spectrum as seen in Figure 28.34. We zoomed in around 0.

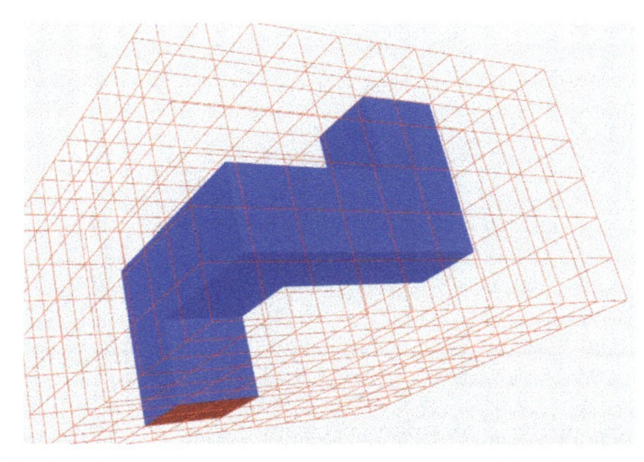

Figure 28.31 Same structure and mesh under a different viewing angle.

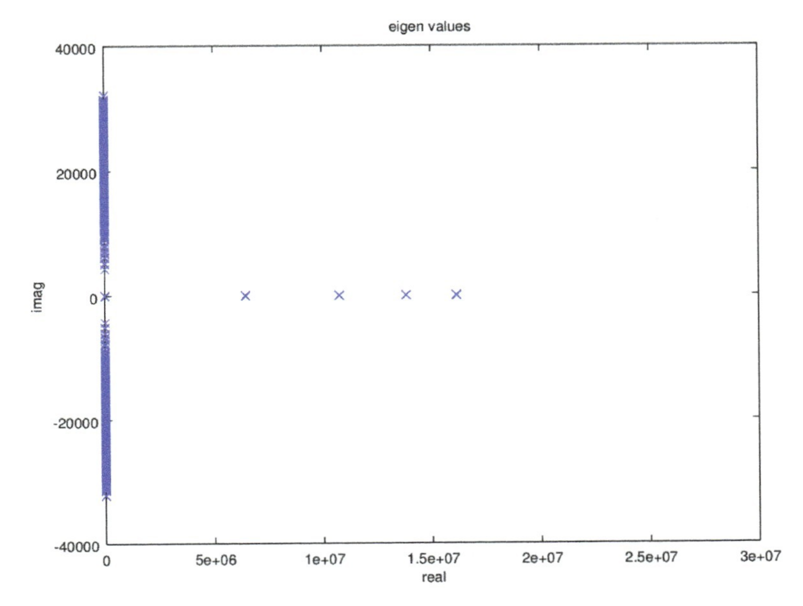

Figure 28.32 Eigenvalue spectrum (full range).

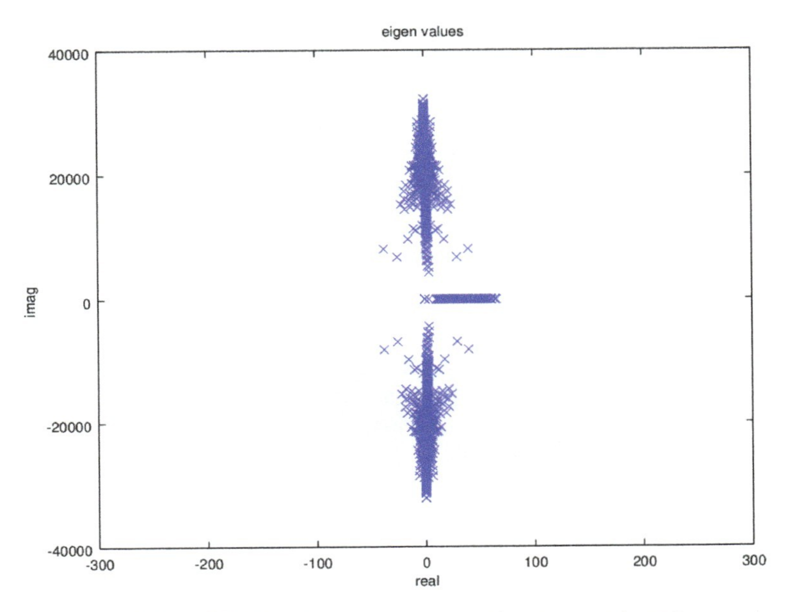

Figure 28.33 Eigenvalue spectrum (zoom around zero).

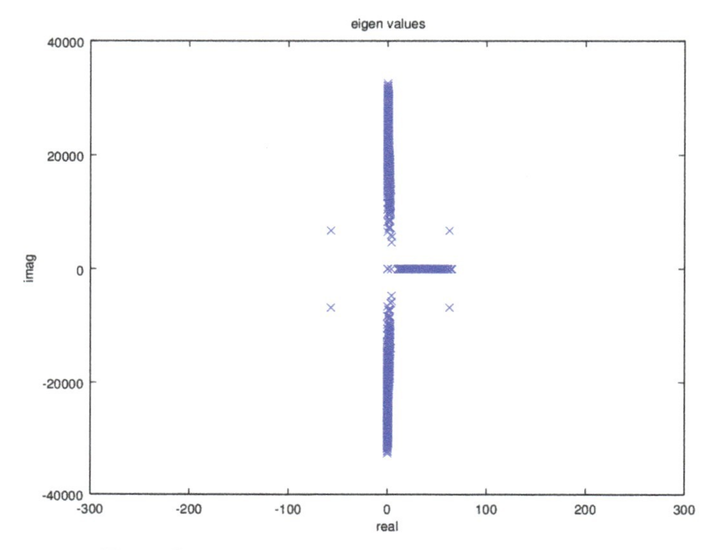

Figure 28.34 Eigenvalue spectrum (zoom around zero).

28.6 Results for a Metal Loop

Further experiments were done using the one-loop one-metal-layer inductor case. The structure is shown in Figure 28.35 The results are shown in Figure 28.36. There is a sign convention difference such that the stable real parts are at the negative side of the real axis. In Figure 28.37 a zoom is presented. The eigenvalues corresponding to unstable response have now positive real parts. The computed impulse response is shown in Figure 28.38.

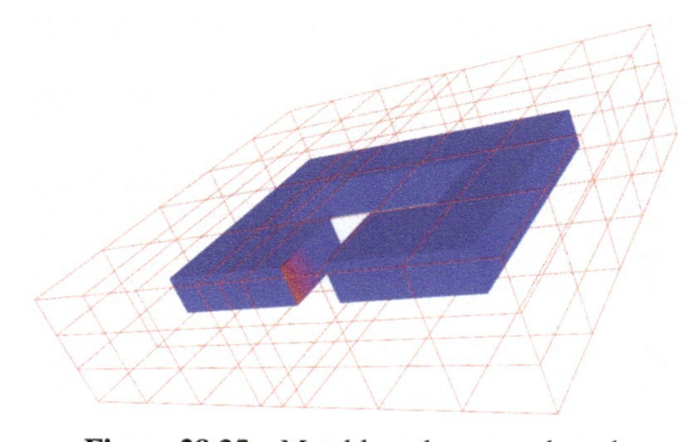

Figure 28.35 Metal loop layout and mesh.

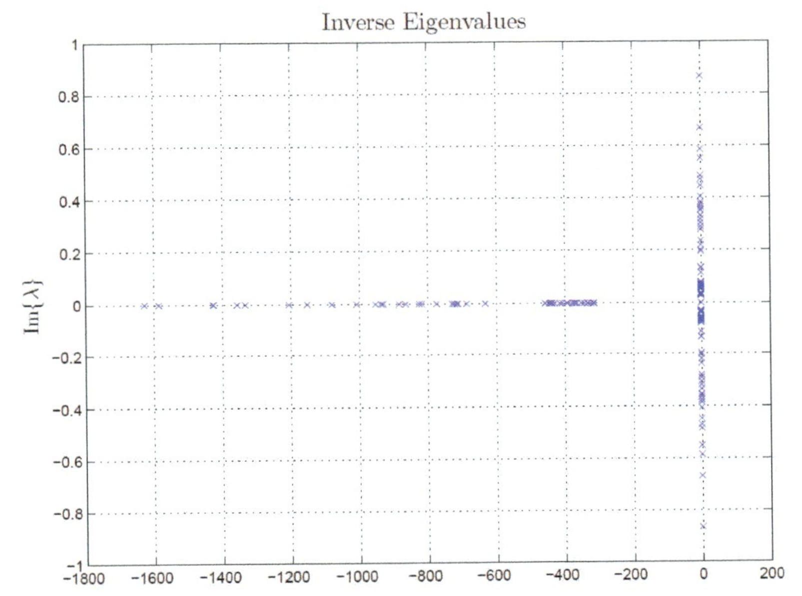

Figure 28.36 Eigenvalue spectrum results for the inductor case.

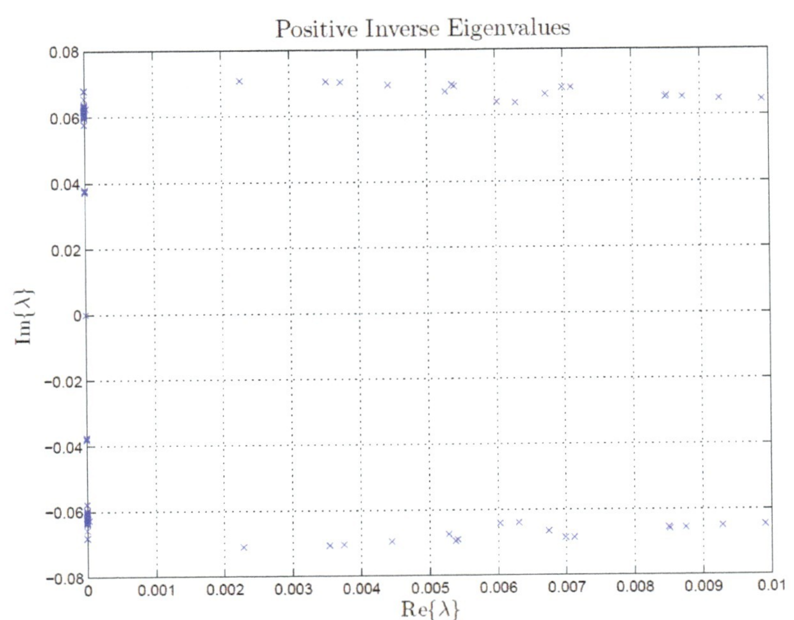

Figure 28.37 Zoom of the results for the inductor case.

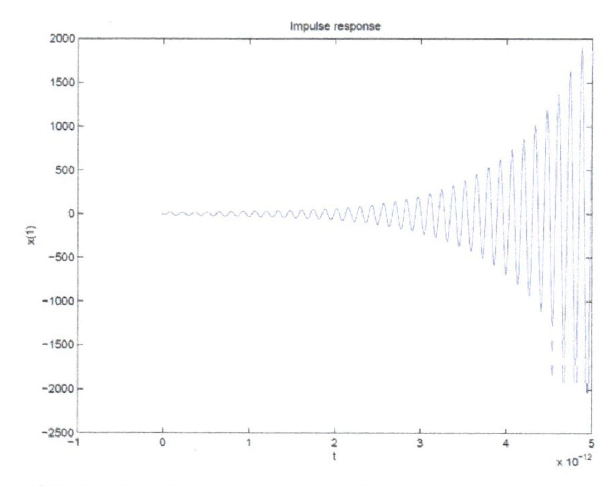

Figure 28.38 Impulse response of the inductor showing the instability.

28.7 Results for a Twisted Bar

A set of experiments were performed using the twisted bar structure using different simulation domains. One experiment was performed changing the mesh density. For the finer mesh example the spectrum of Figure 28.39 was obtained. The result is very much in agreement with Figure 28.33. Next consider stretching of the simulation domains in Figures 28.40–28.42. The spectrum results for the three cases are found in Figures 28.43–28.45.

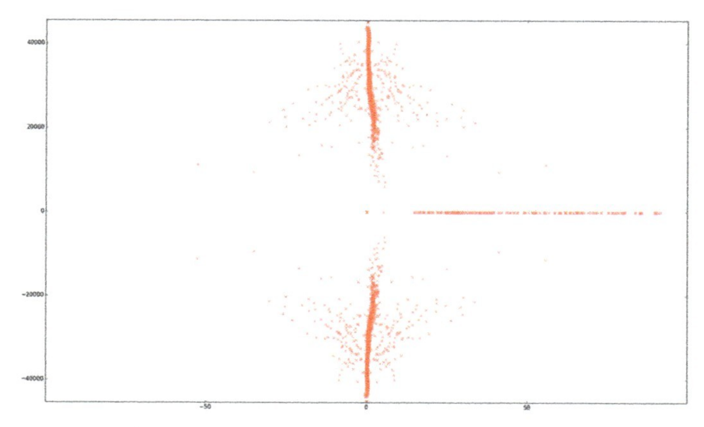

Figure 28.39 Spectrum results for a finer mesh.

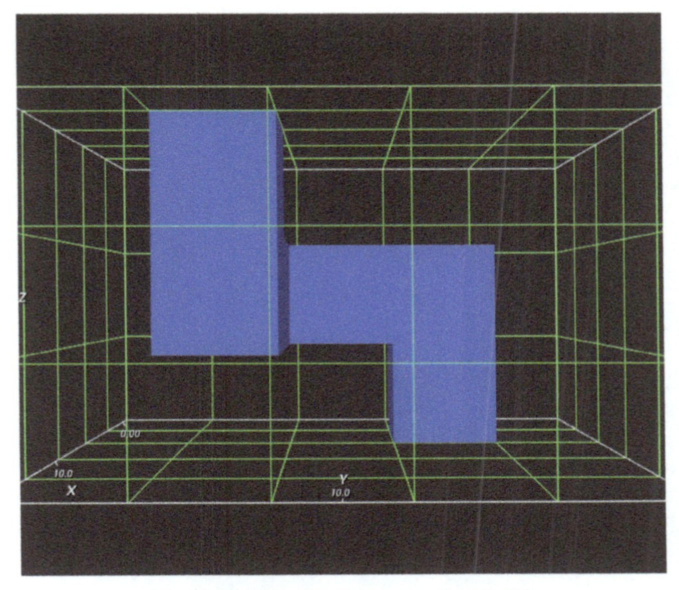

Figure 28.40 Domain size: 20x20x12.

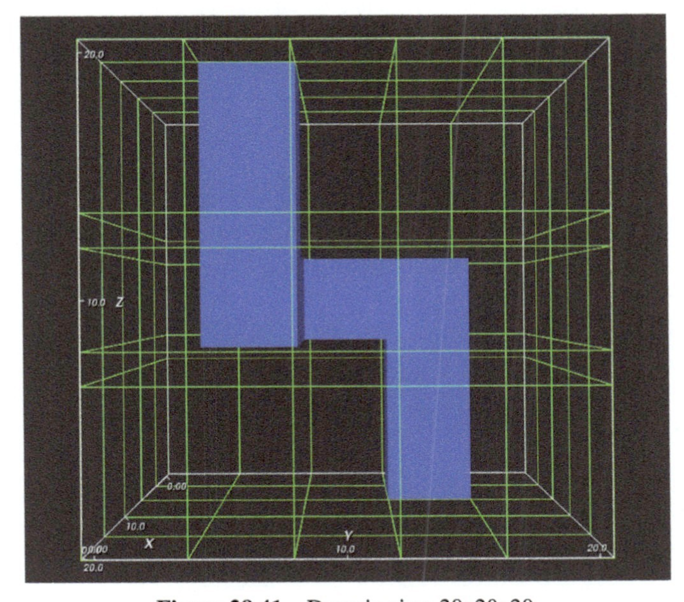

Figure 28.41 Domain size: 20x20x20.

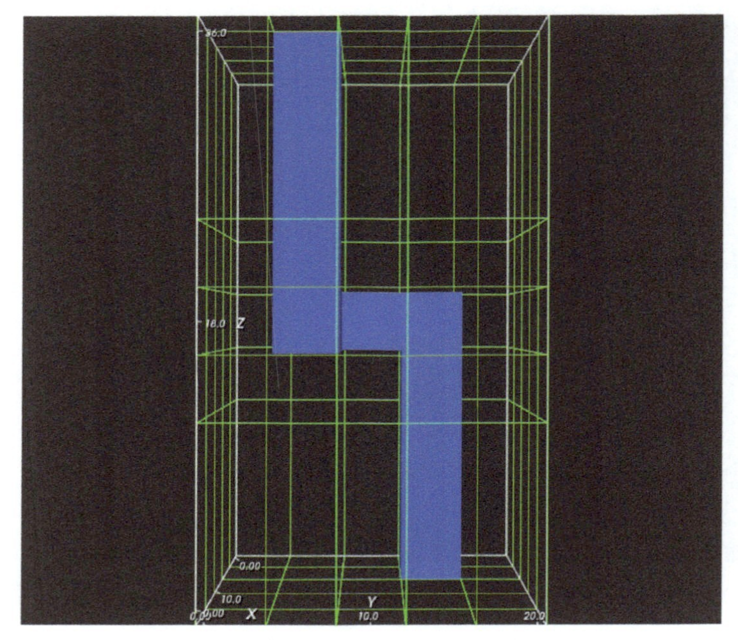

Figure 28.42 Domain size: 20x20x36.

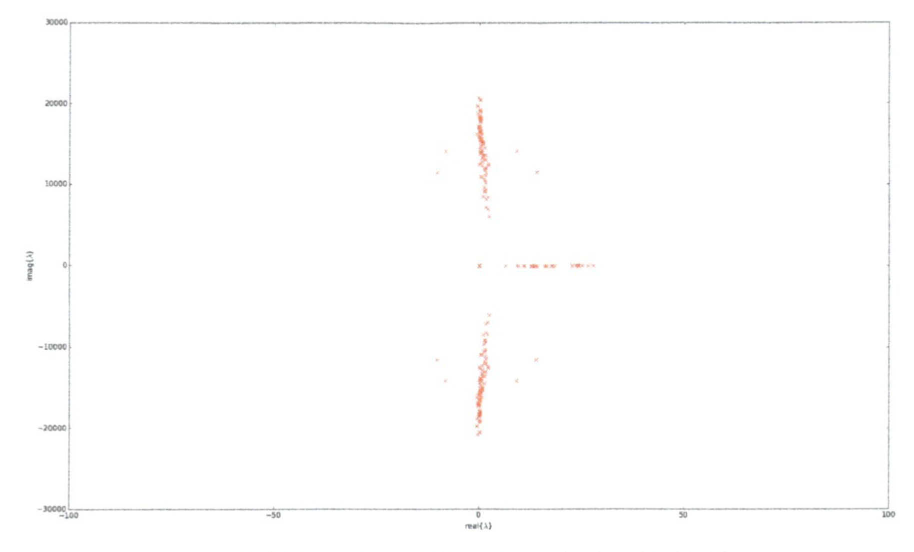

Figure 28.43 Spectrum for the domain size: 20x20x12.

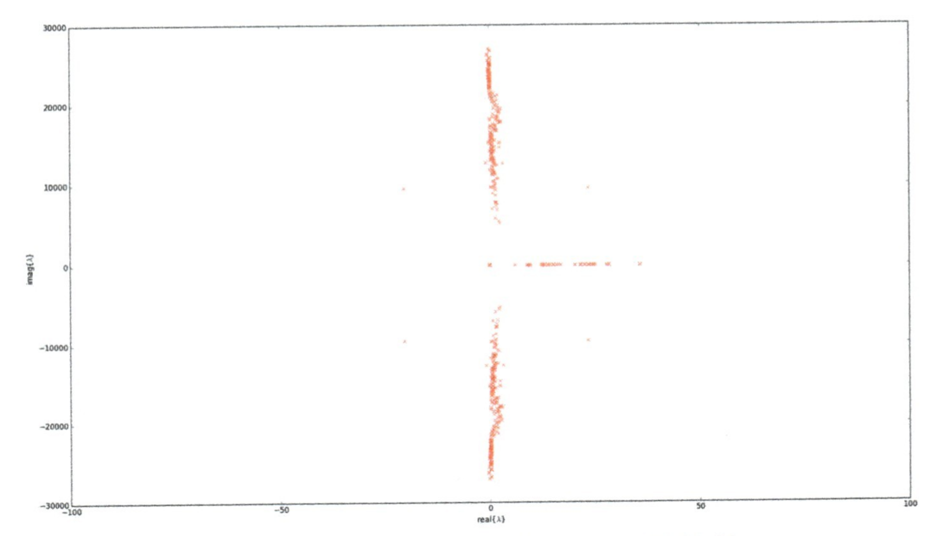

Figure 28.44 Spectrum for the domain size: 20x20x20.

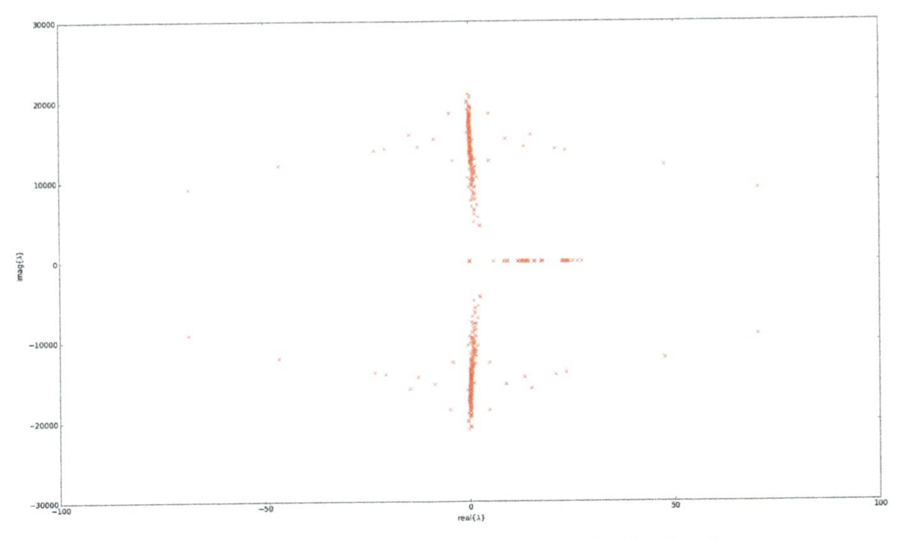

Figure 28.45 Spectrum for the domain size: 20x20x36.

By using a refined mesh the instability problem does not disappear. The number of eigenvalues with unstable real parts grows with the amount of metal/oxide surface nodes. The experiments point into the direction that the metal/oxide interface and its jump in permittivity play a key role in the instability response. The symmetric distribution of the eigenvalues is not merely a bug or discretization error. It suggests that there is a physical connection.

28.8 Corner Example

We make an even simpler example and test the stability problem here. Consider a simple corner of metal embedded in oxide. The structure is shown in Figure 28.46 and Figure 28.47. The results using A_x, A_y and A_z with zoom-in to the spectrum around zero is shown in Figure 28.48. Since we have a planar 'device' it suffices to include A_x and A_y only, since we expect that $A_z = 0$. The result is shown in Figure 28.49. With a zoom-in at the spectrum we find the Figure 28.50. As is seen, the spectrum still shows large negative real-part eigenvalues.

Figure 28.46 2D view of the metal corner embedding in oxide. The green block is oxide with $\varepsilon_{\text{oxide}} = 5$. The yellow part is oxide with $\varepsilon_{\text{oxide}} = 1$.

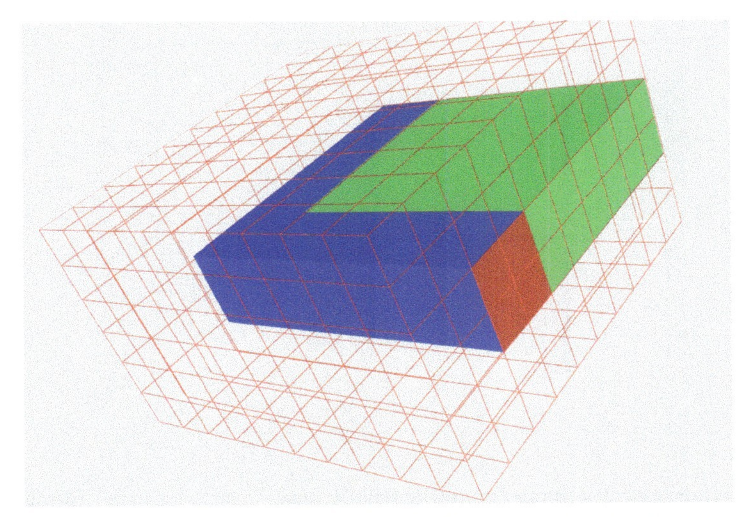

Figure 28.47 3D view of the metal corner embedding in Oxide. The background oxide ($\varepsilon_{\text{oxide}} = 1$) is not shown but present in all layers.

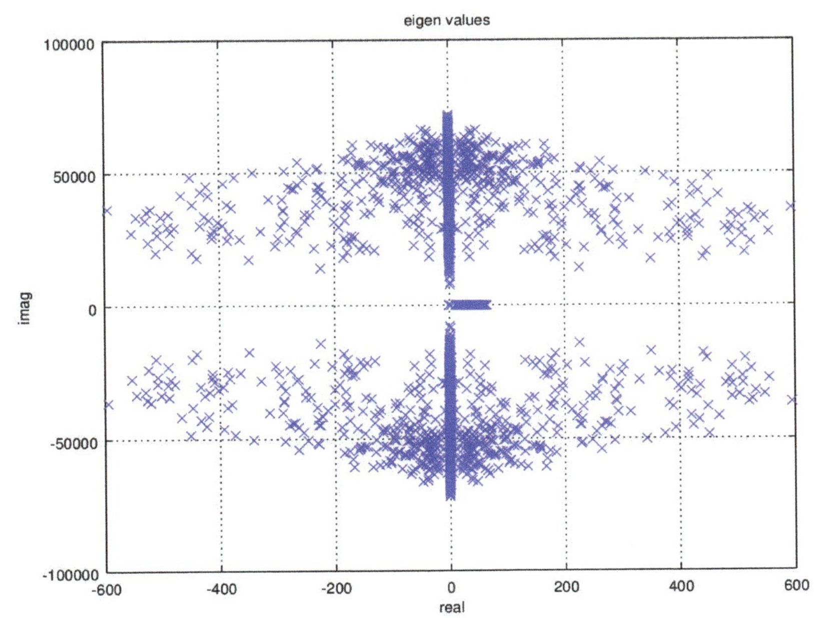

Figure 28.48 Zoomed view of the spectrum with Ax, Ay and Az included in the calculation.

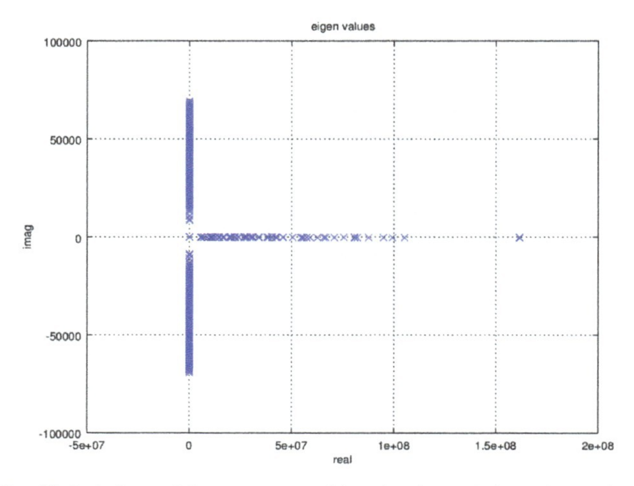

Figure 28.49 Global view of the spectrum with only A_x and A_y included in the calculation.

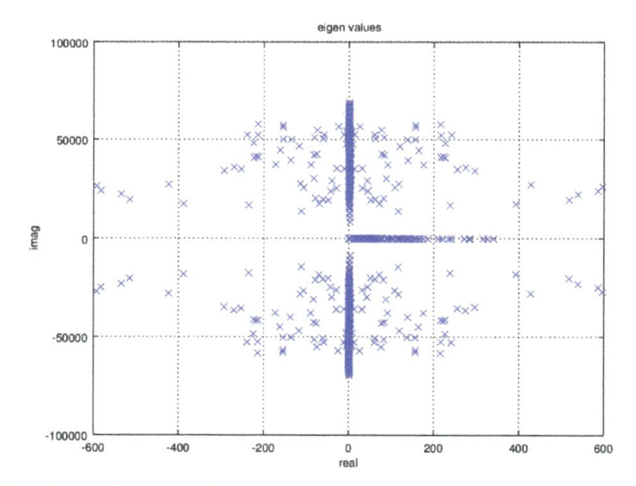

Figure 28.50 Zoomed view at the spectrum with only A_x and A_y included in the calculation.

When inspecting the Π – part of the A-matrix we find that the diagonal values are very small e.g. O(10^{-10}) whereas the off-diagonal part is O(10^{-1}). So the ratio diagonal/off-diagonal is O(10^{-9}). This number is very small because $K = 10^{-7}$. By modifying the scaling scenario we obtain a much better balancing between the diagonal and off-diagonal part of the Π – part of the matrix. Using the time scaling constant $\tau = 10^{-14}$ seconds, the ratio diagonal/off-diagonal =O(10^{-1}). The be more precise: the diagonal is O(10^{-3}) and off-diagonal is O(10^{-2}). In Figure 28.51 we present the spectrum using the time scaling of 10^{-15} sec. Now $K = 10^3$. In fact, K scales quadratically with the time-scaling constant. In Figure 28.52 we perform a zoom of the spectrum around zero. We observe that the negative real-part eigen values have moved towards the imaginary axis.

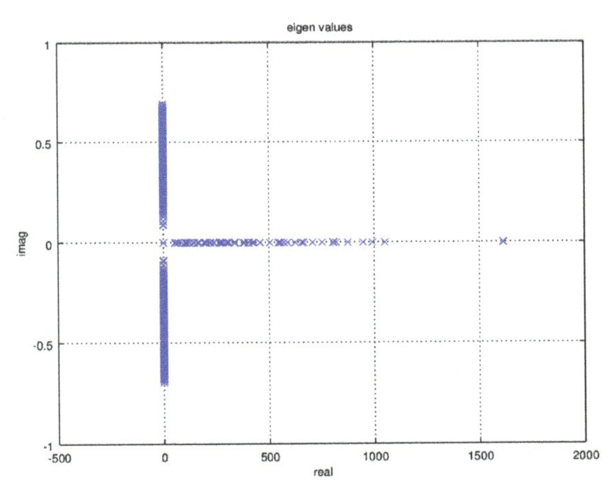

Figure 28.51 Spectrum for $\tau = 10^{-15}$. With an adapted time scaling we obtain a spectrum that corresponds to a much more stable equation.

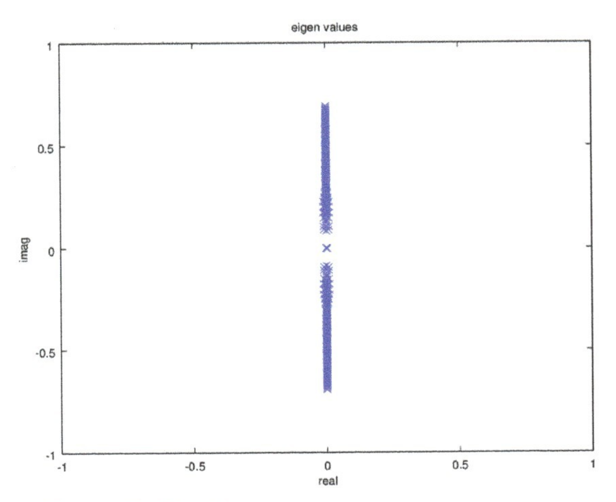

Figure 28.52 Zoom in to the spectrum around zero.

28.9 Returning to the Original Problem

While exploring various variations on the the theme of parameter choices, device designs we will now return to the original structure as shown in Figure 28.53. The original wire of Figure 28.53 is folded in the x, y and z directions and has the inclusion of the A_x A_y A_z components of the vector potential and the calculation of the spectrum is done in the Lorenz gauge. The time scaling factor is chosen to be $\tau = 1.428675 \times 10^{-12}$ sec. (The original value is $\tau = 1.428675 \times 10^{-10}$ seconds.) We find the spectrum as shown in Figure 28.54: Figure 28.55 shows a zoom-in of the spectrum around zero.

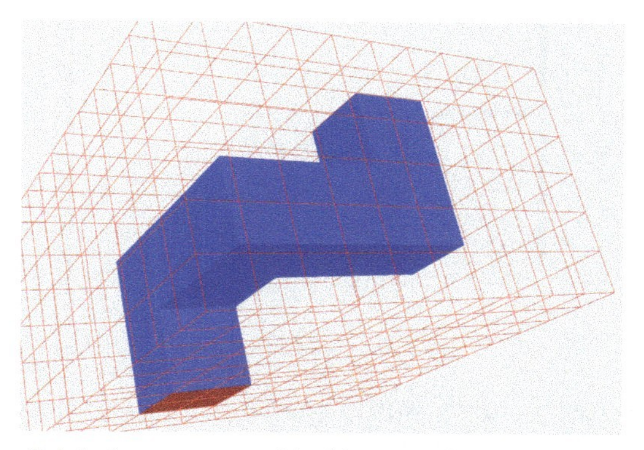

Figure 28.53 Original structure used in this appendix. $\varepsilon_{\text{metal}} = 1$ and $\varepsilon_{\text{oxide}} = 5$.

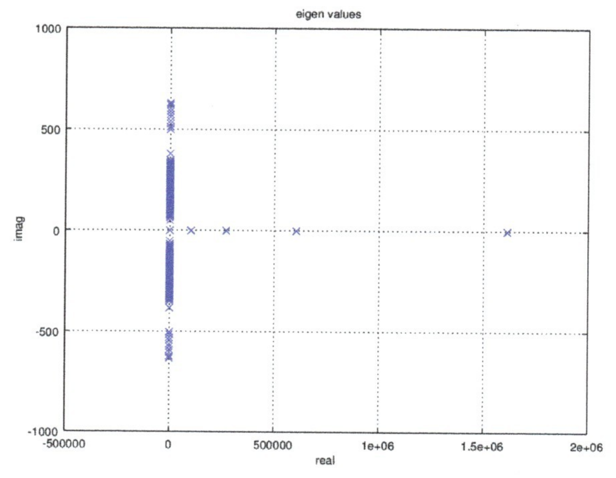

Figure 28.54 Full-range spectrum using the new time scaling factor.

Using the time scaling value $\tau = 1.428675 \times 10^{-14}$ seconds we obtain the result shown in Figure 28.56. We have obtained very excellent results e.g. the result corresponds to stable transient response.

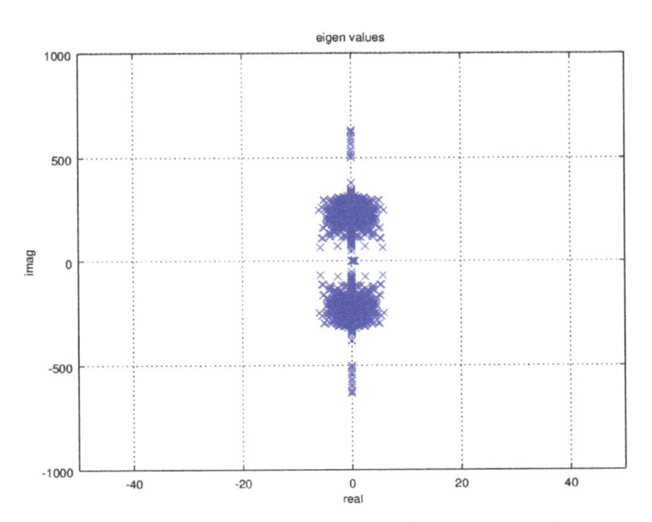

Figure 28.55 Spectrum zoom in around Re(λ) is 0. The spread is limited to (–10, 10) using time-scaling 10^{-12} seconds.

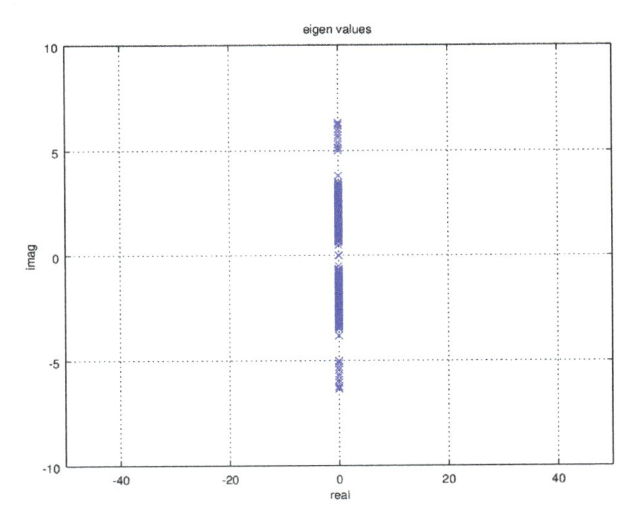

Figure 28.56 Spectrum zoom in around Re(λ) is 0. The time-scaling 10^{-14} seconds.

For comparison we also present the zoom-in results for the time scaling 10^{-13} seconds. The results is found in Figure 28.57. Note that some spreading around zero both in the negative and positive real part plane is observed in going from 10^{-14} to 10^{-13} seconds. This quickly get larger when the scaling value $\tau = 10^{-12}$ seconds is used. The lesson extracted from these simulations is that the time-scaling choice is essential in avoiding or delaying at least the unstable behavior of the transient response.

Next we apply the Coulomb gauge. We consider the spectrum for time-scaling $\tau = 10^{-14}$ seconds. The result is shown in Figures 28.58–28.60.

The problem of the negative real-part eigenvalues appears to be triggered by a bad choice of the time-scaling constant. In order to set up a stable transient calculation one must adapt the scaling scenario to the time step. In the MAGWEL code this method was implemented for the simulation of ESD protection structures Schoenmaker et al. [2014], Galy and Schoenmaker [2014], Chen et al. [2015]. At that time we observed failure of convergence and adapted the scaling scenario to solve the convergence issue.

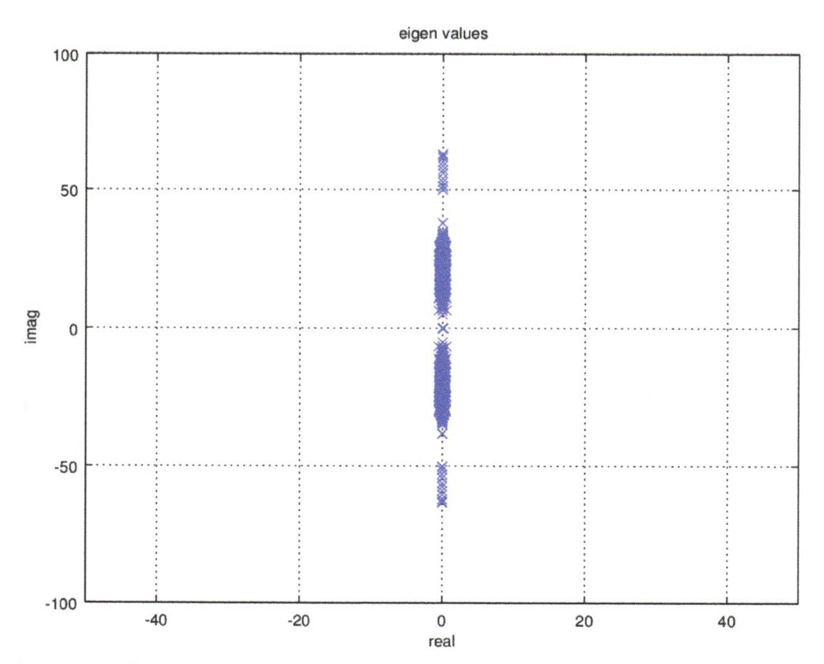

Figure 28.57 Spectrum zoom in around Re(λ) is 0. The time-scaling 10^{-13} seconds.

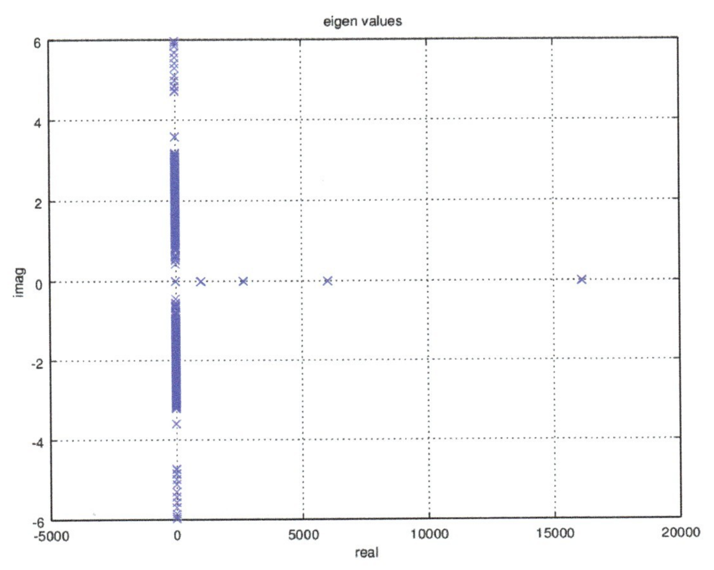

Figure 28.58 Result (no zoom) in the Coulomb gauge for the time-scaling $\tau = 10^{-14}$ sec. $\varepsilon_{\text{oxide}} = 5$, $\varepsilon_{\text{metal}} = 1$ and mesh as shown in Figure 28.53.

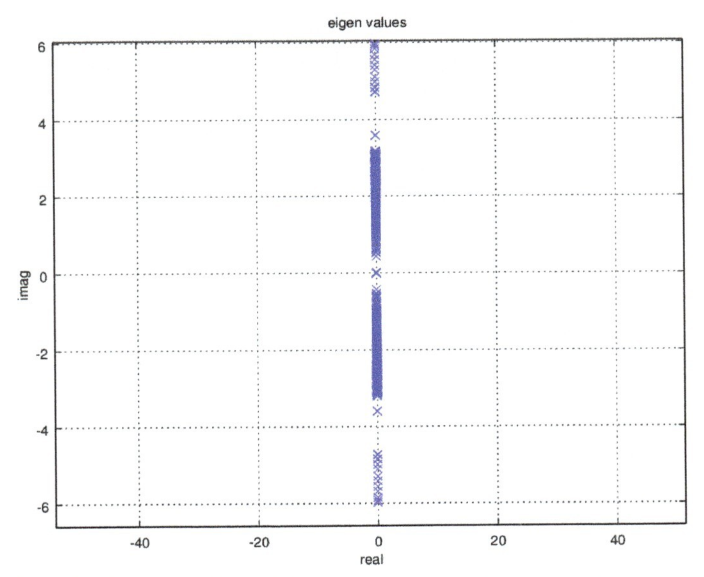

Figure 28.59 Result (zoom) in the Coulomb gauge for the time-scaling $\tau = 10^{-14}$ sec. $\varepsilon_{\text{oxide}} = 5$, $\varepsilon_{\text{metal}} = 1$ and mesh as shown in Figure 28.53.

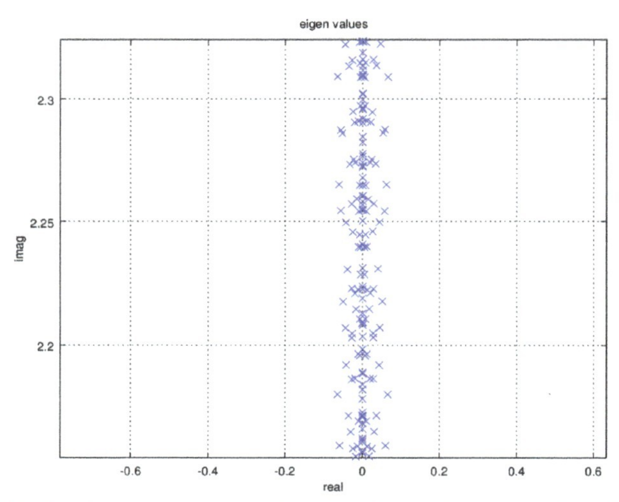

Figure 28.60 Result (extra zoom) in the Coulomb gauge for the time-scaling $\tau = 10^{-14}$ sec. $\varepsilon_{\text{oxide}} = 5$, $\varepsilon_{\text{metal}} = 1$ and mesh as shown in Figure 28.53.

28.10 Revisiting the Equations

We consider the twisted bar as shown in Figure 28.53. The size of the state space is 3601, meaning that there are 3601 degrees of freedom. We found that the eigenvalue spectrum shows a balloon structure around the imaginary axis. A zoom-in around the imaginary axis is shown in Figure 28.61. We have next re-visited (once more) the implementation of the various terms in of the Maxwell-Ampere equation. This equation reads with $\mu_r = 1$:

$$\nabla \times (\nabla \times \mathbf{A}) = \mu_0 \mathbf{J}_c + \frac{\partial}{\partial t}(-\mu_0 \epsilon (\nabla V + \mathbf{\Pi})) \tag{28.5}$$

In order to make the curl-curl operator more laplacian-like we add the gauge condition to this equation:

$$\nabla \times (\nabla \times \mathbf{A}) - \nabla (\nabla \cdot \mathbf{A}) = \mu_0 \mathbf{J}_c + \frac{\partial}{\partial t}(-\mu_0 \epsilon (\nabla V + \mathbf{\Pi})) + \mu_0 \xi \nabla \left(\epsilon \frac{\partial V}{\partial t}\right) \tag{28.6}$$

There a two terms that contain a mixture of a spatial and a time derivative, e.g.

$$-\mu_0 \frac{\partial}{\partial t}(\epsilon \nabla V) \tag{28.7}$$

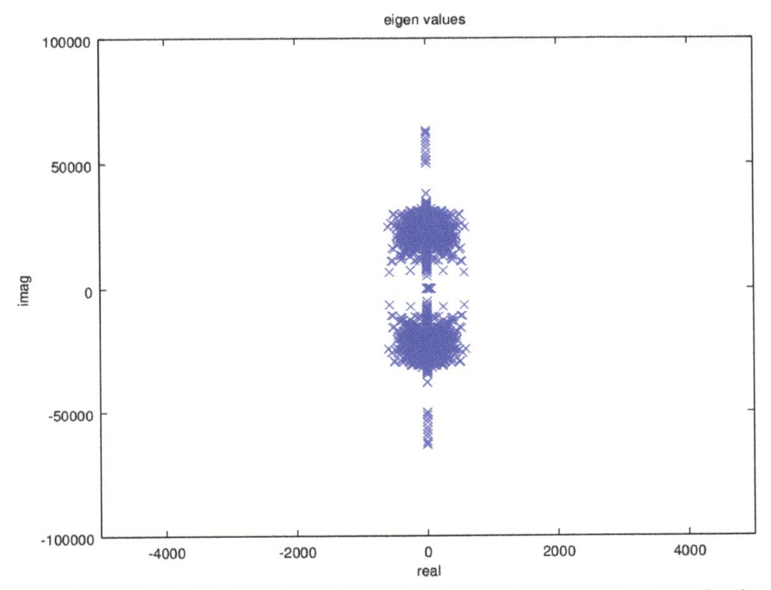

Figure 28.61 Zoom-in to the eigenvalue spectrum around the real axis.

and

$$\mu_0 \xi \nabla \left(\epsilon \frac{\partial V}{\partial t} \right) \tag{28.8}$$

It turns out that these terms need a different discretization based on the origin of appearance in the Maxwell-Ampere equation. The term (28.7) needs to be discretized as is done for the term

$$\frac{\partial}{\partial t} \left(-\mu_0 \epsilon \mathbf{\Pi} \right) \tag{28.9}$$

whereas the term (28.7) needs to be discretized as is done for the term

$$\nabla(\nabla \cdot \mathbf{A}) \tag{28.10}$$

Applying these new insights we adapted the solver and then obtain the eigenvalue spectrum as shown in Figure 28.62. The zoom-in is shown in Figure 28.63 for the structure of Figure 28.62. In this simulation we have not modified the scaling scenario. The computation was done in the Lorenz gauge. Of the 3601 eigenvalues the following onces have wrong-sign real part.

```
-5.7092e-01 + 1.0865e+05i
-5.7092e-01 - 1.0865e+05i
-5.9641e-01 + 1.0847e+05i
-5.9641e-01 - 1.0847e+05i
-6.6423e-01 + 1.0834e+05i
-6.6423e-01 - 1.0834e+05i
-1.3098e+03 + 1.8515e+04i
-1.3098e+03 - 1.8515e+04i
-1.5856e+00 + 2.3426e+04i
-1.5856e+00 - 2.3426e+04i
-2.0324e+02 + 2.3089e+04i
-2.0324e+02 - 2.3089e+04i
-2.0374e+02 + 2.3323e+04i
-2.0374e+02 - 2.3323e+04i
-1.5292e+00 + 2.2822e+04i
-1.5292e+00 - 2.2822e+04i
-1.5856e+00 + 2.3426e+04i
-1.5856e+00 - 2.3426e+04i
```

This is 0.5% of all the eigenvalues. All other eigenvalues have correct positive real part. So the balloon structure of eigenvalues has disappeared. Finally, we show some typical fields in Figures 28.64–28.66.

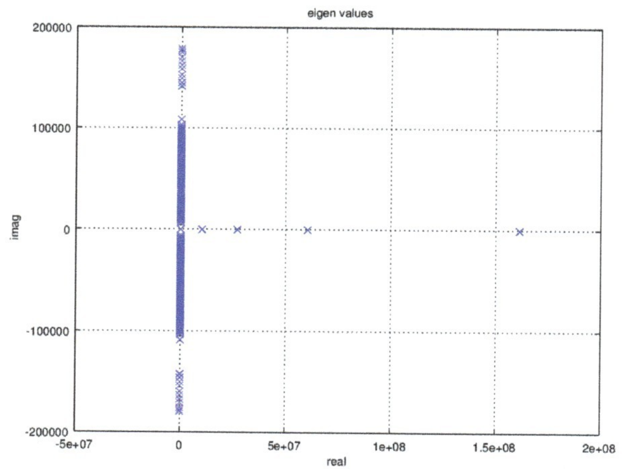

Figure 28.62 Spectrum of all eigenvalues for the structure in Figure 28.53. The Lorenz gauge is used and the new implementation of mixed space-time terms is applied.

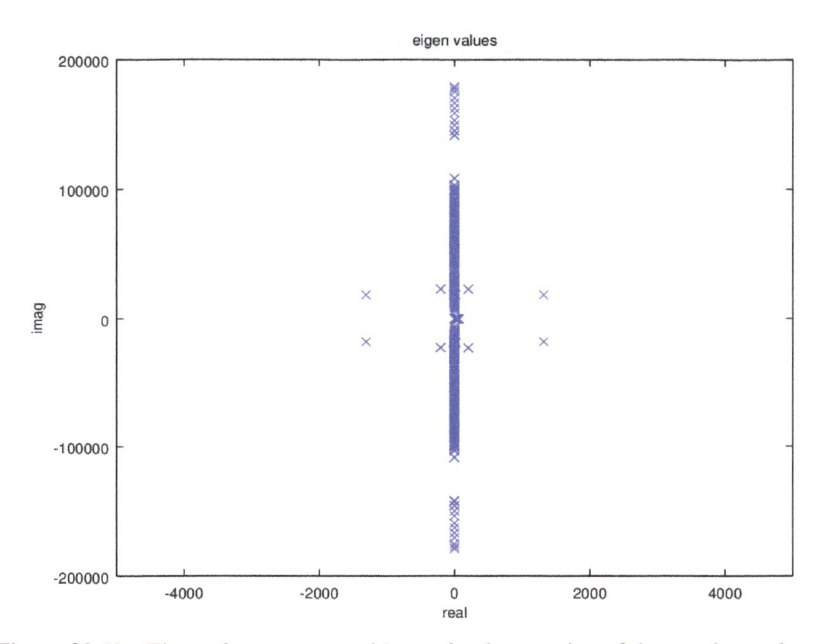

Figure 28.63 Eigenvalue spectrum with new implementation of the transient voltage terms in the Maxwell-Ampere equation. (Zoom in around zero).

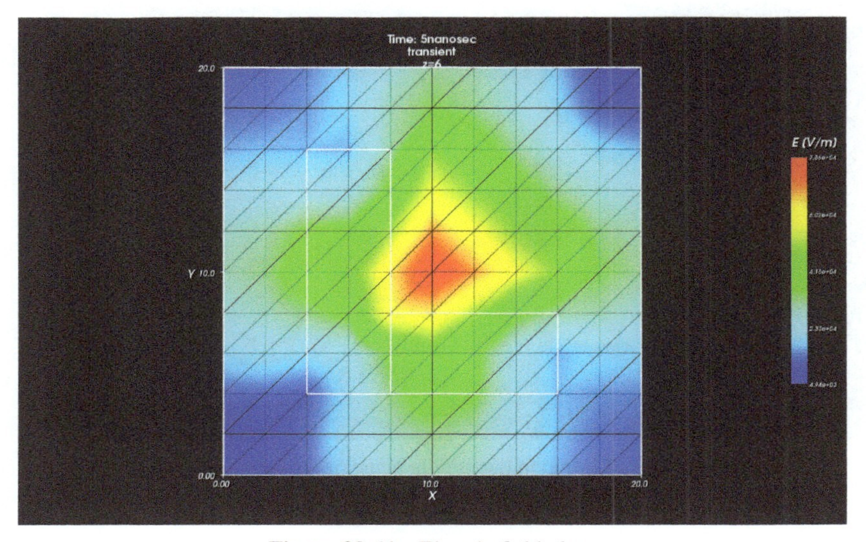

Figure 28.64 Electric field plot.

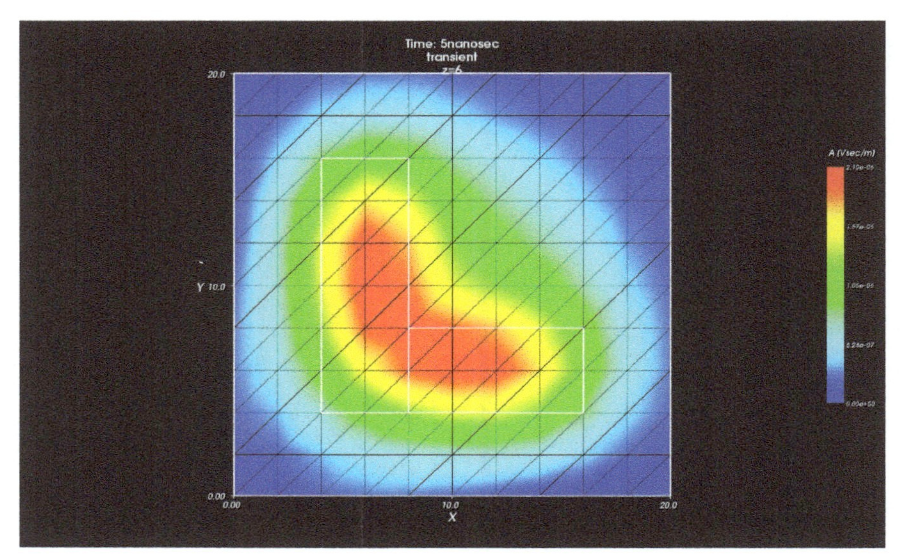

Figure 28.65 A-field in the middle plane of the structure.

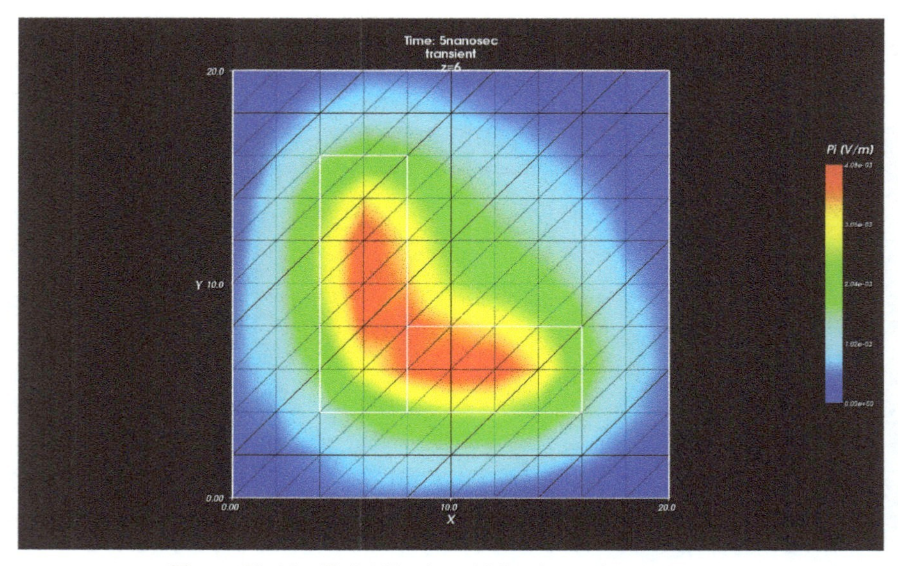

Figure 28.66 Π-field in the middle plane of the structure.

28.11 Redoing the Corner Structure

We also revisited the corner structure as shown in Figure 28.67. The result of the old implementation is shown in Figure 28.68. There are 720 eigenvalues with a wrong sign for the real part. The calculation was done with A_x, A_y and A_z components of the vector potential included, leading to 2629 eigenvalues. Thus the erroneous amount is 27%. With the improved implementation we find that there are 32 wrong-sign real-part eigenvalues for A_x, A_y, A_z included. This amounts to 1.2% of all eigenvalues. For the calculation using Ax, Ay only the results is shown in Figure 28.69. For the calculation using using A_x, A_y and A_z the result is shown in Figure 28.70.

Figure 28.67 Π-field in the middle plane of the structure.

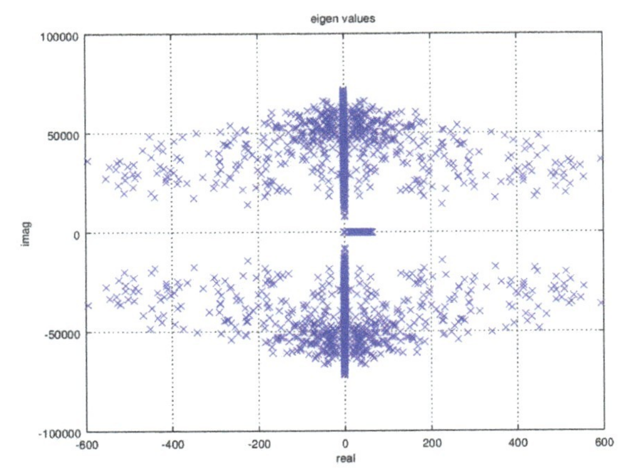

Figure 28.68 Results for the corner using the old implementation.

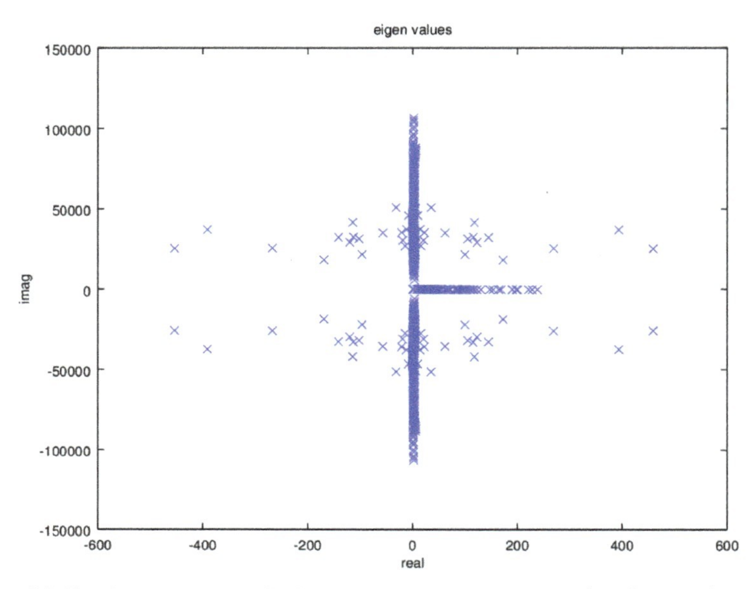

Figure 28.69 Spectrum zoom-in for the calculation including A_x, A_y (no A_z) containing the wrong-sign real part eigenvalues.

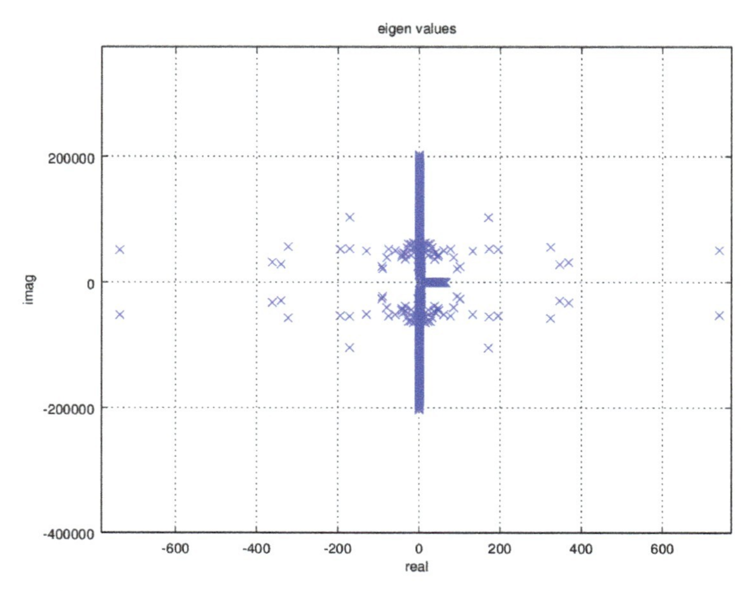

Figure 28.70 Eigenvalue spectrum for the calculation including A_x, A_y and A_z containing the wrong-sign real part eigenvalues.

We continued experimenting with the twisted bar and corner structures. For the twisted-bar structure we returned to the coarse mesh and obtained the following results. The problem size is 392. So there are a total of 392 eigenvalues. $\varepsilon_{\text{oxide}} = 5$ and $\varepsilon_{\text{Alum}} = 1$. The Alum conductance, $\sigma = 10^7$ S/m. A zoom-in of the spectrum is shown in Figure 28.71. There are 4 wrong-sign eigen values with values (–2.7617e-02 + 5.0476e+04i), (–2.7617e-02 – 5.0476e+04i), (–3.8202e-02 + 1.8259e+04i), (–3.8202e-02 – 1.8259e+04i). A further zoom-in is given in Figure 28.72. We also computed the spectrum without the magnetic fields. The result is shown in Figure 28.73. There are no negative real-part eigenvalues. Therefore we conclude that the wrong-sign eigenvalues are triggered by the Maxwell-Ampere equation. For the corner structure, we take for the $\varepsilon_{\text{oxide}} = 5$ and conductance $\sigma_{\text{oxide}} = 1$ S/m while keeping the conductance of the metal Alum = 10^7 S/m. Thus we add a small conductance to the oxides. The result is shown in Figure 28.74.

In this case all wrong sign eigenvalues have real part of the order of 10^{-9} whereas the corresponding imaginary part is around 10^4. Thus with adding a small conductivity to the oxide we can remove the unstable eigenvalues.

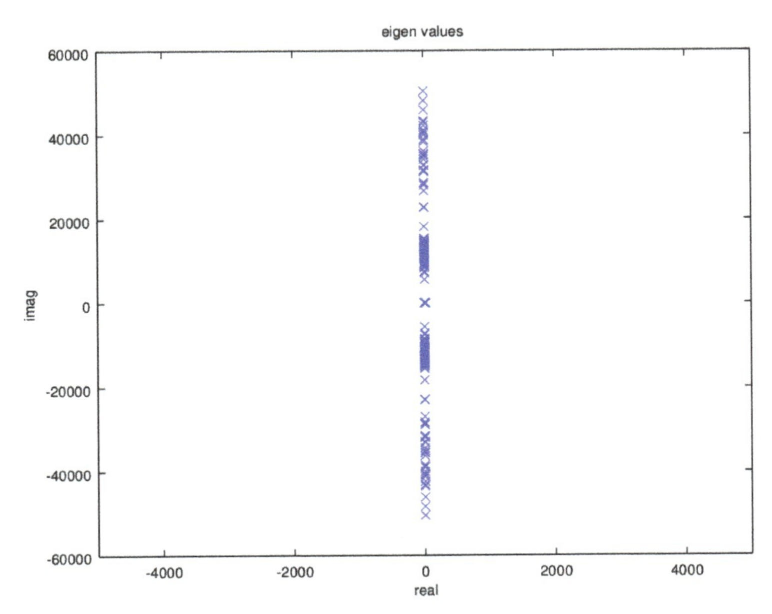

Figure 28.71 Zoom-in to the spectrum for the bar structure using a course mesh.

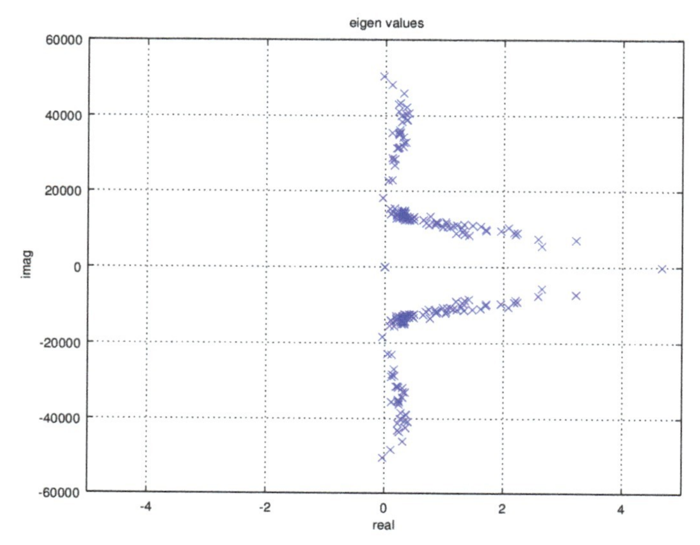

Figure 28.72 Further zoom-in to the spectrum for the bar structure using a course mesh.

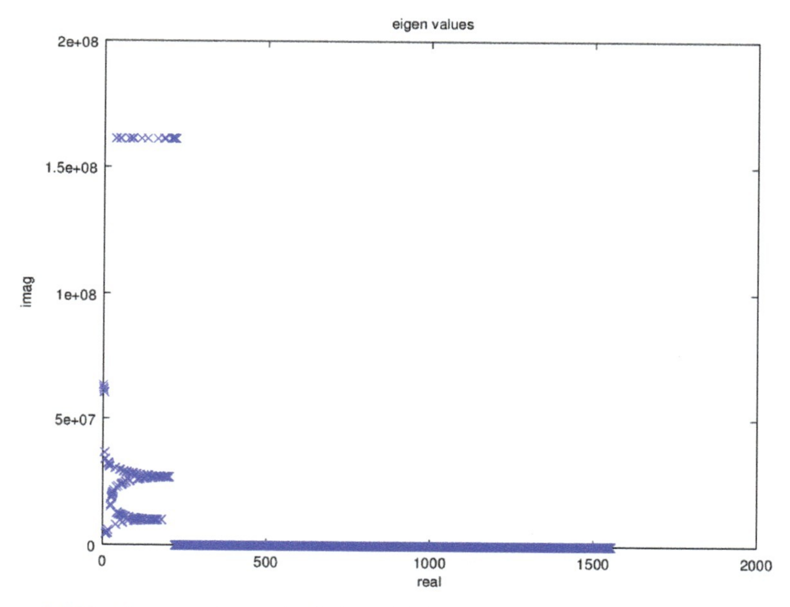

Figure 28.73 Spectrum for a calculation with the E-field switched on but the vector potential is not activated.

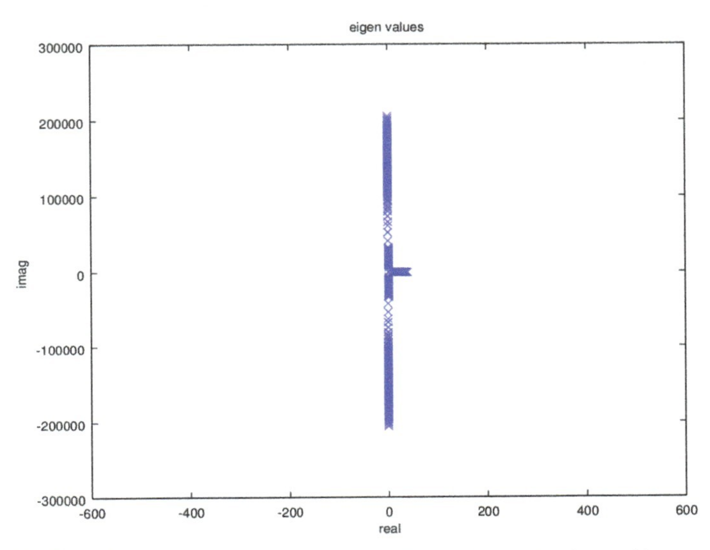

Figure 28.74 Eigenvalue spectrum zoom-in with lowly conducting oxide: $\sigma_{\mathrm{Ox}} = 1$ S/m.

The number of wrong-sign real part eigenvalues in the spectrum gets reduced dramatically if an alternative implementation of the mixed space-time derivative term is implemented. Since the number of such eigenvalues is now very limited we suspect that the remaining onces are related to the boundaries of the simulation domain. All the calculations were done without exploiting the inflation of the eigenvalue balloons by changing the time-scaling constant. From all experiments performed so far, we come to the following conclusions.

- The erroneous eigenvalues are induced by the Maxwell-Ampere equation.
- All erroneous eigenvalues can be removed by added a tiny conductance to the insulator.

It should be noted that the Maxwell-Ampere is essentially a wave equation. From Figure 28.70, we find that for every solution

$$X(t) = X_1^k \sin(\lambda_k t) + X_2^k \cos(\lambda_k t) \tag{28.11}$$

there exists also the solution:

$$X(t) = X_1^k \sin(-\lambda_k t) + X_2^k \cos(-\lambda_k t) \tag{28.12}$$

In wave mechanics this just reflects the properties of waves that can be captured by Green functions. The latter appear in advanced and retarded versions.

The physical Green' s function is of course the retarded one, but mathematically there are also the advanced Green's functions. So the damping is triggered by the conductance (in metal) and if there are solution (eigenvectors) that have no component in the metal then they are exclusively present in the insulating regions with the consequence that they are wave-like. In order to test this idea we consider the corner (again and replace the metal by insulator with $\varepsilon = 1$). Furthermore we set all conductances equal zero. Then there is no damping mechanism in the problem and the spectrum should be symmetric with respect to the real axis (which corresponds to the fact that the solution is real) and symmetric with respect to the imaginary axis (which corresponds to the fact that for each retarded solution there is an advanced solution). We see in Figure 28.75 that the spectrum is indeed fully symmetric. Now the question arises why there are eigenvalues found away from the imaginary axis. Since there is no damping we would expect that all eigenvalues have real part equal to zero. Is it a discretization effect? If it is an unavoidable discretization phenomenon then it must disappear in the infinite volume limit and/or zero grid distance limit. To further analyze the symmetry considerations we proceed with considering a simple U-shape conductor.

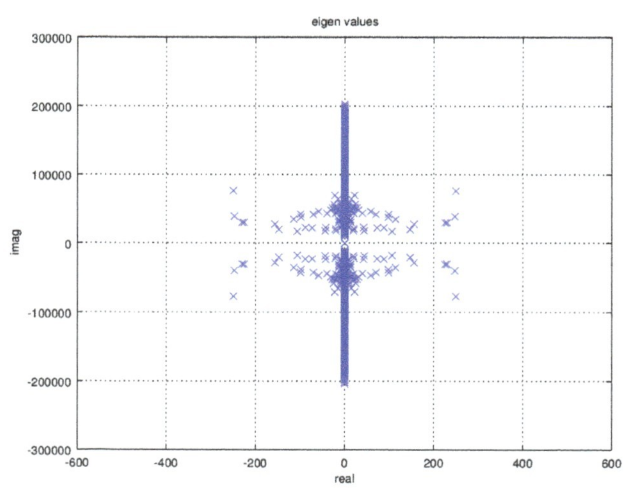

Figure 28.75 Eigenvalue spectrum for a system without conducting material.

28.12 Simple Test Structure for the Stability Problem

We present an extremely small test structure which demonstrates the stability problem of the transient Maxwell-Ampere system. The test structure is a U shape of metal. The contacts are at the surface of the simulation domain and there is no ground plane. Furthermore, the relative permittivity in the structure is equal to one throughout the structure. Figure 28.76 shows a 2D view of the structure.

In Figure 28.77 we show the 3D structure.

Figure 28.76 2D view in the mid plane of the structure including the mesh.

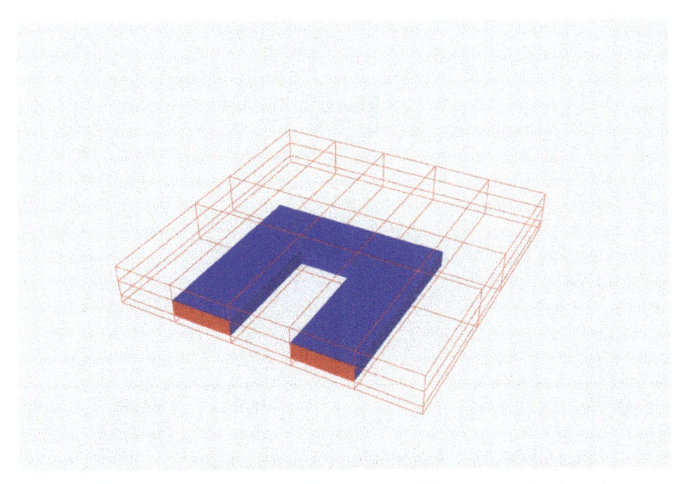

Figure 28.77 3D view of the structure. The mesh is also shown.

Detailed description of the test structure. The embedding material is non-conductive and has relative permittivity $\varepsilon_r = 1$. The lowest layer has height 5 micron, the middle layer has height 10 micron and the top layer has height 10 micron. The U-shape has coordinates:

- $200 < X < 400$ and $0 < Y < 400$ (in microns) for the left leg
- $600 < X < 800$ and $0 < Y < 400$ (in microns) for the right leg
- $200 < X < 800$ and $400 < Y < 600$ (in microns) for the top bar
- $5 < Z < 15$

The simulation domain has size $1000 \times 1000 \times 25\ \ \mu\text{m}^3$ in the X, Y and Z directions. The conductivity of the metal is 10^7 S/m. The mesh has $6 \times 4 \times 3 = 72$ nodes and there are 162 links. Since there are 2 contacts containing each 4 nodes, the DOF count for V is $96 - 4 \times 2 = 88$. Since all links on the edge of the simulation domain are non-DOF, the DOF count for A is 68. The DOF count for Π is also 68. Therefore the dimension of X is $88 + 68 + 68 = 224$. In Figure 28.78, we show the global spectrum. At first sight everything looks fine.

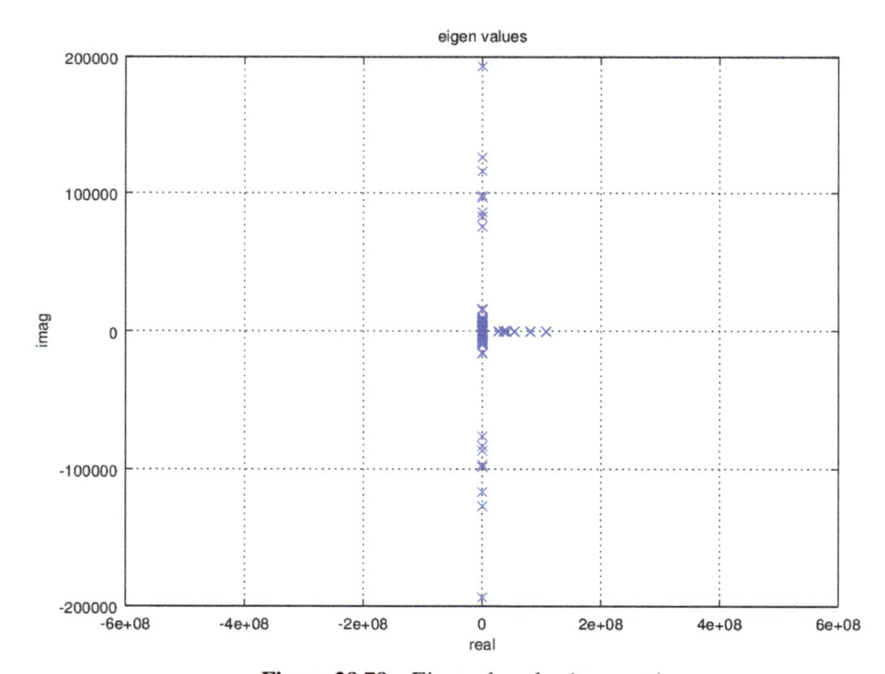

Figure 28.78 Eigenvalue plot (no zoom).

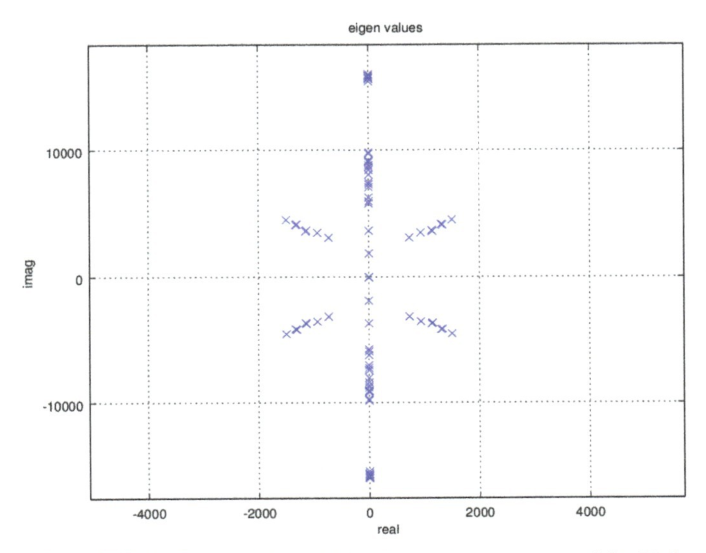

Figure 28.79 Zoomed view of the eigenvalue spectrum of the U-shape.

In Figure 28.79 we show the eigenvalue spectrum where we have zoomed in to the eigenvalues around zero.
The unstable eigenvalues are:

```
-6.7723e-02 + 1.9330e+05i
-6.7723e-02 - 1.9330e+05i
-8.4967e-02 + 1.2678e+05i
-8.4967e-02 - 1.2678e+05i
-6.6898e-02 + 1.1638e+05i
-6.6898e-02 - 1.1638e+05i
-8.2804e-02 + 9.8427e+04i
-8.2804e-02 - 9.8427e+04i
-8.3988e-02 + 9.6842e+04i
-8.3988e-02 - 9.6842e+04i
-5.8395e-02 + 8.6275e+04i
-5.8395e-02 - 8.6275e+04i
-8.1934e-02 + 8.3133e+04i
-8.1934e-02 - 8.3133e+04i
-5.8027e-02 + 7.5884e+04i
-5.8027e-02 - 7.5884e+04i
-1.5908e-02 + 8.4158e+03i
```

```
-1.5908e-02 - 8.4158e+03i
-1.1372e-02 + 8.0620e+03i
-1.1372e-02 - 8.0620e+03i
-4.2141e-03 + 7.4597e+03i
-4.2141e-03 - 7.4597e+03i
-9.3715e-03 + 7.2916e+03i
-9.3715e-03 - 7.2916e+03i
-8.4787e-03 + 7.0699e+03i
-8.4787e-03 - 7.0699e+03i
-8.7671e-03 + 6.2601e+03i
-8.7671e-03 - 6.2601e+03i
-2.7133e-02 + 5.9582e+03i
-2.7133e-02 - 5.9582e+03i
-5.5854e-02 + 5.7790e+03i
-5.5854e-02 - 5.7790e+03i
-1.5024e+03 + 4.5037e+03i
-1.5024e+03 - 4.5037e+03i
-1.3098e+03 + 4.1479e+03i
-1.3098e+03 - 4.1479e+03i
-1.3305e+03 + 4.1133e+03i
-1.3305e+03 - 4.1133e+03i
-1.6936e-01 + 3.6365e+03i
-1.6936e-01 - 3.6365e+03i
-1.1498e+03 + 3.6021e+03i
-1.1498e+03 - 3.6021e+03i
-7.3287e+02 + 3.0992e+03i
-7.3287e+02 - 3.0992e+03i
-1.1368e+03 + 3.6724e+03i
-1.1368e+03 - 3.6724e+03i
-9.3809e+02 + 3.4897e+03i
-9.3809e+02 - 3.4897e+03i
```

We used default scaling settings, e.g. length scaling $\lambda = 1.19527 \times 10^{-05}$ meter. Time scaling $\tau = 1.42867 \times 10^{-10}$ seconds.

Turn off A_z. Since the current is only in the XY-plane, there is almost no source for A_z. The candidate source for the A_z component originates from the displacement current. In the quasi-magneto static approximation, e.g. when there is no displacement current in the metal then the only source for the A_z

component is the displacement current in the oxide and the induced current in the metal. Therefore we turn off the z-component of the magnetic field. There are now 176 DOFs. The unstable eigenvalues now are:

```
-6.5046e-03 + 1.0362e+04i
-6.5046e-03 - 1.0362e+04i
-2.0517e-03 + 1.0235e+04i
-2.0517e-03 - 1.0235e+04i
-6.2108e+01 + 7.3040e+03i
-6.2108e+01 - 7.3040e+03i
-1.2708e-03 + 7.4246e+03i
-1.2708e-03 - 7.4246e+03i
```

In Figure 28.80 we show a zoom in of the spectrum.

Repairing the instability. The key observation leading to a damped system is that metal gives a term proportional to the electric field in the current continuity equation, e.g. using Gauss' law, we have in metal:

$$\nabla \cdot \left(\sigma \mathbf{E} + \frac{\partial}{\partial t} \epsilon \mathbf{E} \right) = 0 \tag{28.13}$$

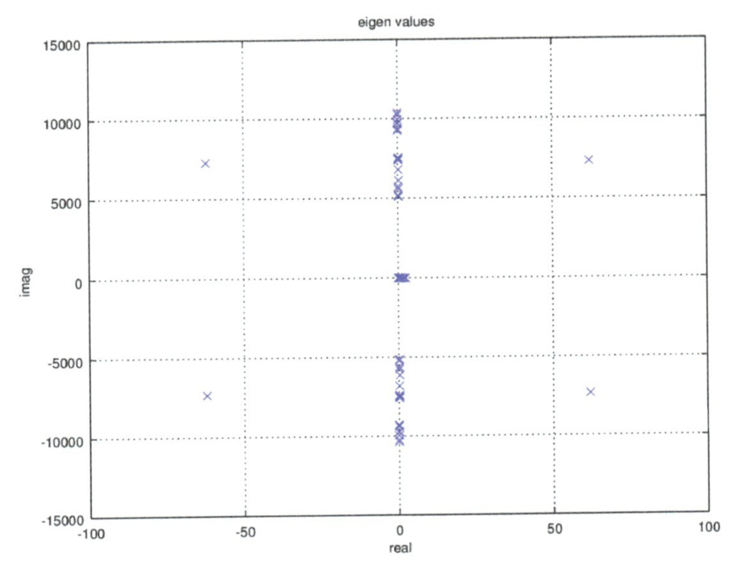

Figure 28.80 Spectrum corresponding the $A_z = 0$.

When $\sigma = 0$ there is no formal guarantee anymore that the system should be damped. Let is now have a look at the Maxwell-Ampere equation. Using a gauge condition in the following form:

$$\nabla \cdot (\frac{1}{\mu}\mathbf{A}) = -\epsilon \frac{\partial V}{\partial t} \tag{28.14}$$

We may rewrite the Maxwell-Ampere equation as:

$$\epsilon \frac{\partial}{\partial t}\mathbf{\Pi} = \mathbf{J}_c + \frac{\partial V}{\partial t}(\nabla\epsilon) + \nabla\left(\nabla \cdot \left(\frac{1}{\mu}\mathbf{A}\right)\right) - \nabla \times \left(\frac{1}{\mu}\nabla \times \mathbf{A}\right) \tag{28.15}$$

The term Jc introduces a contribution $-\sigma\Pi$, such that a damping is guaranteed in conductors. For insulators there is no damping provided. However, if we add a term $-\delta\Pi$ to the right-hand side of (28.15) then this will damp the equation. In other words the unstable eigenvalues will be removed for some value of δ. Thus in insulators we can insert an additional term δ unequal zero such that (28.15) becomes:

$$\epsilon \frac{\partial}{\partial t}\mathbf{\Pi} = \mathbf{J}_c - \delta\mathbf{\Pi} + \frac{\partial V}{\partial t}(\nabla\epsilon) + \nabla\left(\nabla . \left(\frac{1}{\mu}\mathbf{A}\right)\right) - \nabla \times (\frac{1}{\mu}\nabla \times \mathbf{A}) \tag{28.16}$$

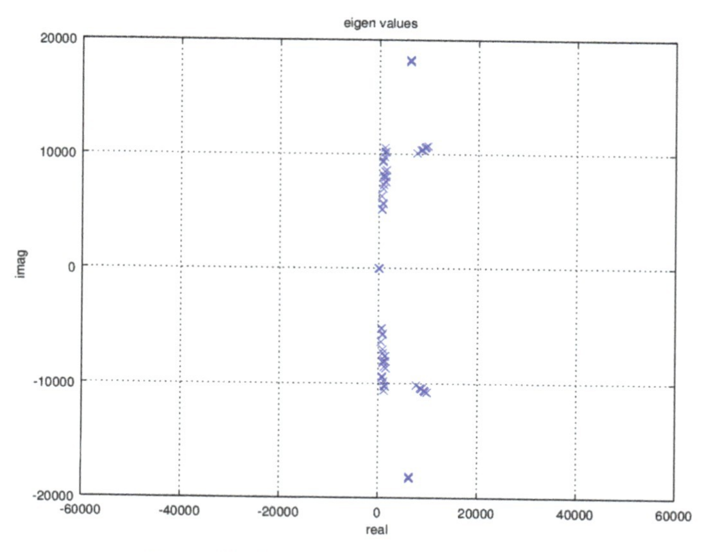

Figure 28.81 Damping Pi using delta= 0.01.

In the Figures 28.81–28.83, we show some results using (28.16) for the test structure. Using an arbitrary number like $\delta = 0.01$ is hard to justify. We can define δ as a material constant that in the frequency regime translates to an imaginary permittivity. With $\epsilon = \epsilon_r + i\epsilon_i$ we may consider $\delta = \tan(\frac{\epsilon_i}{\epsilon_r})$. In Figure 28.82 the result is shown using a loss tangent $\delta = 2000$. This is a big number but the set up of the test structure is also rather artificial, e.g. the mesh si very coarse and there are only a few nodes. In Figure 28.83 the result is shown using a loss tangent $\delta = 1000$. Some first sign of instability pop up. A brute-force cure for instability can be done but further work is needed to correlate it to mesh size.

28.13 Results for a Single Line

The next structure that is considered consists of a single line. We set $\varepsilon_{\text{metal}} = 0$ and $\varepsilon_{\text{oxid}} = 1$. The structure and its mesh are shown in the Figures 28.84 and 28.85. We compute the eigenvalues and zoom in to the spectrum around zero (see Figure 28.86). The problem size is 128 with three fields A_x, A_y and A_z. Using this very simple example we found that there was an error dealing with the computation of the matrix element corresponding to the

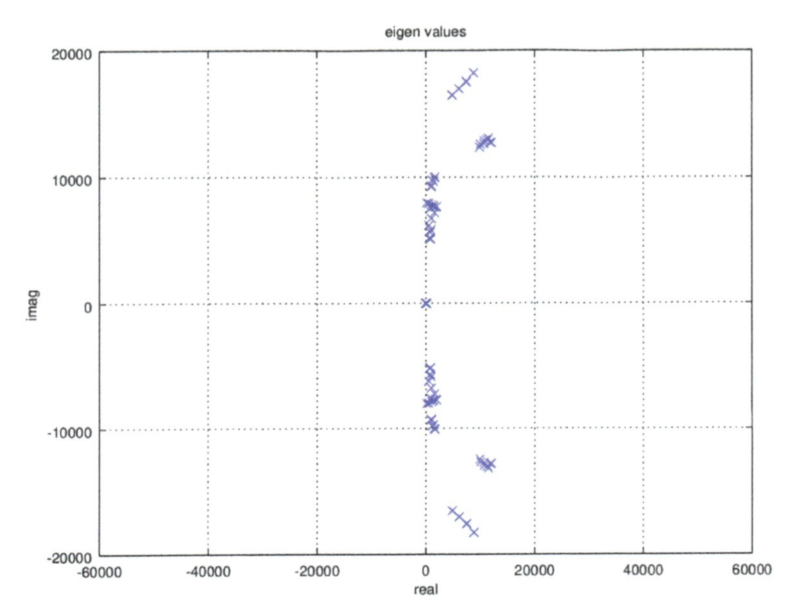

Figure 28.82 Using Pi damp with loss = 2000.

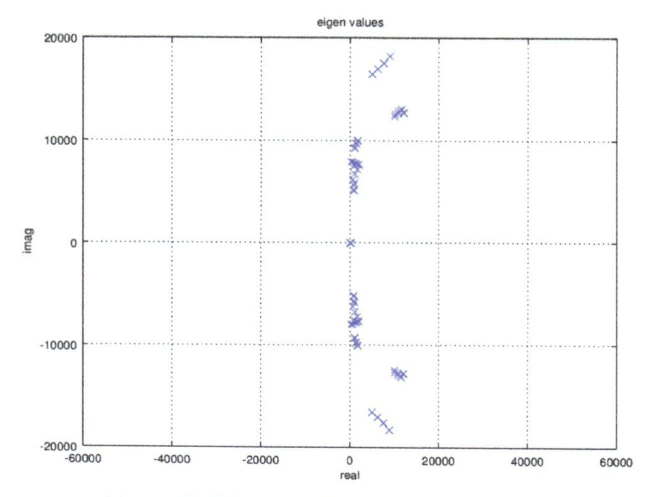

Figure 28.83 Using Pi damp with loss = 1000.

Figure 28.84 The bar of metal.

derivative of the Maxwell-Ampere equation with respect to the voltage. We have implemented the computation of the link area differently and applied a weighted average of the volume integral over the permittivity. For the test structure of Figure 28.76 this gives now the following 'wrong-sign' eigenvalues.

```
-7.7653e-05 + 5.0289e+03i
-7.7653e-05 - 5.0289e+03i
-8.3629e-06 + 4.4777e+03i
-8.3629e-06 - 4.4777e+03i
-3.0232e-07 + 4.4802e+03i
-3.0232e-07 - 4.4802e+03i
-2.1242e-06 + 4.2647e+03i
-2.1242e-06 - 4.2647e+03i
-3.8719e-06 + 4.2647e+03i
-3.8719e-06 - 4.2647e+03i
-1.5545e-04 + 2.7464e+03i
-1.5545e-04 - 2.7464e+03i
-2.5810e-05 + 2.7510e+03i
-2.5810e-05 - 2.7510e+03i
-5.9052e-07 + 2.5161e+03i
-5.9052e-07 - 2.5161e+03i
-4.9273e-06 + 2.4169e+03i
-4.9273e-06 - 2.4169e+03i
```

Comparing this result with the results of Figure 28.79 we find a substantial improvement.

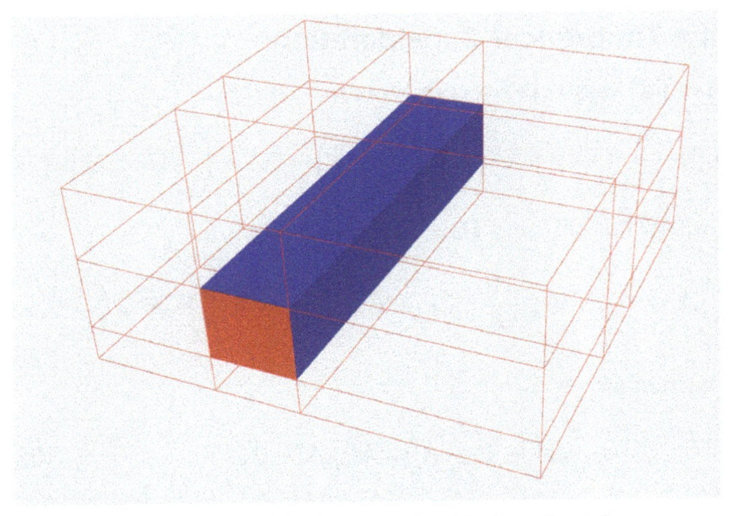

Figure 28.85 3D view + mesh of the bar of metal.

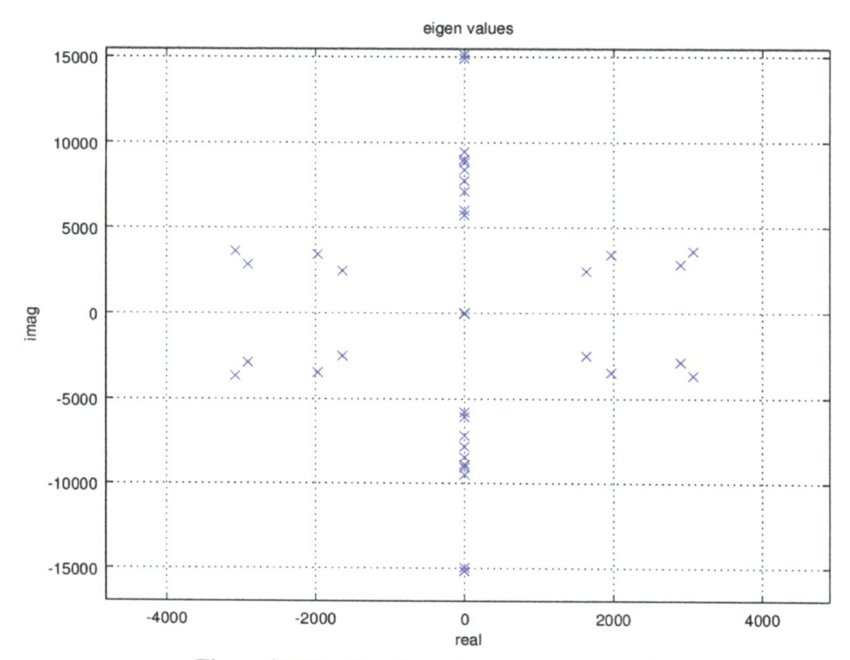

Figure 28.86 The eigenvalue spectrum around zero.

28.14 Some Theoretical Considerations

Let us return to the Maxwell-Ampere Equation (28.15).

$$\epsilon \frac{\partial}{\partial t}\mathbf{\Pi} = \mathbf{J}_c + \frac{\partial V}{\partial t}(\nabla\epsilon) + \nabla\left(\nabla\cdot\left(\frac{1}{\mu}\mathbf{A}\right)\right) - \nabla\times\left(\frac{1}{\mu}\nabla\times\mathbf{A}\right) \quad (28.17)$$

Since $\mathbf{J}_c = \sigma(-\nabla V - \mathbf{\Pi})$ and $\mathbf{\Pi} = \partial_t \mathbf{A}$ we obtain

$$\epsilon \frac{\partial^2}{\partial t^2}\mathbf{A} + \sigma\frac{\partial}{\partial t}\mathbf{A} = -\sigma\nabla V + \frac{\partial V}{\partial t}(\nabla\epsilon) + \nabla\left(\nabla\cdot\left(\frac{1}{\mu}\mathbf{A}\right)\right) - \nabla\times\left(\frac{1}{\mu}\nabla\times\mathbf{A}\right) \quad (28.18)$$

This can be written as:

$$\epsilon \frac{\partial^2}{\partial t^2}\mathbf{A} + \sigma\frac{\partial}{\partial t}\mathbf{A} = M_{op}\mathbf{A} + \mathbf{J}_s \quad (28.19)$$

$$\mathbf{J}_s = -\sigma\nabla V + \frac{\partial V}{\partial t}(\nabla\epsilon) \quad (28.20)$$

$$M_{op} = \nabla\left(\nabla\cdot\left(\frac{1}{\mu}\cdot\right)\right) - \nabla\times\left(\frac{1}{\mu}\nabla\times\cdot\right) \simeq \left(\frac{1}{\mu}\right)\nabla^2 \tag{28.21}$$

In here M_{op} is a spatial differential operator and $\mathbf{J}_s$ is a source term. For a planar structure, as is the case here, the component A_z decouples from the equations system. Moreover the source term for this component is zero. The second order spatial derivative will lead to wave-like solutions. Consider the very simple one-DOF equation:

$$\epsilon\frac{\partial^2 x}{\partial t^2} + \sigma\frac{\partial x}{\partial t} + k^2 x = 0 \tag{28.22}$$

There are solutions of the type $x(t) = x_0\exp(\lambda t)$. Inserting this solution gives:

$$\epsilon\lambda^2 + \sigma\lambda + k^2 = 0 \tag{28.23}$$

The solutions of this equation for $\sigma > 0$ are:

$$\lambda_{1,2} = -\frac{\sigma}{2\epsilon}\left(1 \pm \sqrt{1 - \frac{4\epsilon k^2}{\sigma^2}}\right) \tag{28.24}$$

This can not lead to an unstable eigenvalue since the argument of the square root is a number less than one. If the argument is less than zero we get wave-like solutions. Our observation critically depends on the assumption that the "Laplace operator" M_{op} gives rise to $k^2 \geq 0$. Unstable-solutions eigenvalues can arise if M_{op} gives rise to negative eigenvalues for the imposed boundary conditions. It should also be noted that if $\sigma = 0$ then the eigenvalues become pure imaginary.

28.15 The Impact of the Meshing

In this section we consider a structure with contacts at the edge of the simulation domain. The structure is shown in Figure 28.87 and Figure 28.88. The Manhattan meshing gives rise to to an eigenvalue spectrum which has no negative real-parts. However, when using $2D$ – Delaunay meshing, we find that the spectrum has the following severe negative real-part eigenvalues.

```
-1.9237e+10 + 0.0000e+00i
-4.9211e+00 + 4.2624e+03i
-4.9211e+00 - 4.2624e+03i
```

Figure 28.87 Test structure: 2D view.

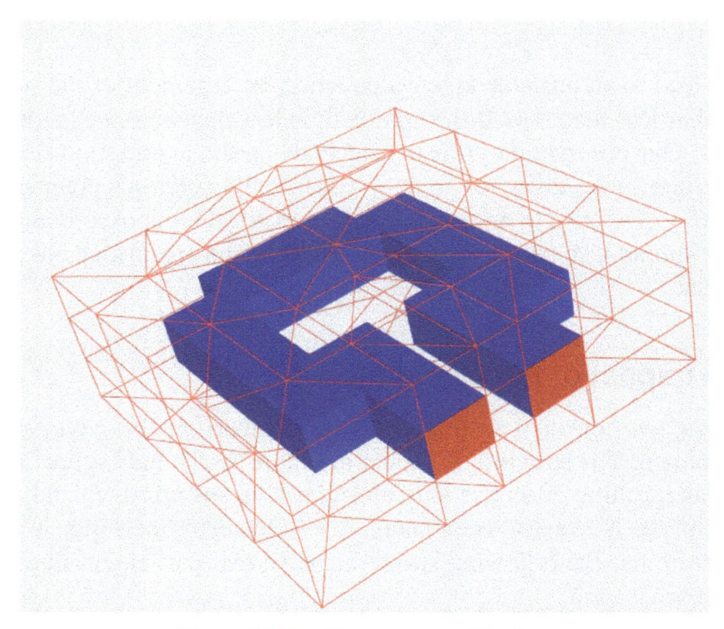

Figure 28.88 Test structure: 3D view.

As is seen in Figure 28.87, some cells have obtuse angles. This will lead to negative dual areas. One may modify the meshing algorithm by assigning a dual volume to each node in each cell by starting from the center of gravity for the surfaces of the cell and the cell volume. Using this modified method of obtaining dual volumes and dual areas, the negative real-part eigenvalues are removed again.

28.16 Final Summary of Stability Study

We went through a rather detailed discussion of the various research tracks to resolve the stability problem. The goal of this exposure has been to give the reader a flavor of how we have attempted to resolve the issue, The road to the solution was 'windy' and 'thorny'. Several aspects obscured the route to the correct final implementation. For example, the impact of meshing, a subtle misjudgement of how a term must be discretized or what is theoretically a correct expectation contributes to resolve the problem. Nevertheless, we finally arrived at an implementation that is stable and the success critically depends on the semi-definiteness of the implementation of the Laplace operator. As a side product we presented a view tricks to make the system stable by insertion of loss tangents.

29

Summary of the Numerical Techniques

In this chapter we will summarize and highlight the most essential steps that underly the coding schemes that were used in order to realize the simulations in the part III.

29.1 Equations

The four Maxwell's equation read

$\nabla \times \mathbf{E} = -\dot{\mathbf{B}}$	Induction law, Maxwell-Faraday's law
$\nabla \times \mathbf{H} = \mathbf{J} + \dot{\mathbf{D}}$	Maxwell-Ampère's law
$\nabla \cdot \mathbf{D} = \rho$	Gauss' law
$\nabla \cdot \mathbf{B} = 0$	Gauss' law for magnetism

where $\mathbf{E}$, $\mathbf{D}$ are the electric field strength and the displacement, and $\mathbf{H}$, $\mathbf{B}$ the magnetic field strength and induction, respectively. Moreover ρ and $\mathbf{J}$ are the electric charge density and current density, respectively. A dot on a variable denotes the partial derivative with respect to time : $\dot{X} = \partial_t X$. From Maxwell's equation one obtains the continuity law

$$\nabla \cdot \mathbf{J} + \dot{\rho} = 0 \quad \text{continuity law} \tag{29.1}$$

In what follows we assume linear isotropic materials, i.e.

$$\begin{aligned} \mathbf{D} &= \epsilon \mathbf{E} \\ \mathbf{B} &= \mu \mathbf{H} \end{aligned} \tag{29.2}$$

where ϵ is the dielectric constant and μ the permeability. It should be noted that materials of different types can be stacked or blocks of different materials can be placed next to each other. This results into abrupt jumps in the overall permitity ϵ and permeability μ. However, generally we assume that the parameters depend on the space coordinate. Magwel's simulator devEM employs as unknowns the scalar potential $\mathbf{E} = -(\mathrm{grad}\, V + \dot{\mathbf{A}})$ and vector potentials $\mathbf{B} = \nabla\times\mathbf{A}$. To avoid second order partial differential equations (PDEs) in time, one may introduce the so called quasi-canonical momentum $\mathbf{\Pi} = \dot{\mathbf{A}} = \partial_t \mathbf{A}$.

With Ohmic law $\mathbf{J} = \sigma\,\mathbf{E}$, $\mathbf{D} = \epsilon\,\mathrm{e}$ and Maxwell-Ampère law one obtains

$$\begin{aligned}\frac{1}{\mu}\,\nabla\times\,\nabla\times\mathbf{A} &= \frac{1}{\mu}\,(\nabla\,\nabla\cdot\mathbf{A} - \Delta\,\mathbf{A}) \\ &= -\sigma\,(\nabla\,V + \mathbf{\Pi}) - \epsilon\,\frac{\partial}{\partial t}\,(\nabla\,V + \mathbf{\Pi})\end{aligned} \tag{29.3}$$

From the continuity law (29.1) we get a second equation

$$\begin{aligned}-\,\nabla\cdot\mathbf{J} - \dot{\rho} &= -\nabla\cdot\mathbf{J} - \nabla\cdot\dot{\mathbf{D}} \\ &= \sigma\,\nabla\cdot(\nabla\,V + \mathbf{\Pi}) + \epsilon\,\frac{\partial}{\partial t}\,(\nabla\cdot(\nabla V + \mathbf{\Pi})) = 0\end{aligned} \tag{29.4}$$

From the Equations (29.3) and (29.4) one obtains 4 equations for the scalar and vector potentials $(V, \mathbf{A})$.

Since in ideal isolators $\mathbf{J} = 0$ and $\rho = 0$ is valid and moreover linearity of the materials can be assumed, i.e. $\mathbf{D} = \epsilon\,\mathbf{E}$ and $\mathbf{B} = \mu\,\mathbf{H}$, we obtain with $\nabla\cdot\mathbf{D} = 0$

$$\begin{aligned}&\epsilon\,\nabla\cdot(\nabla\,V + \mathbf{\Pi}) = 0 \\ &\frac{1}{\mu}\,\nabla\times\nabla\times\,\mathbf{A} = \frac{1}{\mu}\,(\nabla\,\nabla\cdot\mathbf{A} - \Delta\,\mathbf{A}) = -\epsilon\,\frac{\partial}{\partial t}\,(\nabla\,V + \mathbf{\Pi})\end{aligned} \tag{29.5}$$

From the Equations (29.5) one obtains 4 equations for the scalar and vector potentials $(V,\, \mathbf{A})$.

As already discussed above, the induction can be neglected in most cases, i.e. $\mathbf{\Pi} \approx 0$. However, we write the equations here without simplification. From Gauss' law we get

$$-\epsilon\,\nabla\cdot(\nabla\,V + \mathbf{\Pi}) = \rho, \quad \rho = q\,(p - n + N_D - N_A) \tag{29.6}$$

where n, p are the concentrations of free electrons/holes, respectively, and N_D, N_A the donator/exceptor concentrations. From the Maxwell-Ampère law one obtains

$$\frac{1}{\mu}\,\nabla\times\nabla\times\,\mathbf{A} = \mathbf{J}_p + \mathbf{J}_n - \epsilon\,\frac{\partial}{\partial t}\,(\nabla V + \mathbf{\Pi}) \tag{29.7}$$

where $\mathbf{J}_n$, $\mathbf{J}_p$ are the currents densities of electrons/holes, respectively. The current densities of the electrons $\mathbf{J}_n$ and holes $\mathbf{J}_p$ are given by

$$\mathbf{J}_n = -q\,\mu_n\left(n\,(\nabla V + \mathbf{\Pi}) - \frac{k\,T}{q}\cdot\nabla n\right)$$
$$\mathbf{J}_p = -q\,\mu_p\left(p\,(\nabla V + \mathbf{\Pi}) + \frac{k\,T}{q}\cdot\nabla p\right) \tag{29.8}$$

Herein q is the elementary charge, μ_n, μ_p the mobilities of electrons and holes, k the Boltzmann constant and T the absolute temperature. Finally $V_T = \frac{k\,T}{q}$ is the thermal voltage.

The first term in (29.8) is the drift term whereas the second term corresponds to the diffusion current. IN homogeneously doped semiconductors one may say that the drift current is an Ohmic current corresponding to a conductance $\sigma = q(\mu_p p + \mu_n n)$.

The densities of electrons and holes read

$$n = n_i\,\exp\frac{V - \Phi^n}{V_T}$$
$$p = n_i\,\exp\frac{\Phi^p - V}{V_T} \tag{29.9}$$

where Φ^n, Φ^p are the quasi-Fermi potentials for electrons/holes, respectively. The continuity equation holds for the electrons and holes separately, i.e.

$$\nabla\cdot\mathbf{J}_n - q\,\frac{\partial n}{\partial t} = -q\,(G - R) = -q\,U(n,p),$$
$$\nabla\cdot\mathbf{J}_p + q\,\frac{\partial p}{\partial t} = q\,(G - R) = q\,U(n,p) \tag{29.10}$$

with generation/recombination terms G and R. Often the net generation rate $U(n,p) = G - R$ is introduced. The unknown physical quantities are the concentrations of electrons and holes (n, p) and the scalar potential V and vector potential A, i.e. 6 scalar unknowns. Alternatively, one can choose as unknowns the quasi-Fermi potentials Φ^n, Φ^p instead of the carriers concentrations. The latter formulation is of practical advantage since the carrier concentration can vary by orders of magnitude.

Since the equations are underdetermined, the gauge conditions are required, i.e. it should be noted that the Maxwell-Ampère equation is complete with the following gauge condition

$$\frac{1}{\mu}\,\nabla(\nabla\cdot\mathbf{A}) + \xi\,\epsilon\nabla\,(\partial_t V) = 0 \tag{29.11}$$

However this is usually implemented by demanding

$$\frac{1}{\mu}\nabla \cdot \mathbf{A} + \xi\, \epsilon \partial_t V = 0 \tag{29.12}$$

For $\xi = 0$ one obtains the Coulomb and for $\xi = 1$ the Lorenz gauge. The system of PDEs is of second order in time. Introducing the quasi-canonical momentum $\Pi = \partial_t A$, a first order system is obtained. For the coupling of a circuit with a device simulator it is preferred to have a first order system, since circuit simulators are typically based on first order ordinary differential-algebraic equations in time. The interfaces presuppose therefore a first order system.

Remark 29.1 *For numerical stability a good scaling of the field equations is mandatory. In devEM the scalar potentials are normalized to* $V_T = \frac{kT}{q}$*, where* k *is Boltzmann's constant,* T *absolute temperature and* q *the elementary charge.*

29.2 Boundary conditions

Metal-semiconductor (Schottky-) contact. At the metal-semiconductor contact charge neutrality is presupposed, i.e. $p - n + N_D - N_A = 0$. Therefore the Fermi potentials are equal $\Phi^n = \Phi^p$ and moreover continuous at the contact, i.e. $\Phi^n = \Phi^p = V_{metal}$. Employing the short hand $N = N_D - N_A$ and (29.10) we get the abrupt voltage drop at the contact $\delta V = \Phi - V = V_{metal} - V$

$$\delta V = V_T \ln\left(-\frac{N}{2n_i}\left(1 + \sqrt{1 - \frac{4n_i^2}{N^2}}\right)\right) \quad p - dope, \quad N < 0$$

$$\delta V = -V_T \ln\left(\frac{N}{2n_i}\left(1 + \sqrt{1 - \frac{4n_i^2}{N^2}}\right)\right) \quad n - dope, \quad N > 0$$

Isolator-semiconductor contact. The continuity of the scalar potential V is assumed. The continuity equation for electrons and holes at the semiconductor side determines the potentials uniquely.

Metal-isolator contact. The scalar potential V is the sole unknown. The surface charge at the contact is obtained from the gradient in normal direction.

Metal-semiconductor-isolator contact. From δV, given by the metal-semiconductor contact above, one obtains by averaging

$$V_{insul} = V_{metal} - \frac{1}{2}\delta V$$

the contact voltage on the isolator side.

29.3 Spatial Discretization

Semiconductor Equations As already mentioned, it is preferred to obtains after space discretization a system of first order ordinary differential-algebraic equations in time for compatibility reasons with circuit simulators. Therefore the quasi-canonical momentum $\mathbf{\Pi} = \partial_t \mathbf{A}$ is introduced.

In what follows we introduce the following notation i and j for numbering the space grid points of adjacent grid nodes. Moreover, let Δw_i be a finite volume element, associated with i, and $\sigma_{ij} = \pm 1$ the orientation of a link connecting node i and node j, positive when oriented from inside to outside of the volume.

The links between nodes i and j are denoted with $\langle ij \rangle$ and have an associated length h_{ij}, correspondingly d_{ij} for the dual surface. Every link has an *intrinsic* orientation vector of length 1 and is denoted by $\mathbf{e}_{ij}$. A projection of a vector onto a link $\langle ij \rangle$ is marked with index i and j, e.g. $\mathbf{e}_{ij} \cdot \mathbf{A} = A_{ij}$.

When applying the finite-volume method, each node generates a balance equation corresponding to elaborating the divergence of a flux over the surface of the dual volume element of each node. As a result, a each surface element naturally gets a normal vector pointing away from the node under consideration. This vector $\mathbf{n}$, that is also found on a each link, can be parallel or anti-parallel to $\mathbf{e}$. The resulting sign is denoted as s, e.g. $s_{ij} = \mathbf{n} \cdot \mathbf{e} = \pm 1$.

Scharfetter-Gummel stabilisation. The Scharfetter-Gummel assumes that the current and field strength are nearly constant along a link $\langle i, j \rangle$. From the continuity equation for electrons one obtains

$$n\,(\nabla V + \mathbf{\Pi}) - V_T \nabla n = -c$$

A similar equation is valid for the holes. For ease of presentation we assume the x coordinate along the link, i.e. the linear differential equation

$$n\,E_x + V_T \, \frac{\mathrm{d}n}{\mathrm{d}x} = c$$

with boundary conditions $n(0) = n_i$ and $n(\Delta x) = n_{i+1}$. One obtains the solution

$$\begin{aligned} n(x) &= n_i \exp\left(-\frac{E_x}{V_T}\,x\right) \\ &\quad + \frac{n_{i+1} - n_i \exp\left(-\frac{E_x}{V_T}\,\Delta x\right)}{1 - \exp\left(-\frac{E_x}{V_T}\,\Delta x\right)} \left(1 - \exp\left(-\frac{E_x}{V_T}\,x\right)\right) \\ p(x) &= p_i \exp\left(\frac{E_x}{V_T}\,x\right) \\ &\quad + \frac{p_{i+1} - p_i \exp\left(\frac{E_x}{V_T}\,\Delta x\right)}{1 - \exp\left(\frac{E_x}{V_T}\,\Delta x\right)} \left(1 - \exp\left(\frac{E_x}{V_T}\,x\right)\right) \end{aligned} \tag{29.13}$$

where $\Delta x = x_{i+1} - x_i$. The field strength is approximated by

$$E_x \approx -\left(\frac{\Delta V}{\Delta x} + \mathbf{\Pi}\right)$$

For the current densities of electrons and holes one obtains the formulas

$$\begin{aligned} \mathbf{J}_n &= q\,\mu_n\,E_x\,n + q\,\mu_n\,V_T\,\tfrac{\mathrm{d}n}{\mathrm{d}x} \\ \mathbf{J}_p &= q\,\mu_p\,E_x\,p - q\,\mu_p\,V_T\,\tfrac{\mathrm{d}p}{\mathrm{d}x} \end{aligned}$$

Introducing the Bernoulli function

$$B(x) = \frac{x}{\exp x - 1}$$

and evaluation of the current densities on an intermediate grid point $\Delta x/2$ the current densities read

$$\begin{aligned} \mathbf{J}_n(\Delta x/2) &= q\,\mu_n\,\frac{V_T}{\Delta x}\left[B\left(\frac{-E_x\,\Delta x}{V_T}\right)\,n_{i+1} - B\left(\frac{E_x\,\Delta x}{V_T}\right)\,n_i\right] \\ \mathbf{J}_p(\Delta x/2) &= q\,\mu_p\,\frac{V_T}{\Delta x}\left[-B\left(\frac{E_x\,\Delta x}{V_T}\right)\,p_{i+1} + B\left(\frac{-E_x\,\Delta x}{V_T}\right)\,p_i\right] \end{aligned}$$

Gauss law. The discretization of

$$\epsilon\,\nabla\cdot\left(\nabla V + \frac{\partial}{\partial t}\mathbf{A}\right) + q\,(p - n + N_D - N_A) = 0$$

leads by integration over a volume element ΔV

$$\epsilon \oint_{\partial \Delta V} (\nabla V + \mathbf{\Pi})\, \mathrm{d}\mathbf{S} + \int_{\Delta V} \rho \, \mathrm{d}V = 0, \quad \rho = q\,(p - n + N_D - N_A)$$

and a summation over all nodes j, incident with i via a branch $\langle ij \rangle$

$$\begin{aligned} \sum_j \epsilon \frac{d_{ij}}{h_{ij}} \left(V_i - V_j - \sigma_{ij} \Pi_{ij}\, h_{ij}\right) - q p_i(\Phi_i^p, V_i) \Delta w_i \\ + q n_i(\Phi_i^n, V_i)\, \Delta w_i - q(N_D - N_A)\, \Delta w_i \; = 0 \end{aligned} \tag{29.14}$$

where Δw_i is a volume element. The required concentrations are obtained from (29.9). The electric field strength along a link $\langle ij \rangle$ - positive when directed from inside to outside of a volume - is obtained by

$$-E_{ij} = \nabla V + \frac{\partial}{\partial t} \mathbf{A} = \nabla V + \mathbf{\Pi} = \frac{V_j - V_i + \sigma_{ij} \Pi_{ij} h_{ij}}{h_{ij}}$$

Continuity equation. For the discretization of the continuity equations we employ for the current densities

$$\begin{aligned} \mathbf{J}_n &= -q\, \mu_n \left(n \left(\nabla V + \frac{\partial}{\partial t} \mathbf{A} \right) - \frac{k\,T}{q} \cdot \nabla n \right) \\ &= -q\, \mu_n \left(n \left(\nabla V + \mathbf{\Pi} \right) - V_T \nabla n \right) \\ \mathbf{J}_p &= -q\, \mu_p \left(p \left(\nabla V + \frac{\partial}{\partial t} \mathbf{A} \right) + \frac{k\,T}{q} \cdot \nabla p \right) \\ &= -q\, \mu_p \left(p \left(\nabla V + \mathbf{\Pi} \right) + V_T \nabla p \right) \end{aligned}$$

the Scharfetter-Gummel stabilization, i.e.

$$\begin{aligned} J_{ij}^n &= -\, q\, V_T\, \mu_n \frac{d_{ij}}{h_{ij}} \left(n_i B(-X_{ij}) - n_j B(X_{ij}) \right) \\ J_{ij}^p &= q\, V_T\, \mu_p \frac{d_{ij}}{h_{ij}} \left(p_i B(X_{ij}) - p_j B(-X_{ij}) \right) \end{aligned}$$

where $B(x)$ is the Bernoulli function and

$$X_{ij} = \frac{V_j - V_i + \sigma_{ij} h_{ij} \Pi_{ij}}{V_T}$$

The discretization of the continuity equations by an integration over a finite volume ΔV read

$$\begin{aligned}
&\int_{\Delta V} \left(\frac{\partial n}{\partial t} - \frac{1}{q} \nabla \cdot \mathbf{J}_n - (G - R) \right) \mathrm{d}V \\
&\quad = \int_{\Delta V} \left(\frac{\partial n}{\partial t} - (G - R) \right) \mathrm{d}V - \oint_{\partial \Delta V} \frac{1}{q} \mathbf{J}_n \, \mathrm{d}S \\
&\quad \approx \Delta w_i \, \frac{\partial}{\partial t} n_i + \sum_j V_T \, \mu_n \frac{d_{ij}}{h_{ij}} \left(n_i B(-X_{ij}) - n_j B(X_{ij}) \right) \\
&\qquad + \left(R(p_i, n_i) - G(p_i, n_i) \right) \Delta w_i = 0 \\
&\int_{\Delta V} \left(\frac{\partial p}{\partial t} + \frac{1}{q} \nabla \cdot \mathbf{J}_p - (G - R) \right) \mathrm{d}V \\
&\quad = \int_{\Delta V} \left(\frac{\partial p}{\partial t} - (G - R) \right) \mathrm{d}V + \oint_{\partial \Delta V} \frac{1}{q} \mathbf{J}_p \, \mathrm{d}S \\
&\quad \approx \Delta w_i \, \frac{\partial}{\partial t} p_i + \sum_j V_T \, \mu_p \frac{d_{ij}}{h_{ij}} \left(p_i B(X_{ij}) - p_j B(-X_{ij}) \right) \\
&\qquad + \left(R(p_i, n_i) - G(p_i, n_i) \right) \Delta w_i = 0 \qquad (29.15)
\end{aligned}$$

where ΔV is the volume of a considered finite volume element.

Maxwell-Ampère law. Let $\mathbf{J}_c = \mathbf{J}_p + \mathbf{J}_n$. The Maxwell-Ampère law reads

$$\frac{1}{\mu} \nabla \times \nabla \times \mathbf{A} = \mathbf{J}_c - \epsilon \frac{\partial}{\partial t} \left(\nabla V + \mathbf{\Pi} \right)$$

Employing the gauge condition

$$\frac{1}{\mu} \nabla \left(\nabla \cdot \mathbf{A} \right) + \xi \varepsilon \nabla \left(\frac{\partial}{\partial t} V \right) = 0 \qquad (29.16)$$

where $0 \leq \xi \leq 1$ is a free parameter.

Remark 29.2 *The Coulomb gauge is obtained with $\xi = 0$ and the Lorenz gauge with $\xi = 1$ as special cases.*

Integration over a surface area $\partial \Delta V$ of a finite volume element ΔV gives

$$\begin{aligned}
\oint_{\partial \Delta V} \frac{1}{\mu} \nabla \times \nabla \times \mathbf{A} \, \mathrm{d}S = & \oint_{\partial \Delta V} \mathbf{J}_c \, \mathrm{d}S + \oint_{\partial \Delta V} \frac{1}{\mu} \nabla \left(\nabla \cdot \mathbf{A} \right) \mathrm{d}S \\
& - \oint_{\partial \Delta V} \epsilon \frac{\partial}{\partial t} \mathbf{\Pi} \, \mathrm{d}S - \oint_{\partial \Delta V} (1 - \xi) \, \epsilon \frac{\partial}{\partial t} \nabla V \mathrm{d}S
\end{aligned}$$

where the normal vector S is directed from inside to outside. The latter term vanishes when the Lorenz gauge is employed and will not be treated subsequently.

Discretization of the Maxwell-Ampère system Just as for nodal variables, the entities associated to other geometrical objects, such as links or surfaces, we must first limit ourselves to a finite subset of links. A domain restriction is required. Suppose, we have a finite domain Ω selected. Furthermore, a grid is built using the nodes that were identified in the foregoing section. Next, we focus on all the links that connect these nodes. Again, some links will be found on the surface of Ω, to be precise, $\langle ij\rangle \in \partial\Omega$. The construction of the equations of motion (and/or constrain equations) requires special care, because the finite-integration methods around such links we bring us outside Ω and that falls outside the region for which we compute information. In order to avoid this situation we generalize the finite-volume method with a *finite-surface method.* We will first consider the situation when $\langle ij\rangle$ is an internal link, i.e. the link is not at the surface of the simulation domain.

Let us start with the Maxwell-Ampère law

$$\frac{1}{\mu}\nabla \times \nabla \times \mathbf{A} = \mathbf{J}_c - \epsilon \frac{\partial}{\partial t}\left(\nabla V + \mathbf{\Pi}\right) \tag{29.17}$$

and consider for each link its dual surface. We will take the integral of this equation over the dual surface. Furthermore, we multiply the results with the length $L = h_{ij}$ of the link $\langle ij\rangle$ under consideration and obtain

$$\begin{aligned}\epsilon L \frac{\partial}{\partial t}\int_{\Delta S} \mathrm{d}\mathbf{S}\cdot\mathbf{\Pi} = &-L \int_{\Delta S} \mathrm{d}\mathbf{S}\cdot\nabla\times\left(\frac{1}{\mu}\nabla\times\mathbf{A}\right)\\ &+L\int_{\Delta S}\mathrm{d}\mathbf{S}\cdot\frac{1}{\mu}\nabla\left(\nabla\cdot\mathbf{A}\right)\\ &-L\int_{\Delta S}\mathrm{d}\mathbf{S}\cdot\sigma\nabla V - L\int_{\Delta S}\mathrm{d}\mathbf{S}\cdot\sigma\mathbf{\Pi}\end{aligned} \tag{29.18}$$

The discretization of each term will now be discussed. Starting at the left-hand side, we define a link variable Π_{ij} for the link going from node i to node j. The surface integral is approximated by taking $\mathbf{\Pi}$ constant over the dual area. Thus

$$\epsilon L \frac{\partial}{\partial t}\int_{\Delta S}\mathrm{d}\mathbf{S}\cdot\mathbf{\Pi} \simeq \epsilon L\,\Delta S_{ij}\,\frac{d\Pi_{ij}}{dt} \tag{29.19}$$

We can assign to each link a volume being $\Delta v_{ij} = L\,\Delta S_{ij}$.

Remark 29.3 *Note that $\Delta v_{ij} \neq \Delta w_{ij}$, since Δv_{ij} is the volume corresponding to the area of a dual surface multiplied with the length of a primary link whereas Δw_{ij} is a dual volume.*

The first term on the right-hand side is dealt with using Stokes theorem twice in order to evaluate the circulations

$$-L \int_{\Delta S} \mathrm{d}\mathbf{S} \cdot \nabla \times \left(\frac{1}{\mu}\nabla \times \mathbf{A}\right) = -L \oint_{\partial(\Delta S)} \mathrm{d}\mathbf{l} \cdot \left(\frac{1}{\mu}\nabla \times \mathbf{A}\right) \quad (29.20)$$

The circumference $\partial(\Delta S)$ consists of N segments. Each segment corresponds to a dual link that pierces through a *primary* surface. Therefore, we may approximate the right-hand side of (29.20) as

$$-L \oint_{\partial(\Delta S)} \mathrm{d}\mathbf{l} \cdot \left(\frac{1}{\mu}\nabla \times \mathbf{A}\right) = -L \sum_{k=1}^{N} \Delta l_k \frac{1}{\mu_k} (\nabla \times \mathbf{A})_k \quad (29.21)$$

where the sum goes over all primary surfaces that were identified above as belonging to the circulation around the starting link. Note that we also attached an index on μ. This will guarantee that the correct value is taken depending in which material the segment Δl_k is located.

Next we must obtain an appropriate expression for $(\nabla \times \mathbf{A})_k$. For that purpose, we consider the primary surfaces. In particular, an approximation for this expression is found by using

$$(\nabla \times \mathbf{A})_k \simeq \frac{1}{\Delta S_k} \int_{\Delta S_k} \mathrm{d}\mathbf{S} \cdot \nabla \times \mathbf{A} = \frac{1}{\Delta S_k} \oint_{\partial(\Delta S_k)} \mathrm{d}\mathbf{l} \cdot \mathbf{A} \quad (29.22)$$

The last contour integral is evidently replaced by the collection of primary links variables around the primary surface. As a consequence, the first term at the right-hand side of (29.18) becomes

$$-L \sum_{k=1}^{N} \Delta l_k \frac{1}{\mu_k} \frac{1}{\Delta S_k} \left(\sum_{l=1}^{N'} \Delta l_{<kl>} A_{<kl>} \right) \quad (29.23)$$

where we distinguished the link labeling from node labeling (ij) to surface labeling $\langle kl \rangle$. So we obtain

$$-L \int_{\Delta S} \mathrm{d}\mathbf{S} \cdot \nabla \times \left(\frac{1}{\mu}\nabla \times \mathbf{A}\right) = -L \sum_{k=1}^{N} \Delta l_k \frac{\alpha}{\mu_k} \frac{1}{\Delta S_k} \left(\sum_{l=1}^{N'} \Delta l_{<kl>} A_{<kl>} \right) \quad (29.24)$$

Next we consider the second term of (29.18). Now we use the fact that each link has a specific *intrinsic* orientation from 'start' to 'end' or from 'front' to 'back' or whatever that was earlier set equal to $\mathbf{e}$.

$$L \int_{\Delta S} \mathrm{d}\mathbf{S} \cdot \frac{1}{\mu} \nabla (\nabla \cdot \mathbf{A}) \simeq \int_{\Delta S} \mathrm{d}\mathbf{S} \cdot \frac{1}{\mu} (\nabla \cdot \mathbf{A})_{back} - \int_{\Delta S} \mathrm{d}\mathbf{S} \cdot \frac{1}{\mu} (\nabla \cdot \mathbf{A})_{front} \tag{29.25}$$

The two terms in (29.25) are now discretized as

$$\begin{aligned} \int_{\Delta S} \mathrm{d}\mathbf{S} \cdot \frac{1}{\mu} (\nabla \cdot \mathbf{A}) &= \frac{\Delta S}{\mu \Delta v} \int_{\Delta v} \mathrm{d}v \nabla \cdot \mathbf{A} \\ &= \frac{\Delta S}{\mu \Delta v} \oint_{\partial(\Delta v)} \mathrm{d}\mathbf{S} \cdot \mathbf{A} \\ &= \frac{\Delta S}{\mu \Delta v} \sum_{j}^{n} \Delta S_{ij} A_{ij} \end{aligned} \tag{29.26}$$

where the sum is now from the front or back node to their corresponding neighbor nodes. The boundary conditions enter this analysis in a specific way. Suppose the front or back node is on the surface of the simulation domain. Then the closed surface integral around such a node will require a dual area contribution from a dual area outside the simulation domain. These surfaces are by definition not considered.

However, we can go back to the gauge condition and use

$$\int_{\Delta S} \mathrm{d}\mathbf{S} \cdot \frac{1}{\mu} (\nabla \cdot \mathbf{A}) = -\Delta S \, \epsilon \frac{\partial V}{\partial t}. \tag{29.27}$$

At first sight this looks weird: First we insert the gauge condition to get rid of the singular character of the curl-curl operation and now we 'undo' this for nodes at the surface. This is however fine because for Dirichlet boundary conditions for $\mathbf{A}$ their are no closed circulations around primary surfaces and there is no uniqueness problem and therefore the operator is well defined.

The last two terms are rather straightforeward: For the third term we consider ∇V constant over the dual surface. Thus we obtain

$$-L \int_{\Delta S} \mathrm{d}\mathbf{S} \cdot \sigma \nabla V = (V_{front} - V_{back}) \left(\sum \Delta S_i \sigma_i \right). \tag{29.28}$$

The variation of σ is taken into account by looking at each volume contribution separately.

The fourth term of Equation (29.18) can be dealt with in a similar manner

$$-L \int_{\Delta S} \mathrm{d}\mathbf{S} \cdot \sigma \mathbf{\Pi} = L\, \Pi_{ij} \left(\sum \Delta S_i \sigma_i \right). \tag{29.29}$$

This brings us to the end of the discretization discussion.

References

Al-Mohy, A. H. and Higham, N. J.: [2011]; Computing the Action of the Matrix Exponential, with an Application to Exponential Integrators. SIAM Journal on Scientific Computing 33, 488–511.

Bertazzi, F.; Cappelluti, F.; Guerrieri, S. D.; Bonani, F. and Ghione, G.: [2006a]; Self-Consistent Coupled Carrier Transport Full-Wave EM Analysis of Semiconductor Traveling-Wave Devices. IEEE Trans. on Microwave Theory and Techniques 54, 1611–1617.

Bertazzi, F.; Cappelluti, F.; Guerrieri, S. D.; Bonani, F. and Ghione, G.: [2006b]; Self-Consistent Coupled Carrier Transport Full-Wave EM Analysis of Semiconductor Traveling-Wave Devices. IEEE Trans. on Microwave Theory and Techniques 54, 1611–1617.

Beylkin, G.; Keiser, J. M. and Vozovoi, L.: [1998]; A New Class of Time Discretization Schemes for the Solution of Nonlinear PDEs. J. Comput. Phys. 147, 362–387.

Bossavit, A.: [2003]; The sommerville mesh in yee-like schemes. In Scientific computing in electrical engineering, Series Maths. Series Mathematics in Industry (Editor S. H. W.H.A. Schilders, E.J.W. ter Maten), volume 4, (pp. 128–136). Springer-Verlag.

Bossavit, A.: [2005]; Discretization of electromagnetic problems: The generalized finite differences approach. In Handbook of numerical analysis (Editors W. Schilders and E. ter Maten), volume XIII, (pp. 105–197). Elsevier North-Holland.

Burnett, D.: [1987]; Finite Element Analysis. Addison-Wesley Pubishing Company.

Butcher, P.; March, N. H. and Tosi, M. P.: [1993]; Physics of Low-Dimensional Semiconductor Structrures. Plenum Press, New York.

Campbell, S. L.; Ipsen, I. C. F.; Kelley, C. T. and Meyer, C. D.: [1996]; GMRES and the Minimal Polynomial. BIT 36, 32–43.

Cendes, J.: [1991]; Vector finite elements for electromagnetic field computation. IEEE Transactions on Magnetics 27, 3958–3966.

Chameleon-RF: [2006]; Comprehensive high-accuracy modeling of electromagnetic effects in complete nanoscale RF blocks.
URL: *http://www.chameleon-rf.org*

Chen, Q.; Schoenmaker, W.; Banagaaya, N.; Schilders, W. and Wong, N.: [2011a]; EM-TCAD solving from 0–100 THz: A new implementation of an electromagnetic solver. In Solid-State Device Research Conference (ESSDERC), 2011 Proceedings of the European. IEEE.

Chen, Q.; Schoenmaker, W.; Chen, G.; Jiang, L. and Wong, N.: [2013]; Numerically Efficient Formulation for Time-Domain Electromagnetic-Semiconductor Co-Simulation for Fast-Transient Systems. IEEE Transactions on Computer-Aided Design of Integrated Circuits and Systems 32, 802–806.

Chen, Q.; Schoenmaker, W.; H.Weng, S.; Cheng, C. K.; Chen, G. H.; Jiang, L. J. and Wong, N.: [2012]; A fast time-domain EM-TCAD coupled simulation framework via matrix exponential. In Proc. IEEE Intl. Conf. on Computer Aided-Design (ICCAD).

Chen, Q.; Schoenmaker, W.; Meuris, P. and Wong, N.: [2011b]; An Effective Formulation of Coupled Electromagnetic-TCAD Simulation for Extremely High Frequency Onward. Computer-Aided Design of Integrated Circuits and Systems, IEEE Transactions on 30, 866–876.

Chen, Q.; Schoenmaker, W.; Weng, S.-H.; Cheng, C.-K.; Chen, G.-H.; Jiang, L.-J. and Wong, N.: [2015]; A fast time-domain EM-TCAD coupled simulation framework via matrix exponential with stiffness reduction. Int. J. Circ. Theor. Appl. Published online in Wiley Online Library (wileyonlinelibrary.com).

Chua, L. O. and Lin, P. M.: [1975]; Computer-Aided Analysis of Electronic Circuits, chapter 8. Prentice-Hall, New Jersey.

CODESTAR: [2003]; Compact modeling of on-chip passive structures at high frequencies.
URL: *http://www.magwel.com/codestar*

Collin, R.: [1960]; Field Theory of Guided Waves. Mc-Graw Hill, New York.

Datta, S.: [1995]; Electronic Transport in Mesoscopic Systems. Cambridge University Press, UK.

Davis, T. A.; Gilbert, J. R.; Larimore, S. and Ng, E.: [2004a]; Algorithm 836: COLAMD, an approximate column minimum degree ordering algorithm. ACM Transactions on Mathematical Software 30, 377–380.

Davis, T. A.; Gilbert, J. R.; Larimore, S. I. and Ng, E. G.: [2004b]; A column approximate minimum degree ordering algorithm. ACM Trans. Math. Softw. 30, 353–376.

devEM: [2003]; MAGWEL.
URL: *http://www.magwel.com*

Dittrich, T.; Haenggi, P.; Ingold, G.-L.; Kramer, B.; Schoen, G. and Zwerger, W.: [1997]; Quantum Transport and Dissipation. WILEY-VCH, Weinheim, Germany.

Drude, P.: [1900a]; Zur Elektronentheorie der Metalle, I. Teil. Annalen der Physik 1, 566–613.

Drude, P.: [1900b]; Zur Elektronentheorie der Metalle, II. Teil. Annalen der Physik 3, 369–402.

Duff, I. and Koster, J.: [2001]; On algorithms for permuting large entries to the diagonal of a sparse matrix. SIAM J MATRIX ANAL APPL 22, 973–996.

Edwards, S.; Yallup, K. and De Meyer, K.: [1988]; Two-dimensional Numerical Analysis of the Floating Region in SOI MOSFET's. IEEE Trans. on Electron Devices 35, 1012–1020.

Einziger, J.: [2004]; Planar inductance. Patent # WO 2004/012213.

Enders, P.: [2009]; Underdeterminacy and redundance in Maxwell's equations equations. Origin of gauge freedom – Transversality of free electromagnetic waves – Gaugefree canonical treatment without constraints. Electronic Journal of Theoretical Physics 6, 135–166.

Endes, P.: [2009]; Underteminacy and Redundance in Maxwell's Equations. Electronic Journal of Theoretical Phys. EJTP 6, 135–166.

Ezawa, Z. F.: [2000]; Quantum Hall Effects – Field Theoretical Approach and Related Topics. World Scientific Publishing Co. Pte. Ltd., Singapore.

Fach, N.; Hese, J. V.; Libbrecht, F.; Sercu, J.; Vandormael, L.; Herreman, M.; Kok, P.; Dhaene, T. and Blomme, K.: [1984]; Momentum – Agilent Technologies (now Keysight Technologies).
URL: *http://www.agilent.com/*

Fahs, B.; Gamand, P. and Berland, C.: [2010]; Low-phase-noise LC-VCO using high-Q inductor. Electronics Letters 46, 140–141.

Feynman, R.; Leighton, R. and Sands, M.: [1964]; The Feynman Lectures on Physics, volume 2. Addison-Wesley Publishing Company, New York.

Forghieri, A.; Guerri, R.; Ciampolini, P.; Gnudi, A. and Rudan, M.: [1988]; A New Discretization Strategy of the Semiconductor Equations Comprising Momentum and Energy Balance. IEEE Trans. on Computer-Aided Design 7, 231–242.

Fowler, R.: [1936]; Statistical Mechanics. New York, The MacMillan Company.

Frankel, T.: [1997]; The Geometry of Physics. University Press, Cambridge.

Galy, P.; Jimenez, J.; Schoenmaker, W.; Meuris, P. and Dupuis, O.: [2011]; ESD RF protections in advanced CMOS technologies and its parasitic capacitance evaluation. In ICICDT, IEEE International Conference on IC Design and Technology, Kaohsiung, Taiwan.

Galy, P. and Schoenmaker, W.: [2014]; In-depth Electromagnetic Analysis of ESD Protection for Advanced CMOS Technology During Fast Transient and High-Current Surge. *Electron Devices*, IEEE Transactions on 61, 1900–1906.

Gear, C.: [1971]; Simultaneous Numerical Solution of Differential-Algebraic Equations. IEEE Trans. Circuit Theory CT-18, 89–95.

Grondin, R. O.; El-Ghazaly, S. M. and Goodnick, S.: [1999]; A review of global modeling of charge transport in semiconductors and full-wave electromagnetics. IEEE Trans. on Microwave Theory and Techniques 47, 817–829.

Hammerstad, E. O. and Jensen, O.: [1980]; Accurate Models for Microstrip Computer-Aided Design. IEEE MTT-S Digest International Microwave Symposium (pp. 408–409).

HFSS: [1990]; Ansoft (now ANSYS).
URL: *http://www.ansoft.com/products/hf/hfss*

Huang, K.: [1963]; Statistical Mechanics. John Wiley & Sons, Inc., New York.

ICESTARS: [2008–2011]; Integrated Circuit/EM Simulation and Design Technologies for Advanced Radio Systems-on-Chip. FP7, Grant Agreement No. FP7-ICT-2007-1.

Itoh, T.: [1987]; Planar Transmission Line Structures. IEEE Press, New York.

Jackson, J.: [1975]; Classical Electrodynamics. John Wiley & Sons, Inc., New York.

Kapora, S.; Stuber, M.; Schoenmaker, W. and Meuris, P.: [2009]; Substrate Noise Isolation Characterization in 90nm CMOS Technology. In User track IEEE Design Automation Conference (DAC).

Kharchenko, S. A. and Yeremin, Y. A.: [1995]; Eigenvalue translation based preconditioners for the GMRES(k) method. Numerical Linear Algebra with Applications 2, 51–77.

Kubo, R.: [1957]; Statistical-Mechanical Theory of Irreversible Processes. I. General Theory and Simple Applications to Magnetic and Conduction Problems. J. Phys. Soc. Japan 12, 570.

Landau, L. and Lifshitz, E. M.: [1962]; The Classical Theory of Fields. Addison-Wesley Publishing Company, Inc.

Lee, J.; Sun, D. and Cendes, Z.: [1991]; Tangential vector finite elements for electromagnetic field computation. IEEE Transactions on Magnetics 27, 4032–4035.

Lipkinov, K.; Morel, J. and Shashkov, M.: [2004]; Mimetic finite difference methods for diffusion equations on non-orthogonal non-conformal meshes. Journal of Computational Physics 199, 589–597.

Lipnikov, K.; Manzini, G.; Brezzi, F. and Buffa, A.: [2011]; The mimetic finite difference method for the 3D magnetostatic field problems on polyhedral meshes. Journal of Computational Physics 20, 305–328.

Lundstrom, M.: [1999]; Fundamentals of carrier transport, 2nd Edition. Cambridge University Press, Cambridge.

Magnus, W. and Schoenmaker, W.: [1993]; Dissipative motion of an electron-phonon system in a uniform electric field: an exact solution. Phys. Rev. B 47, 1276–1281.

Magnus, W. and Schoenmaker, W.: [1998]; On the use of a new integral theorem for the quantum mechanical treatment of electric circuits. J. Math. Phys. 39, 6715–6719.

Mahan, G. D.: [1981]; Many-Particle Physics. Plenum Press, New York.

Maxwell, J.: [1954]; A Treatise on Electricity and Magnetism, volume 1. Dover Publications Inc., New York.

Meng, L.; Yam, C.; Koo, S.; Chen, Q.; Wong, N. and Chen, G.: [2012]; Dynamic Multiscale Quantum Mechanics/Electromagnetics Simulation Method. J. Chem. Theory and Comput. (JCTC) 8, 1190–1199.

Meuris, P.; Schoenmaker, W. and Magnus, W.: [2001a]; Strategy for Electromagnetic Interconnect Modeling. IEEE Trans. Computer-Aided Design of Circuits and Integrated Systems 20, no. 6, 753–762.

Meuris, P.; Schoenmaker, W. and Magnus, W.: [2001b]; Strategy for electromagnetic interconnect modeling. IEEE Trans. Computer-Aided Design of Circuits and Integrated Systems 20, 753–762.

Morse, P. M. and Feshbach, H.: [1953]; Methods of Theoretical Physics, PART I. McGraw-Hill Book Company, Inc.

Nastos, N. and Papananos, Y.: [2006]; RF Optimization of MOSFETs Under Integrated Inductors. IEEE Trans. on Microwave Theory and Techniques 54, 2106–2117.

Nedelec, J.: [1980]; Mixed finite elements in r^3. Numer. Math. 35, 315–341.

Nie, Q.; Zhang, Y.-T. and Zhao, R.: [2006]; Efficient semi-implicit schemes for stiff systems. Journal of Computational Physics 214, 521–537.

Niknejad, A. M. and Meyer, R. G.: [1998]; Analysis, Design and Optimization of Spiral Inductors and Transformers for Si RF ICs. IEEE Journ. of Solid-State Circuits 33, 1470–1481.

Noether, E.: [1918]; Invariante Variationsprobleme. Nachr. D. König. Gesellsch. D. Wiss. Zu Göttingen, Math-phys. Klasse (pp. 235–257).

Opera: [1984]; COBHAM.
URL: *http://operafea.com*

Orfanidis, S.: [2002]; Electromagnetic Waves and Antennas, volume http://www.ece.rutgers.edu/ orfanidi/ewa/. ECE Department Rutgers University 94 Brett Road Piscataway, NJ 08854-8058.

Palusinski, O. A.: [2005]; Transient analysis-2. Lecture note of Computer Aided Engineering for Integrated Circuits, University of Arizona.
URL: *www2.engr.arizona.edu/ ece570/session8.pdf*

Ricobene, C.; Wachutka, G. and Baltes, H.: [1991]; Operation of Vertical and Lateral Dual Collector Magnetotransistors Studied by Exact 2D-Simulation. In proceedings of Simulation of Semiconductor Devices and Processes, SISDEP 1991 (Editor Z. W. Fichtner, D. Aemer). Hartung-Gorre.

Rudan, M.; Reggiani, S.; Gnani, E.; Baccarani, G.; Corvasce, C.; Barlini, D.; Ciappa, M.; Fichtner, W.; Denison, M.; Jensen, N.; Groos, G. and Stecher, M.: [2006]; Theory and Experimental Validation of a New Analytical Model for the Position-Dependent Hall Voltage in Devices with Arbitrary Aspect Ratio. IEEE Transactions on Electron Devices 53, 314–322.

Saad, Y.: [1992]; Analysis of some Krylov subspace approximations to the matrix exponential operator. SIAM Journal on Numerical Analysis.

Saad, Y.: [1996]; Iterative methods for sparse linear systems. PWS, Boston.

Scharfetter, D. L. and Gummel, H. K.: [1969]; Large scale analysis of a silicon read diode oscillator. IEEE Trans. on Electron Devices 16, 64–77.

Schilders, W.; Polak, S. and van Weij, J.: [1985]; Singular Perturbation Theory and its Application to the Computation of Electromagnetic Fields. IEEE Trans. on Magn. 21, 2211–2216.

Schoenmaker, W.; Chen, Q. and Galy, P.: [2014]; Computation of self-induced magnetic field effects including the lorentz force for fast-transient phenomena in integrated-circuit devices. *Computer-Aided Design of Integrated Circuits and Systems*, IEEE Transactions on 33, 893–902.

Schoenmaker, W.; Magnus, W. and Meuris, P.: [2002a]; Ghost Fields in Classical Gauge Theories. Phys. Rev. Lett. 88, no. 18, 181602–01 181602–04.

Schoenmaker, W.; Magnus, W.; Meuris, P. and Maleszka, B.: [2002b]; Renormalization group meshes and the discretization of TCAD equations. IEEE Trans. Computer-Aided Design of Circuits and Integrated Systems 21.

Schoenmaker, W.; Matthes, M.; Smedt, B. D.; Baumanns, S.; Tischendorf, C. and Janssen, R.: [2012]; Large Signal Simulation of Integrated Inductors on Semi-Conducting Substrates. In Proc. Design, Automation and Test in Europe (DATE), (pp. 1221–1226).

Schoenmaker, W. and Meuris, P.: [2002a]; Electromagnetic Interconnects and Passives Modeling: Software Implementation Issues. IEEE Trans. Computer-Aided Design of Circuits and Integrated Systems 21, no. 5, 534–543.

Schoenmaker, W. and Meuris, P.: [2002b]; Electromagnetic interconnects and passives modeling: software implementation issues. IEEE Trans. Computer-Aided Design of Circuits and Integrated Systems 21, 534–543.

Schoenmaker, W.; Meuris, P.; Galy, P. and Jiminez, J.: [2010a]; On the inclusion of Lorentz force effects in TCAD simulations. In Proc. 37th European Solid-State Device Research Conference, ESSDERC-2007, Sevilla, Spain.

Schoenmaker, W.; Meuris, P.; Janssens, E.; Schilders, W. and Ioan, D.: [2007a]; Modeling of Passive-Active Device Interactions. In Proc. 37th European Solid-State Device Research Conference, ESSDERC-2007, Munich Germany.

Schoenmaker, W.; Meuris, P.; Janssens, E.; Schilders, W. and Ioan, D.: [2007b]; Modeling of Passive-Active Device Interactions. In Proceedings ESSDERC 2007, Munich Germany (Editors D. S. Landsiedel and R. Thewis), (pp. 163–166).

Schoenmaker, W.; Meuris, P.; Janssens, E.; Schilders, W. and Ioan, D.: [2007c]; Modeling of Passive-Active Device Interactions. In Proc. 37th European Solid-State Device Research Conference, ESSDERC-2007, Munich Germany.

Schoenmaker, W.; Meuris, P.; Janssens, E.; van der Kolk, K.-J.; van der Meijs, N. and Schilders, W.: [2007d]; Maxwell Equations on Unstructured Grids Using Finite-Integration Methods. In proceedings of the 12th International Conference in Simulation of Semiconductor Processes and Devices SISPAD 2007, (pp. 333–335). IEEE.
URL: *http://in4.iue.tuwien.ac.at/pdfs/sispad2007/pdfs/333.pdf*

Schoenmaker, W.; Meuris, P.; Jiminez, J. and Galy, P.: [2011]; On the inclusion of Lorentz force effects in TCAD simulations. Solid-State Electronics 103, 65–66.

Schoenmaker, W.; Meuris, P.; Pflanzl, W. and Steinmair, A.: [2010b]; Evaluation of Electromagnetic Coupling between Microelectronic Device Structures using Computational Electrodynamics. In Scientific computing in electrical engineering 2008, Series Maths. Series Mathematics in Industry (Editors J. Roos and L. Costa). Springer Verlag Berlin Heidelberg 2010.

Schuhmann, R. and Weiland, T.: [1998]; A stable interpolation technique for fdtd on nonorthogonal grids. International Journal on Numerical Modelling 11, 299–306.

Sonneveld, P.: [1989]; CGS, a fast Lanczos-type solver for nonsymmetric linear systems. SIAM J. Sci. Statist. Comput. 10, 36–52.

Stratton, R.: [1962]; Diffusion of hot and cold electrons in semiconductor devices. Phys. Rev. 126, 2002–2014.

't Hooft, G.: [1971]; Renormalizable Lagrangians for Massive Yang-Mills Fields. Nucl. Phys. B35, 167–188.

Tesson, O.: [2008]; High Quality Monolithic 8-Shaped Inductors for Silicon RF IC Design. In IEEE Topical Meeting on Silicon Monolithic Integrated Circuits in RF Systems (SiRF 2008). IEEE.

Trottenberg, U. and T. Clees, T.: [2009]; Multigrid Software for Industrial Applications – From MG00 to SAMG. Notes on Numerical Fluid Mechanics and Multidisciplinary Design 100, 423–436.

van der Vorst, H.: [1989]; Bi-CGSTAB, A fast and smoothly converging variant of Bi-CG for the solution of nonsymmetric linear systems. SIAM J. Sci. Statist. Comput. 13, 631–644.

Von Klitzing, K.; Dorda, G. and Pepper, M.: [1980]; New method for high-accuracy determination of the fine-structure constant based on quantized Hall resistance. Phys. Rev. Lett. 45, 494–497.

Wang, G. F.; Dutton, R. W. and Rafferty, C. S.: [2002a]; Device-Level Simulation of Wave Propagation Along Metal-Insulator-Semiconductor Interconnects. IEEE-J-MTT 50, 1127–1136.

Wang, G. F.; Dutton, R. W. and Rafferty, C. S.: [2002b]; Device-Level Simulation of Wave Propagation Along Metal-Insulator-Semiconductor Interconnects. IEEE Trans. on Microwave Theory and Techniques 50, 1127–1136.

Weiland, T.: [1977]; A discretization method for the solution of maxwells equations for six-component fields. Electronics and Communications AEU 31, 116–120.

Weiland, T.: [1996]; Time Domain Electric Field Computation with Finite Difference Methods. International Journal of Numerical Modelling: Electronic Networks, Devices and Fields 9, 295–319.

Weng, S.-H.; Chen, Q. and Cheng, C.-K.: [2011]; Circuit simulation using matrix exponential method. In ASIC (ASICON), 2011 IEEE 9th International Conference on, (pp. 369–372).

Weng, S. H.; Chen, Q. and Cheng, C. K.: [2012a]; Time-domain analysis of large-scale circuits by matrix exponential method with adaptive control. IEEE Trans. on CAD and Int. Circuits 31, 1180–1193.

Weng, S. H.; Chen, Q.; Wong, N. and Cheng, C. K.: [2012b]; Circuit simulation using matrix exponential method for stiffness handling and parallel processing. In Prof. IEEE Intl. Conf. on Computer Aided-Design (ICCAD), (pp. 407–414).

Weyl, H.: [1918]; Gravitation und Elektrizität. Sitzungsberichte der Preussischen Akademie der Wissenschaften 26, 465–480.

Willis, K. J.; Ayubi-Moak, J. S.; Hagness, S. C. and Knezevic, I.: [2009]; Global modeling of carrier-field dynamics in semiconductors using EMC-FDTD. Journal of Computational Electronics 8, 153–171.

Willis, K. J.; Hagness, S. C. and Knezevic, I.: [2011]; Multiphysics simulation of high-frequency carrier dynamics in conductive materials. J. Appl. Phys 10, 063714.1–063714.15.

Wilson, K.: [1974]; Confinement of Quarks. Phys. Rev. D 10, 2445–2459.

Xu, C.; Li, H.; Suaya, R. and Banerjee, K.: [2009]; Compact AC Modeling and Analysis of Cu, W, and CNT based Through-Silicon Vias (TSVs) in 3-D ICs. In Proc. IEEE Electron Device Meeting (IEDM), (pp. 521–524).

Yam, C.; Meng, L.; Chen, G.; Chen, Q. and Wong, N.: [2011]; Multiscale quantum mechanics/electromagnetics simulation for electronic devices. Phys. Chem. Chem. Phys. 13, 14365–14369.

Yeh, P. C.; Chiou, H. K.; Lee, C. Y.; Yeh, J.; Tang, D. and Chern, J.: [2008]; An Experimental Study on High-Frequency Substrate Noise Isolation in BiCMOS Technology. Electron Device Letters, IEEE 29, 255–258.

Index

'l Hopital's rule 494
't Hooft gauge 95

A
accumulation layer 31
action functional 18, 19
action integral 87, 88, 131
affine connection 100, 101, 103, 105
amber 7, 97
Ampère's law 11, 12, 18, 303
angular momentum 15
AV solver 321, 322, 396, 427
axial gauge 95

B
backward Euler 330, 355, 396, 419
ballistic transport 92
band gap 31, 36
basis, contravariant 109
basis, covariant 109
BDF 215, 216, 364, 399
BEOL 383, 392
Bernoulli function 35, 139, 300, 479
Bessel function 493
Bianci identities 44
Biot-Savart's law 97, 438
Boltzmann constant 23, 356, 397, 412
Boltzmann transport equation 24, 31, 33, 359
Boltzmann transport theory 31
boson 79, 96
boundary conditions 59, 121, 176, 199
boundary conditions, surface impedance 440, 441
box/surface-integration method 134
branch 74, 78, 86, 354
Brillouin zone 31

C
canonical momentum 88, 94, 330
canonical quantization 18
capacitor 37, 74, 82, 86
carrier heating 27, 35
carrier mean energy 34
carrier temperature 33
Cartesian coordinates 113
Cartesian grid 129, 130, 132, 492
Cartesian grids 132, 133
causality 29, 90, 188
causality principle 29
cell variable 39
CGS 326, 329
characteristic impedance 50, 52, 53, 240

charge accumulation 31
charge conservation 98, 134, 355, 468
charge density 87, 214, 304, 358
charge distribution 37, 56, 91, 181
chemical potential 25, 30, 31
circuit 50, 67, 75, 81
circuit modeling 4, 86, 107
circulation 112, 345, 445, 572
classical electrodynamics 18
classical fields 97
classical Hall effect 92
closed curve 11, 107, 108, 112
closed surface 107, 108, 206, 573
CMOS 36
CMOS technology 379, 392
coarse graining 24, 73
coax 50, 51, 233, 256
coaxial cable,
square 229, 231
COLAMD 298, 307, 315, 404
computational physics 3
computer calculations 97
conducting media 27, 57, 182
conduction band 30
conductive rod 233, 235
conductivity 27, 29, 36, 295
conductivity tensor 29
conductor 57, 79, 190, 215
confinement potential 93, 94
confining potential 92
conjugated variables 30
connected regions 17, 76, 78, 107
conservation laws 12, 17, 119
conservation of electric charge 12
conservation of linear momentum 14
constitutive equations 14, 23, 32, 36
constitutive relation 28, 293, 341, 481
constraint 42, 89, 303, 328
continuity equation 12, 137, 222
contravariant vector 99
convective currents 33
Cooper pairs 79, 96
coordinate covariance 97, 99
coordinate curve 111, 112
coordinate differentials 99
coordinate surfaces 109
coordinate system 98, 102, 177, 221
coordinate systems,
Euclidean 101, 104
coordinate transformation 99, 100
Coulomb blockade 78
Coulomb gauge 42, 89, 172, 490
Coulomb potential 89
coupled inductors 281
Courant limit 187
covariance 97, 98, 99, 100
covariant derivative 101, 102, 105
covariant vector 99
crossing wires 229, 230
cryogenic temperatures 92
crystal defects 25
crystal lattice 28
curl-curl operator 132, 398, 482, 538
current density 77, 188, 343, 360
current distribution 42, 44, 481
current-voltage characteristics 28
currents in metals 24, 26, 34
curvature 44, 98, 103, 481
curvature of space-time 98
curvature local 101
curvilinear
coordinates 109, 110, 113

cyclic coordinate 109, 110
cyclotron frequency 93
cylindrical symmetry 95

D

DAE 215, 216, 411, 421
decay time 29
degrees of freedom 17, 23, 201, 206
degrees of freedom field 4
degrees of freedom, of a link 129, 200, 206
delta function 38, 167
depletion layers 31
dielectric 37, 72, 82, 181
dielectric material 36, 37, 295, 478
dielectric media 37, 38, 42
dielectric medium 45, 478
dielectric polarization 37
differential geometry 107, 328
differential operator 4, 132, 300, 491
diffusion 31, 34, 399, 565
diffusion equation 35, 139, 317
dipole charges 37
dipole moment 38, 39
Dirac string 213
direct tunneling 36
discretization 10, 115, 179, 197
discretization Scharfetter-Gummel scheme 138, 179, 330
displaced Maxwellian distribution 28, 34
displacement 40, 85, 101, 563
displacement current 12, 170, 400, 415
dissipative processes 14
distance between two near-by points 100
distribution function 25, 27, 28, 327
drift-diffusion model 299, 343, 354, 359
Drude's model 27, 28

E

EDA 192, 340
edge state 92, 93
effective mass 30
Einstein 97, 99
Einstein relation 32
Einstein general theory of relativity 97
Einstein theory of gravity 97
elastic collision 26
elastic scattering 14
electric charge 8, 10, 13, 79
electric charge density 10, 563
electric circuit 12, 73, 75, 81
electric current 10, 14
electric current density 10, 14
electric displacement 39
electric field 7, 9, 71, 82
electric flux 12, 40
electric monopole 38
electric permittivity 10
electric susceptibility 40
electrical conductivity 27, 34
electrical polarization 43
electrodynamics 9, 18, 97, 98
electrodynamics, geometrical interpretation 104, 292, 482
electrodynamics, geometry of 97, 98

electromagnetic energy 13, 45, 74, 84
electromagnetic field 12, 17, 87, 353
electromagnetic field tensor 14, 105
electromagnetic potentials 67, 96
electromagnetic radiation 18
electromagnetic waves 91, 208, 437
electromagnetism 3, 19, 354, 409
electromotive force 11, 76
electron mobility 28
EM-TCAD 296, 317, 356, 409
EM-TCAD-LF 356
EMC 332, 379, 382, 393
EMLF 379, 382, 383, 392
energy balance equation 34, 72
energy flux 33, 63
energy relaxation times 34
energy transport model 35
entropy 23
eigenvalue spectrum 507, 517, 520, 548
EQS 282
equations of motion 88, 91, 202, 571
equivalent circuit 50, 190
error control 420, 428, 430
ESD 342, 367, 383, 392
Euclidean coordinate system 101, 104
Euclidean geometry 97, 100, 101
Euler-Lagrange equations 19, 20, 21
EV solver 174, 285, 321, 324
EV solving 322, 324

F

Faraday's law 12, 44, 97, 327
FDTD 116, 210, 297
FEM 116, 297
FEOL 383, 392
Fermi's Golden Rule 26
field equations 40, 44, 329, 566
fill-in 164, 168, 307, 404
finite difference method 117, 119
finite surface method 129, 183, 190, 354
finite volume method 116, 129, 354
finite-element method 116, 121, 297, 327
first Maxwell equation 20, 39, 41, 84
FIT 116, 281
floating domains 455, 462, 463, 467
Fourier expansion 90
fourth Maxwell equation 18, 20, 84
FSM 116, 190, 354, 410
functional 18, 19, 73
functional differentiation 131
FVM 116, 190, 309, 355

G

gain 232, 293, 404, 522
Galerkin method 122, 124
gate length 36
gate voltage 92
gauge 18, 60
gauge condition 42, 55, 56
gauge conditions 87, 89, 164, 191
gauge covariance 97

gauge covariant variables 104
gauge field 19
gauge invariance 19, 87
gauge invariance,
principle of 98
gauge slider 299
gauge theories 97, 98, 103
gauge transformation 18, 19,
41, 87
Gauss' law 11, 20, 38, 44, 135
Gauss' theorem 76, 83, 108, 110
general relativity 98
general theory of relativity 97, 98
generalized coordinates 24, 88
generalized Coulomb gauge 172
generalized momenta 24
geometrical interpretation 44, 98,
104, 482
geometry 97, 107, 328, 460
geometry,
non-Euclidean 97
ghost field 55, 91, 284, 456
ghost method 455
Gibbs' theory of assemble 28
GMRES 301, 404, 421, 426
gravitational field 98, 102
gravitational forces 3
gravity 97, 98, 103, 561
Green's function 4, 90,
115, 548
Green's function free space 90
Green's function poles 90
Green's theorem 108
grid generation 5
grid
Cartesian 129, 130,
132, 492
grid links 478
grid nodes 330, 567

H

Hall coefficient 347, 348,
353, 359
Hall effect 92, 342, 358, 393
Hall resistance 92
Hall voltage 91, 92
Hamiltonian 18, 19, 26, 92
harmonic modes 45
harmonic oscillator functions 93
Helmholtz equation 47, 49
Helmholtz operator 46
Helmholtz' theorem 17, 41,
107, 112
Hermite functions 93
heterojunction 91
HFSS 266, 280
Hilbert space 28
hole diffusivity 32
hole mobility 32, 349
holes 30, 79, 343, 356
homogeneous magnetic
field 91
hydrodynamic model 33, 34
hysteresis 43

I

ill-posedness 458
ILUT 307, 312, 315, 404
impedance 50, 211, 290, 437
implicit method 187
incompressible stationary
flow 41
inductance 83, 229, 264, 500
inductor 78, 183, 233, 258
inductor
octa-shaped 332
inelastic scattering 14
insulator 36, 58, 304, 441
integral theorem 107

integrating factor 265, 477, 493, 495
interacting electromagnetic field 20
interaction Hamiltonian 26
interaction Lagrangian 96
interfaces 31, 163, 305, 457
internal energy 30
intrinsic semiconductors 30
irreversible processes 578

J

Jacobian 110, 311, 396, 406

K

Kirchhoff 73, 74, 86, 298
Kirchhoff's laws 73, 77, 78, 86
Kramers-Kronig relations 29
Krylov subspace method 301, 312, 419, 427
Kubo's theory 28

L

Lagrange multiplier 131
Lagrangian 17, 88, 96
Lagrangian density 19, 20, 88, 89
Lagrangian
  electromagnetic field 19
Landau gauge 91, 93
Laplace's equation 42, 304, 482
Laplacian 55, 133, 319, 482
large signal 188, 327, 336, 339
large signal response 188, 355, 396, 410
lattice temperature 34
Legendre polynomials 37
line integral 12, 63, 112, 491
linear momentum 14
linear multi-step method 411
linear responses 14
link 18, 168, 174, 218
link variables 66, 201, 288, 571
Liouville's theorem 25
LNA 332
longitudinal components 46
longitudinal current 92
longitudinal electric field 49
longitudinal magnetic field 49
longitudinal polarization 91
longitudinal resistance 92
Lorentz force 341, 345, 353, 358
Lorentz force
  self-induced 342, 348, 349, 358
Lorenz gauge 191, 398, 491, 511

M

macroscopic field equations 44
macroscopic Maxwell equations 23, 44
magnetic field 8, 43, 208, 238
magnetic flux 12, 44, 84, 95
magnetic induction 9, 188, 372, 445
magnetic media 41
magnetic moment 42
magnetic moment density 42
magnetic monopoles 8, 11, 44, 188
magnetic
  permeability 10, 356, 481
magnetic susceptibility 43
magnetization 36, 37
magnetohydrodynamics 17
MAGWEL editor 216, 217, 236, 271
MAGWEL solver 53, 180, 215, 254

massive scalar particles 86
Matrix fill-in 144, 265
matter 44, 85, 86, 430
Maxwell equations 3, 7, 10, 17
Maxwell equations
differential form of 12, 87
Maxwell equations
integral form 10
Maxwell stress tensor 14
Maxwell-Ampère law 485, 486, 492, 493
mechanical energy 13
metal-oxide-semiconductor
field-effect transistor 81
metals 18, 25, 146, 375
MEXP 354, 364, 370, 374
mimetic 298, 299
mobility 22, 30, 293, 356
molecular charge distribution 33
moment expansion 25, 303
Momentum 15, 25, 84, 315
momentum flux 27
momentum relaxation time 21, 26
momentum representation 80
MQS 262
multiply connected 68, 87, 88, 90
multiply connected region 68, 90
multiply connected surface 87
multipole moments 31

N

N-port system 191, 214, 269
Newton-Raphson
matrix 110, 283, 310
Newton-Raphson
method 110, 358
no-ghost approach 114
Noether's theorem 299
non-equilibrium state 26
non-linear response 14
non-singular Lagrangian
density 79
numerical simulation 10, 87, 217
numerical stability 110, 488

O

Ohm's law 14, 21, 68, 286
ohmic losses 39
one-electron
Hamiltonian 82
OPERA simulations 229
Opera
VectorField 227
orientable surface 87
orientation 12, 88, 178, 489
oxide thickness 30

P

parallel transport 98, 102, 103, 105
parallel-plate capacitor 40, 74
path integral 18
Pauli's exclusion
principle 26, 27, 28
PEEC 116, 190
perfect conductor 79, 80, 81
periodic boundary conditions 92
permeability 43, 265, 347, 563
perturbative methods 45
phase space 24, 28
phenomenological theory 97
phonons 25
photon modes 18
plane waves 92
plasma physics 23
plotmtv 237
polarization 37, 43, 72, 91
potential 17, 25, 51, 56

potential difference 18, 51, 77, 86
potential well 91, 93
Poynting vector 13, 14, 62, 69
Poynting's theorem 13, 69
pseudo-canonical
momentum 190, 194,
358, 371
Pythagoras' theorem 100

Q

Q factor 264, 275, 332, 500
quadrupole moment 38
quantum Hall effect 91, 92
quantum mechanical probability
density 19
quantum mechanics 98, 18, 410
quantum-Liouville equation 28

R

radiation 62, 86, 91, 332
radiation field 18, 86, 87
recombination
generation 330, 455
resistance 28, 80, 264, 349
resistivity 78, 229, 348
retarded Green function 90
Riemann geometry 98, 100, 101

S

S parameters 50, 184, 233, 335
S-matrix 52, 184, 233, 288
SAMG 338
scalar 17, 96, 99
scalar field 19, 88, 113, 229
scalar field charged 96
scalar potential 17, 98
scaling 36, 142, 222, 442
scaling parameters 24, 225,
399, 443
scattering 27, 53, 212, 289
scattering parameters 212, 234,
288, 289
scattering processes 14, 26
Scharfetter-Gummel
discretization 179, 330,
477, 492
Scharfetter-Gummel
scheme 35, 138, 299, 477
Schrödinger equation 3, 18, 93
SCR 373, 380, 385, 390
self-interaction 96
semiconducting
materials 30, 304, 327, 351
semiconductors 23, 30, 169, 342
series impedance 50
shift operator 46, 117
shunt admittance 50
SIBC 440, 442, 445, 453
silicon wire 349, 366, 368, 425
simply connected region 10, 17,
76, 112
singular Lagrangian 88, 89
singular matrix 130, 416, 458, 459
singular operator 172, 318, 328
skin depth 239, 440, 443, 446
solenoid 95, 111
solver combo 366
spatial discretization 119, 197,
399, 567
spectral analysis 311, 315,
401, 407
speed of light 91, 190, 443
spherical harmonics 37
spiral inductor 183, 231, 291, 332
stability analysis 503
State-Space Matrices 216
statistical physics 23
step function 90, 170, 334, 508
stiffness 125, 409, 422, 437

Stokes' theorem 10, 107, 203, 572
strip line 233, 239, 241, 249
superconductor 79, 96
surface element 12, 66, 174, 567
surface impedance 435, 437, 440, 449
surface integral 85, 206, 484, 571

T

tangent vector 111
TCAD 59, 65, 295, 430
TCAD boundary conditions 62, 234, 289
TE modes 49, 219
telegraph equation 50
TEM modes 47, 48, 65
temperature 25, 32, 299, 412
temporal discretization 214
temporal gauge 94, 95, 298, 319
tensor 14, 98, 102, 105
tensor equations 99, 101
thermal conductance 33, 34, 380
thermal conductivity 33, 34
thermal equilibrium 25, 30, 31, 32
thermal flux 33
time reversal invariance 23
TLP 379, 391, 392, 393
TM modes 49, 50
topology 76, 109, 332, 364
toroidal region 76
transient regime 187, 328, 341, 367
transistor 36, 379, 389, 393
translational invariance 45, 46, 92, 93
transmission line 45, 50, 379, 463
transmission line theory 50, 51
transmission lines 45, 233
transverse components 47, 50, 109, 113
transverse potential 49
trial function 122, 123, 124, 127
TSV 296, 393
tunneling 36, 78, 79, 189
two-dimensional electron gas 91

U

uniform electric field 34, 579
unifying theory 3

V

vacuum 8, 40, 223, 401
vacuum expectation value 96
valence band 30
variational principle 19
VCO 332, 333, 404, 429
vector calculus 113
vector field 17, 37, 39, 55
vector field
conservative 110
vector field
irrotational 109
vector field
longitudinal part 107
vector field
non-conservative 110
vector field
transverse part 107
vector potential 17, 20, 60, 176
velocity field 41
Virtuosa 291
volume integral 76, 110, 469, 556
von Klitzing resistance 92

W

wave equation 18, 56, 329, 547
wave function 19, 93, 98, 104
wave guide 45, 64, 74, 437
weak decay 3
weight 126, 173, 479, 490
Whitney elements 127
Wiedemann-Franz law 34

About the Author

Wim Schoenmaker spent about one third of his carrier in physics working on several aspects of gauge theories. The other two-third of his professional life he worked in the field of microelectronics, starting in IMEC as a researcher in technology computer-aided design. When being confronted with numerous design problems in microelectronics that were induced by the increasing demand of high-frequency applications, he developed the numerical methods that are discussed in this book by exploiting his expertise in (lattice) gauge theories by adapting these methods for performing numerical simulations. Encouraged by the successful outcomes of these calculations, he co-founded the company *MAGWEL* that provides electronic design automation tools for microelectronic industry. He is presently the Chief-Technology Officer and member of the Board of Directors of MAGWEL. Wim Schoenmaker is the principal or co-author of $\sim$175 peer-reviewed journal and conference contributions. He is the co-author of one book, several book chapters and three patents.